Pavement Design

Pavement Design

Materials, Analysis, and Highways

M. Rashad Islam
Colorado State University
Pueblo, Colorado

Rafiqul A. Tarefder
University of New Mexico
Albuquerque, New Mexico

McGraw Hill

New York Chicago San Francisco
Athens London Madrid
Mexico City Milan New Delhi
Singapore Sydney Toronto

Library of Congress Control Number: 2020936569

Pavement Design: Materials, Analysis, and Highways

1 2 3 4 5 6 7 8 9 CD 25 24 23 22 21 20

ISBN 978-1-260-45891-6
MHID 1-260-45891-1

Sponsoring Editor
Ania Levinson

Editing Supervisor
Stephen M. Smith

Production Supervisor
Pamela A. Pelton

Acquisitions Coordinator
Elizabeth M. Houde

Project Manager
Poonam Bisht, MPS Limited

Copy Editor
Mohammad Taiyab Khan, MPS Limited

Proofreader
Nicole Schlutt

Indexer
Edwin Durbin

Art Director, Cover
Jeff Weeks

Composition
MPS Limited

About the Authors

M. Rashad Islam, Ph.D., P.E., is an assistant professor of civil engineering at Colorado State University, Pueblo. He also serves as an ABET evaluator and a journal reviewer for ASCE, TRB, ASTM, and Elsevier. Dr. Islam has more than 100 publications in pavement engineering.

Rafiqul A. Tarefder, Ph.D., P.E., is a professor of civil engineering at the University of New Mexico, where he developed the pavement engineering graduate program. He also serves as an associate editor of ASCE's *International Journal of Geomechanics*. Dr. Tarefder has more than 400 refereed journal and conference publications, and is a registered Professional Engineer in New Mexico and Idaho.

Contents

Preface

The authors are delighted to introduce this pavement design book to civil engineering students. It focuses on the most recent invention in pavement design, the pavement mechanistic-empirical (ME) design approach. Practical design examples are included in this textbook to help students understand the pavement mechanics and models used in this new approach. Fundamentals of Engineering (FE) exam–style questions are also included so that this book can be helpful for the FE examination as well. The authors believe that this text will be valuable for understanding pavement engineering.

Being the first edition, this book might contain some errors or confusing explanations. Please send any suggestions on improving it to books.mrislam@gmail.com. The authors will address these, with appropriate acknowledgment, in the next edition. Thank you.

M. Rashad Islam, Ph.D., P.E.
Rafiqul A. Tarefder, Ph.D., P.E.

CHAPTER 1
Introduction

1.1 Background

Pavement design is one of the most important parts of transportation engineering. To carry traffic from one place to another place comfortably, economically, and safely, an engineering design of pavements is essential. In this textbook, the required background knowledge about pavement materials, structural design of pavements, pavement design software, geometric design of pavements, and pavement maintenance procedures is discussed. Students are expected to be competent in pavement engineering after mastering this book.

A highway pavement is a structure consisting of layers of natural and processed materials above the natural ground (often called subgrade). A pavement's primary function is to distribute the vehicle loads from the top of the pavement to a larger area of the subgrade without causing any damage to the subgrade. The pavement structure should be able to provide an acceptable riding quality, satisfactory skid resistance, favorable light-reflecting characteristics, and low noise. The aim is to ensure that the transmitted wheel loads are sufficiently reduced, so that they do not exceed the capacity of all the layers of pavement including the subgrade. This chapter gives an overview of pavement types, layers and their functions, and pavement design methods.

A pavement is expected to meet the following requirements:

- Sufficient thickness to distribute the wheel-induced stresses to a reduced value on the subgrade soil
- Structurally adequate to keep the cracking and deformation within tolerable limits
- Structurally strong to withstand all types of stresses imposed upon it
- Adequate coefficient of friction to prevent skidding of vehicles
- Smooth surface to provide comfort to road users even at the expected speed
- Produces least noise from moving vehicles
- Dust and waterproof surface for avoiding reduced visibility
- Drains water laterally or vertically without washing layer particles
- Long service life with a desirable level of comfort considering the economy

1.2 Pavement Types

Two types of pavements are generally recognized: flexible pavement and rigid pavement, as shown in Fig. 1.1. A combination of these two pavements is also possible, and is termed *composite pavement*. Simply, a layer of asphalt layer can be placed on top of a concrete layer to create a composite pavement.

1.2.1 Flexible Pavements

Flexible pavements are usually surfaced with asphalt materials. These pavements are called flexible because the pavement structures can flex or bend under a traffic loading. A flexible pavement structure requires several layers of materials because these layers are not stiff enough to distribute the wheel load to a large area (Fig. 1.2). Beneath the asphalt layer, a crushed aggregate base layer is commonly seen. Below the base layer, a subbase layer is also used based on the subgrade strength. The natural subgrade soil can be improved by compaction or mixing of some improved soil, asphalt millings, low-quality aggregate based on the availability of these materials, and degree of improvement required.

FIGURE 1.1 Flexible and rigid pavements.

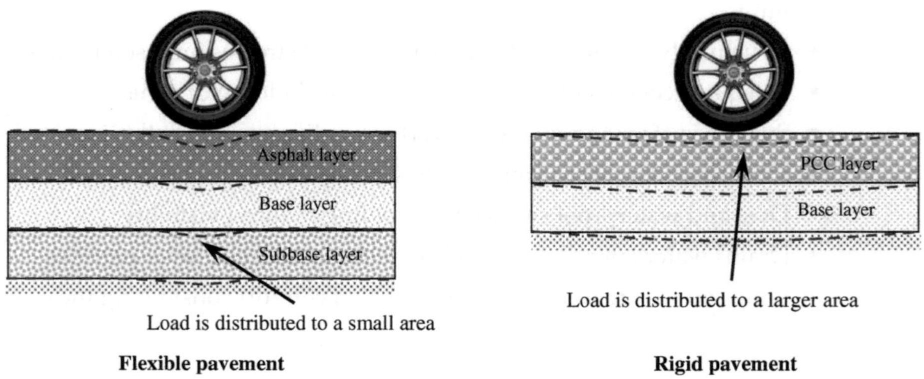

FIGURE 1.2 Deformation behavior of flexible and rigid pavements.

1.2.2 Rigid Pavements

Rigid pavements are composed of reinforced or non-reinforced portland cement concrete (PCC) surface course. Such pavements are stiffer than flexible pavements due to the high modulus of elasticity [typically 3,000–4,000 ksi (21–28 GPa) for PCC and 500–1,000 ksi (3.4–6.9 GPa) for asphalt layer] of the PCC material. These pavements can have reinforcing steel to reduce thermal cracking or eliminate joints. Each of these pavement types distributes load over the subgrade in a different fashion. Rigid pavement, because of PCC's high elastic modulus, tends to distribute the load over a relatively wide area of a subgrade (see Fig. 1.2). The concrete slab itself supplies most of a rigid pavement's structural capacity. On the other hand, a flexible pavement having a low modulus distributes loads over a smaller area. It requires a thicker pavement, which is achieved through a combination of thin layers due to field compaction difficulty of constructing a thicker layer.

Compared to flexible pavements, rigid pavements are placed either directly on the prepared subgrade or on a single layer of granular or stabilized material called base course. On a rigid pavement, a load is distributed by the slab action, in which the pavement behaves like an elastic plate resting on an elastic medium. Rigid pavements should be analyzed by plate theory instead of layer theory, assuming an elastic plate resting on an elastic foundation. Plate theory assumes the concrete slab as a medium-thick plate that is plane before loading and remains plane after loading. Bending of the slab due to wheel load and temperature variation causes tensile and flexural stress within the pavement layers.

Rigid pavements can be classified into four types:

1. Jointed plain concrete pavement (JPCP)
2. Jointed reinforced concrete pavement (JRCP)
3. Continuously reinforced concrete pavement (CRCP)
4. Prestressed concrete pavement (PCP)

Jointed Plain Concrete Pavement (JPCP)

Jointed plain concrete pavement (Fig. 1.3) uses plain concrete slabs without any reinforcement and it has transverse and longitudinal joints between slabs. Dowel bars are typically used at transverse joints to assist in load transfer. Epoxy-coated bars with a diameter of 1.0 or 1.5 in. (2.50 or 3.75 cm) and length of 18 in. (45 cm) are widely used in JPCP. Currently, fiber reinforced polymer (FRP) dowel bars are being used to avoid corrosion (FHWA-HRT-06-106) (FHWA, 2009). Dowel bars are placed in male-female fashion so that no tensile stress develops when a slab contracts. More clearly, dowel bars are tightly bonded to one slab and axially move freely in another slab. This action just transfers the load from one slab to another. However, no tensile stress is developed during the contraction of the slabs. Transverse joint spacing is selected such that temperature and moisture stresses do not produce intermediate cracking between joints. This typically results in a spacing no longer than about 20 ft (6 m). Tie bars are typically used at longitudinal joints or between an edge joint and a curb or shoulder for holding faces of rigid slabs in contact to maintain aggregate interlock. Tie bars are not load-transferring device, but they transfer some loads. Tie bars are about 0.5 in. (12.5 mm) in diameter and between 24 and 40 in. (0.6 and 1.0 m) long.

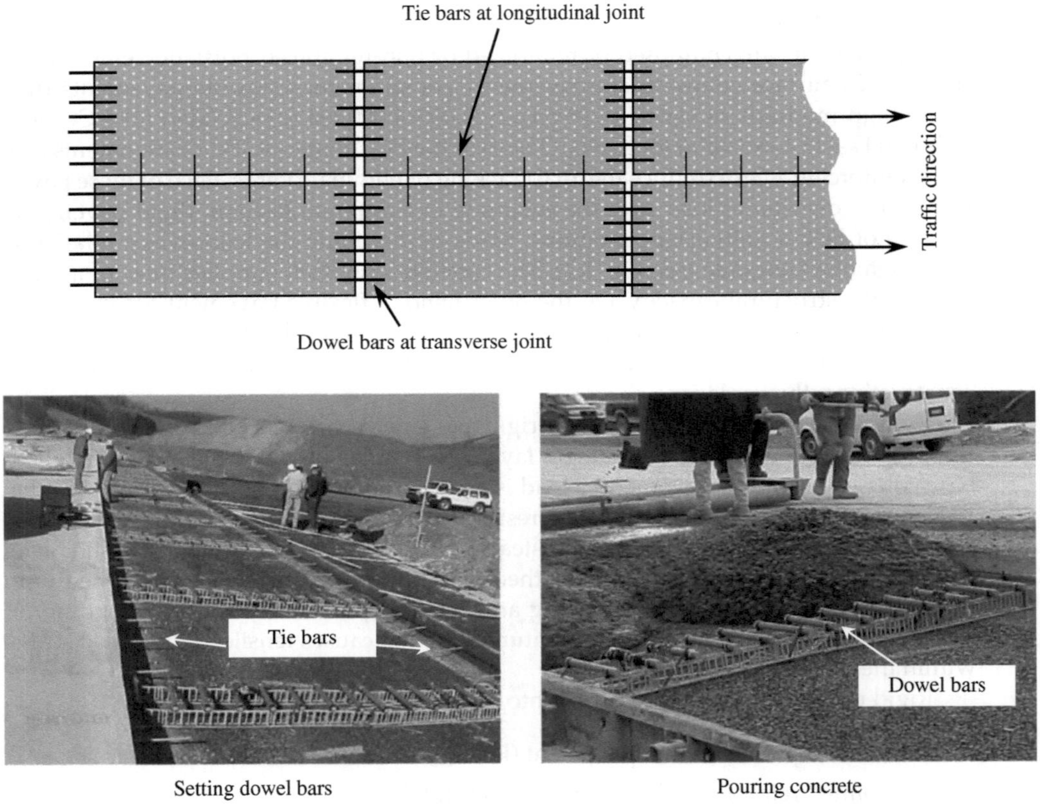

FIGURE 1.3 Jointed plain concrete pavement (JPCP) in Route 219, Elkins, WV. (*Courtesy of FHWA (2009). FHWA-HRT-06-106. Design and Evaluation of Jointed Plain Concrete Pavement with Fiber Reinforced Polymer Dowels. McLean, VA: Office of Research and Technology Services, Federal Highway Administration.*)

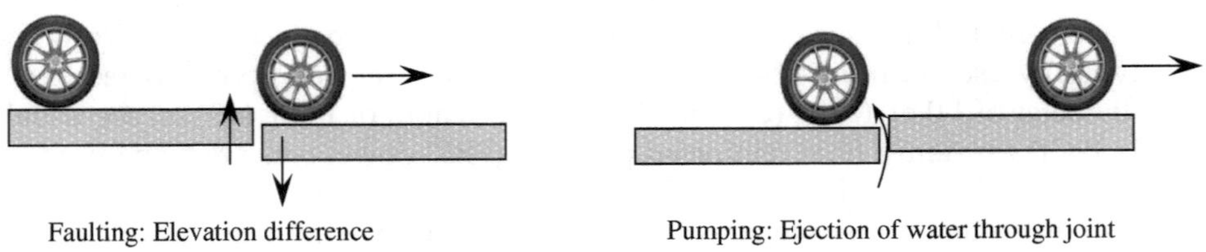

FIGURE 1.4 Faulting and pumping phenomena.

In JPCP pavements, if no dowel bars are provided or if inadequate amount of dowel bars are provided, then load transfer across the joint causes substantially higher stresses and deflections due to joint loading than those due to interior loading (Fig. 1.4). A dowel bar transfers a portion of the applied wheel load from the loaded slab across the joint to the adjacent unloaded slab. Load transfer through dowel bars significantly reduces stresses and deflections due to joint loading and minimizes faulting and pumping. Faulting is a difference in elevation across the joint of two slabs, while pumping is defined as the expulsion of subgrade material through joints and along the edges of the pavement.

Jointed Reinforced Concrete Pavement (JRCP)

Jointed reinforced concrete pavement is similar to the JPCP with the exception that some reinforced is used in the concrete slabs to control cracking with contraction joints. Using reinforced slabs, transverse joint spacing can be provided longer than that for JPCP and ranges from about 25 ft (7.5 m) to 50 ft (15 m). Temperature and moisture stresses are expected to cause cracking between joints. Reinforcing steel or a steel mesh is used to avoid these cracks or hold these cracks tightly together. Dowel bars are typically used at transverse joints to assist in load transfer.

Continuous Reinforced Concrete Pavement (CRCP)

Continuous reinforced concrete pavement completely eliminates the transverse joints, except as required at end-of-day construction and at bridge approaches and transitions to other pavement structures, by providing continuous reinforcement as shown in Fig. 1.5. In newly constructed CRCP, volumetric change occurs due to cement hydration,

Figure 1.5 Continuous reinforced concrete pavement (CRCP) construction. (*Courtesy of FHWA (2016). FHWA-HIF-16-026. Continuously Reinforced Concrete Pavement Manual: Guidelines for Design, Construction, Maintenance, and Rehabilitation. McLean, VA: Office of Research and Technology Services, Federal Highway Administration.*)

Figure 1.6 Prestressed concrete panel being placed over a base. (*Courtesy of FHWA (2009). FHWA-HIF-09-008. Concrete Pavement Technology Program (CPTP) TechBrief. McLean, VA: Office of Research and Technology Services, Federal Highway Administration.*)

thermal effects, and external drying. This volumetric change is restrained by the underneath base layer creating tensile stresses in the CRCP. Due to this developed tensile stresses, full-depth transverse cracks form and divide the pavement into short, individual slabs. However, CRCP provides long-term and high load transfer across the transverse cracks, resulting in a smooth and quiet ride.

Prestressed Concrete Pavement (PCP)

Prestressed concrete pavement is built using precast concrete pavement slabs that are fabricated at manufacturers' plant and transported to and installed at the project site. PCP is very applicable in busy areas where high traffic volume requires stronger pavement and construction time is very challenging considering traffic controls. Panels (Fig. 1.6) are manufactured in sizes to match the width of one, two, or three lanes of the pavement permitting one or multiple lanes of an existing pavement to be reconstructed at one time. Panels are pretensioned in the longer direction during fabrication, and post-tensioned together in groups longitudinally (in the direction of traffic). Thus, the installed slabs act as a continuous slab.

1.3 Design Philosophy of Pavements

The design philosophy of a pavement is quite different from the other classical civil engineering structures such as steel structures, reinforced concrete structures, and so on. These classical structures are designed such that they do not show any crack, significant deformation, or collapse during the service life. For example, a concrete beam in a building is not expected to show any crack. However, a pavement structure is designed with the consideration that there might be many cracks or some permanent deformation, and or roughness, during its service life, as shown in Fig. 1.7. However, these

FIGURE 1.7 Cracked pavements in service.

distress severities are expected to be within some tolerable limits set by the designers. The threshold values of distresses are determined considering the riding quality, safety, and economy. This means even after showing some distresses, pavement structures provide some services for some time.

1.4 Major Pavement Design Methods

Two major pavement design guides used in the United States are the American Association of State Highway and Transportation Officials (AASHTO) 1993 pavement design guide and the AASHTOWare pavement mechanistic-empirical (ME) design guide, also known as pavement ME design. They are briefly described in this section.

1.4.1 The AASHTO 1993 Pavement Design Guide

The *AASHTO Guide for Design of Pavement Structures* (AASHTO, 1993) was developed based on field performance data collected from the American Association of State Highway Officials (AASHO) road test project during 1956–1960 at Ottawa, IL, shown in Fig. 1.8. That project focused on the performance of pavement structures of known thickness under moving loads of known magnitude and frequency. It consisted of six two-lane loops along the alignment of Interstate 80 (I-80). The pavement structure within each loop was varied so that the interaction of vehicle loads with pavement structure could be investigated. The results from the AASHO road test were used to develop some regression equations for a pavement design guide, first issued in 1961 as the *AASHO Interim Guide for the Design of Rigid and Flexible Pavements*, with major updates issued in 1972, 1986, and 1993. The 1993 version (Fig. 1.9*a*) was in widespread use in the world until the AASHTOWare pavement ME design came out in 2008. The design philosophy of the AASHTO (1993) method is to limit vertical stress on subgrade within a tolerable limit. It does not consider the performance of the pavement such as cracking, rutting, and smoothness with its service life.

The AASHTO 1993 design method is purely empirical. The design equations, methodology, reliability, etc. were developed based on the road test data from Ottawa, IL. However, the climate, materials, and traffic behavior in that test section are not similar

FIGURE 1.8 AASHO road test in 1958–1960 in Ottawa, IL.

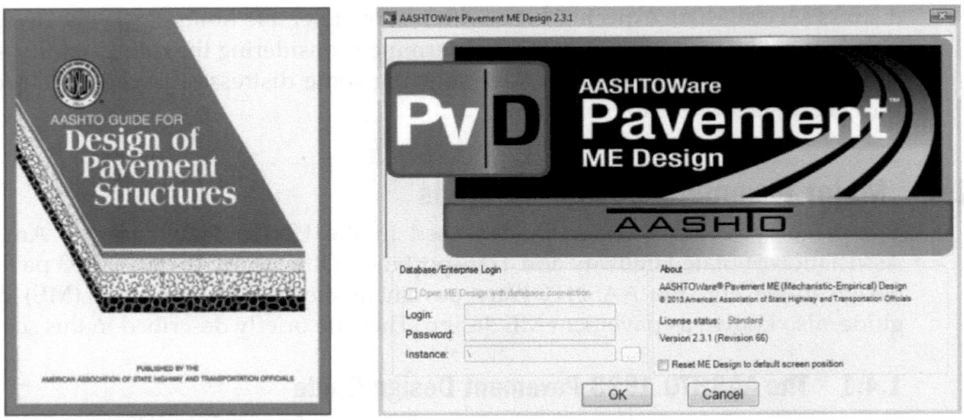

(a) AASHTO 1993 guide (b) AASHTOWare pavement ME design software

FIGURE 1.9 Pavement design guides.

to other areas in the United States or in any other countries. In fact, the climate varies city to city, material behavior varies pavement site to pavement site, traffic behavior (speed, distribution, load spectra, etc.) changes highway to highway. In addition, the material and traffic condition change with time. More specifically, the climate, materials, and traffic in a city in Illinois are not similar to another city (e.g., New York City). Even the climate, materials, and traffic in a city in Illinois in 1950 and at present are not the same. Therefore, the AASHTO 1993 method is not appropriate to other pavement sites.

1.4.2 The AASHTOWare Pavement Mechanistic-Empirical (ME) Design Guide

To overcome the limitations of the AASHTO 1993 pavement design guide, a new mechanistic-empirical (ME) method was started to develop about the year 2000. The advancements in computational tools and invention of performance models enable pavement designers to predict certain distress more accurately. These performance models use mechanistic pavement responses (such as stress and strain) while conducting analysis and design. The AASHTOWare pavement ME design (also termed *pavement ME design*) was officially implemented in 2008 as the *Mechanistic-Empirical Pavement Design Guide* (MEPDG) under NCHRP project 1-37A (AASHTO, 2015; NCHRP, 2004). This guide has been improved to a new version with the new name of the AASHTOWare pavement ME design software. Different sequences of the AASHTOWare pavement ME design software are listed below:

- First released in 2004 as an experiment with the name of *Design Guide 2002* (DG 2002)
- Revised in 2008 with the revised name of *Mechanistic-Empirical Pavement Design Guide* (MEPDG)
- Further revised in 2011 and renamed as *Design, Analysis and Rehabilitation for Windows* (DARWin ME)
- Further revised in 2013 and renamed as AASHTOWare pavement ME design (Fig. 1.9*b*)

The latest version of the AASHTOWare pavement ME design software is available at https://me-design.com/MEDesign/. The AASHTOWare pavement ME design software approach considers the performance of pavement with its service life. More specifically, the AASHTOWare pavement ME design software first analyzes a trial pavement section for possible stress-strain for an applied load using the software built-in numerical analysis program. Using the stress-strain data, the software predicts the amount of cracking, rutting, and smoothness during its service life. If the trial pavement section is found to predict less or equal to the threshold amount of distress during its service life, it is considered adequate. If otherwise, the pavement section or layers are revised and trial continues. After several trials, an optimum section is reached considering demand and economy.

The input levels available in the AASHTOWare pavement ME design software method depend on the availability of the input parameters at hand. Three levels are available for materials and traffic parameters:

- *Level 1*. Level 1 is used for pavement design with the greatest accuracy. All or most of the input parameters such as material density, modulus, strength, traffic volume, traffic distribution, axle load spectra, climate, etc. are measured directly for the site or project.
- *Level 2*. Level 2 input parameters represent measured regional average values. The input parameters and the calibration coefficients of the software are estimated from correlations or regression equations developed based on local or state-level conditions.
- *Level 3*. Most of the input parameters and the calibration coefficients of the software are ME default values, which are based on global or regional default values. This input level is the least accurate, and commonly used in noncrucial pavements.

In the next few chapters, materials properties required for pavement design are discussed. For better understanding of these materials properties, knowledge of the

AASHTOWare pavement ME design input levels is essential. The input level to be used in the AASHTOWare pavement ME design method depends on the availability of the input parameters at hand.

1.4.3 Other Design Methods

Some other design methods such as Asphalt Institute (AI) method and local methods are also available. These methods are not discussed here as these methods are obsolete and have no trace of being used now-a-days. Noncritical local roads or parking lots might be designed using these methods.

1.4.4 International Design Methods

Pavement design methodologies in different countries of the world are different. Some of the design methodologies followed in certain countries are discussed in App. A.

1.5 Other Design Considerations

Besides the structural design of pavement, some other essential components of design need to be considered: for example, drainage design to save pavement material from moisture damage, and geometric design for passengers' comfort and safety. Drainage design includes design and selection of drainage materials section, installation, and slopping (grading) of pavement. Geometric design includes designing vertical and horizontal curves for passengers' comfort and safety. Safety measure also includes pavement marking, rumble strips, barriers, etc. The designs of these components are discussed in later chapters of this textbook.

1.6 Summary

This chapter is the introduction to pavements, types, design methods, and philosophies. Pavement structure consists of layers of materials and compacted in place to provide a smooth surface to drive traffic comfortably, efficiently, economically, and safely. Different layers of pavement serve different purposes. The main purpose of all layers is to distribute the tire load to a larger area of natural soil.

Pavements are classified into two broad categories: flexible and rigid. The combination of flexible and rigid pavements is also possible and known as composite pavement. Flexible pavements use asphalt materials in the surface layer, which flexes with load and temperatures, whereas rigid pavements use PCC slab as a surface layer, which are less flexible under loads. Rigid pavements can be further classified into four types: JPCP, JRCP, CRCP, and PCP.

Pavement structures are designed with the consideration that there might be cracks, some permanent deformations, roughness, etc. during its service life. Two major pavement design guides used in the United States are the AASHTO 1993 design guide, which is empirical in nature, and the AASHTOWare pavement ME design guide, which is mechanistic-empirical in nature. The AASHTO 1993 method is decreasing its popularity and the AASHTOWare pavement ME design is rocketing up its popularity.

1.7 Fundamentals of Engineering (FE) Exam-Style Questions

FE1.1 A pavement is NOT expected to meet the requirement of:

A. Sufficient thickness to distribute the wheel-induced stresses to a reduced value on the subgrade
B. Structurally adequate to keep the cracking and deformation within tolerable limits
C. Structurally strong to withstand all types of stresses imposed upon it
D. Adequate coefficient of friction to prevent skidding of vehicles
E. Optimum traffic signal timing for efficient passage of vehicles

Solution E

Traffic signal timing is not covered in pavement design.

FE1.2 A pavement has 3.0-in. (75-mm) asphalt layer underlain by a 4.0-in. (100-mm) PCC layer. The pavement is a:

A. Flexible pavement
B. Rigid pavement
C. Composite pavement
D. Asphalt pavement

Solution C

Flexible pavement consists of asphalt layer only. Rigid pavement consists of PCC layer only. Composite pavement consists of both asphalt and PCC layers.

FE1.3 The AASHTOWare pavement ME design method is:

A. Empirical in nature
B. Mechanistic in nature
C. Mechanistic-empirical in nature
D. Scientific in nature

Solution C

The AASHTOWare pavement ME design method is mechanistic-empirical in nature. The developed stress-strain in pavement is determined using mechanics, but the amount of distress is predicted using empirical equations.

1.8 Practice Problems

1.1 List some requirements of pavements.

1.2 Classify the types of pavements.

1.3 Define and state the purpose of tie bars and dowel bars.

1.4 Differentiate the faulting and pumping phenomena.

1.5 Discuss the major two pavement design methods.

1.6 What are the three input levels of the AASHTOWare pavement ME design method?

CHAPTER 2

Stress-Strain in Pavement

2.1 Background

Flexible or rigid pavement, undergoes deformation upon applying loading. The deformation behavior and the resulting stress-strain at different layers of pavement are discussed in this chapter. Determining the stress-strain or deformation is essential to analyze and design pavement. Damages in different layers of pavement depend on the repeated stress-strain developed in pavement layers due to the applied repeated wheel load. The cracking or deformation is the consequence of the resulting damages. Deformation behavior of pavement is not similar to that of other civil engineering structures.

Let us consider a simply supported beam used in classical civil engineering mechanics shown in Fig. 2.1. Two concentrated loads of equal magnitude (P) are acting on the beam. The beam deforms as shown and a horizontal tensile strain (ε_h or ε_t) develops at the bottom of the beam. The supports A and B are rigid and do not undergo any deformation. The properties of the beam such as modulus of elasticity also do not change with time and season. In fact, in all classical civil engineering structures, supports are considered rigid, do not undergo any remarkable deformation, and do not change their properties with time or season.

The solution of this beam, that is, the resulting deflection, stress-strain, etc., can be obtained from knowledge of the strength of materials. Now, let us consider the scenario for a flexible pavement. If we consider the loading of a single-axle (single axle on each end) load on flexible pavement, it may look like as shown in Fig. 2.2.

Each wheel on each end of the axle applies load on the pavement surface. The support of the asphalt slab is the underneath unbound aggregate base layer which provides continuous support. The asphalt slab is not like a rigid beam; it is more like a flexible member that can bend any point upon loading. The support base layer is also not a rigid support. It also undergoes deformation remarkably upon loading. Horizontal tensile strain (ε_h or ε_t) develops at the bottom of asphalt layer and vertical strain (ε_v) develops in all layers including the asphalt layer. Thus, the deformation behavior of the asphalt slab is different from the traditional beam or slab. Another difference in flexible pavement compared to the classical beam/slab is that the materials properties such as modulus change with time. For example, asphalt softens upon heating and hardens upon cooling. Thus, its modulus changes with day-night or summer-winter temperature fluctuations. In addition, its modulus changes with frequency of loading as well. The higher the frequency (fast loading), the higher the modulus and vice versa. For underneath unbound layers, their modulus does not change with temperature variations unless they are asphalt treated. However, the moduli of unbound aggregate layers

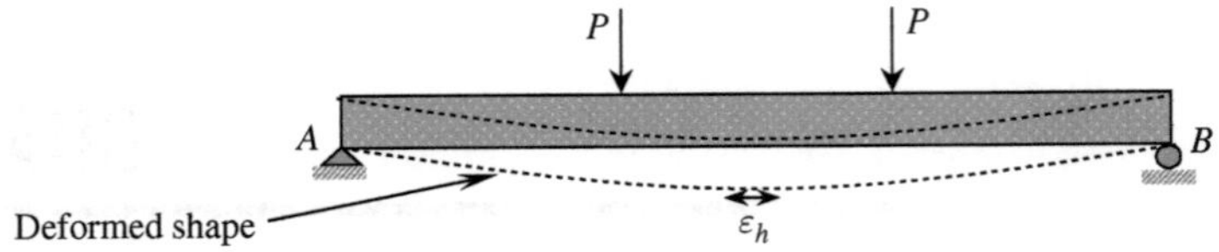

FIGURE 2.1 A classical structural member used in civil engineering.

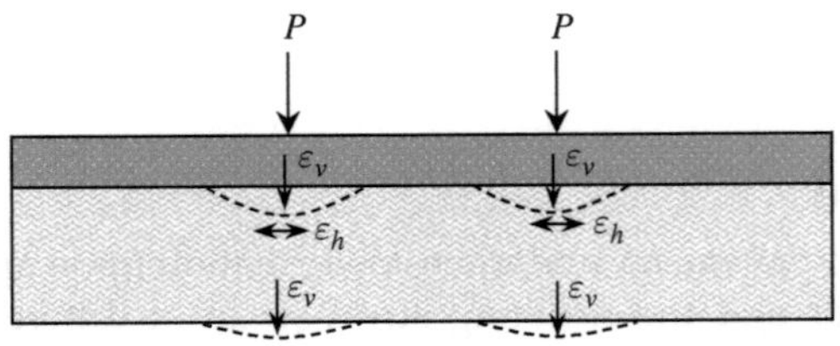

FIGURE 2.2 Loading on a flexible pavement.

or soil layers change with the moisture content and magnitude of loading. Due to seasonal variations in moisture content or capillary rise of underground water or infiltration of surface water, modulus of unbound aggregate layers or soil layers decreases with increase in moisture content. Thus, the reaction capacity of unbound aggregate layers or soil layers changes throughout the year. The difference in mechanical properties between classical and pavement structures is summarized in Table 2.1.

2.2 Stress-Stain in Flexible Pavement

Analytical determinations of stress-strain in pavement layers are impossible. This is because the accurate mechanical and nonmechanical properties of pavement layers are not known. For example, asphalt layer is very often considered elastic or viscoelastic. However, it might have some plastic properties as well, which is still under research. Also, the loading on pavement is considered static although dynamic loading is being considered in researches nowadays. In fact, the loading on pavement is dynamic with impact action as there are so many large surface roughness in pavement that make the tire load as impact-type load. The axle types and the distances among wheel vary from vehicle to vehicle. In addition, there are no well-developed computational tools (equations) which can deal with all types of conditions in pavement. For learning purpose, only few idealized cases whose solutions are available in the literature are discussed in this chapter. These computational solutions available in the literature are rarely used while designing pavement. For accurate analysis, finite-element analysis is performed nowadays. The following assumptions are used in the computational solutions available in the literature:

- Each layer is homogeneous and behaves linear-elastic.
- The material is infinite and weightless.
- Each layer is fully bonded with adjacent layers and has a finite thickness.

Parameter	Classical structures	Flexible pavement	Rigid pavement
Modulus	Constant for all members	For asphalt: • Decreases with temperature • Increases with frequency of loading	For concrete: • Increases with moisture • Decreases with temperature
		For base/soil: • Changes with magnitude of loading • Changes with moisture	For base/soil: • Changes with magnitude of loading • Changes with moisture
Support type	Discrete	Continuous	Continuous
Loading magnitude	Assumed constant throughout its life	Changes every moment in terms of magnitude and point of application.	Changes every moment in terms of magnitude and point of application.
Loading mode	Static	Dynamic with impact action	Dynamic with impact action
Serviceability	Limited serviceability; the structure becomes unusable even if a small amount of cracking is seen	Provides service even after severe amount of distresses	Provides service even after some amount of distresses

TABLE 2.1 Difference in Mechanical Properties between Classical and Pavement Structures

2.2.1 Single-Layer Elastic Analysis

Different theories are available in the literature to determine the stress-strain at some depth from the surface. The simplest way to determine the stress-strain under wheel load is to consider the pavement system as a homogeneous half-space. Half-space means the domain with infinitely large area and an infinite depth with a top plane. Boussinesq (1885) theory was developed based on a homogeneous linear-elastic half-space. However, the real pavement does not satisfy this assumption as it has several layers of distinctly different properties. However, this theory is a good starting point for learning stresses-strains in pavement layers.

Stress below a Concentrated Load

Boussinesq (1885) solved the problem of stresses produced at any point in a homogeneous and elastic medium as the result of a point load applied on the surface of an infinitely large half-space. Let us consider a semi-infinite half-surface, as shown in Fig. 2.3, where a concentrated load of P is acting at the surface. A point of concern is located at a depth of z and radial distance of r.

In Fig. 2.3, P is the applied concentrated load, z is any vertical distance, r is any radial distance (zero along the centerline of the wheel), σ_z is the vertical stress, σ_r is the

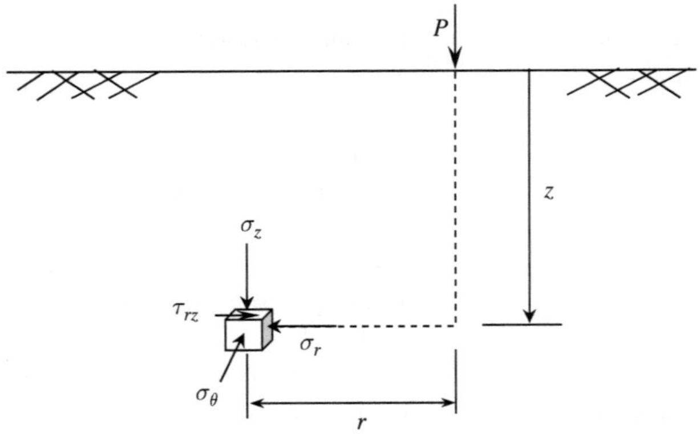

FIGURE 2.3 Stresses under a concentrated load for a single-layer elastic system.

radial stress, σ_θ is the tangential stress, and τ_{zr} is the horizontal shear stress in the radial direction. Arrows show the positive sign convention.

Boussinesq's solutions for stresses at a point caused by the point load P are presented in Eqs. (2.1) to (2.4):

$$\sigma_z = \frac{3P}{2\pi z^2\left[1+\left(\frac{r}{z}\right)^2\right]^{5/2}} \tag{2.1}$$

$$\sigma_r = \frac{P}{2\pi}\left[\frac{3r^2 z}{(r^2+z^2)^{5/2}} - \frac{1-2\mu}{r^2+z^2+z\sqrt{r^2+z^2}}\right] \tag{2.2}$$

$$\sigma_\theta = \frac{P}{2\pi}(1-2\mu)\left[\frac{z}{(r^2+z^2)^{3/2}} - \frac{1}{r^2+z^2+z\sqrt{r^2+z^2}}\right] \tag{2.3}$$

$$\tau_{zr} = \frac{P}{2\pi}\frac{3rz^2}{(r^2+z^2)^{5/2}} \tag{2.4}$$

where μ is Poisson's ratio of the material.

The resulting strains are presented in Eqs. (2.5) to (2.8):

$$\varepsilon_z = \frac{1}{E}[\sigma_z - \mu(\sigma_r + \sigma_\theta)] \tag{2.5}$$

$$\varepsilon_r = \frac{1}{E}[\sigma_r - \mu(\sigma_z + \sigma_\theta)] \tag{2.6}$$

$$\varepsilon_\theta = \frac{1}{E}[\sigma_\theta - \mu(\sigma_r + \sigma_z)] \tag{2.7}$$

$$\gamma_{zr} = \frac{2\tau_{zr}(1+\mu)}{E} = \frac{\tau_{zr}}{G} \tag{2.8}$$

where ε_z = Vertical strain
ε_r = Radial strain
ε_θ = Tangential strain
γ_{zr} = Horizontal shear strain in the radial direction

If Δ_z and Δ_h are the vertical and horizontal deflections, respectively, then:

$$\varepsilon_z = \frac{\partial \Delta_z}{\partial z} \tag{2.9}$$

$$\text{Or, } \int \partial \Delta_z = \int \varepsilon_z \partial z \tag{2.10}$$

On integrating,

$$\Delta_z = \frac{P}{2\pi E}\left[(1+\mu)z^2(r^2+z^2)^{-3/2} + 2(1-\mu^2)(r^2+z^2)^{-1/2}\right] \tag{2.11}$$

It should be noted that at the surface (i.e., $z = 0$), the vertical deflection is:

$$\Delta_z = \frac{P(1-\mu^2)}{\pi E r} \tag{2.12}$$

Similarly,

$$\varepsilon_r = \frac{\partial \Delta_h}{\partial r} \tag{2.13}$$

$$\text{Or, } \int \partial \Delta_h = \int \varepsilon_r \partial r \tag{2.14}$$

On integrating,

$$\Delta_h = \frac{P(1+\mu)(1-2\mu)}{2\pi r E}\left[z(r^2+z^2)^{-1/2} - 1 + \frac{1}{(1-2\mu)}(r^2 z)(r^2+z^2)^{-3/2}\right] \tag{2.15}$$

Example 2.1: Responses Due to Concentrated Load A wheel applies 4.5-kip load on a semi-infinite elastic space, as shown in Fig. 2.4. The modulus of elasticity of the material is 12 ksi and Poisson's ratio is 0.40. Assume the load is applied on the surface as a concentrated force.

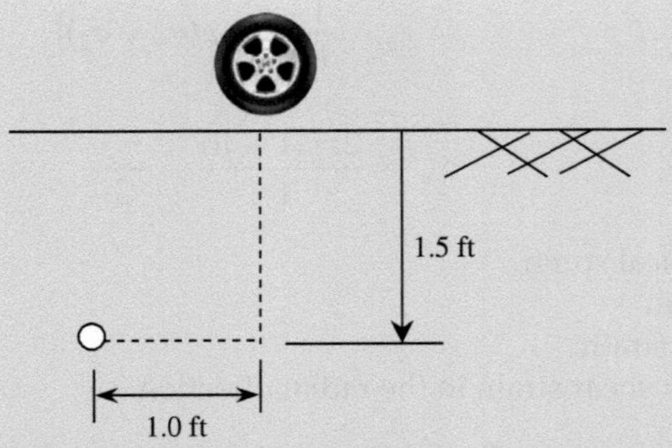

FIGURE 2.4 Profile of the semi-infinite elastic space for Example 2.1.

Calculate the following at a depth of 1.5 ft and a radial offset of 1.0 ft:

a. Vertical stress
b. Radial stress
c. Tangential stress
d. Horizontal shear stress in the radial direction
e. Vertical strain
f. Radial strain
g. Tangential strain
h. Vertical deformation
i. Horizontal deformation

Solution

Applied load, $P = 4{,}500$ lb
Modulus of elasticity, $E = 12{,}000$ psi
Poisson's ratio, $\mu = 0.40$
Depth, $z = 1.5$ ft $= 18$ in.
Radial offset, $r = 1.0$ ft $= 12$ in.

a. Vertical stress

$$\sigma_z = \frac{3P}{2\pi z^2\left[1+\left(\frac{r}{z}\right)^2\right]^{5/2}} = = \frac{3(4{,}500\text{ lb})}{2\pi(18\text{ in.})^2\left[1+\left(\frac{12\text{ in.}}{18\text{ in.}}\right)^2\right]^{5/2}} = 2.645\text{ psi}$$

b. Radial stress

$$\sigma_r = \frac{P}{2\pi}\left[\frac{3r^2z}{(r^2+z^2)^{5/2}} - \frac{1-2\mu}{r^2+z^2+z\sqrt{r^2+z^2}}\right]$$

$$= \frac{(4{,}500)}{2\pi}\left[\frac{3(12)^2(18)}{(12^2+18^2)^{5/2}} - \frac{1-2(0.40)}{12^2+18^2+18\sqrt{12^2+18^2}}\right]$$

$$= 1.008 \text{ psi}$$

c. Tangential stress

$$\sigma_\theta = \frac{P}{2\pi}(1-2\mu)\left[\frac{z}{(r^2+z^2)^{3/2}} - \frac{1}{r^2+z^2+z\sqrt{r^2+z^2}}\right]$$

$$= \frac{(4{,}500)}{2\pi}[1-2(0.40)]\left[\frac{18}{(12^2+18^2)^{3/2}} - \frac{1}{12^2+18^2+18\sqrt{12^2+18^2}}\right]$$

$$= 1.088 \text{ psi}$$

d. Horizontal shear stress in the radial direction

$$\tau_{zr} = \frac{P}{2\pi}\frac{3rz^2}{(r^2+z^2)^{5/2}} = \frac{4{,}500}{2\pi}\frac{3(12)(18)^2}{(12^2+18^2)^{5/2}} = 1.763 \text{ psi}$$

e. Vertical strain

$$\varepsilon_z = \frac{1}{E}[\sigma_z - \mu(\sigma_r + \sigma_\theta)] = \frac{1}{12{,}000}[2.645 - 0.40(1.008 + 0.088)] = 0.0001897$$

f. Radial strain

$$\varepsilon_r = \frac{1}{E}[\sigma_r - \mu(\sigma_z + \sigma_\theta)] = \frac{1}{12{,}000}[1.008 - 0.40(2.645 + 0.088)] = -0.000007$$

g. Tangential strain

$$\varepsilon_\theta = \frac{1}{E}[\sigma_\theta - \mu(\sigma_r + \sigma_z)] = \frac{1}{12{,}000}[0.088 - 0.40(2.645 + 1.008)] = -0.0001145$$

h. Vertical deformation

$$\Delta_z = \frac{P}{2\pi E}\left[(1+\mu)z^2(r^2+z^2)^{-3/2} + 2(1-\mu^2)(r^2+z^2)^{-1/2}\right]$$

$$\Delta_z = \frac{4{,}500}{2\pi(12{,}000)}\left[(1+0.40)(18)^2(12^2+18^2)^{-3/2} + 2(1-0.40^2)(12^2+18^2)^{-1/2}\right]$$

$$= 0.0073 \text{ in.}$$

i. Horizontal deformation

$$\Delta_h = \frac{P(1+\mu)(1-2\mu)}{2\pi r E}\left[z(r^2+z^2)^{-1/2} - 1 + \frac{1}{(1-2\mu)}(r^2 z)(r^2+z^2)^{-3/2}\right]$$

$$\Delta_h = \frac{4{,}500(1+0.40)(1-2(0.40))}{2\pi(12)(12{,}000)}\left[(18)(12^2+18^2)^{-1/2} - 1 + \frac{1}{(1-2(0.40))}(12^2(18))(12^2+18^2)^{-3/2}\right]$$

$$= 0.00155 \text{ in.}$$

Answers

a. Vertical stress: = 2.645 psi

b. Radial stress: 1.008 psi

c. Tangential stress: 0.088 psi

d. Horizontal shear stress in the radial direction: 1.763 psi

e. Vertical strain: 0.0001897

f. Radial strain: –0.000007

g. Tangential strain: –0.0001145

h. Vertical deformation: 0.0073 in.

i. Horizontal deformation: 0.00155 in.

Note: Strain has no unit. It is very often expressed as in./in. or m/m, or so.

Stress below a Uniform Circular Load

Let us consider a surface has a uniform distributed stress of p on a circular area of radius a. Note that tire-imprint area is most often considered a circle. Two types of cases may be considered bearing in mind the rigidity of the load application method, as shown in Fig. 2.5:

1. For flexible-plate load application such as load applied by rubber tire
2. For rigid-plate load application such as load applied during plate loading test

For flexible-plate load application such as load applied by rubber tire, stress expressions for points on the centerline of the circular load (i.e., $r = 0$) are as follows:

$$\sigma_z = p\left[1 - \frac{z^3}{\left[a^2+z^2\right]^{3/2}}\right] \tag{2.16}$$

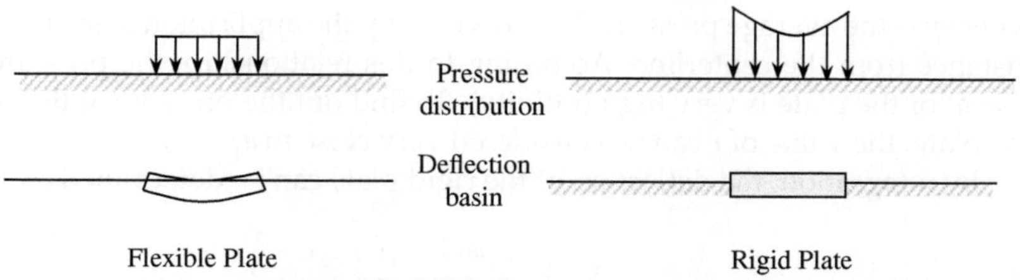

FIGURE 2.5 Deflection behavior under flexible and rigid plates.

$$\sigma_r = \frac{p}{2}\left[(1+2\mu) - \frac{2z(1+\mu)}{\sqrt{a^2+z^2}} + \frac{z^3}{(a^2+z^2)^{3/2}}\right] \tag{2.17}$$

$$\sigma_r = \sigma_\theta \tag{2.18}$$

$$\tau_{zr} = 0 \tag{2.19}$$

The corresponding strains for the flexible-plate load application can be calculated as follows:

$$\varepsilon_z = \frac{p(1+\mu)}{E}\left[1 - 2\mu + \frac{2\mu z}{(a^2+z^2)^{1/2}} - \frac{z^3}{(a^2+z^2)^{3/2}}\right] \tag{2.20}$$

$$\varepsilon_r = \frac{p(1+\mu)}{2E}\left[1 - 2\mu - \frac{2(1-\mu)z}{(a^2+z^2)^{1/2}} + \frac{z^3}{(a^2+z^2)^{3/2}}\right] \tag{2.21}$$

The vertical deflection under the centerline of the circular load for the flexible-plate load application is given by:

$$\Delta_z = \frac{pa(1+\mu)}{2E}\left[\frac{a}{(a^2+z^2)^{1/2}} + \frac{1-2\mu}{a}\left((a^2+z^2)^{1/2} - z\right)\right] \tag{2.22}$$

If $z = 0$ (i.e., at the surface), the vertical deflection under the centerline of the flexible-plate load application is given by:

$$\Delta_{zo} = \frac{2pa(1-\mu^2)}{E} \tag{2.23}$$

For rigid-plate load application such as load applied by plate test, pressure distribution is nonuniform and can be expressed as follows:

$$p(r) = \frac{pa}{2\left(a^2 - r^2\right)^{1/2}} \tag{2.24}$$

where p is the average pressure (load divided by the application area) and r is the radial distance from the centerline. According to this relationship, the pressure at the edge ($r = a$) of the plate is very high (infinity). To find out the pressure at the edge ($r = a$) of the plate, the value of r can be considered very close to a.

By integration, the deflection of the rigid plate can be determined as:

$$\Delta_{zo} = \frac{\pi pa(1-\mu^2)}{2E} \tag{2.25}$$

With comparison of the surface deflection [Eqs. (2.23) and (2.25)], the deflection produced by the rigid plate is 79% of that produced by the flexible plate. This is reasonable as the pressure at the center of plate is less at the center of loading for rigid plate.

Example 2.2: Responses Due to Circular Load in Flexible Pavement A wheel applies a 65-psi load on a 6-in.-diameter area on a semi-infinite elastic space, as shown in Fig. 2.6. The modulus of elasticity of the material is 12 ksi and Poisson's ratio is 0.40. Assume the load is applied on the surface as a uniform circular load using flexible plate.

Calculate the following at a depth of 1.5 ft under the centerline of the load:

a. Vertical stress
b. Radial stress
c. Tangential stress
d. Vertical strain
e. Radial strain
f. Vertical deflection under the centerline

Also, calculate the vertical deformation at the surface just below the loading.

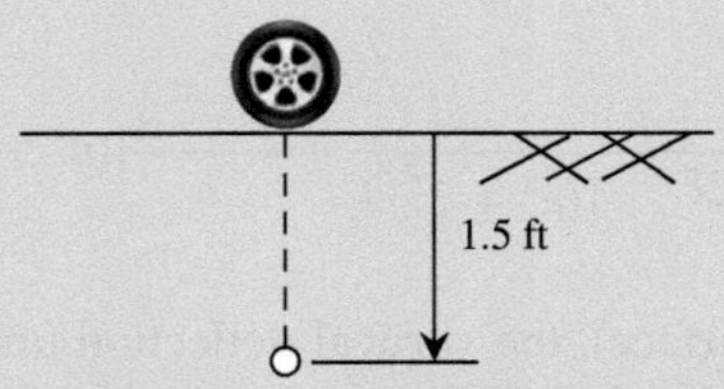

Figure 2.6 Profile of the semi-infinite elastic space for Example 2.2.

Solution

Applied pressure, $p = 65$ psi
Modulus of elasticity, $E = 12{,}000$ psi
Poisson's ratio, $\mu = 0.40$
Depth, $z = 1.5$ ft $= 18$ in.
Radius of loading area, $a = 3$ in.

a. Vertical stress

$$\sigma_z = p\left[1 - \frac{z^3}{\left[a^2 + z^2\right]^{3/2}}\right] = (65\text{ psi})\left[1 - \frac{(18\text{ in.})^3}{\left[(3\text{ in.})^2 + (18\text{ in.})^2\right]^{3/2}}\right] = 2.62\text{ psi}$$

b. Radial stress

$$\sigma_r = \frac{p}{2}\left[(1 + 2\mu) - \frac{2z(1+\mu)}{\sqrt{a^2 + z^2}} + \frac{z^3}{(a^2 + z^2)^{3/2}}\right]$$

$$= \frac{65\text{ psi}}{2}\left[(1 + 2(0.40)) - \frac{2(18\text{ in.})(1 + 0.40)}{\sqrt{(3\text{ in.})^2 + (18\text{ in.})^2}} + \frac{(18\text{ in.})^3}{\left((3\text{ in.})^2 + (18\text{ in.})^2\right)^{3/2}}\right]$$

$$= -0.0705\text{ psi}$$

c. Tangential stress

$$\sigma_\theta = \sigma_r = -0.0705\text{ psi}$$

d. Vertical strain

$$\varepsilon_z = \frac{p(1+\mu)}{E}\left[1 - 2\mu + \frac{2\mu z}{(a^2 + z^2)^{1/2}} - \frac{z^3}{(a^2 + z^2)^{3/2}}\right]$$

$$= \frac{(65\text{ psi})(1 + 0.40)}{12{,}000\text{ psi}}\left[1 - 2(0.40) + \frac{2(0.40)(18\text{ in.})}{\left((3\text{ in.})^2 + (18\text{ in.})^2\right)^{1/2}} - \frac{(18\text{ in.})^3}{\left((3\text{ in.})^2 + (18\text{ in.})^2\right)^{3/2}}\right]$$

$$= 0.0002228$$

e. Radial strain

$$\varepsilon_r = \frac{p(1+\mu)}{2E}\left[1 - 2\mu - \frac{2(1-\mu)z}{(a^2 + z^2)^{1/2}} + \frac{z^3}{(a^2 + z^2)^{3/2}}\right]$$

$$= \frac{(65\text{ psi})(1 + 0.40)}{2(12{,}000\text{ psi})}\left[1 - 2(0.40) - \frac{2(1 - 0.40)(18\text{ in.})}{\left((3\text{ in.})^2 + (18\text{ in.})^2\right)^{1/2}} + \frac{(18\text{ in.})^3}{\left((3\text{ in.})^2 + (18\text{ in.})^2\right)^{3/2}}\right]$$

$$= -0.0000908$$

f. Vertical deflection under the centerline

$$\Delta_z = \frac{pa(1+\mu)}{2E}\left[\frac{a}{\left(a^2+z^2\right)^{1/2}} + \frac{1-2\mu}{a}\left(\left(a^2+z^2\right)^{1/2} - z\right)\right]$$

$$= \frac{(65\text{ psi})(3\text{ in.})(1+0.40)}{2(12{,}000\text{ psi})}\left[\frac{3\text{ in.}}{\left((3\text{ in.})^2+(18\text{ in.})^2\right)^{1/2}} + \frac{1-2(0.40)}{3\text{ in.}}\left(\left((3\text{ in.})^2+(18\text{ in.})^2\right)^{1/2} - 18\text{ in.}\right)\right]$$

$$= 0.0021\text{ in.}$$

Vertical deformation at the surface just below the loading:

$$\Delta_{zo} = \frac{2pa(1-\mu^2)}{E} = \frac{2(65\text{ psi})(3\text{ in.})\left(1-(0.40)^2\right)}{12{,}000\text{ psi}} = 0.0273\text{ in.}$$

Answers

a. *Vertical stress: 2.62 psi*

b. *Radial stress: – 0.0705 psi*

c. *Tangential stress: –0.0705 psi*

d. *Vertical strain: 0.0002228*

e. *Radial strain: – 0.0000908*

f. *Vertical deflection under the centerline: 0.0021 in. Vertical deformation at the surface just below the loading: 0.0273 in.*

Example 2.3: Responses Due to Circular Load in Rigid Pavement A plate load test was conducted using a 0.3-m-diameter rigid plate applying a load of 80 kN on a subgrade, as shown in Fig. 2.7. A deformation of 2.5 mm was measured after applying this load. Poisson's ratio of the subgrade is 0.45. Determine the elastic modulus of the subgrade.

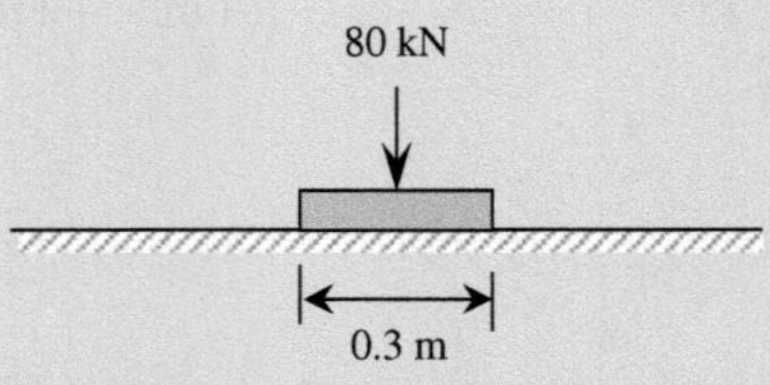

FIGURE 2.7 Plate load test for Example 2.3.

Solution

Applied load, $P = 80{,}000$ N
Radius of plate, $a = 0.3\text{ m}/2 = 0.15$ m
Poisson's ratio, $\mu = 0.45$
Surface deflection, $\Delta_{zo} = 2.5\text{ mm} = 0.0025$ m

Applied average pressure, $p = \frac{P}{\pi a^2} = \frac{80{,}000\text{ N}}{\pi(0.15\text{ m})^2} = 1{,}131{,}766\frac{\text{N}}{\text{m}^2}$

Now,

$$\Delta_{zo} = \frac{\pi pa(1-\mu^2)}{2E}$$

$$E = \frac{\pi pa(1-\mu^2)}{2\Delta_{zo}}$$

$$= \frac{\pi\left(1{,}131{,}766\,\frac{\text{N}}{\text{m}^2}\right)(0.15\text{ m})(1-0.45^2)}{2(0.0025\text{ m})}$$

$$= 85{,}066{,}666\text{ N/m}^2$$

$$= 85\text{ MN/m}^2$$

$$= 85\text{ MPa}$$

Answer *85 MPa*

2.2.2 Two-Layer Elastic Analysis

There are a few methods of solving layered elastic media. Method of equivalent thicknesses is an approximate method for solving the elastic layered system problem and consists of translating multiple layers of different moduli into an equivalent single layer, hence known as the method of equivalent thicknesses. For a system of two layers, such as the one shown in Fig. 2.8, the top layer with thickness h can be translated into an equivalent thickness h_e, with a modulus E_2. For two-layer system of equal Poison's ratio (i.e., $\mu_1 = \mu_2$), the equivalent layer thickness of the top layer is given by Odemark (1949):

$$h_e = 0.9h\left(\frac{E_1}{E_2}\right)^{1/3} \tag{2.26}$$

where 0.9 is an approximation factor. This allows utilizing the single-layer solutions in computing pavement responses in the lower layer.

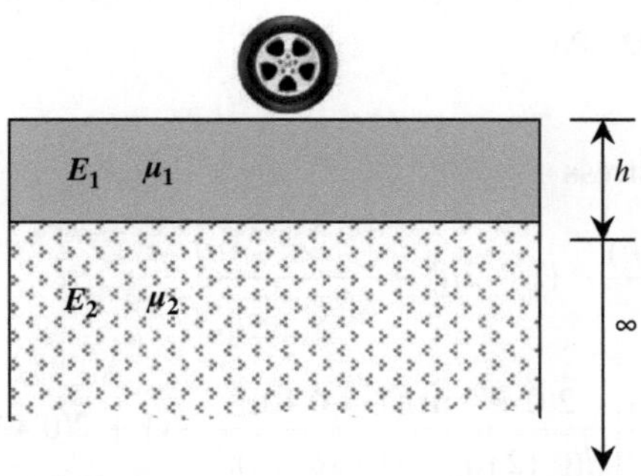

FIGURE 2.8 Schematics of two-layer system.

Example 2.4: Responses in Two-Layer System A flexible pavement surface layer is 0.25 m thick resting on a semi-infinite subgrade layer. The load consists of a circular tire with a 0.12-m radius carrying a uniform pressure of 650 kPa. The layer moduli are 1,200 and 120 MPa, respectively, and μ is 0.4 for both layers.

Calculate the following at the bottom of the surface layer at the centerline of the load:

a. Vertical stress
b. Radial stress
c. Vertical deformation

Solution

Equivalent thickness of the top layer in terms of the modulus of the bottom layer as:

$$h_e = 0.9h\left(\frac{E_1}{E_2}\right)^{1/3} 0.9(0.25\text{ m})\left(\frac{1{,}200\text{ MPa}}{120\text{ MPa}}\right)^{1/3} = 0.485\text{ m}$$

This means the top of the 0.25-m-thick layer with a modulus of 1,200 MPa is equivalent to a 0.485-m-thick layer with a modulus of 120 MPa.

Applied pressure, $p = 650{,}000$ Pa
Modulus of elasticity, $E = 120{,}000{,}000$ Pa
Poisson's ratio, $\mu = 0.40$
Depth, $z = 0.485$ m
Radial offset, $r = 0$
Radius of loading area, $a = 0.12$ m

a. Vertical stress

$$\sigma_z = p\left[1 - \frac{z^3}{\left[a^2 + z^2\right]^{3/2}}\right]$$

$$= 650{,}000\text{ Pa}\left[1 - \frac{(0.485\text{ m})^3}{\left((0.12\text{ m})^2 + (0.485\text{ m})^2\right)^{3/2}}\right]$$

$$= 55{,}425\text{ Pa}$$

b. Radial stress

$$\sigma_r = \frac{p}{2}\left[\frac{2z(1+\mu)}{\sqrt{a^2 + z^2}} - (1 + 2\mu) - \frac{z^3}{(a^2 + z^2)^{3/2}}\right]$$

$$= \frac{650{,}000\text{ Pa}}{2}\left[\frac{2(0.485\text{ m})(1 + 0.40)}{\sqrt{(0.12\text{ m})^2 + (0.485\text{ m})^2}} - (1 + 2(0.40)) - \frac{(0.485\text{ m})^3}{\left((0.12\text{ m})^2 + (0.485\text{ m})^2\right)^{3/2}}\right]$$

$$= 1{,}075\text{ Pa}$$

c. Vertical deformation

$$\Delta_z = \frac{2pa(1-\mu^2)}{E}$$

$$= \frac{2(650{,}000\,\text{Pa})(0.12\,\text{m})(1-(0.40)^2)}{120{,}000{,}000\,\text{Pa}}$$

$$= 0.0011\ \text{m}$$

$$= 1.1\ \text{mm}$$

Answers

a. *Vertical stress: 55,425 Pa*

b. *Radial stress: 1,075 Pa*

c. *Vertical deformation: 1.1 mm*

There are some other methods which use charts to determine vertical stresses, vertical deflection, etc. However, these charts are valid for $\mu = 0.50$ only, which is not realistic for pavement materials. The commonly used μ values for portland cement concrete, asphalt concrete, aggregate base, subbase, and subgrade are 0.15, 0.35, 0.40, 0.45, and 0.45, respectively (Islam et al., 2016).

2.2.3 Multilayer Elastic Analysis

A multilayer elastic system consists of multiple finite-thickness layers resting on a subgrade of infinite thickness. It is an idealized representation of multiple pavement layers, such as asphalt concrete friction and leveling layers, base layers, and subbase layers, each having different elastic properties. Elastic response solutions to this system were developed by extending the Burmister analytical approach for the two-layer system to multiple layers. A variety of software implement such solutions, such as ELSYM, DAMA, KENLAYER, and EVERSTRESS. These programs have, to a great extent, similar features and input requirements and can handle up to five layers, including an infinite depth subgrade. They accept multiple circular tires and compute stresses at any location in the layered system. The calculations are made in the radial coordinate system, and then translated to a Cartesian coordinate system, with its origin in the middle of a tire imprint.

Granular materials exhibit a stress-dependent behavior; that is, their resilient (i.e., elastic) modulus is a function of the stress state at any given location. Coarse-grained layers (e.g., base, subbase) exhibit an exponential dependence to bulk stress, while fine-grained/cohesive layers (e.g., subgrade) exhibit an exponential dependence to deviatoric stress. Layer elastic analysis can handle this nonlinearity using a piecewise linear iterative algorithm. Seed moduli are initially assigned; the layered analysis is conducted to compute initial stresses; and in subsequent iterations, the moduli are updated on the basis of the calculated stresses.

2.3 Stress-Stain in Rigid Pavement

Stresses in concrete pavement are the result of the interaction of a number of factors, which can be grouped into three main categories:

1. Curling stress
2. Traffic-induced stress
3. Friction-induced stress

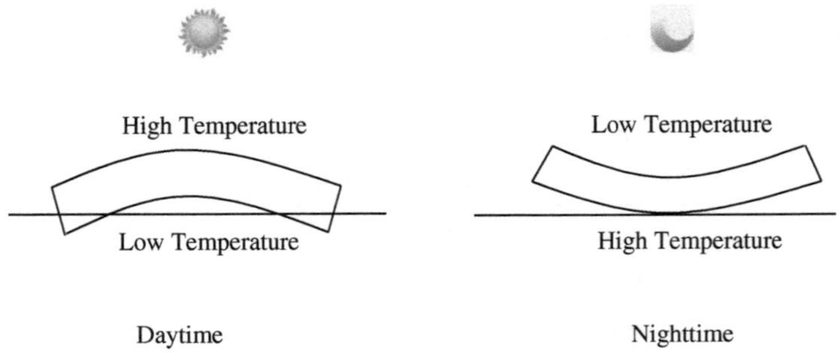

FIGURE 2.9 Expansion/contraction of concrete slab during the day and at night.

2.3.1 Curling Stress

Curling stress occurs due to the temperature variations at the top and at the bottom of the concrete slab. The temperature at the top of the slab is greater than that at the bottom of the slab during the day. At night, the top of the concrete slab is cooler than the bottom, as shown in Fig. 2.9. During the day, the top of the concrete expands with respect to the neutral axis as the top is hotter than the bottom. This means the concrete slab tends to move up (concave downward). However, the weight of the slab tends to pull it down which results in compressive stress at the top of the slab and tensile stress at the bottom of the slab. At night, the scenario is opposite. The bottom of the concrete slab expands and the edges tend to go up. While the weight of the slab tends to balance it, tensile stress develops at the top of the slab and compressive stress develops at the bottom of the slab.

Concrete slab is a three-dimensional member. Its bending behavior is thus not similar to a concrete beam. Concrete beam bends about only one direction. However, concrete slab bends about two directions.

Concrete slab is analyzed using the plate theory, which considers the plate has two axes (x and y). The thickness of the plate is so small and is discarded. When the plate bends about the x-axis, stress develops about both the x and y axes. However, strain develops about the x-axis only. Similarly, when the plate bends about the y-axis, stress develops about both the x and y axes. However, strain develops about the y-axis only.

Now, let us consider the bending about the x-direction only. Again to remind you, stress exists about both the x and y axes and strain develops about the x-axis only.

From Hooke's law, strain about the x-axis (ε_x) is:

$$\varepsilon_x = \frac{1}{E}(\sigma_x - \mu\sigma_y) \tag{2.27}$$

where E = Modulus of elasticity of the plate or slab
μ = Poisson's ratio of the plate or slab
σ_x = Stress about the x-axis in the plate or slab
σ_y = Stress about the y-axis in the plate or slab

Similarly, strain about y-axis (ε_y) is:

$$\varepsilon_y = \frac{1}{E}(\sigma_y - \mu\sigma_x) \tag{2.28}$$

As the slab bends only about the x-axis, $\varepsilon_y = 0$

$$\frac{1}{E}(\sigma_y - \mu\sigma_x) = 0 \tag{2.29}$$

$$\sigma_y = \mu\sigma_x \tag{2.30}$$

Then,

$$\varepsilon_x = \frac{1}{E}[\sigma_x - \mu(\mu\sigma_x)] \tag{2.31}$$

$$\varepsilon_x = \frac{1}{E}(\sigma_x - \mu^2\sigma_x) \tag{2.32}$$

$$\varepsilon_x = \frac{\sigma_x}{E}(1 - \mu^2) \tag{2.33}$$

$$\sigma_x = \frac{E\varepsilon_x}{(1 - \mu^2)} \tag{2.34}$$

σ_x indicates the stress along the bending direction and σ_y is the stress perpendicular to the bending direction. Similar expressions can be written for bending in the y-direction only. If the temperature difference between the top and bottom of concrete slab is ΔT, the temperature difference between the neutral axis and the top/bottom of concrete slab is $\Delta T/2$ considering the linear variation in temperature throughout the slab depth. The real concrete slab in pavement has nonlinear temperature distribution. The developed thermal strain at the top or bottom is:

$$\varepsilon_x = \varepsilon_y = \alpha\left(\frac{\Delta T}{2}\right) = \frac{\alpha\Delta T}{2} \tag{2.35}$$

where α is the coefficient of thermal expansion and contraction of the concrete slab. The thermal stress about the x-axis is:

$$\sigma_x = \frac{E\varepsilon_x}{(1 - \mu^2)} = \frac{E}{(1 - \mu^2)}\left(\frac{\alpha\Delta T}{2}\right) = \frac{E\alpha\Delta T}{2(1 - \mu^2)} \tag{2.36}$$

$$\sigma_y = \mu\sigma_x = \mu\frac{E\alpha\Delta T}{2(1 - \mu^2)} \tag{2.37}$$

Similar expressions can be derived if there is bending about the y-axis only.

Now, consider the bending about the both axes, x and y. Combining bending in both the x and y axes, the total stress about x- or y-axis ($\sigma_{x/y}$) is obtained by superimposing the stresses from the two stresses as follows:

$$\sigma_{x/y} = \frac{E\alpha\Delta T}{2(1-\mu^2)} + \mu\frac{E\alpha\Delta T}{2(1-\mu^2)} = \frac{E\alpha\Delta T}{2(1-\mu^2)}(1+\mu) = \frac{E\alpha\Delta T}{2(1-\mu)} \tag{2.38}$$

Bradbury (1938) extended this pure bending formulation to slabs of finite dimensions by weighing the contribution of stresses from bending in the two axes using the variables C_x and C_y, resulting in the following stress expressions of the total stress about the x-axis:

$$\sigma_x = \frac{E\alpha\Delta T}{2(1-\mu^2)}(C_x + \mu C_y) \tag{2.39}$$

where C_x and C_y are the correction factors for a finite slab. Similarly, the total stress about the y-axis:

$$\sigma_y = \frac{E\alpha\Delta T}{2(1-\mu^2)}(C_y + \mu C_x) \tag{2.40}$$

The value of C (C_x and C_y) can be obtained from Fig. 2.10. It requires L/l (L_x/l for C_x and L_y/l for C_y) as an input; the output is the value of C (C_x or C_y). L_x is the length along the x-axis and L_y is the length along the y-axis, and l is termed as the radius of relative stiffness and expressed using Eq. (2.41).

$$l = \left(\frac{Eh^3}{12(1-\mu^2)k}\right)^{0.25} \tag{2.41}$$

where h = Thickness (depth) of the concrete slab
k = Modulus of subgrade reaction

The k-value is defined as the reaction pressure of the subgrade support upon unit deformation of the concrete slab. More clearly, upon applying wheel load on concrete slab, it deforms. The underneath subgrade reacts to this applied pressure or resulting deformation. The amount of reactive pressure per unit deformation of the concrete slab is known as the modulus of subgrade reaction (k). There is no direct laboratory procedure for determining k-value. It can be determined by field tests or by correlation with other tests. The value of k is in terms of MPa/m (psi/in. or pci) and ranges from about 13.5 MPa/m (50 pci) for weak support to over 270 MPa/m (1,000 pci) for strong support. Typically, the modulus of subgrade reaction is estimated from other strength/stiffness tests; however, in situ values can be measured using the plate bearing test.

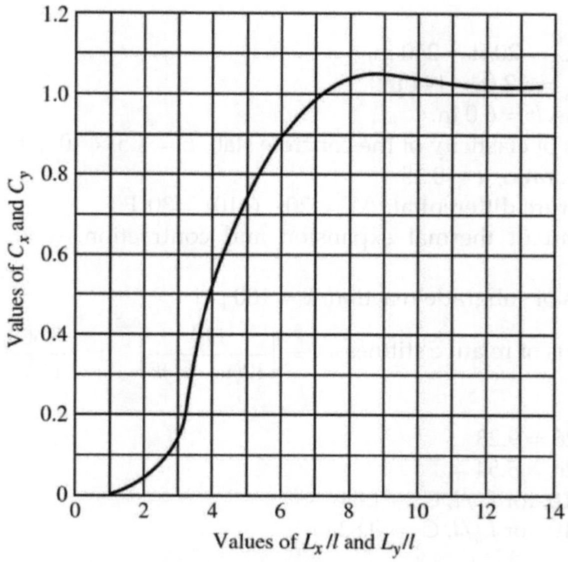

FIGURE 2.10 Stress correction factors for finite slabs. *(From Bradbury, R. D. (1938). Reinforced Concrete Pavements. Washington, DC: Wire Reinforcement Institute.)*

In interior region of a concrete slab, stresses along x and y axes exist. At the free edge of a slab no stress exists. Example 2.5 clarifies the curling stress variations in a concrete slab.

Example 2.5: Curling Stress in Concrete Slab A concrete slab, shown in Fig. 2.11, has a length of 20 ft, width of 12 ft, and thickness of 6.0 in. It is subjected to an increase in temperature of 20°F at its upper surface and a temperature decrease of 10°F at its lower surface. The modulus of elasticity of the concrete slab is 4.5×10^6 psi, Poisson's ratio of the concrete slab is 0.15, and the coefficient of thermal expansion and contraction is 4.5×10^{-6} per degree Fahrenheit. Assume the modulus of subgrade reaction is 180 pci. Determine the stresses at locations 1 (interior), 2 (edge), 3 (edge), and 4 (corner).

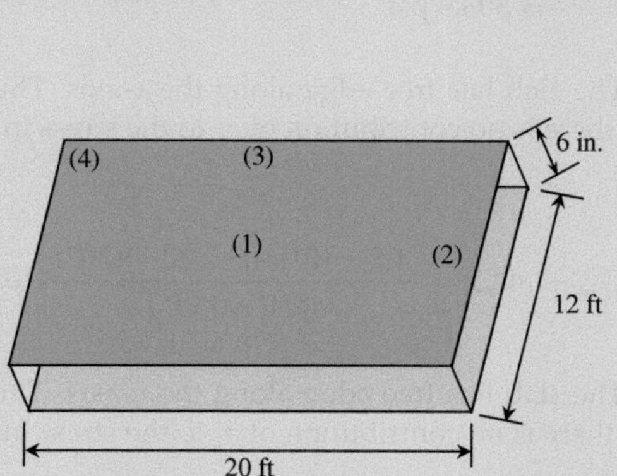

FIGURE 2.11 A concrete slab for Example 2.5.

Solution

Length, $L_x = 20$ ft $= 240$ in.
Width, $L_y = 12$ ft $= 144$ in.
Thickness, $h = 6.0$ in.
Modulus of elasticity of the concrete slab, $E = 4.5 \times 10^6$ psi
Poisson's ratio, $\mu = 0.15$
Temperature differential, $\Delta T = 20 - (-10) = 30°F$
Coefficient of thermal expansion and contraction, $\alpha = 4.5 \times 10^{-6}$ per degree Fahrenheit
Modulus of subgrade reaction, $k = 180$ pci

The radius of relative stiffness, $l = \left(\dfrac{Eh^3}{12(1-\mu^2)k}\right)^{0.25} = \left(\dfrac{4{,}500{,}000\,\text{psi}(6.0\,\text{in.})^3}{12(1-0.15^2)(180\,\text{pci})}\right)^{0.25} = 26$ in.

$L_x/l = 240/26 = 9.23$
$L_y/l = 144/26 = 5.54$
From Fig. 2.10, for L_x/l, $C_x = 1.04$.
From Fig. 2.10, for L_y/l, $C_y = 0.82$.

At Point 1 Total stress about the x-axis:

$$\sigma_x = \frac{E\alpha\Delta T}{2(1-\mu^2)}(C_x + \mu C_y)[0,1]$$
$$= \frac{(4.5\times10^6)(4.5\times10^{-6})(30°F)}{2(1-0.15^2)}[1.04 + 0.15(0.82)]$$
$$= 361.4 \text{ psi}$$

Total stress about the y-axis:

$$\sigma_y = \frac{E\alpha\Delta T}{2(1-\mu^2)}(C_y + \mu C_x)$$
$$= \frac{(4.5\times10^6)(4.5\times10^{-6})(30°F)}{2(1-0.15^2)}[0.82 + 0.15(1.04)]$$
$$= 303.3 \text{ psi}$$

At Point 2 The slab has free edge along the x-axis. Thus, there is no stress in the x-direction and there is no contribution of σ_x to the stress in the y-direction.

$C_x = 0$

$$\sigma_y = \frac{E\alpha\Delta T}{2(1-\mu^2)}(C_y + \mu C_x) = \frac{(4.5\times10^6)(4.5\times10^{-6})(30°F)}{2(1-0.15^2)}(0.82 + 0.15(0)) = 254.8 \text{ psi}$$

At Point 3 The slab has free edge along the y-axis. Thus, there is no stress in the y-direction and there is no contribution of σ_y to the stress in the x-direction.

$C_y = 0$

$$\sigma_x = \frac{E\alpha\Delta T}{2(1-\mu^2)}(C_x + \mu C_y) = \frac{(4.5\times10^6)(4.5\times10^{-6})(30°F)}{2(1-0.15^2)}(1.04 + 0.15(0)) = 323.2 \text{ psi}$$

At Point 4 Being the corner, the slab is free along both x and y axes. Thus, there is no stress in the x-direction and in the y-direction.

$\sigma_x = 0$

$\sigma_y = 0$

Answers

At Point 1: 361.4 psi, 303.3 psi

At Point 2: 0 psi, 254.8 psi

At Point 3: 323.2 psi, 0 psi

At Point 4: 0 psi, 0 psi

2.3.2 Traffic-Induced Stress

The wheel-induced stress can be of different types, such as stress due to corner loading, stress due to interior loading, stress due to edge loading, etc. The net stresses at different times and levels of loading are determined combining the wheel-induced stress and curling stress. Hand calculations of stresses at different times of a day and year for millions of vehicles are impossible; this is true for flexible pavement as well. Therefore, computer simulation of the stress-strain is essential.

Concentrated Corner Loading

Corner of a slab may be the most critical point under loading, considering the small area resisting the load. Cantilever beam analysis is performed to determine the maximum tensile stress at the top of the slab. It is considered that the subgrade reaction is negligible to resist the load. This assumption is very conservative and gives an approximate solution of the corner loading. Figure 2.12 shows the schematics of a concentrated corner loading in a slab. For the concentrated load P acting at the corner A, let us consider that the failure tends to occur along the line B-B at a distance x from the loading. Then, the maximum stress at B-B line can be calculated using the classical strength of materials formula ($\sigma = \frac{Mc}{I}$, where σ, M, c, and I are the developed stress, developed moment, distance from the neutral axis to the top/bottom fiber, and moment of inertia of the section, respectively) as:

$$\sigma_{\text{corn}} = \frac{Mc}{I} = \frac{(Px)\left(\frac{h}{2}\right)}{\frac{(2x)h^3}{12}} = \frac{3P}{h^2} \tag{2.42}$$

where h is the thickness of the slab.

The computed bending stress is independent of the location of the slab section considered. The subgrade reaction would reduce this developed stress. This is why it is conservative, ignoring the subgrade reaction. However, if the load acts at the very corner, the subgrade reaction can be reasonably ignored.

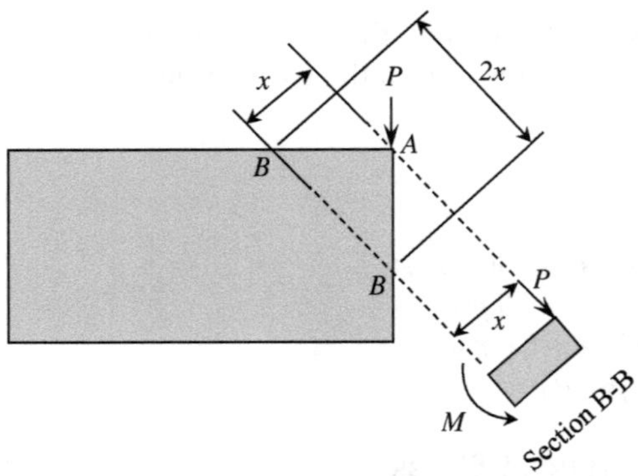

Figure 2.12 Schematics due to a concentrated corner loading.

Example 2.6: Stress Due to a Concentrated Corner Loading Compute the maximum tensile stress on a 0.25-m-thick slab of a JPCP under a corner point load of 35 kN.

Solution

Given:

Applied concentrated load, $P = 35$ kN

Thickness of the slab, $h = 0.25$ m

Therefore, the maximum corner stress $\sigma_{\text{corn}} = \dfrac{3P}{h^2} = \dfrac{3(35\,\text{kN})}{(0.25\,\text{m})^2} = 1{,}680\dfrac{\text{kN}}{\text{m}^2}$

Answer *1,680 kPa*

Circular Corner Loading

For a uniform circular load that is applied at the corner of a slab, as shown in Fig. 2.13, Ioannides et al. (1985) proposed the following solutions for the developed maximum stress and deflection. The maximum stress is developed at a distance of $2.38\sqrt{al}$ from the corner of the slab, where a is the radius of the circular area and l is the radius of the relative stiffness (Westergaard, 1926).

$$\text{Maximum stress, } \sigma_{\text{corn}} = \frac{3P}{h^2}\left[1 - \left(\frac{c}{l}\right)^{0.72}\right]$$

$$\text{Maximum deflection, } \Delta_{\text{corn}} = \frac{P}{kl^2}\left[1.205 - 0.69\left(\frac{c}{l}\right)\right]$$

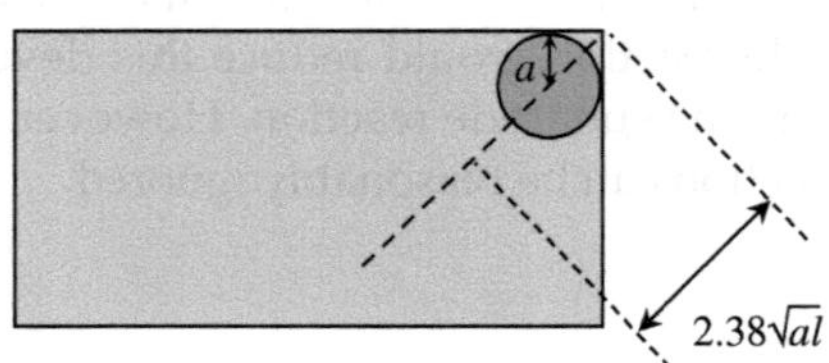

Figure 2.13 Schematics due to a circular corner loading.

where k = Modulus of subgrade reaction
$c = 1.772a$

Example 2.7: Stress Due to a Circular Corner Loading Determine the maximum tensile stress and the corner deflection under a circular load of 12-in. diameter carrying a 10-kip load, given a slab thickness of 8 in., a modulus of subgrade reaction of 100 pci, a concrete modulus of 3,000 ksi, and a Poisson's ratio of 0.15.

Solution

Given:

Applied load, $P = 10{,}000$ lb
Radius of circular area, $a = 12$ in./2 = 6 in.
Thickness of the slab, $h = 8$ in.
Modulus of subgrade reaction, $k = 100$ pci
Concrete modulus, $E = 3{,}000$ ksi = 3,000,000 psi
Poisson's ratio, $\mu = 0.15$
$c = 1.772a = 1.772$ (6 in.) = 10.632 in.

$$\text{Radius of relative stiffness, } l = \left(\frac{Eh^3}{12(1-\mu^2)k}\right)^{0.25} = \left(\frac{(3{,}000{,}000\,\text{psi})(8\,\text{in.})^3}{12(1-0.15^2)(100\,\text{pci})}\right)^{0.25} = 33.83\text{ in.}$$

$$\text{Maximum stress, } \sigma_{\text{corn}} = \frac{3P}{h^2}\left[1-\left(\frac{c}{l}\right)^{0.72}\right] = \frac{3(10{,}000\,\text{lb})}{(8\,\text{in.})^2}\left[1-\left(\frac{10.632\,\text{in.}}{33.83\,\text{in.}}\right)^{0.72}\right] = 265\text{ psi}$$

Maximum deflection:

$$\Delta_{\text{corn}} = \frac{P}{kl^2}\left[1.205 - 0.69\left(\frac{c}{l}\right)\right]$$

$$= \frac{(10{,}000\,\text{lb})}{(100\,\text{pci})(33.83\,\text{in.})^2}\left[1.205 - 0.69\left(\frac{10.632\,\text{in.}}{33.83\,\text{in.}}\right)\right] = 0.0863\text{ in.}$$

Answers *265 psi, 0.0863 in.*

Edge Loading

When the wheel is close to the edge, the full wheel may be inside the slab or part of the wheel may be in the current slab and the other part may be in the adjacent slab. Two cases are commonly considered: circular loading and semicircular loading (Fig. 2.14).

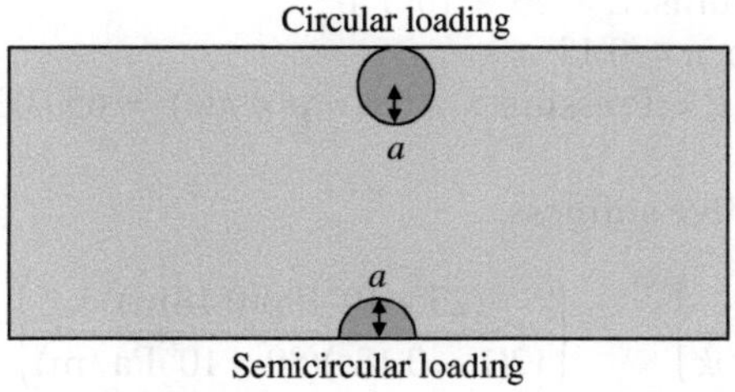

FIGURE 2.14 Schematics of edge loading.

Ioannides et al. (1985) proposed the following solutions for calculating the developed maximum stress for edge loading:

$$\sigma_{e(\text{circle})} = \frac{3(1+\mu)P}{\pi(3+\mu)h^2}\left[\ln\left(\frac{Eh^3}{100ka^4}\right) + 1.84 - \frac{4\mu}{3} + \frac{1-\mu}{2} + \frac{1.18(1+2\mu)a}{l}\right] \tag{2.43}$$

$$\sigma_{e(\text{semicircle})} = \frac{3(1+\mu)P}{\pi(3+\mu)h^2}\left[\ln\left(\frac{Eh^3}{100ka^4}\right) + 3.84 - \frac{4\mu}{3} + \frac{(1+2\mu)a}{2l}\right] \tag{2.44}$$

Ioannides et al. (1985) proposed the following solutions for calculating the deflection for edge loading:

$$\Delta_{e(\text{circle})} = \frac{\sqrt{2+1.2\mu P}}{\sqrt{Eh^3k}}\left[1 - \frac{(0.76+0.4\mu)a}{l}\right] \tag{2.45}$$

$$\Delta_{e(\text{semicircle})} = \frac{\sqrt{2+1.2\mu P}}{\sqrt{Eh^3k}}\left[1 - \frac{(0.323+0.17\mu)a}{l}\right] \tag{2.46}$$

where $\sigma_{e(\text{circle})}$ = Maximum stress for circular loading, psi or Pa
$\sigma_{e(\text{semicircle})}$ = Maximum stress for semicircular loading, psi or Pa
$\Delta_{e(\text{circle})}$ = Maximum deflection for circular loading, in. or m
$\Delta_{e(\text{semicircle})}$ = Maximum deflection for semicircular loading, in. or m
P = Applied load on the area, lb or N

Example 2.8: Stress Due to an Edge Loading Determine the stresses in the edge point of a 0.18-m-thick slab under a uniform stress of 650 kPa distributed over a circular area with a radius of 0.15 m. It is provided that the modulus of subgrade reaction of 80 MPa/m, an elastic modulus of concrete of 25 GPa, and a Poisson's ratio of concrete of 0.18.

Solution

Given:
Applied pressure, p = 650,000 Pa
Radius of circular area, a = 0.15 m
Thickness of the slab, h = 0.18 m
Modulus of subgrade reaction, k = 80 MPa/m
Concrete modulus, $E = 25 \times 10^9$ Pa
Poisson's ratio, μ = 0.18
Applied load, P = Pressure × Area = $p \times \pi(a)^2$ = 650,000 Pa × $\pi(0.15\text{ m})^2$ = 45,945 N

Radius of relative stiffness,

$$l = \left(\frac{Eh^3}{12(1-\mu^2)k}\right)^{0.25} = \left(\frac{(25\times10^9\text{ Pa})(0.18\text{ m})^3}{12(1-0.18^2)(80\times10^6\text{ Pa/m})}\right)^{0.25} = 0.6294\text{ m}$$

$$\text{Stress, } \sigma_{e(\text{circle})} = \frac{3(1+\mu)P}{\pi(3+\mu)h^2}\left[\ln\left(\frac{Eh^3}{100ka^4}\right) + 1.84 - \frac{4\mu}{3} + \frac{1-\mu}{2} + \frac{1.18(1+2\mu)a}{l}\right]$$

$$= \frac{3(1+0.18)(45{,}945\,\text{N})}{\pi(3+0.18)(0.18\,\text{m})^2}\left[\begin{array}{l}\ln\left(\dfrac{(25\times 10^9\,\text{Pa})(0.18\,\text{m})^3}{100\left(80\times 10^6\,\dfrac{\text{Pa}}{\text{m}}\right)(0.15\,\text{m})^4}\right) + 1.84 - \dfrac{4(0.18)}{3} \\ + \dfrac{1-0.18}{2} + \dfrac{1.18(1+2(0.18))(0.15\,\text{m})}{0.6294\,\text{m}}\end{array}\right]$$

$= 3{,}002{,}808$ Pa
$= 3.0$ MPa

$$\text{Deformation, } \Delta_{e(\text{circle})} = \frac{\sqrt{2+1.2\mu P}}{\sqrt{Eh^3k}}\left[1 - \frac{(0.76+0.4\mu)a}{l}\right]$$

$$= \frac{\sqrt{(2+1.2(0.18)(45{,}945\,\text{N}))}}{\sqrt{(25\times 10^9\,\text{Pa})((0.18\,\text{m}))^3\left(80\times 10^6\,\dfrac{\text{Pa}}{\text{m}}\right)}}\left[1 - \frac{(0.76+0.4(0.18))(0.15\,\text{m})}{0.6294\,\text{m}}\right]$$

$= 7.4 \times 10^{-7}$ m
$= 7.4 \times 10^{-4}$ mm

Answers *3.0 MPa, 7.4 × 10⁻⁴ mm*

2.3.3 Friction-Induced Stress

Due to temperature change, concrete slab undergoes thermal expansion and contraction. The expansion and contraction can be the result of uniform temperature changes or due to post-construction concrete shrinkage. The thermal expansion and contraction are resisted by the friction between the slab and the underneath soil/base layer. The amount of frictional force depends on the relative movement of the slab. The centerline of the slab does not move and the frictional force is zero at the centerline. The edges move the most and thus the frictional force is the maximum at the edges. Let us consider a slab that is expanding due to increase in temperature, as shown in Fig. 2.15. The slab is intending to expand and the subgrade friction is resisting the expansion of the slab.

Let the length of the slab be L, the width be unity, the thickness be h, the unit weight of concrete be γ_c, and the frictional coefficient be f (typically ranges 1.5–2.0). The tensile stress (σ_c) developed is the maximum at the center due to the thermal expansion in two opposite directions and can be determined as follows:

Tensile force developed in the slab $= \sigma_c\,(h \times 1.0) = \sigma_c h$

Frictional force up to the centerline = Frictional coefficient × Weight of slab $= f \times (L/2 \times h \times \gamma_c) = ½\, Lh\gamma_c f$

Equating the above two:

$$\sigma_c h = \frac{1}{2} Lh\gamma_c f \tag{2.47}$$

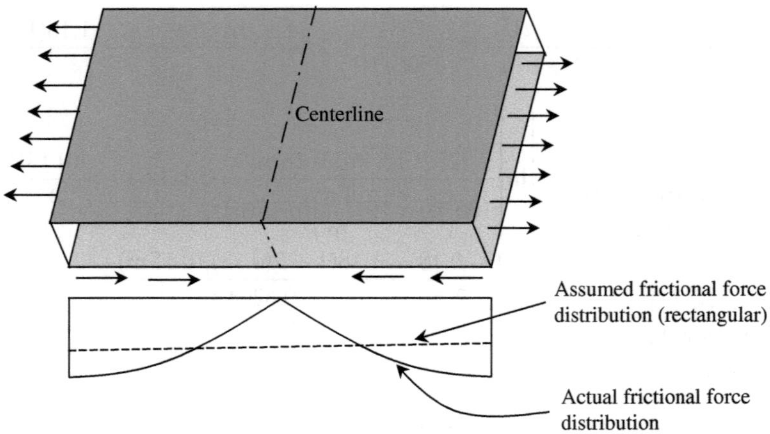

FIGURE 2.15 Subgrade friction in a concrete slab of unit width.

$$\sigma_c = \frac{L\gamma_c f}{2} \tag{2.48}$$

This stress in slab is developed due to the subgrade friction. If the developed stress exceeds the tensile capacity of the slab, thermal crack occurs.

Example 2.9: Stress Due to Subgrade Friction A plain concrete slab is supported by a subgrade. The concrete has the unit weight of 150 pcf. The frictional coefficient between the concrete slab and the subgrade is 1.5. If the length of the concrete slab is 25 ft, what is the developed frictional stress in the slab?

Solution

Unit weight, $\gamma_c = 150$ pcf
Length, $L = 25$ ft
Frictional coefficient, $f = 1.5$

Frictional stress, $\sigma_c = \frac{L\gamma_c f}{2} = \frac{(25\,\text{ft})(150\text{ pcf})(1.5)}{2} = 2{,}813\text{ psf}$

Answer *2.8 ksf*

2.3.4 Joint Opening

Joint opening is very important if dowel bars are not used to transfer the wheel load from one slab to another. For this type of rigid pavement, load transfer is relied only through the aggregate-slab interlock. If slabs contract due to uniform temperature reductions or due to post-construction shrinkage, the joints open and their vertical load transfer efficiency dramatically decreases. Practically, plain joints with openings larger than 1 mm are not effective in transferring vertical loads across adjacent slabs (Papagiannakis and Masad, 2008). The joint opening (ΔL) can be approximately estimated using the following equation (Darter and Barenberg, 1977):

$$\Delta L = CL(\alpha\, \Delta T + \varepsilon) \tag{2.49}$$

where α = Coefficient of thermal expansion and contraction of the concrete slab
ΔT = Temperature in placement minus the lowest mean monthly temperature
ε = Drying shrinkage coefficient of concrete (varies from 0.5×10^{-4} to 2.5×10^{-4})
C = Adjustment factor due to slab-subbase friction (0.65 for stabilized base and 0.80 for granular subbase)
L = Joint spacing or the slab length

Example 2.10: Joint Opening A concrete slab is to be placed with the condition that the maximum joint opening allowable is 0.8 mm. The pavement may experience a temperature difference between construction and the coldest winter month of 35°C. The coefficient of thermal expansion and contraction of the concrete slab is 9.5×10^{-6} per degree Celsius. The drying shrinkage coefficient of concrete is 1.5×10^{-4}. Assume the adjustment factor due to the slab-subbase friction is 0.65.

Compute the maximum feasible slab length.

Solution

Given:
Coefficient of thermal expansion and contraction, $\alpha = 9.5 \times 10^{-6}$ per degree Celsius
Temperature difference, $\Delta T = 35°C$
Drying shrinkage coefficient, $\varepsilon = 1.5 \times 10^{-4}$
Adjustment factor, $C = 0.65$
Joint opening, $\Delta L = 0.80$ mm $= 0.0008$ m
Slab length, L = ?

We know:

$$\Delta L = CL(\alpha\,\Delta T + \varepsilon)$$

$0.0008 = 0.65 \times L \times (9.5 \times 10^{-6}$ per degree Celsius $\times$ 35°C $+ 1.5 \times 10^{-4})$
Therefore, $L = 2.55$ m.

Answer *2.55 m*

2.4 Stress in Dowels

Dowels are used in the transverse joints to transfer load from one slab to another. The design of dowel bars is primarily dependent on the developed bearing stress between the dowel bars and the concrete. The allowable bearing stress (f_b) is determined using Eq. 2.50 (ACI, 1956).

$$f_b = \left(\frac{4-d}{3}\right) f_c' \tag{2.50}$$

where d = Diameter of the dowel (in.) (no other unit is possible)
f_c' = Compressive strength of the concrete (units of f_b and f_c' are the same)

If the load is applied to one dowel, the deflection in the dowel bar can be determined considering a cantilever beam (Friberg, 1940), as shown in Eq. (2.51).

$$y_o = \frac{P_t(2+\beta z)}{4\beta^3 E_d I_d} \tag{2.51}$$

where y_o = Deformation of the dowel at the face of the joint
P_t = Load on the dowel bar
z = Opening of the joint
E_d = Modulus of elasticity of the dowel
I_d = Moment of inertia of the dowel, can be calculated as $I_d = \frac{\pi d^4}{64}$
d = Diameter of the dowel
β = Relative stiffness of a dowel, can be calculated as $\beta = \left(\frac{Kd}{4E_d I_d}\right)^{1/4}$
K = Modulus of dowel support ranging from 300,000 to 1,500,000 pci (82–410 GN/m^3).

The developed bearing stress σ_b is proportional to the deformation and can be calculated as:

$$\sigma_b = Ky_o = \frac{KP_t(2+\beta z)}{4\beta^3 E_d I_d} \tag{2.52}$$

This developed stress (σ_b) must be equal or greater than the allowable bearing stress (f_b). The above analysis considers a single dowel bar. If the applied load affects a group of dowel bars, the dowel bars getting the maximum load and its magnitude must be determined. According to Friberg (1940) load affects dowels up to a distance of 1.8*l* away from its location, where *l* is the radius of relative stiffness of the slab. It is commonly assumed that the load distributed across a group of dowels in a linear fashion allows calculation of the load carried by each dowel. If the dowel bars are 100% efficient, then each slab carries 50% of the load applied at the proximity of the joint. The following examples clarify the computation.

Example 2.11: Single Dowel Bar A 1.95-cm-diameter dowel bar is transferring a vertical load of 3,000 N across a 0.4-cm-wide joint. Given: modulus of dowel support of 150,000 MPa/m, modulus of elasticity of the dowel of 180,000 MPa, and f_c' of 28 MPa. Compute the dowel bar deflection at the edge of the joint. Can the concrete handle this stress?

Solution

Given:
Load on the dowel bar, P_t = 3,000 N
Opening of the joint, z = 0.4 cm = 0.004 m
Modulus of elasticity of the dowel, $E_d = 180{,}000 \times 10^6$ Pa
Diameter of the dowel, d = 1.95 cm = 0.0195 m = 0.0195 × 39.37 in. = 0.77 in.
Modulus of dowel support, $K = 150{,}000 \times 10^6$ Pa/m
Concrete strength, f_c' = 28 MPa

$$\text{Moment of inertia of the dowel, } I_d = \frac{\pi d^4}{64} = \frac{\pi d^4}{64} = \frac{\pi(0.0195\,\text{m})^4}{64} = 7.1 \times 10^{-9}\,\text{m}^4$$

$$\text{Relative stiffness, } \beta = \left(\frac{Kd}{4E_d I_d}\right)^{1/4} = \left(\frac{(150{,}000\times 10^6)(0.0195\,\text{m})}{4(180{,}000\times 10^6)(7.1\times 10^{-9})}\right)^{1/4} = 27.5$$

$$\text{Deflection, } y_o = \frac{P_t(2+\beta z)}{4\beta^3 E_d I_d}$$

$$= \frac{(3{,}000)(2+(27.5)(0.004))}{4(27.5)^3(180{,}000\times 10^6)(7.1\times 10^{-9})} = 5.95\times 10^{-5}\text{ m} = 0.0595\text{ mm}$$

$$\text{The allowable bearing stress, } f_b = \left(\frac{4-d}{3}\right)f_c'$$

$$= \left(\frac{4-0.77\text{ in.}}{3}\right)(28\times 10^6\text{ Pa}) = 30.1\times 10^6\text{ Pa}$$

The developed bearing stress:
$\sigma_b = Ky_o = 150{,}000\times 10^6(5.95\times 10^{-5}\text{ m}) = 8{,}925{,}000\text{Pa} = 8.9\times 10^6\text{ Pa}.$

Comparing this developed stress (8.9 MPa) to the allowed bearing stress of concrete (30.1 MPa) dictates that the concrete can indeed handle the applied stress.

Answers *0.0595 mm, concrete can handle the applied stress*

Example 2.12: Forces in a Group of Dowels A set of 0.75-in.-diameter dowel bars are placed 12 in. on center in a 12-ft-wide concrete slab, as shown in Fig. 2.16. Two loads (6,000 lb each) are applied as shown. Assume the full efficiency of the dowel bars. The modulus of subgrade reaction is 100 pci; the concrete modulus is 3,000 ksi and Poisson's ratio is 0.15.

Determine the force distribution in the dowel bars.

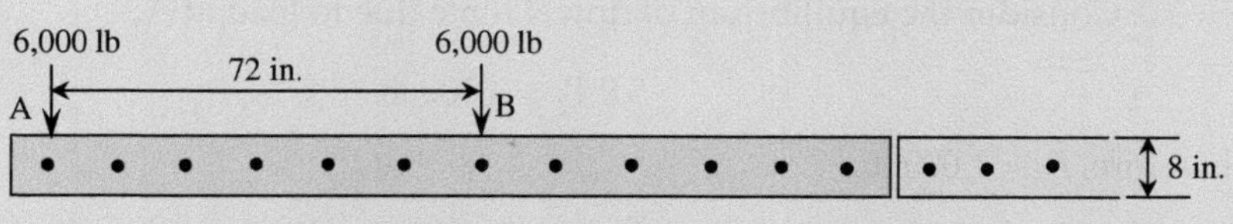

FIGURE 2.16 Dowel bars arrangements for Example 2.12.

Solution
Given:
Thickness of the slab, $h = 8$ in.
Modulus of subgrade reaction, $k = 100$ pci
Concrete modulus, $E = 3{,}000$ ksi
Poisson's ratio, $\mu = 0.15$

$$\text{Radius of relative stiffness, } l = \left(\frac{Eh^3}{12(1-\mu^2)k}\right)^{0.25} = \left(\frac{(3{,}000{,}000\text{psi})(8\text{in.})^3}{12(1-0.15^2)(100\text{pci})}\right)^{0.25} = 33.83\text{ in.}$$

$1.8l = 1.8\ (33.83\text{ in.}) = 60.894 \approx 60\text{ in.}$

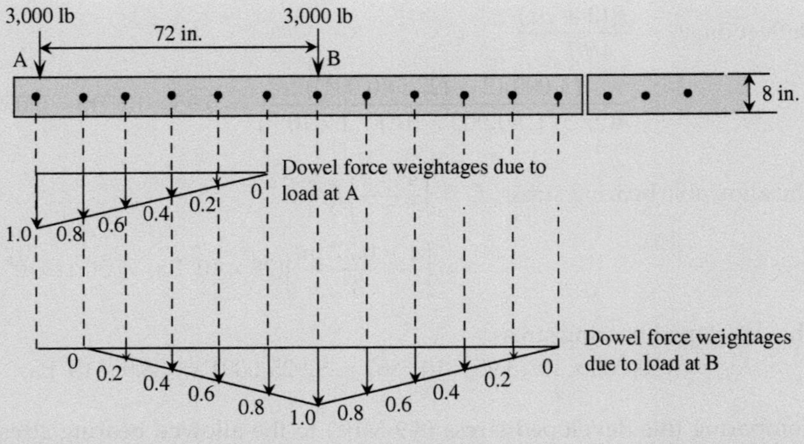

FIGURE 2.17 Force weightages in dowels for loads at A and B.

Let us distribute the force in 60 in. considering the weightage, 1.0 at just below the load and the weightage, 0 at a distance of 60 in. from the loading point (Fig. 2.17). Similar angled triangles rule is used to find out the weightage for different ordinates. For example, ordinate at 12 in. = (1.0/60)12 = 0.20.

As the dowel bars are 100% efficient, 50% of the load will be carried by the next slab.

Sum of the weightage for dowel force due to load at $A = 1.0 + 0.8 + 0.6 + 0.4 + 0.2 + 0 = 3.0$. Consider the equilibrium of dowel force due to load at A:

$$3.0\, P_t = 3{,}000 \text{ lb}$$

Therefore, $P_t = 1{,}000$ lb.

Similarly, the sum of the weightage for dowel force due to load at B = 1.0 + 2(0.8 + 0.6 + 0.4 + 0.2 + 0) = 5.0.

Consider the equilibrium of dowel force due to load at B:

$$5.0\, P_t = 3{,}000 \text{ lb}$$

Therefore, $P_t = 600$ lb.

Now, the force distribution can be calculated as shown in Fig. 2.18.

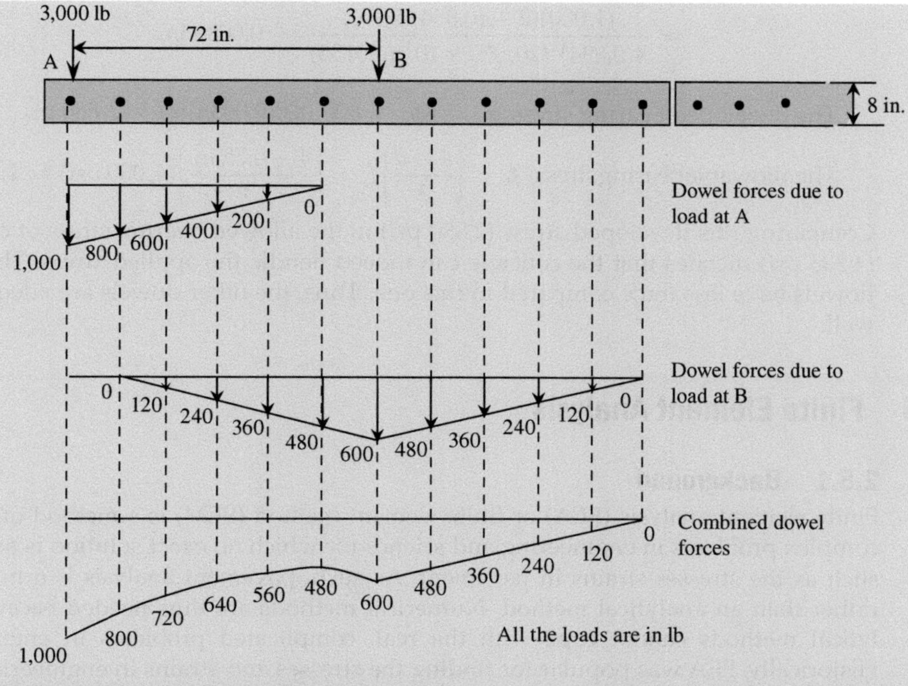

FIGURE 2.18 Dowel forces for loads at A and B.

Example 2.13: Adequacy of a Group of Dowels For Example 2.12, given are a slab thickness of 8 in., joint opening of 0.75 in., a modulus of dowel support of 800,000 pci, modulus of elasticity of the dowel of 20,000 ksi, and f_c' of 4,000 psi.

Can the concrete handle this stress?

Solution

Given:

Maximum dowel force, $P_t = 1{,}000$ lb (the dowel force as obtained from Example 2.12)

Joint opening, $z = 0.75$ in.

Modulus of elasticity of the dowel, $E_d = 20{,}000$ ksi

Diameter of the dowel bar, $d = 0.75$ in.

Modulus of dowel support, $K = 800{,}000$ pci

Concrete strength, $f_c' = 4{,}000$ psi

Moment of inertia of the dowel, $I_d = \dfrac{\pi d^4}{64} = \dfrac{\pi (0.75\text{ in.})^4}{64} = 0.0155\text{ in.}^4$

Relative stiffness of a dowel, $\beta = \left(\dfrac{Kd}{4E_d I_d}\right)^{1/4}$

$$= \left(\frac{(800{,}000\,\text{pci})(0.75\,\text{in.})}{4(20{,}000\times 10^3\,\text{psi})(0.0155\,\text{in.}^4)}\right)^{1/4} = 0.834$$

Deflection, $y_o = \dfrac{P_t(2+\beta z)}{4\beta^3 E_d I_d}$

$$= \frac{(1{,}000)(2 + (0.834)(0.75))}{4(0.834)^3(20{,}000 \times 10^3)(0.0155)} = 0.0036 \text{ in.}$$

The developed bearing stress, $\sigma_b = Ky_o = 800{,}000(0.0036\text{in.}) = 2{,}880$ psi.

Theallowablebearingstress, $f_b = \left(\frac{4-d}{3}\right)f_c' = \left(\frac{4-0.75\text{ in.}}{3}\right)(4{,}000 \text{ psi}) = 4{,}333$ psi.

Comparing this developed stress (2,880 psi) to the allowed bearing stress of concrete (4,333 psi) dictates that the concrete can indeed handle the applied stress. The other dowels have less force compared to this one. Thus, the other dowels are adequate as well.

2.5 Finite Element Analysis

2.5.1 Background

Finite element analysis (FEA) or finite element method (FEM) is a method of solving complex problems in engineering and science for which no exact solution is available, such as the stresses-strains in pavement. As such, pavement analysis is a numerical rather than an analytical method. Numerical methods are thus needed because analytical methods cannot cope with the real, complicated problems in engineering. Historically, FEA was popular for finding the stresses and strains in engineering components considering elastic analysis under load. Nowadays, it can be used for problems involving a wide range of phenomena, including vibrations, heat conduction, fluid mechanics, and electrostatics, and a wide range of material properties, such as linear-elastic (Hookean) behavior and behavior involving deviation from Hooke's law (e.g., viscosity, plasticity, or combination of them).

The FEA is the recommended approach for analyzing the complex behavior of slabs under load/environmental input and subgrade support. The method is based on discretizing the pavement systems (i.e., slabs, subgrade, and reinforcement) into finite elements, defining the stresses and the strains inside each element as a function of its nodal displacements, solving for the nodal displacements of all the elements simultaneously. Finally, FEA computes stresses and strains in all elements from the known displacements. Pavement surface can be modeled as two-dimensional thin plates or three-dimensional solids, depending on the desired level of detail. The underneath base/subbase layers are modeled as a separate layer with different mechanical properties. The bonding between layers can be assumed as required. However, full bonding between any two layers is common. Nonlinear behavior of layers in terms of its modulus, cross-anisotropy, or even anisotropy of all layers are becoming popular nowadays.

2.5.2 The User's View

Many comprehensive general-purpose computer packages are now available that provide very friendly user interface to analyze a structure. The first step of using the FEA is to draw the geometry of the structure. Some computer programs allow the geometry to be imported from a drawing tool such as computer-aided design (CAD). Then, the area is divided into a number of small imaginary elements for analysis purpose only. The degree of division depends on the fineness of the solution desired. Figure 2.19

FIGURE 2.19 A pavement section divided into a number of finite elements.

shows a two-dimensional model of a pavement that has been so divided: the process is called discretization, and the assembly of elements is called a mesh. Finer mesh (smaller element size) produces finer solution.

Elements can be of various shapes, as shown in Fig. 2.20, in two dimensions, quadrilateral or triangular, and in three dimensions, brick-shaped (hexahedral), wedge-shaped (pentahedral), or tetrahedral. This is, of course, not an exhaustive list. The corner points of the elements are called nodes. Depending on the element type, there are nodes at the corners and possibly at the midsides of the elements or even within the element. Nodes on the boundaries of adjacent elements must belong to the elements that meet there. More clearly, every node must be shared by two or more elements. Thus, all elements are bound with each other.

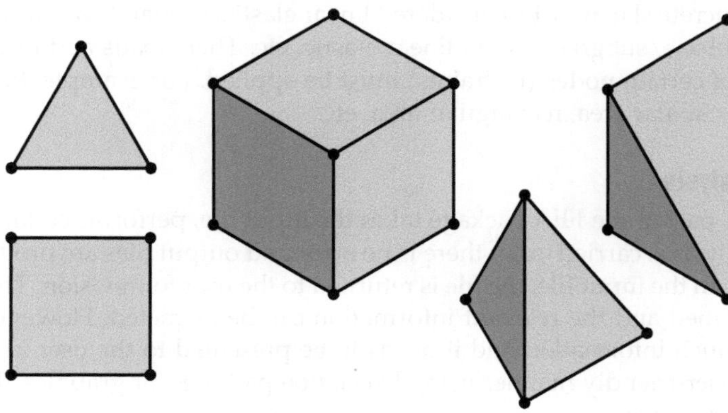

FIGURE 2.20 Various finite elements commonly available.

For the linear elastic analysis, the first found in the analysis is the displacement at the series of nodes. The FEA calculates the displacement at the nodes for the loading applied to the FEA model. The displacement of each point is then defined in terms of the displacements of the nodes of the element to which the point belongs. If we are considering a two-dimensional model, then the displacement of each node consists of two components, a first parallel to a reference x-axis and a second parallel to the y-axis: these are called degrees of freedom. Each node in this case has two degrees of freedom associated with it. For a three-dimensional brick-shaped element, the degrees of freedom would be three. If there are n nodes, then the total number of degrees of freedom to be determined is, in the FEA model, $n \times$ (number of degrees of freedom per node), as compared with an infinite number in the actual component. The computer time and the cost of the analysis naturally increase as the number of degrees of freedom of the model is increased. Having calculated the nodal displacements, the program then goes on to find the corresponding strains and, from the strains, the stresses are computed. All this information is made ready for the user to examine. To summarize, there are generally three stages of FEA:

1. Pre-processing
2. Analysis
3. Post-processing

2.5.3 Pre-Processing

Pre-processing means the creation of the model geometry using the graphics package to build up the model and to display the model on the computer screen. Some idealization is required with the model as compared with the real structure in terms of support types, rigidity, shape, etc. For example, a beam may be modeled as a cable if it has only axial load; half of the pavement can be modeled considering the centerline of the pavement does not move laterally. If the software permits, the geometric information for the model may be taken in by the preprocessor from a CAD package. Then the choice of the mesh and type of element to be used are selected. Too fine mesh may lengthen the simulation time and type of element may affect the results. Appropriate mechanical or physical properties must be allocated to the material of which the model is made. For example, concrete slab may be considered linear elastic, asphalt layer may be viscoelastic, base/subbase/subgrade be no-linear elastic, etc. Then, loads and restriction to the movement of certain nodes (restraints) must be applied. For example, tire load can be applied in a circular area, rectangular area, etc.

2.5.4 Analysis

The analysis part of the FEA package takes the input file, performs certain checks, and then the analysis is carried out if there is no error and output files are produced. If there are no errors in the input file, the file is returned to the user for revision. The output files can be examined and the relevant information can be extracted. However, the output file has so much information and it needs to be presented to the user in a more intelligible and user-friendly manner using Excel, notepad files, or graphics. This is the job of the post-processor.

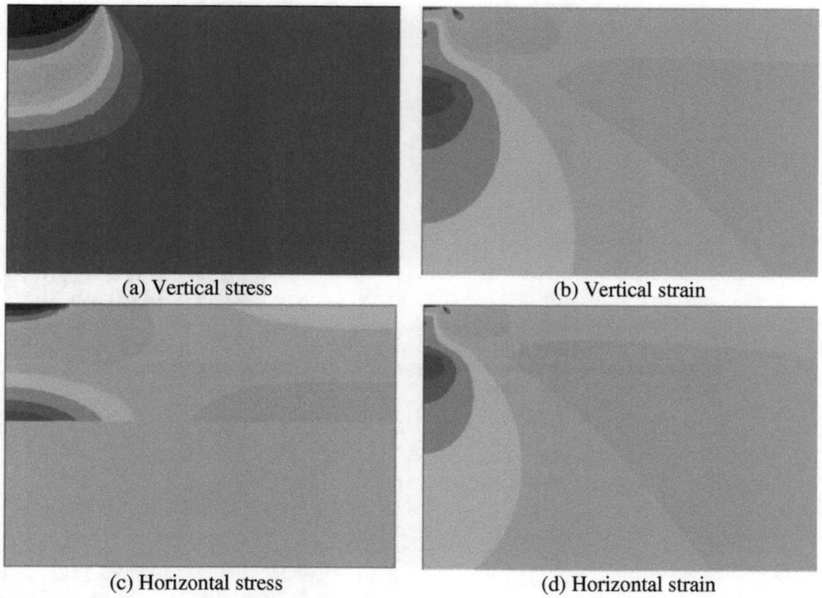

FIGURE 2.21 Contours of stress and strain values for a trial pavement.

2.5.5 Post-Processing

The post-processor takes in the information from the output files and can present it to the user in a range of different graphical and tabular forms. For example, depending on the facilities available, color may be used to indicate the value of some component of stress on the surface of the component, or contour lines of equal stress may be drawn as in Fig. 2.21, or similar forms of display may be produced on sections through the model. As in the pre-processor, the model may be rotated and examined from different viewpoints.

2.6 Numerical Analysis Tools

With the advancement of simulation tools, computerized method is used to determine the stress-strain in different layers of pavement. The AASHTOWare software has the embedded finite element simulation algorithms to determine the stress-strain at different layers of a pavement. Some other simulation tools are listed in Fig. 2.22.

2.7 Summary

Pavement structures are not similar to other classical civil engineering structures. Pavement structures have several layers of distinctly different layers with different properties and behavior. Their properties also may change with temperature, loading magnitude, loading frequency, moisture content, etc. The load application is also different. For example, the loads applied are dynamic in nature in different positions of a pavement. Thus, every load is analyzed separately.

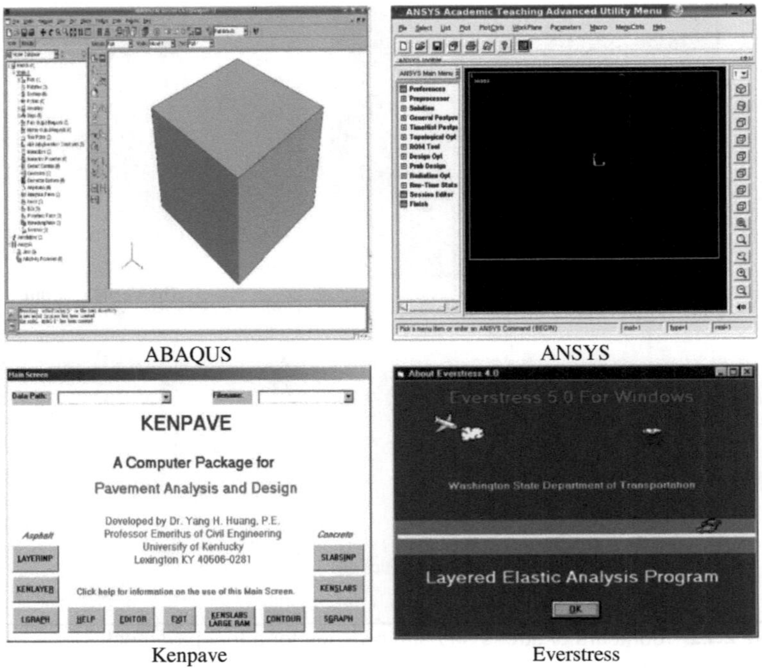

Figure 2.22 Few numerical tools to analyze pavement.

Boussinesq (1885) proposed a solution for stress-strain for a homogeneous elastic half-space for concentrated load and uniform circular load. For uniform circular load, the stress and deflection vary for flexible- and rigid-plate load applications. Odemark (1949) presented the method of equivalent thicknesses for solving the elastic layered system problem, which consists of translating multiple layers of different moduli into an equivalent single layer. For a multilayer system, a variety of software can be used such as ELSYM, DAMA, KENLAYER, and EVERSTRESS.

For concrete pavement, three main categories of stresses occur: curling stress due to temperature changes in the slab, stress due to traffic loading, and stress due to base/subgrade friction. Curling stress occurs due to the temperature variations at the top and at the bottom of concrete slab. In the daytime, when the surface temperature is higher than that of the bottom of the concrete slab, the surface expands and the slab bends upward. At night, when the surface temperature cools down rapidly, the surface of the slab contracts faster than the bottom (i.e., the slab downward). The position of wheel induces stresses at the corner and at the edge of a slab. Uniform thermal expansion and contraction of concrete slab also causes stress due to the friction between the slab and the underneath layer. Stresses also develop in the dowel bars while transferring load from one slab to another. All these stresses are considered while analyzing concrete slab.

Analyzing all these stresses analytically accurately is impossible. Nowadays, finite element software is used to solve stresses for these diversified conditions.

2.8 Fundamentals of Engineering (FE) Exam-Style Questions

FE2.1 Hooke's law is valid for the:

A. One-dimensional body only
B. Linear elastic analysis only
C. Three-dimensional body only
D. Viscoelastic material

Solution B

Hooke's law is valid for the linear elastic analysis only for a one- or multi-dimensional body.

FE2.2 The vertical stress in asphalt pavement:

A. Decreases with the depth but increases with the radial distance
B. Increases with the depth but decreases with the radial distance
C. Decreases with the depth and decreases with the radial distance
D. Increases with the depth and increases with the radial distance

Solution C

FE2.3 In the daytime when the temperature is very high, concrete slab:

A. Bends with concave downward
B. Bends with concave upward
C. Does not bend but curling stress develops
D. Bends without any curling stress

Solution A

In the daytime, the top of concrete expands with respect to the neutral axis as the top is hotter than the bottom. This means the concrete slab tends to move up (concave downward).

2.9 Practice Problems

2.1 A wheel applies a 40-kN load on a semi-infinite elastic space, as shown in Fig. P2.1. The modulus of elasticity of the material is 150 MPa and Poisson's ratio is 0.38. Assume the load is applied on the surface as a concentrated force.

Calculate the following at a depth of 500 mm at the centerline of the wheel:

a. Vertical stress
b. Radial stress
c. Tangential stress
d. Horizontal shear stress in the radial direction
e. Vertical strain
f. Radial strain

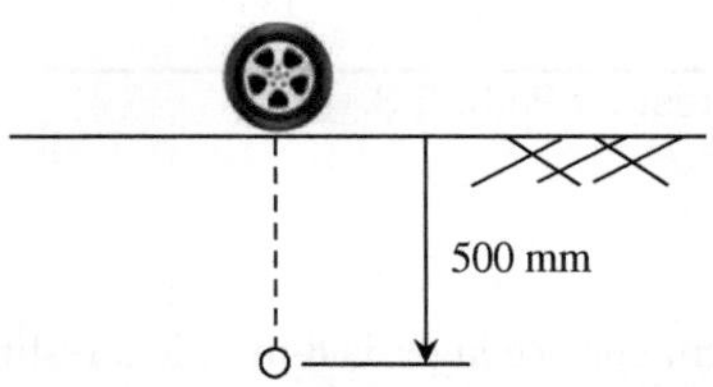

Figure P2.1 Profile of the semi-infinite elastic space for Prob. 2.1.

g. Tangential strain
h. Vertical deformation
i. Horizontal deformation

2.2 A wheel applies a 120-psi load on a 6-in.-diameter area on a semi-infinite elastic space, as shown in Fig. P2.2. The modulus of elasticity of the material is 350 ksi and Poisson's ratio is 0.35. Assume the load is applied on the surface as a uniform circular load using a flexible plate.

Calculate the following at a depth of 1.5 ft under the centerline of the load:

a. Vertical stress
b. Radial stress
c. Tangential stress
d. Vertical strain
e. Radial strain
f. Vertical deflection

Also, calculate the vertical deformation at the surface just below the loading.

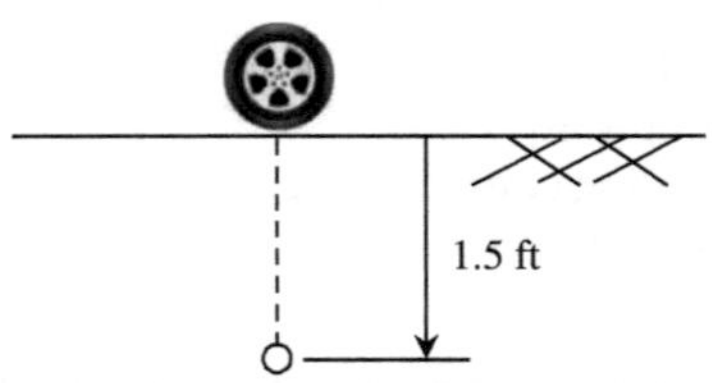

FIGURE P2.2 Profile of the semi-infinite elastic space for Prob. 2.2.

2.3 A plate load test was conducted using an 18-in.-diameter rigid plate applying a load of 12 kip on a subgrade, as shown in Fig. P2.3. A deformation of 0.12 in. was measured after applying this load. Poisson's ratio of the subgrade is 0.40. Determine the elastic modulus of the subgrade.

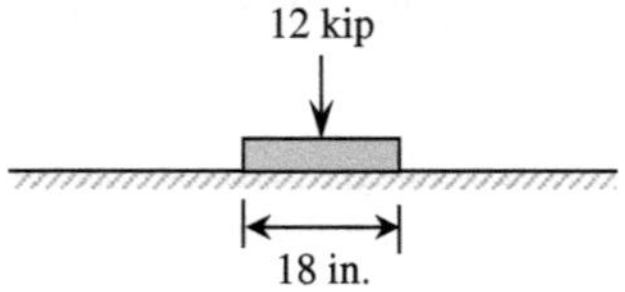

FIGURE P2.3 Plate load test for Prob. 2.3.

2.4 A flexible pavement surface layer is 8-in. thick, resting on a semi-infinite subgrade layer. The load consists of a circular tire with a 6-in. radius carrying a uniform load of 7,500 lb. The layer moduli are 350 and 18 ksi, respectively, and μ is 0.45 for both layers.

Calculate the following at the bottom of the surface layer at the centerline of the load:

a. Vertical stress
b. Radial stress
c. Vertical deformation

2.5 A load is applied on a soil using a rigid plate consisting of a circular 8-in. radius carrying a uniform load of 11,500 lb. The soil's modulus is 35 ksi and μ is 0.4.
Calculate the vertical stress at the surface at the edge of the loading plate.
Calculate the vertical deformation at the surface.

2.6 A concrete slab, as shown in Fig. P2.6, has a length of 10 m, width of 4 m, and thickness of 0.2 m. It is subjected to an increase in temperature of 25°F at its upper surface and a temperature decrease of 10°F at its lower surface. The modulus of elasticity of the concrete slab is 28 GPa, Poisson's ratio of the concrete slab is 0.15, and the coefficient of thermal expansion and contraction is 4.5×10^{-6} per degree Fahrenheit. Assume the modulus of subgrade reaction is 80 MPa/m. Determine the stresses at locations 1 (interior), 2 (edge), 3 (edge), and 4 (corner).

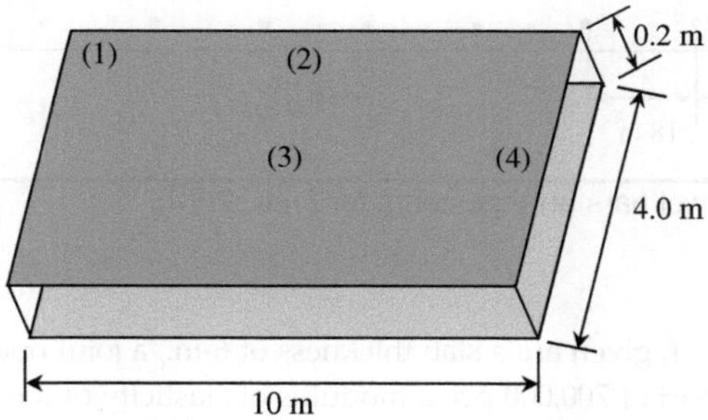

FIGURE P2.6 A concrete slab for Prob. 2.6.

2.7 Compute the maximum tensile stress on an 8-in.-thick slab of a JPCP under a corner point load of 6.9 kip.

2.8 The tensile strength of a 6-in.-thick slab of a JPCP is 450 psi. Calculate the maximum amount of a corner point load that can be applied to this slab.

2.9 Determine the maximum tensile stress and the corner deflection under a circular load of 0.3-m diameter carrying a 55-kN load, given a slab thickness of 0.15 m, a modulus of subgrade reaction of 100 MPa/m, a concrete modulus of 30 GPa, and a Poisson's ratio of 0.15.

2.10 Determine the stresses in the edge point of a 0.18-m-thick slab under a uniform stress of 650 kPa distributed over a semicircular area with a radius of 0.15 m. Given: a modulus of subgrade reaction of 80 MPa/m, an elastic modulus of concrete of 25 GPa, and a Poisson's ratio of concrete of 0.18.

2.11 A plain concrete slab is supported by a subgrade. The concrete has the unit weight of 25 kN/m³. The frictional coefficient between the concrete slab and the subgrade is 1.5. If the length of the concrete slab is 7 m, determine the developed frictional stress in the slab.

2.12 A 20-ft-long concrete slab is to be placed with the condition that the pavement may experience a temperature difference between construction and the coldest winter month of 35°C. The coefficient of thermal expansion and contraction of the concrete slab is 9.5×10^{-6}

per degree Celsius. The drying shrinkage coefficient of concrete is 1.5×10^{-4}. Assume the adjustment factor due to the slab-subbase friction is 0.80.

Compute the maximum joint opening.

2.13 A 25.4-mm-diameter dowel bar is transferring a vertical load of 12 kN across a 0.1-cm-wide joint. Compute the dowel bar deflection at the edge of the joint. Can the concrete handle this stress?

Given: a modulus of dowel support of 180,000 MPa/m, a modulus of elasticity of the dowel of 230,000 MPa, and f_c' of 35 MPa.

2.14 A set of 1.0-in.-diameter dowel bars are placed 18 in. on center in a 12-ft-wide concrete slab, as shown in Fig. P2.14. Two loads (9,000 lb each) are applied as shown. Assume the full efficiency of the dowel bars. The modulus of subgrade reaction is 500 pci; the concrete modulus is 3,800 ksi and Poisson's ratio is 0.15.

Determine the force distribution in the dowel bars.

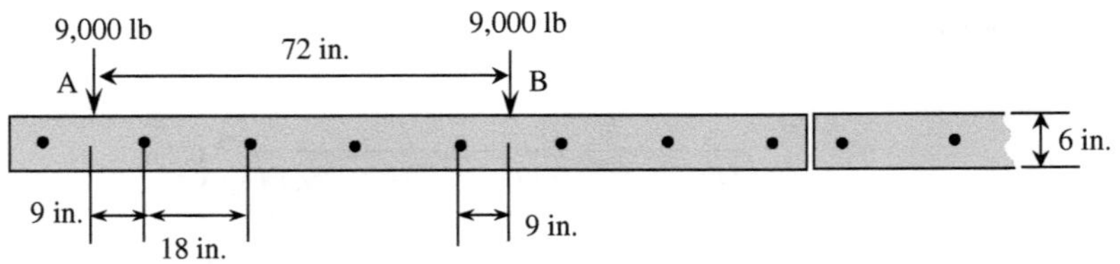

FIGURE P2.14 Dowel bars arrangements for Prob. 2.14.

2.15 For Prob. 2.14, given are a slab thickness of 6-in., a joint opening of 0.25 in., a modulus of dowel support of 700,000 pci, a modulus of elasticity of the dowel of 25,000 ksi, and f_c' of 4,500 psi.

Can the concrete handle this stress?

CHAPTER 3

Soils and Aggregates

3.1 Background

Soils and aggregates are essential ingredients of highway pavements. Soil and aggregate layers of a typical pavement are shown in Fig. 3.1. It shows that the pavement is constructed over the natural soil (subgrade). On the top of subgrade, aggregate subbase and base layers are provided, which consist of crushed-stone coarse aggregate and fine aggregate that are highly compacted at the optimum moisture content. The surface layer, asphalt concrete (AC), or portland cement concrete (PCC) is applied on top of the base layer. The surface, base and subbase layers together provide the structural integrity and strength to the pavement. The base and subbase layers also drain moisture if any moisture infiltrates through the surface layer or occurs due to ice melting or capillary action. As such, soils and aggregates are used in all parts of a pavement, and its strength and performance depend on the quality and properties of those soils and aggregates. Therefore, prior to construction, proper characterization of soils and aggregates is essential to understand its performance during the service life of a pavement.

3.2 Physical Properties

The physical properties of soils and aggregates are determined prior to applications. These properties are not necessarily indicators of material strength or stiffness, but these physical properties help select whether materials to borrow or use existing materials. For example, subgrade cut and fill is a common approach where soft clay or soft soil is encountered. Materials characterization to determine physical properties includes, but not limited to, the following:

- Sieve analysis
- Atterberg limits
- Soil classification
- Proctor test
- Flat and elongated particles
- Fine aggregate angularity
- Coarse aggregate angularity
- Clay content
- Los Angeles (LA) abrasion

Pavement layers

Exposed aggregate base layer

FIGURE 3.1 Typical layers of a pavement.

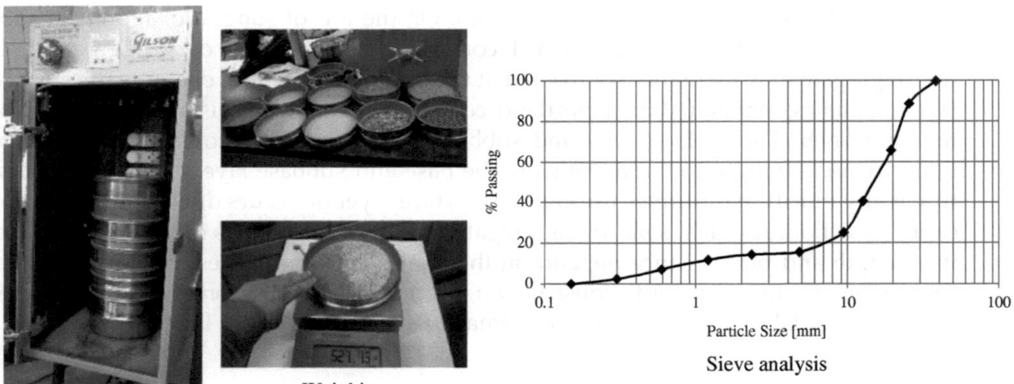

Sieve analysis

Weighing

Shaking

FIGURE 3.2 Sieve analysis procedure and the sieve analysis curve.

- Soundness
- Deleterious materials

3.2.1 Sieve Analysis

Sieve Analysis Procedure

A typical sieve analysis involves separating soils or aggregates by a nested column of sieves with wire mesh (screen), as shown in Fig. 3.2 and Table 3.1. A representative weighed specimen is poured into the top sieve which has the largest screen openings. The column is typically placed in a mechanical shaker. The shaker shakes the column for a certain amount of time. Next, the material retained on each sieve is weighed. The weight of the specimen of each sieve is then divided by the total weight to calculate the percentage retained on each sieve as shown in Eq. (3.1).

$$\%\ \text{Retained} = \frac{\text{Amount retained}}{\text{Total amount}} \times 100 \tag{3.1}$$

Sieve number	Size (mm)
1.5 in.	37.5
1.0 in.	25.0
¾ in.	19.0
½ in.	12.5
No. 4	4.75
No. 10	2.00
No. 20	0.85
No. 40	0.425
No. 100	0.15
No. 200	0.075

TABLE 3.1 Some Common Sieves and Their Sieve Openings

Then, the cumulative percent retained is calculated by summing up the "% Retained" in the corresponding sieve plus that in the larger sieves. The "% Finer" or "% Passing" for certain sieve is calculated as 100 minus the "Cumulative % Retained" for that sieve. The results of sieve analysis are presented in a graphical form to identify the type of gradation of the aggregate. A common practice is to plot "percent passing against sieve size."

Maximum Aggregate Size

ASTM C 125 defines the maximum aggregate size in one of the following two ways:

- *Maximum size.* The traditional definition of the maximum size is the smallest sieve through which 100% of the aggregate specimen particles pass. Superpave defines the maximum aggregate size as one sieve larger than the nominal maximum aggregate size (NMAS).
- *Nominal maximum aggregate size (NMAS).* The traditional definition of the NMAS is the largest sieve that retains some of the aggregate particles but generally not more than 10% by weight. Superpave defines NMAS as one sieve size larger than the first sieve to retain more than 10% of the material. Each stock of aggregate is represented by its NMAS. For example, a 19-mm aggregate batch represents a batch of aggregate whose NMAS is 19 mm.

Example 3.1: Aggregate Sizes After the sieve analysis, the masses of materials retained on each sieve are shown in Table 3.2.

a. Draw the sieve analysis curve (% Passing versus Sieve size).
b. Determine the NMAS based on the Superpave criteria.
c. Determine the maximum aggregate size based on the Superpave criteria.
d. Determine the NMAS based on the traditional criteria.
e. Determine the maximum aggregate size based on the traditional criteria.

Sieve number	Sieve size (mm)	Mass retained (g)
1.0 in.	25	0
0.75 in.	19	3
0.375 in.	9.5	26
No. 4	4.75	55
No. 8	2.38	155
No. 16	1.19	189
No. 30	0.6	315
No. 50	0.297	255
No. 100	0.149	110
No. 200	0.075	78
Pan	Pan	22

Table 3.2 Sieving Data for Example 3.1

Solution

a. Draw the sieve analysis curve (% Passing versus Sieve size).

- First, calculate the percent retained in each sieve as follows:

$$\%\text{Retained} = \frac{\text{Amount retained}}{\text{Total amount}} \times 100$$

- Then, the "Cumulative % Retained" is calculated by summing up the "% Retained" in the corresponding sieve plus that in the larger sieves. For example, "Cumulative % Retained" in No. 8 = 12.83 + 4.55 + 2.15 + 0.25 + 0 = 19.78, as shown in Table 3.3.
- "% Passing" is calculated as 100 – "Cumulative % Retained".
- Plot the "% Passing" versus "Sieve size" curve, as shown in Fig. 3.3.

b. Determine the NMAS based on the Superpave criteria.

Superpave defines the NMAS as one sieve size larger than the first sieve to retain more than 10% of the material. Here the first sieve to retain more than 10% of the material is No. 8 (2.4 mm). Therefore, NMAS = 4.75 mm.

c. Determine the maximum aggregate size based on the Superpave criteria.

Superpave defines the maximum aggregate size as one sieve larger than the nominal maximum size. As, NMAS = 4.75 mm, the maximum size is 0.375 in. (9.5 mm).

d. Determine the NMAS based on the traditional criteria.

The traditional definition of the NMAS is the largest sieve that retains some aggregate but generally not more than 10% by weight. Therefore, the NMAS is 0.75 in. (19 mm) as it retains 0.25% materials.

Sieve number	Sieve size (mm)	Mass retained (g)	% Retained	Cumulative % retained	% Passing
1.0 in.	25	0	0.00	0.00	100.0
0.75 in.	19	3	0.25	0.25	99.8
0.375 in.	9.5	26	2.15	2.40	97.6
No. 4	4.75	55	4.55	6.95	93.0
No. 8	2.38	155	12.83	19.78	80.2
No. 16	1.19	189	15.65	35.43	64.6
No. 30	0.6	315	26.08	61.51	38.5
No. 50	0.297	255	21.11	82.62	17.4
No. 100	0.149	110	9.11	91.72	8.3
No. 200	0.075	78	6.46	98.18	1.8
Pan	Pan	22	1.82	100.0	0.0
		Total = 1,208			

TABLE 3.3 Analysis of Sieving Data for Example 3.1

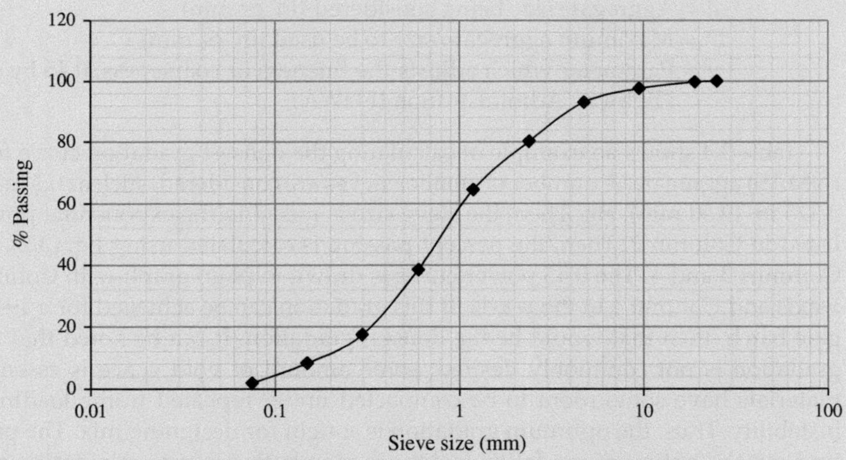

FIGURE 3.3 Sieve analysis curve for Example 3.1.

e. Determine the maximum aggregate size based on the traditional criteria.

The traditional definition of the maximum size is the smallest sieve through which 100% of aggregate specimen particles pass. Therefore, the maximum size is 1.0 in. (25 mm), as it is the smallest size which passes all aggregates.

Desired Gradation

Material gradation has a solid effect on pavement performances. The optimum gradation is a complicated question. It depends on the material (AC or PCC), its desired characteristics, anticipated loading, environment, and mix properties. The optimum gradation produces the maximum density. This would involve a particle arrangement where smaller particles are inserted among the larger particles, which reduces the void space between particles. This creates more particle-to-particle contact, which in AC would increase stability and reduce water infiltration. In PCC, the reduced void space reduces the amount of cement paste required. However, some minimum amount of void space is necessary to:

- Provide adequate volume for the binder.
- Promote rapid drainage and resistance to frost action for base and subbases.

Therefore, although it may not be the optimum aggregate gradation, a maximum density gradation does provide a common reference. A widely used equation to describe a maximum density gradation is:

$$P = \left(\frac{d}{D}\right)^n \times 100 \tag{3.2}$$

where P = % finer (% passing) than the sieve
d = Aggregate size being considered (in. or mm)
D = Maximum aggregate size to be used (in. or mm)
n = Parameter which adjusts the fineness or coarseness, 0.45 by the Federal Highway Administration (FHWA)

Table 3.4 shows an example of calculating the densest gradation curve for a 19-mm (NMAS) aggregate. A number of smaller sieves are considered, such as 0.5 in. (12.5 mm), 0.375 in. (9.50 mm), etc. Then, the sieve size is raised to the exponential power 0.45, as listed in Column 2. Then, the percent passing is calculated using Eq. (3.2), as listed in Columns 3 and 4. The 0.45 power curve is drawn in plain graph with Column 2 in the x-axis and Column 4 in the y-axis. If this gradation can be achieved for a 19-mm aggregate batch, then that would be the densest gradation. It can be noted that the densest gradation is not commonly desired; some amount of void space is essential so that materials have some room to be compacted under repeated traffic loading avoiding instability. Thus, the optimum gradation is sought for designing mix. The procedure to find out the optimum gradation is discussed in both concrete mix design and asphalt mix design chapters as each of these mix designs has different requirements.

Gradation Types

Several common terms are used to classify aggregate gradation, as shown in Fig. 3.4. These are not precise technical terms, but terms that refer to gradations that share common characteristics.

Dense or well-graded. Dense gradations are near the 0.45 power curve, but may not be right on it. It consists of materials of all sizes. Coarse aggregates generate voids in the mix which are filled up by smaller aggregates. Dense-graded mix produces the best quality mix.

Gap graded. Gap graded refers to a gradation that contains only a small percentage of aggregate particles in the mid-size range. The curve is flat in the mid-size range. Some PCC mix designs use gap-graded aggregate to provide a more economical mix

Particle size (mm)	(size, mm) 0.45	% Passing Calculation	% Passing
Column 1	Column 2	Column 3	Column 4
19.0	3.762	$P = \left(\frac{19.0}{19.0}\right)^{0.45} \times 100$	100
12.50	3.116	$P = \left(\frac{12.5}{19.0}\right)^{0.45} \times 100$	82.8
9.50	2.754	$P = \left(\frac{9.5}{19.0}\right)^{0.45} \times 100$	73.2
2.00	1.366	$P = \left(\frac{2.00}{19.0}\right)^{0.45} \times 100$	36.3
0.30	0.582	$P = \left(\frac{0.30}{19.0}\right)^{0.45} \times 100$	15.5
0.075	0.312	$P = \left(\frac{0.075}{19.0}\right)^{0.45} \times 100$	8.3

Table 3.4 Calculations for 0.45 Exponential Power Curve Using a 19-mm NMAS

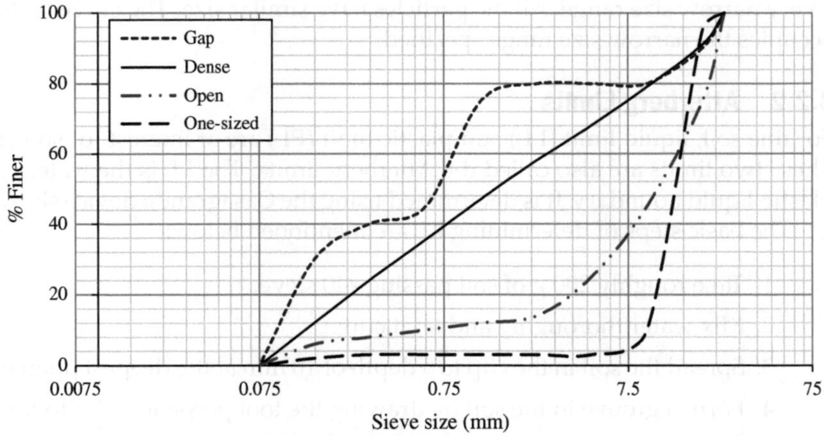

Figure 3.4 Different types of aggregate gradations.

since a reduced percentage of sand can be used for a given workability. Gap-graded AC mixes can be prone to segregation during placement.

Open graded. Open gradation contains only a small percentage of aggregate particles in the small range. This results in more air voids because there are not enough small

(a) Mixing soil and water (b) Placing in the cup (c) Leveling

(d) Forming the groove (e) Lifting and dropping (f) Groove closed by ½ in.

FIGURE 3.5 Different stages of liquid-limit testing.

particles to fill the voids between the larger particles. The gradation curve is near vertical in the mid-size range, and flat and near-zero in the small-size range.

One-sized or uniformly graded. One-sized gradation contains most of the particles in a very narrow size range. All the particles have similar size. The curve is steep and only occupies the narrow size range specified.

3.2.2 Atterberg Limits

For fine soil, liquid limit (LL) and plastic limit (PL) are of interest for soil classification. These two limits are also called the Atterberg limits. The LL is the water content at the plastic-liquid boundary. It is determined using the Casagrande device shown in Fig. 3.5.

The basic steps of determining LL are mentioned below:

1. Take roughly 200 g of soil passing #40 sieve.
2. Mix water thoroughly and let it cure.
3. Spread the soil in the cup to a depth of 10 mm at the deepest point of the device.
4. Form a groove in the soil by drawing the tool perpendicular to the surface.
5. Lift and drop the cup about two drops per second.
6. Record the number of blows required to close the groove.
7. After several trials, determine the water content which needs 25 blows to close the groove by ½ in. (12.5 mm). This water content is called the LL.

PL is the water content at the plastic-semisolid boundary. It is determined by rolling a soil specimen to a thread of 1/8 in. (3.2 mm) diameter as shown in Fig. 3.6. The basic steps are as follows:

(a) Mixing soil and water (b) Taking small specimen (c) Rolling with palm

(d) Crumbled before reaching the diameter of 3.2 mm (too dry) (e) Did not crumble after reaching the diameter of 3.2 mm (too wet) (f) Crumbled just after reaching the diameter of 3.2 mm (exactly on PL)

Figure 3.6 Different stages of plastic-limit testing.

1. Take about 20 g of soil passing #40 sieve.
2. Thoroughly mix small amount of water and cure it for at least 16 minutes.
3. Take about 2 g of wet soil.
4. Roll by hand between palm and a glass plate.
5. If the thread breaks before reaching the diameter of 1/8 in. (3.2 mm), mix more water and repeat steps 1 through 4.
6. If the thread diameter can be reached easily, break it into several pieces and reroll.
7. Determine the water content at which the diameter can be barely reached. This water content is called the PL.

Plasticity index (PI) is the difference between the LL and the PL, i.e., PI = LL – PL.

3.2.3 Soil Classification

Soil can be classified in many different ways, such as the Unified Soil Classification System (USCS), the MIT Soil Classification, the International Classification, the American Association of State Highway and Transportation Officials (AASHTO) Soil Classification, etc. The following two parameters are found to be useful for soil classification after years of researches and experience with wide-ranging soils all over the world:

- Particle size distribution of soils
- Liquid limit and the plasticity index

General classification	Granular materials (35% or less passing 0.075 mm)							Silt-clay materials (More than 35% passing 0.075 mm)			
Group classification	A-1			A-2							A-7
	A-1-a	A-1-b	A-3	A-2-4	A-2-5	A-2-6	A-2-7	A-4	A-5	A-6	A-7-5 A-7-6
Sieve analysis % passing 2.00 mm (No 10) 0.425 mm (No 40) 0.725 mm (No 200)	50 max 30 max 15 max	50 max 25 max	51 min 10 max	35 max	35 max	35 max	35 max	36 min	36 min	36 min	36 min
Characteristics of fraction passing Liquid limit Plastic index	6 max		N.P.	40 max 10 max	41 min 10 max	40 max 11 min	41 min 11 min	40 max 10 max	41 min 10 max	40 max 11 min	40 min 11 min
Usual types of significant constituent material	Stone fragment Gravel and sand		Fine sand	Silty or clayey gravel and sand				Silty soils		Clayey soils	
General rating	Excellent to good							Fair to poor			

Source: From AASHTO M 145 (2017) by the American Association of State Highway and Transportation Officials, Washington, DC. Used with permission.

Table 3.5 AASHTO Soil Classification Chart

The soil in nature consists of a mixture of particles of different sizes and cohesions. Thus, every soil consists of clay, silt, sand, and gravel in some proportions. Hence, to classify a given soil, it is necessary to determine its grain (or particle) size distribution and the Atterberg limits.

For roads and highways, the AASHTO Soil Classification system is commonly used. It classifies soil from A-1 to A-8, where A-1 is the best engineering soil and A-8 is the worst engineering soil (organic soil). A-8 soil is never recommended for engineering purpose, although it has a good value in agriculture and gardening. The AASHTO Soil Classification chart is shown in Table 3.5.

At the beginning, the soil passing the 0.075-mm sieve (No. 200 sieve) is looked at. If 35% or less than 35% of soil passes through the No. 200 sieve, then it is a granular soil, either A-1, A-2, or A-3. If more than 35% passes through the No. 200 sieve, then it is silt or clay materials, and falls under the AASHTO soil classes A-4 to A-7. As mentioned earlier, A-8 is an organic soil, and is excluded from the chart as it is not suitable for engineering purposes. A-1 group consists of stone fragments, gravel, and sand with a maximum of 25% fines. A-3 group consists of a fine sand with a maximum of 10% fines. Silty or clayey soil belongs to A-2, whereas pure silty soils with minimum of 36% fines fall within the A-4 and A-5 groups. A-6 and A-7 groups consist of clayey soils with minimum of 36% fines.

For A-1-a group, the percentages of soil passing through the 2-mm and 425-μm sieves are the additional criteria. For classifying as the A-1-b and A-3 groups, the percentage of soil passing through the 425-μm sieve is the additional criterion. For all other groups, liquid limit and plasticity index of the soil are the additional criteria.

In AASHTO classification system, soil type is further grouped by computing an index called group index (GI). GI is used to describe the relative performance of soils, when used for highway construction, and is defined by the equation:

$$\text{GI} = 0.2\,(F - 35) + 0.005\,(F - 35)\,(\text{LL} - 40) + 0.01\,(\text{F} - 15)\,(\text{PI} - 10) \tag{3.3}$$

where F = Percentage passing through the 0.075 mm or No. 200 sieve
LL = Liquid limit
PI = Plasticity index

The group index is rounded off to the nearest whole number and appended in parentheses. If the computed group index is either zero or negative, the number zero is used as the group index.

Example 3.2: AASHTO Soil Classification A specimen of soil was tested in the laboratory, and results of the laboratory tests were as follows:

Liquid limit = 39
Plastic limit = 19

The sieve analysis data is presented in Table 3.6.
Determine the AASHTO classification of this soil (symbol plus group index).

Solution

Step 1. Passing No. 200 sieve is more than 35%; therefore, soil is within A-4 to A-7.
Step 2. LL = 39 and PI = 39 – 19 = 20; therefore, soil is A-6 as per the chart.
Step 3.
$$\begin{aligned}\text{GI} &= 0.2\,(F - 35) + 0.005\,(F - 35)\,(\text{LL} - 40) + 0.01\,(F - 15)\,(\text{PI} - 10)\\ &= 0.2\,(45 - 35) + 0.005\,(45 - 35)\,(39 - 40) + 0.01\,(45 - 15)\,(20 - 10)\\ &= 2 - 0.05 + 3\\ &= 4.95\end{aligned}$$

Sieve size	Percent passing
No. 4	100
No. 10	85
No. 40	75
No. 200	45

Table 3.6 Sieve Analysis Data for Example 3.2

Answer *The soil type is A-6(5).*

3.2.4 Proctor Test

Proctor test is performed to determine the optimum moisture content (OMC) and the maximum dry density (MDD) of soils. The dry density of a soil increases with the increase in moisture content. After the optimum moisture content, the dry density decreases with the increase in moisture content. Compaction is quantified considering soil's dry density, which can be expressed as shown in the equation:

$$\gamma_d = \frac{\gamma}{1+w} \tag{3.4}$$

where γ_d = Dry density of soil
γ = Moist density of soil
w = Moisture content of soil

The unit weight of moist soil can be determined using the equation:

$$\gamma = \frac{W}{V} \tag{3.5}$$

where W = Weight of moist soil
V = Volume of moist soil

At least three specimens of the soil are compacted at different water contents, and a curve is drawn with axes of dry density and water content. The resulting plot usually has a distinct peak, as shown in Fig. 3.7. Such inverted "V" curves are obtained for cohesive soils (or soils with fines) and are known as compaction curves.

The relation between the moisture content and the dry unit weight for a saturated soil is the zero air-voids line. It is not feasible to expel all the air completely by compaction, no matter how much compactive effort is used and in whatever manner. As water is added to a soil at a low moisture content, it becomes easier for the particles to move past one another during the application of compacting force. The particles come closer, the voids are reduced, and this causes the dry density to increase. As the water content increases, the soil particles develop larger water films around them. This increase in dry density continues until a stage is reached where water starts occupying the space that could have been occupied by the soil grains. Thus, the water at this stage hinders the closer packing of grains and reduces the dry unit weight. The MDD occurs at an OMC and their values can be obtained from the plot.

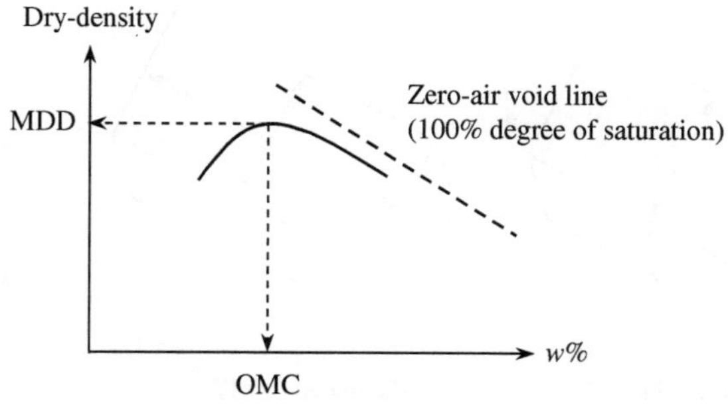

FIGURE 3.7 Compaction test result.

Laboratory compaction test is conducted in the laboratory using the ASTM D 698 test protocol, known as the standard Proctor test. The soil is usually compacted inside a standard mold in three equal layers, each receiving a number of blows from a standard weighted hammer at a specified height. This process is then repeated for various moisture contents and the dry density is determined for each specimen. The graphical relationship of the dry density to moisture content is then plotted to establish the compaction curve. The peak of the curve gives the OMC and MDD.

Example 3.3: Proctor Test A compaction test was conducted in a soils laboratory, and the standard Proctor compaction procedure (ASTM D 698) was used. The weight of a compacted soil specimen plus mold was determined to be 4,805 g. The volume and weight of the mold were 1/30 ft^3 and 3,060 g, respectively. The water content of the specimen was 8%. Compute both the wet and dry unit weights of the compacted specimen.

Solution

Weight of wet soil = 4,805 – 3,060 g = 1,745 g

Volume of soil = 1/30 ft^3

Unit weight of wet soil, $\gamma = \frac{W}{V} = \frac{1{,}745\ \text{g}}{\frac{1}{30}\ \text{ft}^3} = 52{,}350\frac{\text{g}}{\text{ft}^3} = 115\frac{\text{lb}}{\text{ft}^3}$ [1 g = 0.0022 lb]

Moisture content, $w = 8\% = 0.08$

Unit weight of dry soil, $\gamma_d = \frac{\gamma}{1+w} = \frac{115\ \text{pcf}}{1+0.08} = 107\ \text{pcf}$

Answers

Wet unit weight = 115 lb/ft^3

Dry unit weight = 107 pcf

3.2.5 Flat and Elongated Particles

Flat and elongated particles (ASTM D 4791) are a consensus aggregate property requirement in the Superpave asphalt mix design process and it is performed on coarse aggregate larger than No. 4 sieve (4.75 mm). ASTM D 4791 is used to determine the flat and elongated particles and defined as follows (Fig. 3.8):

- A flat particle is defined as one where the ratio of the middle dimension to the smallest dimension of the particle exceeds 3:1.
- An elongated particle is defined as one where the ratio of the largest dimension to the middle dimension of the particle exceeds 3:1.

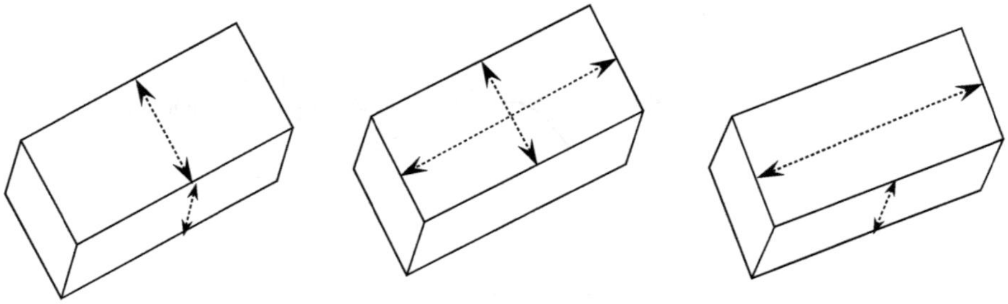

FIGURE 3.8 Concept of flakiness test for coarse aggregates.

- Particles are classified as flat and elongated if the ratio of the largest dimension to the smallest dimension exceeds 5:1.

Elongated particles are undesirable because they have a tendency to break during construction and under traffic. The maximum percentage of flat and elongated particles is commonly used as 10% by weight for total equivalent single axle of load of 0.3 million or more.

3.2.6 Fine Aggregate Angularity

Fine aggregate angularity (FAA) is also a consensus aggregate property requirement in the Superpave asphalt mix design process and is defined as the percent air voids present in a loosely compacted aggregates smaller than No. 8 sieve (2.38 mm). This property ensures enough fine aggregate internal friction and rutting resistance. In the test by the AASHTO T 304, a specimen of fine aggregate is poured into a small calibrated cylinder by flowing through a standard funnel. By determining the weight of fine aggregate in the filled cylinder of known volume, void content can be calculated as the difference between the cylinder volume and fine aggregate volume collected in the cylinder. The uncompacted void content is calculated using the equation:

$$\text{Air void} = \frac{V - \left(\dfrac{M}{G}\right)}{V} \times 100 \tag{3.6}$$

where V = Volume of the cylindrical measure (cm^3, cc, or mL, no other unit is possible)
M = Specimen mass (g, no other unit is possible)
G = Dry bulk specific gravity of the fine aggregate (discussed in the next subsection)

This value is also known as the FAA. The higher the FAA, the higher the angularity and rough surface. The fine aggregate bulk specific gravity is used to compute the fine aggregate volume. The required minimum value for fine aggregate angularity is a function of traffic level and depth within pavement. These requirements apply to the final aggregate blend, although estimates can be made on the individual aggregate stockpiles. Higher void contents mean more fractured faces. The minimum percentage of air voids in loosely compacted fine aggregate is commonly used as 40% to 45% by weight for total equivalent single axle of load of 0.3 million or more.

Example 3.4: Air Void of a Loosely-Filled Soil A 100-cm^3 container is loosely filled with 225 g of fine soil with specific gravity of 2.45. Determine the air void of the loosely filled soil.

Solution

Given:

Volume of the cylindrical measure, $V = 100\ \text{cm}^3$
Specimen mass, $M = 225\ \text{g}$
Specific gravity, $G = 2.45$

The uncompacted void content is calculated as:

$$\text{Air void} = \frac{V - \left(\frac{M}{G}\right)}{V} \times 100$$

$$= \frac{100\ \text{cm}^3 - \left(\frac{225\ \text{g}}{2.45}\right)}{100\ \text{cm}^3} \times 100$$

$$= 8.16\%$$

Answer Air void = 8.2%

Note: Air void is commonly rounded to the nearest 0.1%.

3.2.7 Coarse Aggregate Angularity

Coarse aggregate angularity (CAA) ensures a high degree of aggregate internal friction and mix rutting resistance. It is defined as the percent by weight of aggregates larger than No. 4 sieve (4.75 mm) with one or more fractured faces. ASTM D 5821 procedure involves manually counting particles to determine fractured faces. A fractured face is defined as any fractured surface that occupies more than 25% of the area of the outline of the aggregate particle visible in that orientation. The required minimum values for coarse aggregate angularity are a function of traffic level and position within the pavement and vary between 50% and 100%.

3.2.8 Clay Content

Clay content is the percentage of clay material contained in the aggregate fraction finer than No. 4 sieve (4.75 mm). It is measured following the AASHTO T 176 test protocol. In this test, a specimen of fine aggregate is placed in a graduated cylinder, as shown in Fig. 3.9, with a flocculating solution and agitated to loosen clayey fines present in and coating the aggregate. The flocculating solution forces the clayey material into suspension above the granular aggregate. After a period that allows sedimentation, the cylinder height of suspended clay and sedimented sand is measured. The sand equivalent value is computed as a ratio of the sand to clay height readings expressed as a percentage. The required clay content values (40–50%) for fine aggregate are expressed as a minimum sand equivalent and are a function of traffic level. The ratio of the height of sand over the height of clay is considered the SE, as shown in Eq. (3.7). The larger is the sand equivalent value, the cleaner (less fine dust or clay-like materials) is the aggregate.

Figure 3.9 Sand equivalent testing.

$$SE = \frac{\text{Sand height}}{\text{Clay height}} \times 100 \tag{3.7}$$

Example 3.5 In a sand equivalent test, a clay layer of 13 mm is deposited on the 23 mm of sand particles. Calculate the SE of the specimen.

Solution

Sand height = 23 mm
Clay height = 23 + 13 mm = 36 mm

$$SE = \frac{\text{Sand height}}{\text{Clay height}} \times 100 = \frac{23 \text{ mm}}{36 \text{ mm}} \times 100 = 64\%$$

Answer The SE is 64%.

Note: Sand equivalent is commonly rounded to the nearest 1%.

3.2.9 Los Angeles (LA) Abrasion

Los Angeles (LA) abrasion test is a toughness test. This test simulates the resistance of coarse aggregate to abrasion and mechanical degradation during handling,

FIGURE 3.10 Los Angeles abrasion test apparatus. (*Courtesy of Test Mark Industries, 995 North Market St., East Palestine, Ohio.*)

construction, and in-service. The percent loss of materials from an aggregate blend during the LA abrasion test (AASHTO T 96) is measured. The test is performed by subjecting the coarse aggregate and some steel spheres inside a large drum, as shown in Fig. 3.10. After rotating the drum, the weight of the aggregate that is retained on a No. 12 (1.70 mm) sieve is subtracted from the original weight to obtain a percentage of the total aggregate weight that has broken down and passed through the No. 12 (1.70 mm) sieve. The test result is percent loss, which is the weight percentage of coarse material lost during the test as a result of the mechanical degradation. Maximum loss values typically range from approximately 35% to 45%.

3.2.10 Soundness

Soundness test estimates the resistance of aggregates to weathering while in-service. It can be performed on both coarse and fine aggregates. This test measures the percent loss of materials from an aggregate blend during the sodium or magnesium sulfate soundness test according to the AASHTO T 104 standard. The test is performed by alternately exposing an aggregate specimen to repeated immersions in saturated solutions of sodium or magnesium sulfate each followed by oven drying. One immersion and drying is considered one soundness cycle. During the drying phase, salts precipitate in the permeable void space of the aggregate. Upon reimmersion, the salt rehydrates and exerts internal expansive forces that simulate the expansive forces of freezing water. The test result is total percent loss over various sieve intervals for a required number of cycles. Maximum loss values range from approximately 10% to 20% for five cycles.

3.2.11 Deleterious Materials

Deleterious materials are defined as the weight percentage of contaminants, clay lumps, and friable particles such as shale, wood, mica, and coal in the blended aggregate. The test is performed by wet sieving aggregate size fractions over prescribed sieves following the AASHTO T 112 standard. The weight percentage of material lost as a result of wet sieving is reported as the percent of clay lumps and friable particles. A wide range of maximum permissible percentage of clay lumps and friable particles is evident. Values range from 0.2% to 10%, depending on the exact composition of the contaminant. Some of the adverse effects of several deleterious substances are listed below:

- Organic impurities delay setting and hardening of concrete and reduce strength gain.
- Coal, lignite, clay lumps, and friable particles increase popouts and reduce durability.
- Fines passing No. 200 sieve (smaller than 0.075 mm) weaken the bond between asphalt and aggregates.

3.3 Mechanical Properties

In pavement design, each layer of pavement is treated in mechanistic fashion. More clearly, the stress-strain in each layer of pavement is determined using mechanistic analysis. Hence, the stiffnesses of soil and aggregate layers are essential design input parameters. In general, the more resistant to deformation a subgrade/base is, the more load it can support before reaching a critical deformation value. There are other factors involved when evaluating subgrade materials (such as shrink/swell in the case of certain clays and ash) which are minors. The methods of stiffness characterization are discussed here.

3.3.1 Resilient Modulus

Resilient modulus (M_R) of unbound layer is the most desired input parameter in the AASHTOWare pavement ME design software. This test is conducted according to the AASHTO T 307 test sequence for base/subbase/subgrade materials. This method measures the elastic modulus of untreated base, subbase, and soil materials. A repeated axial cyclic stress of fixed magnitude is applied to a laboratory-prepared cylindrical specimen. During testing, the specimen is subjected to a dynamic cyclic stress and a static confining pressure by means of a pressure chamber. The total recoverable axial deformation of the specimen is measured and used to calculate the M_R value as shown in the equation:

$$M_R = \frac{S_{\text{cyclic}}}{\varepsilon_r} \tag{3.8}$$

where S_{cyclic} is the applied cyclic axial stress and ε_r is the resilient axial strain as presented in Eqs. (3.9) and (3.10).

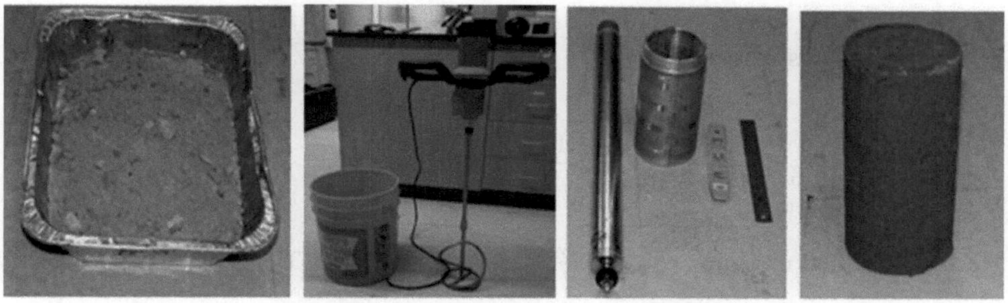

(a) Soil (b) Mixing with water (c) Compaction (d) Prepared sample

FIGURE 3.11 Specimen preparation.

$$S_{\text{cyclic}} = \frac{P_{\text{cyclic}}}{A} \tag{3.9}$$

where P_{cyclic} is the applied cyclic load and A is the cross-sectional area of the specimen.

$$\varepsilon_r = \frac{e_r}{L} \tag{3.10}$$

where e_r is the resilient axial deformation due to S_{cyclic} and L is the original specimen length.

The specimen diameters are typically 6.0, 4.0, and 2.5 in. for the base, subbase, and subgrade materials, respectively. The height of the specimens is taken at least twice of the diameter. Specimen is compacted using dynamic impact compaction by the modified Proctor effort. Prior to the testing, the specimen is capped with gypsum to prepare a uniform top surface. A prepared specimen and a test-ready specimen are shown in Fig. 3.11.

Cyclic haversine–shaped load is applied in a triaxial pressure chamber. Each test sequence (1 s) has 0.1 s of load pulse and 0.9 s of rest period. The average stress-strain at the last five cycles is considered for M_R calculation for each loading condition. The test setup is shown in Fig. 3.12.

Example 3.6: Resilient Modulus In a resilient modulus test, a cylindrical specimen of 100-mm diameter and 200-mm height is used. The applied total axial force is 2,100 N. The resulting axial deformation in the specimen is 0.39 mm. Determine the resilient modulus of the specimen.

Solution

Specimen height, $L = 200 \text{ mm} = 0.2 \text{ m}$

Specimen diameter, $D = 100 \text{ mm} = 0.1 \text{ m}$

Specimen cross-sectional area, $A = \dfrac{\pi D^2}{4} = \dfrac{\pi(0.1 \text{ m})^2}{4} = 0.007854 \text{ m}^2$

$$\text{Applied stress, } S_{\text{cyclic}} = \frac{P_{\text{cyclic}}}{A} = \frac{2{,}100 \text{ N}}{0.007854 \text{ m}^2} = 267{,}380 \text{ Pa}$$

$$\text{Resulting strain, } \varepsilon_r = \frac{e_r}{L} = \frac{0.39 \text{ mm}}{200 \text{ mm}} = 0.00195$$

$$\text{Resilient modulus, } M_R = \frac{S_{\text{cyclic}}}{\varepsilon_r} = \frac{267{,}380 \text{ Pa}}{0.00195} = 137{,}117{,}950 \text{ Pa} = 137 \text{ MPa}$$

Answer *The resilient modulus is 137 MPa.*

Note: M_R is commonly rounded to nearest hundreds of psi or nearest MPa.

3.3.2 California Bearing Ratio (CBR) in Laboratory

The California bearing ratio (CBR) test compares the bearing capacity of a material with that of a well-graded crushed stone. Thus, a high-quality crushed stone material should have a CBR at 100%. It is primarily intended for, but not limited to, evaluating the strength

FIGURE 3.12 Resilient modulus test setup.

FIGURE 3.13 CBR test equipment and its accessories. (*Courtesy of Indiamart.*)

of cohesive materials having maximum particle sizes less than 19 mm (0.75 in.). It was developed by the California Division of Highways about the year 1930 and was subsequently adopted by numerous design agencies. This test can be performed on in situ base/subbase/soil (ASTM D 4429), or laboratory-compacted base/subbase/soil (ASTM D1883 or AASHTO T 193).

In the laboratory, the basic CBR test involves applying a load to a small penetration piston (Fig. 3.13) at a rate of 1.3 mm (0.05 in.) per minute and recording the total load at penetrations ranging from 0.64 mm (0.025 in.) up to 7.62 mm (0.30 in.).

Values obtained are inserted into the following equation to obtain the CBR value:

$$\text{CBR}(\%) = 100\left(\frac{x}{y}\right) \tag{3.11}$$

where x = Resistance pressure on the piston for 0.1 in. (2.54 mm) or 0.2 in. (5.08 mm) penetration

y = Standard pressure required for a well-graded crushed stone

- 1,000 psi (6.9 MPa) for 0.1 in. (2.54 mm) penetration
- 1,500 psi (10.3 MPa) for 0.2 in. (5.08 mm) penetration

The M_R can be obtained empirically from the CBR value using the equation:

Testing setup

Penetration after testing

Figure 3.14 In situ CBR testing setup.

$$M_R(\text{psi}) = 2{,}555(\text{CBR})^{0.64} \tag{3.12}$$

3.3.3 California Bearing Ratio (CBR) in Field

The CBR can also be determined for in situ condition. It is similar in nature to laboratory testing. The main difference is the soil compaction and loading frame. In field, the in situ soil being tested thus does not require any soil preparation. The loading frame is prepared by a heavy vehicle, as shown in Fig. 3.14.

Example 3.7: CBR and Resilient Modulus While conducted the CBR test on an aggregate layer, a load of 1,750 lb was required to penetrate 0.1 in. The diameter of the penetrating piston was 1.95 in.

Determine the CBR and the resilient modulus of the aggregate layer.

Solution

Given:

Piston diameter, $D = 1.95$ in.

$$\text{Piston cross-sectional area, } A = \frac{\pi D^2}{4} = \frac{\pi(1.95\text{ in.})^2}{4} = 2.99\text{ in.}^2$$

$$\text{Applied stress, } x = \frac{\text{Load}}{\text{Area}} = \frac{1{,}750\text{ lb}}{2.99\text{ in.}^2} = 585.3\text{ psi}$$

Known: Stress required for a crushed stone to penetrate by 1.0 in., $y = 1{,}000$ psi

$$\text{California bearing ratio, CBR}(\%) = 100\left(\frac{x}{y}\right) = 100\left(\frac{585.3\text{ psi}}{1{,}000\text{ psi}}\right) = 58.53 \approx 59$$

$$\text{Resilient modulus, } M_R(\text{psi}) = 2{,}555(\text{CBR})^{0.64} = 2{,}555\,(59)^{0.64} = 34{,}733\text{ psi}$$

Answers *The CBR is 59 and the resilient modulus is 34,700 psi.*

3.3.4 *R*-Value

The resistance value (*R*-value) test is a material stiffness test conducted by following the AASHTO T 190 or the ASTM D 2844 standard. Materials tested are assigned an *R*-value. The *R*-value test was developed by the California Division of Highways and first reported in the late 1940s. During this time rutting (or shoving) in the wheel tracks was a primary concern and the *R*-value test was developed as an improvement of the CBR test.

The test procedure to determine *R*-value requires that the laboratory-prepared specimens are fabricated to a moisture and density conditions representative of the worst possible in situ condition of a compacted subgrade. The *R*-value is calculated from the ratio of the applied vertical pressure to the developed lateral pressure, and is essentially a measure of the material's resistance to plastic flow, as shown in Eq. (3.13). The testing apparatus used in the *R*-value test is called a stabilometer and is shown in Fig. 3.15.

$$R = 100 - \left\{ \frac{100}{\left(\frac{2.5}{D}\right)\left[\left(\frac{P_V}{P_H}\right) - 1\right] + 1} \right\} \tag{3.13}$$

where R = Resistance value (0 to 100)
P_V = Applied vertical pressure, 160 psi
P_H = Transmitted horizontal pressure at P_V = 160 psi
D = Displacement of stabilometer fluid necessary to increase horizontal pressure from 5 to 100 psi

FIGURE 3.15 *R*-value test equipment.

The M_R can be obtained empirically from the R-value using the equation:

$$M_R(\text{psi}) = 1{,}155 + (555)(R\text{-value}) \quad (3.14)$$

3.3.5 Dynamic Cone Penetration (DCP)

The DCP test provides a measure of a material's in situ resistance to penetration following the ASTM D 6951. DCP testing execution is shown in Fig. 3.16. It consists of a rod with a standard sliding weight called hammer attached to the top and a disposable cone tip to penetrate the soil on the bottom. The weight of the hammer is 8 kg (17.6 lb) and it slides on a 16-mm driving rod. The tip has an included angle of 60 degrees and a diameter at the base of 20 mm. The hammer is lifted up and dropped from a standard height of 2.26 feet (575 mm), which causes the cone at the bottom of the device to be forced into the ground. The weight is dropped multiple times until there are enough blows to determine the soil characteristics or the cone has reached a depth of interest. With each blow the new depth of the device is recorded. The depths and corresponding blow numbers are then plotted in Excel where a best linear fit is applied. The slope is considered the DCP value and is usually measured in millimeters per blow or inches per blow.

The M_R can be obtained empirically from the R-value using the equation:

$$M_R(\text{psi}) = 2{,}555\left(\frac{292}{\text{DCP}^{1.12}}\right)^{0.64} \quad (3.15)$$

where DCP value is in millimeters per blow. DCP index is related to the CBR, as in the equation below:

$$\text{CBR} = \frac{292}{\text{DCP}^{1.12}} \quad (3.16)$$

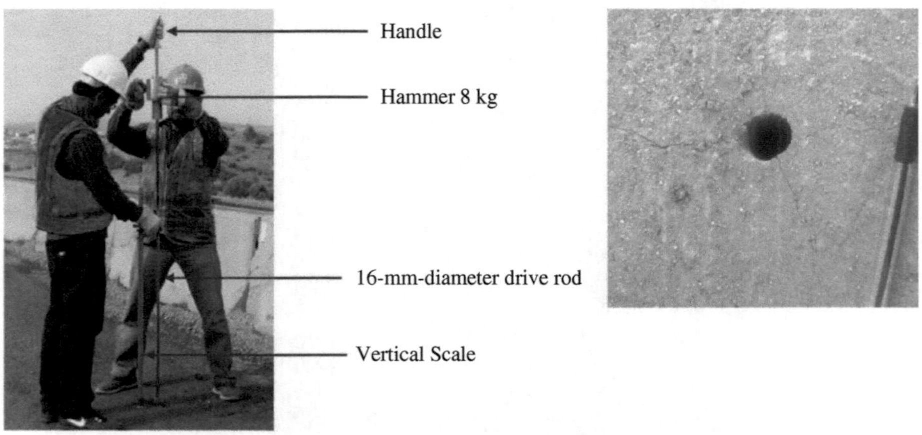

FIGURE 3.16 DCP testing.

Example 3.8: Analysis of DCP Test Data The DCP test readings on a pavement site are listed in Table 3.7. Determine the DCP, M_R and CBR values.

Blow	Penetration reading (mm)
0	2.5
1	15.6
2	25.2
3	31.6
4	36.8
5	42.0
6	47.2
7	52.4
8	57.6
9	62.8
10	67.8

Table 3.7 Penetration Reading for Example 3.8

Solution

Let us calculate the penetration at each blow of loading (Table 3.8).

At the beginning the DCP value is very high (13.1 mm, 9.6 mm, etc.) and then becomes constant. This may be due to the reason that the surface soil is commonly loose and becomes dense with increase in depth. The average DCP value can be taken as 5.2 mm/blow. You may also consider the mathematical average penetration neglecting the penetration by the first few blows.

Blow	Penetration reading (mm)	Penetration (mm)
0	2.5	0
1	15.6	15.6 – 2.5 = 13.1
2	25.2	25.2 – 15.6 = 9.6
3	31.6	31.6 – 25.2 = 6.4
4	36.8	36.8 – 31.6 = 5.2
5	42.0	42.0 – 36.8 = 5.2
6	47.2	47.2 – 42.0 = 5.2
7	52.4	52.4 – 47.2 = 5.2
8	57.6	57.6 – 52.4 = 5.2
9	62.8	62.8 – 57.6 = 5.2
10	67.8	67.8 – 62.8 = 5.0

Table 3.8 Penetration Calculation for Example 3.8

Resilient modulus, $M_R(\text{psi}) = 2{,}555\left(\frac{292}{\text{DCP}^{1.12}}\right)^{0.64} = 2{,}555\left(\frac{292}{5.2^{1.12}}\right)^{0.64} = 29{,}650\,\text{psi} \approx$ 29,700 psi

$$\text{CBR} = \frac{292}{\text{DCP}^{1.12}} = \frac{292}{5.2^{1.12}} = 46$$

Answers *DCP = 5.2 mm/blow, M_R = 29,700 psi, CBR = 46*

3.3.6 Resilient Modulus from Soil Physical Testing

The M_R can also be determined empirically if the Atterberg limits and percent of soil passing No. 200 sieve (0.075-mm opening) are known. Equation (3.17) shows the relationship that can be used to determine empirically the M_R value.

$$M_R(\text{psi}) = 2{,}555\left(\frac{75}{1 + 0.728(P_{200})(\text{PI})}\right)^{0.64} \tag{3.17}$$

where P_{200} = Passing No. 200 sieve (use the decimal value)
PI = Plasticity index = the difference between liquid limit and plastic limit

Example 3.9: Resilient Modulus from Sieve Analysis The LL and PL of a soil are 38 and 22, respectively. After sieve analysis, it was found that 7.5% of soil passes No. 200 sieve. Calculate the resilient modulus of the soil.

Solution
Given data:
Plasticity index, PI = LL – PL = 38 – 22 = 16
Soil passing No. 200 sieve, P_{200} = 7.5% = 0.075

$$\text{Resilient modulus, } M_R(\text{psi}) = 2{,}555\left(\frac{75}{1 + 0.728(P_{200})(PI)}\right)^{0.64}$$

$$= 2{,}555\left(\frac{75}{1 + 0.728(0.075)(16)}\right)^{0.64}$$

$$= 27{,}096\ \text{psi}$$

Answer *The resilient modulus is 27,100 psi.*

3.3.7 Resilient/Elastic Modulus of Chemically Stabilized Soil

There are some empirical relationships which can be used to determine the resilient/elastic modulus of chemically stabilized soil. For cement-treated aggregate, the modulus of elasticity (E) in psi can be determined as:

$$E = 57{,}000\sqrt{f_c'} \tag{3.18}$$

where f_c' is the compression strength in psi tested in accordance with the AASHTO T 22 standard. For soil cement, the modulus of elasticity (E) in psi can be determined as:

$$E = 1{,}200\, q_u \tag{3.19}$$

where q_u is the unconfined compression strength in psi tested in accordance with the ASTM D 1633 standard. For lime-cement-fly ash, the modulus of elasticity (E) in psi can be determined as:

$$E = 500 + q_u \tag{3.20}$$

where q_u is the unconfined compression strength in psi tested in accordance with the ASTM C 593 standard. For lime-stabilized soils, the resilient modulus (M_R) in psi can be determined as:

$$M_R = 0.124\, q_u + 9.98 \tag{3.21}$$

where q_u is the unconfined compression strength in psi tested in accordance with the ASTM D 5102 standard.

3.4 Resilient Modulus Variations Due to Moisture

The AASHTOWare pavement ME design guide considers the M_R of unbound layer affected by the in situ moisture content. For example, the AASHTOWare pavement ME design guide uses the following model to determine the variation in M_R of unbound layer with the degree of saturation (AASHTO 2015):

$$\log\left(\frac{M_R}{M_{\mathrm{Ropt}}}\right) = a + \frac{b - a}{1 + \exp\left[\ln\dfrac{-b}{a} + k_m(S - S_{opt})\right]} \tag{3.22}$$

where M_{Ropt} = resilient modulus at optimum moisture content; a, b, k_m are regression parameters; $(S - S_{\mathrm{opt}})$ = variation in degree of saturation expressed in decimal. For fine-grained materials, a, b, and k_m are –0.5934, 0.4, and 6.1324, respectively. For coarse-grained materials, a, b, and k_m are –0.3123, 0.3, and 6.8157, respectively. An increase in degree of saturation level leads to the decreases in M_R. For example, M_R decreases by 47% for coarse-grained and by 67% for fine-grained material due to increase in degree of saturation by 40% for a particular soil in New Mexico (Islam and Tarefder, 2015). The degree of saturation (S) can be determined using the following equation:

$$S = \frac{wG_S}{e} \tag{3.23}$$

where e = Void ratio
w = Gravimetric moisture content
G_S = Specific gravity of the material

Remember that degree of saturation (S) and water content (w) are not the same parameters.

Unless provided, the AASHTOWare pavement ME design guide determines the change in saturation level using the soil-water characteristic curve (SWCC) based on the enhanced integrated climate model (EICM). The AASHTOWare pavement ME design software also considers the change in M_R value due to freezing and thawing. The details can be found in the I-37A report by the National Cooperative Highway Research Program (NCHRP, 2004; Part 2, Chapter 3).

Example 3.10: Degree of Saturation A chunk of moist soil weighs 49 lb with a volume of 0.46 ft³. After oven-drying, the weight of the soil decreases to 38 lb. The specific gravity and void ratio of the soil are 2.64 and 1.1, respectively. Calculate the degree of saturation of the soil.

Solution

Moist weight of soil = 49 lb
Dry weight of soil = 38 lb
Soil volume = 0.46 ft³
Specific gravity, $G = 2.64$
Void ratio, $e = 1.1$
Degree of saturation of soil, $S = ?$

$$\text{Moisture content, } w = \frac{\text{Weight of water}}{\text{Dry weight of soil}} \times 100\% = \left(\frac{49\text{ lb} - 38\text{ lb}}{38\text{ lb}}\right) \times 100\% = 28.9\%$$

Now, $Se = wG_S$

$$\text{Therefore, } S = \frac{wG_S}{e} = \frac{0.289(2.64)}{1.1} = 0.69 = 69\%$$

Answer *The degree of saturation of the soil is 69%.*

Example 3.11: Variation of Resilient Modulus with Moisture The resilient modulus of a fine-grained soil is 20 ksi at the optimum moisture content of 15.5%. If the field moisture contents of the soil are 20% and 10% in rainy season and dry season, respectively, calculate the resilient moduli of that soil in rainy season and dry season.

For fine-grained materials, a, b, and k_m are – 0.5934, 0.4, and 6.1324, respectively. The specific gravity of the soil is 2.6 and the void ratio is 1.1.

Solution

Resilient modulus at the optimum moisture content, $M_{Ropt} = 20$ ksi
Regression constant $a = -0.5934$
Regression constant, $b = 0.4$
Regression constant, $k_m = 6.1324$

Specific gravity, $G = 2.6$
Void ratio, $e = 1.1$

$$\text{Degree of saturation at the optimum moisture content, } S_{opt} = \frac{wG_S}{e} = \frac{0.155(2.6)}{1.1} = 0.3664$$

During the rainy season

$$\text{Degree of saturation at 20\% moisture content, } S = \frac{wG_S}{e} = \frac{0.2(2.6)}{1.1} = 0.4727$$

$$\text{Known: } \log\left(\frac{M_R}{M_{Ropt}}\right) = a + \frac{b-a}{1+\exp\left[\ln\frac{-b}{a} + k_m(S - S_{opt})\right]}$$

$$\log\left(\frac{M_R}{20}\right) = -0.5934 + \frac{0.4-(-0.5934)}{1+e^{\left[\ln\frac{-0.4}{-0.5934}+6.1324(0.4727-0.3664)\right]}} = -0.1603$$

$$\frac{M_R}{20} = 10^{-0.1603}$$

Therefore, resilient modulus at 20% moisture content, $M_R = 13.8$ ksi.

During the dry season

$$\text{Degree of saturation at 10\% moisture content, } S = \frac{wG_S}{e} = \frac{0.1(2.6)}{1.1} = 0.2364$$

$$\log\left(\frac{M_R}{20}\right) = -0.5934 + \frac{0.4-(-0.5934)}{1+e^{\left[\ln\frac{-0.4}{-0.5934}+6.1324(0.2364-0.3664)\right]}} = 0.1686$$

$$\frac{M_R}{20} = 10^{0.1686}$$

Therefore, resilient modulus at 10% moisture content, $M_R = 29.5$ ksi.

Answers *13.8 ksi, 29.5 ksi*

3.5 Resilient Modulus Variations Due to Stress Level

The resilient modulus is dependent not only on the moisture content but also on the applied stress level. The resilient modulus should be input in the AASHTOWare pavement ME design software for the average probable stress in field. However, it is always not possible to predict the probable stress in soil or aggregate layer in the field as the traffic-induced stress is mostly uncertain. This is why a generalized model is used in the AASHTOWare pavement ME design software, which is presented in the equation:

$$M_R = k_1 p_a \left(\frac{\theta}{p_a}\right)^{k_2} \left(\frac{\tau_{oct}}{p_a} + 1\right)^{k_3} \tag{3.24}$$

where M_R = Resilient modulus, psi

θ = Bulk stress = $\sigma_1 + \sigma_2 + \sigma_3$

τ_{oct} = Octahedral shear stress = $\frac{1}{3}\sqrt{(\sigma_1-\sigma_2)^2 + (\sigma_1-\sigma_3)^2 + (\sigma_2-\sigma_3)^2}$

σ_1 = Major principal stress

σ_2 = Intermediate principal stress

σ_3 = Minor principal stress/confining pressure

p_a = Normalizing stress

k_1, k_2, k_3 = Regression constants

The laboratory-tested M_R data is fitted using the linear or nonlinear regression analysis. Then the M_R can be determined for any stress possible in pavement.

3.6 Other Properties

Some other properties are required while analyzing and designing pavement.

- *Poisson's ratio.* It is the ratio of the lateral strain to the axial strain and is required to calculate the lateral stress-strain upon applying vertical pressure. For soil and aggregate layers, its effect on the pavement response is not very significant. There is no national test standard available so far and the default values included in the AASHTOWare pavement ME design software are mostly used. Depending on the soil type, this value varies from 0.15 to 0.45.
- *Maximum dry density.* Maximum dry density is determined using the Proctor test following the AASHTO T 180 test standard.
- *Optimum moisture content.* Optimum moisture content is determined using the Proctor test following the AASHTO T 180 test standard.
- *Specific gravity.* Specific gravity is a required parameter to analyze the phase relationship and calculate the overburden pressure. This parameter can be determined using the AASHTO T 100 test standard.
- *Saturated hydraulic conductivity.* Saturated hydraulic conductivity is determined using the AASHTO T 215 test standard.
- *Soil-water characteristic curve (SWCC) parameters.* SWCC parameters are required to determine the variations in ground moisture over the years. There are different methods of determining SWCC parameters such as pressure plate (AASHTO T 99), filter paper (AASHTO T 180), and Tempe cell (AASHTO T 100).

If Level 1 resilient modulus value cannot be obtained by any of the mentioned testing system, Level 2 or 3 data (available in the AASHTOWare pavement ME design software) listed in Table 3.9 may be used.

The performances of different types of soils, such as strength, frost susceptibility, compressibility, drainage performance, etc., are listed in Table 3.10.

3.7 Summary

This chapter discusses the knowledge of soils and aggregates required for analyzing and designing pavement. More specifically, the basic physical properties and mechanical properties of aggregates along with their test standards, brief procedures, and common test values are discussed. The details of the testing procedures can be found in the cited ASTM or AASHTO test standards. More details of how these properties are used in the pavement design can be found in the AASHTOWare pavement ME design manual of practice or the software.

Soil type	Base/subbase for flexible and rigid pavements	Embankment and subgrade for flexible pavements	Embankment and subgrade for rigid pavements
A-1-a	40,000	29,500	18,000
A-1-b	38,000	26,500	18,000
A-2-4	32,000	24,500	16,500
A-2-5	28,000	21,500	16,000
A-2-6	26,000	21,000	16,000
A-2-7	24,000	20,500	16,000
A-3	29,000	16,500	16,000
A-4	24,000	16,500	15,000
A-5	20,000	15,500	8,000
A-6	17,000	14,500	14,000
A-7-5	12,000	13,000	10,000
A-7-6	8,000	11,500	13,000

Source: From AASHTO (2015). Mechanistic-Empirical Pavement Design Guide: A Manual of Practice. Washington, DC: American Association of State Highway and Transportation Officials. Table 10-10. Used with permission.

Table 3.9 Recommended Levels 2 and 3 M_R (psi) at Optimum Moisture for Different Types of Soils

Pavement has several layers of which the surface layer is asphalt or PCC. The underneath base/subbase unbound layers are provided for different reasons, such as to ensure enough depth to decrease the tire-induced stress on the subgrade, structural integrity, drainage requirements, elevation requirements, etc. The unbound layers must satisfy several requirements set by the design agency.

Resilient modulus (M_R) is the primary input for unbound layers. This property is determined by laboratory testing or derived from several other testing such as CBR, R-value, DCP, Atterberg limits and soil gradation, etc. The M_R is also dependent on the moisture variations and the stress level imposed on it. The default M_R values are mostly used if no test data is available. There are several other properties of unbound layers such as Poisson's ratio, soil-water characteristic curve (SWCC) parameters, etc. which are mostly considered the default values.

3.8 Fundamentals of Engineering (FE) Exam-Style Questions

FE3.1 What kind of particles do not compact readily?

A. Angular
B. Rounded
C. Flat and elongated
D. Smooth

Major divisions	Name	Strength when not subject to Frost action	Potential frost action	Compressibility and expansion	Drainage characteristics
Gravel and Gravelly Soils	Well-graded gravels or gravel-sand mixes, little to no fines; GW	Excellent	None to very slight	Almost none	Excellent
	Poorly graded gravels or gravel-sand mixes little or no fines; GP	Good to excellent	None to very slight	Almost none	Excellent
	Silty gravels, gravels and silt mixes; GM	Good to excellent	Slight to medium	Very slight	Fair to poor
	Very silty gravels, gravel-sand silt mixes; GM	Good	Slight to medium	Slight	Poor to impervious
	Clayey gravels, gravels, and clay mixes; GC	Good	Slight to medium	Slight	Poor to impervious
Sand and Sandy Soils	Well-graded sands or gravelly sands, little to no fines; SW	Good	None to very slight	Almost none	Excellent
	Poorly graded sands or gravelly sands, little or no fines; SP	Fair to Good	None to very slight	Almost none	Excellent
	Silty sands, sand-silt mixes; SP	Fair to Good	Slight to high	Very slight	Fair to poor
	Silty sands, sand-silt mixes; SM	Fair	Slight to high	Slight to medium	Poor to impervious
	Clayey sands, sand-clay mixes; SC	Poor to fair	Slight to high	Slight to medium	Poor to impervious

Silts and Clays with the Liquid Limit Less Than 50	Inorganic silts and very fine sand, rock flour, silty or clayey fine sand or clayey silts with slight plasticity; MG, MS, and ML	Poor to fair	Medium to very high	Slight to medium	Fair to poor
	Inorganic clays of low to medium plasticity, gravelly clays, sandy clays, silty clays, lean clays; CG, CL, and CS	Poor to fair	Medium to high	Slight to medium	Practically impervious
	Organic silts and organic silt-clays of low plasticity; MSO and CLO	Poor	Medium to high	Medium to high	Poor
Silts and Clays with Liquid Limit Greater Than 50	Inorganic silts, micaceous or diatomaceous fine sand or silty soils, elastic silts; MH	Poor	Medium to very high	High	Fair to poor
	Inorganic clays of high plasticity, fat clays; CH	Poor to fair	Medium to very high	High	Practically impervious
	Organic clays of medium to high plasticity, organic silts; MHO and CHO	Poor to very poor	Medium	High	Practically impervious
Highly Organic Soils	Peat and other highly organic soils	Not suitable	Slight	Very High	Fair to poor

Source: From AASHTO (2015). Mechanistic-Empirical Pavement Design Guide: A Manual of Practice. Washington, DC: American Association of State Highway and Transportation Officials. Table 8-5. Used with permission.

TABLE 3.10 Summary of Soil Performances as a Pavement Material

Solution A

Angular particles do not compact as readily as rounded particles because their angular surfaces tend to lock up with one another and resist compaction.

FE3.2 A Proctor mold contains 4.30 lb of wet soil with the moisture content of 10%. The mold's volume is 1/30 ft^3. The dry unit weight of the specimen is most nearly:

A. 129 pcf
B. 117 pcf
C. 171 pcf
D. 192 pcf

Solution B

Weight of wet soil, $W = 4.30$ lb
Volume of soil, $V = 1/30$ ft^3
Water content, $w = 10\% = 0.10$

Unit weight of wet soil, $\gamma = \dfrac{W}{V} = \dfrac{4.30 \text{ lb}}{\frac{1}{30}\text{ft}^3} = 129 \text{ pcf}$

Unit weight of dry soil, $\gamma_d = \dfrac{\gamma}{1+w} = \dfrac{129 \text{ pcf}}{1+0.10} = 117.27 \text{ pcf} \approx 117 \text{ pcf}$

FE3.3 A CBR test has been conducted on a soil. A stress of 7.35 MPa is required to penetrate the standard piston by 5.08 mm. Standard penetration pressure for well-graded crushed stone is 10.3 MPa for 5.08-mm penetration. The CBR value of the soil is most nearly:

A. 51
B. 61
C. 71
D. 81

Solution C

CBR = (7.35 MPa/10.3 MPa) × 100 = 71

3.9 Practice Problems

3.1 The resilient modulus of an aggregate base layer is 40 ksi at the optimum moisture content of 7.5%. If the field gravimetric moisture contents of the soil are 10% and 5% in rainy season and dry season, respectively, determine the resilient moduli of that layer in rainy season and dry season.

For coarse-grained materials, a, b, and k_m are –0.3123, 0.3, and 6.8157, respectively. The specific gravity of the aggregate base is 2.64 and the void ratio is 1.4.

3.2 While conducting CBR test in a soil, the results presented in Fig. P3.2 were obtained:

The diameter of the standard piston used in the CBR testing is 1.95 in. Calculate the resilient modulus of the soil.

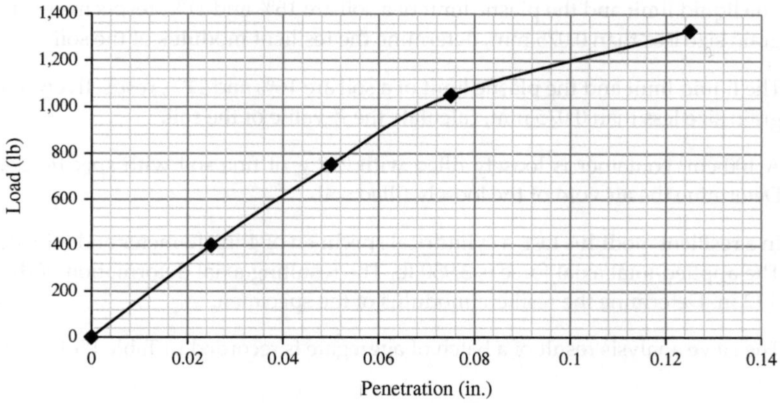

FIGURE P3.2 Test result for Prob. 3.2.

3.3 While conducting the DCP test in a soil, it was found that the average penetration is about 12 mm/blow. Calculate the resilient modulus of the soil.

3.4 The sieve analysis of a soil specimen is shown in Fig. P3.4. The liquid limit and the plastic limit of the soil specimen are 40 and 12, respectively. Calculate the resilient modulus of the soil.

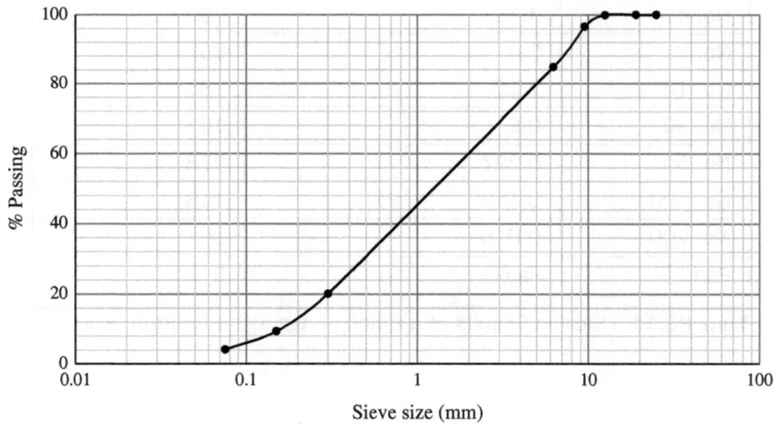

FIGURE P3.4 Sieve analysis of a soil specimen for Prob. 3.4.

3.5 Determine the percentage of the flat and elongated particles in the following coarse aggregate:

Weight of oven dry aggregates: 498.2 g
Weight of flat aggregates: 4.2 g
Weight of elongated aggregates: 11.3 g
Weight of flat and elongated aggregates: 52.1 g
Weight of neither flat nor elongated aggregates: 443.8 g
Flat and elongation limit set by the design agency: 20%

3.6 In an LL test, 20 blows are required to close the groove at 20% moisture content, and 30 blows are required to close the groove at 16% moisture content. Determine the LL of the soil.

3.7 The liquid limit and the plastic limit of a soil are 16% and 11%, respectively. If 6.5% soil has grain size less than 0.075 mm, determine the resilient modulus of the soil.

3.8 The liquid limit and the plastic limit of a soil are 16% and 11%, respectively. If 6.5% soil has grain size less than 0.075 mm, calculate the *R*-value of the soil.

3.9 A 100-cm^3 container is loosely filled with 232 g of fine soil with specific gravity of 2.65. Determine the air void of the loosely filled soil.

3.10 In a resilient modulus test, a cylindrical specimen of 4-in. diameter and 8-in. height is used. The applied total axial force is 3,950 lb. The resulting axial deformation of the specimen is 0.12 in. Determine the resilient modulus of the specimen.

3.11 The sieve analysis result of a batch of aggregate is recorded in Table P3.11.

Sieve	Sieve size (mm)	Mass retained (g)
2.00 in.	50	0
1.50 in.	37.5	8
1.00 in.	25	19
0.75 in.	19	44
0.375 in.	9.5	55
No. 4	4.75	78
No. 8	2.4	255
No. 16	1.19	188
No. 30	0.6	157
No. 50	0.297	278
No. 100	0.149	118
No. 200	0.075	33
Pan	Pan	9

TABLE P3.11 Sieving Results of a Batch of Aggregate for Prob. 3.11

a. What are the NMAS and the maximum aggregate size based on the Superpave definition?
b. What are the NMAS and the maximum aggregate size based on the traditional definition?
c. Draw the sieve analysis curve in a semi-log graph paper.
d. Compare the gradation with the maximum density line.

3.12 The sieve analysis result of a batch of aggregate is recorded in Table P3.12.

Sieve	Sieve size (mm)	Mass retained (g)
2.00 in.	50	0
1.50 in.	37.5	0
1.00 in.	25	0
0.75 in.	19	12
0.375 in.	9.5	145
No. 4	4.75	111
No. 8	2.4	118
No. 16	1.19	180
No. 30	0.6	300
No. 50	0.297	218
No. 100	0.149	11
No. 200	0.075	6
Pan	Pan	4

TABLE P3.12 Sieving Results of a Batch of Aggregate for Prob. 3.12

a. What are the NMAS and the maximum aggregate size based on the Superpave definition?
b. What are the NMAS and the maximum aggregate size based on the traditional definition?
c. Draw the sieve analysis curve in a semi-log graph paper.
d. Compare the gradation with the maximum density line.

3.13 The sieve analysis result of a batch of aggregate is recorded in Table P3.13.

Sieve	Sieve size (mm)	Mass retained (g)
2.00 in.	50	0
1.50 in.	37.5	35
1.00 in.	25	187
0.75 in.	19	180
0.375 in.	9.5	178
No. 4	4.75	323
No. 10	2.4	415
No. 16	1.19	123
No. 40	0.6	211
No. 50	0.297	499
No. 100	0.149	66
No. 200	0.075	490
Pan	Pan	39

TABLE P3.13 Sieving Results of a Batch of Aggregate for Prob. 3.13

a. What are the NMAS and the maximum aggregate size based on the Superpave definition?
b. What are the NMAS and the maximum aggregate size based on the traditional definition?
c. Draw the sieve analysis curve in a semi-log graph paper.
d. Compare the gradation with the maximum density line.

CHAPTER 4

Asphalt Materials

4.1 Background

Asphalt concrete (AC) is the general name of hardened asphaltic material mixed with coarse aggregate, fine aggregate, and some additives. The term "asphalt concrete (AC)" is interchangeably used as "hot-mix asphalt (HMA)" in the pavement industry. However, AC not only is HMA, but also includes some other forms (discussed later in this chapter). AC is the material of most concern for asphalt pavement. In the previous chapter, the physical and mechanical characteristics of soil and aggregates have been discussed in certain pavement projects for suitability. In the current chapter, the asphalt materials, their properties, grading systems, and characterization methods are discussed. All these discussions are targeted at the AASHTOWare pavement ME design procedure in addition to the AASHTO 1993 design system.

In the United States, there are about 4,000 asphalt processing plants which generate about 525 million tons of asphalt with the total value of over $3 billion. There are about 300,000 employees working in this industry. There is a total of 8.7 million miles of pavement (2.5% interstate and 97.5% non-interstate), of which interstate comprises 65% concrete and 35% asphalt pavements. Non-interstate has 94% asphalt and 6% concrete (Rivera et al., 2017). Thus, the importance of asphalt binder is huge. The United States used approximately 130 million barrels (23 million tons) of asphalt binder and road oil in 2011, worth $7.7 billion, according to the U.S. Energy Information Administration. In the recent peak years of 1999 and 2005, nearly 200 million barrels were consumed (EIA, 2011). In 2001, the United States produced almost 35 million tons of asphalt at a rough value of around $6 billion. Roads and highways constitute the largest sole use of asphalt at 85% of the total (Asphalt Institute, 2001). Approximately 83% of asphalt binder used in the United States in 2011 was used for paving purposes (Grass, 2012). In the United States, more than 92% of all paved roads and highways are surfaced with asphalt products. The United States has about 4,000 plants producing asphalt mixtures, with total production of about 452 million tons in 2007 (NAPA, 2012) and about 396 million tons in 2010 (Hansen and Newcomb, 2011). The value of asphalt paving mixtures produced in the United States was estimated at $11.5 billion in 2007 (U.S. Census Bureau, 2007). Asphalt usually accounts for between 4% and 8% of the AC mix by weight and about 30% of the cost of AC pavement structure depending on type and quantity.

4.2 Asphalt Binder

Asphalt is dark brown to black in color, highly viscous hydrocarbon produced from petroleum distillation residue. This distillation can occur naturally, resulting in asphalt

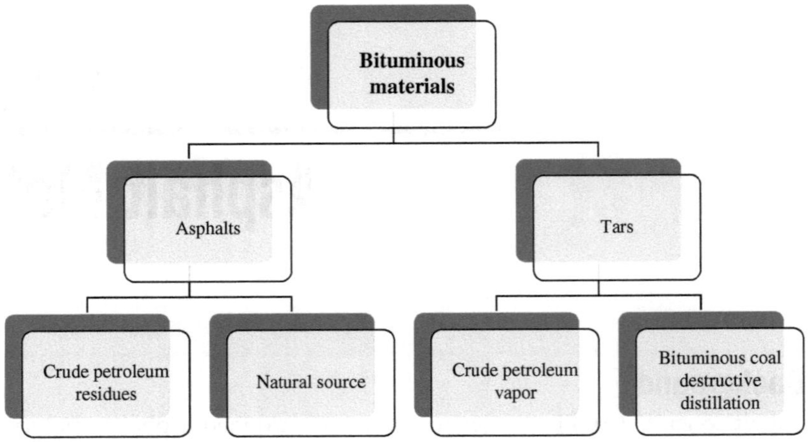

FIGURE 4.1 Classification of asphalt binder.

lakes, or occur in a petroleum refinery using crude oil as shown in Fig. 4.1. Asphalt and tar are both black in color and have outstanding waterproofing and adhesive properties. Asphalt binder is obtained from crude petroleum residue or natural lakes. On the other hand, tar is obtained from petroleum vapor and the destructive distillation of bituminous coal. Asphalt material is primarily used in asphalt pavement road construction, whereas tar is primarily used for waterproofing membranes in roofs and pavement treatment especially in parking lots where fuel spill may dissolve the asphalt. Tar is no longer used for road construction due to health hazards (e.g., eye and skin irritation) and high-temperature susceptibility. Carbon and hydrogen are the main elements present in asphalt binder (Peterson, 1984). Some other minor elements present are sulfur, nitrogen, and oxygen, and occasionally vanadium and nickel are found (Halstead, 1985).

There are three natural asphalt lakes in the world: La Brea Tar Pits in Los Angeles, Lake Guanoco in Venezuela, and Pitch Lake in Trinidad and Tobago. Pitch Lake has the largest natural deposit of asphalt in the world. The lake is about 100 acres and 250 ft deep. This lake contains about 50% to 57% of total asphalt in the world (Wallace and Martin, 1967). Four other types of asphalts are produced in addition to the common asphalt:

1. Asphalt emulsion
2. Asphalt cutback
3. Foamed asphalt
4. Recycled asphalt

4.2.1 Asphalt Emulsion

Asphalt emulsion is the suspension of a small amount of asphalt cement in water, which is assisted by an emulsifying agent (such as soap). The emulsifying agent assists by imparting an electrical charge to the surface of the asphalt cement globules so that they do not coalesce. Emulsions are used because they reduce asphalt viscosity effectively for lower temperature uses.

4.2.2 Cutback Asphalt

Cutback asphalt is simply a combination of asphalt cement and petroleum solvent. Like emulsions, cutbacks are used because they reduce asphalt viscosity for lower temperature uses (tack coats, fog seals, slurry seals, stabilization material). Like emulsified asphalts, after a cutback asphalt is applied the petroleum solvent evaporates leaving behind asphalt cement residue on the surface to which it was applied. It is said that cutback asphalt cures as the petroleum solvent evaporates away.

4.2.3 Foamed Asphalt

Foamed asphalt is formed by combining hot asphalt binder with small amounts of cold water. When the cold water comes into contact with the hot asphalt binder, it becomes steam, which is trapped in small bubbles of asphalt binder. The result is a thin-film, high-volume asphalt foam with approximately 10 times more coating potential than the asphalt binder in its normal liquid state (World Highways, 2001). This high-volume foam state only lasts for a few minutes, after which the asphalt binder resumes its original properties. Foamed asphalt can be used as a binder in soil or base course stabilization, and is often used as the stabilizing agent in full-depth asphalt reclamation.

4.2.4 Recycled Asphalt

Recycled asphalt binder is extracted from recycled asphalt pavement (RAP). Solvent extraction following the AASHTO T 164 or ASTM D 2172 standard uses a chemical solvent (trichloroethylene, 1,1,1-trichloroethane or methylene chloride) to separate the asphalt binder from the aggregate. Asphalt milling is first submerged in trichloroethylene for some time. The asphalt binder/solvent and aggregate are then separated using a centrifuge (Fig. 4.2). The binder/solvent mixture is then processed in order to separate binder and solvent. This solvent extraction method is only sparingly used due to the hazardous nature of the specified solvents.

FIGURE 4.2 Extraction of asphalt binder from RAP by one of the authors.

4.3 Grading of Asphalt Binder

Asphalt binder can be graded broadly into three systems:

1. Penetration grading
2. Viscosity grading
3. Performance grading

4.3.1 Penetration Grading

Penetration grading (in short named as *pen grade*) is primarily developed based on the penetration test on asphalt binder following ASTM D 946. In this test, a standard needle penetrates an asphalt binder specimen when placed under a 100-g (0.22-lb) load for 5 s as shown in Fig. 4.3. The test does not measure any fundamental parameter and characterizes asphalt binder only at 77°F. Penetration grades are classified as a range of penetration units (one penetration unit = 0.1 mm) as 40–50 if the penetration ranges from 4 to 5 mm. Other four types of penetration grading are 60–70, 85–100, 120–150, and 200–300 as listed in Table 4.1. The higher is the penetration, the softer is the asphalt binder. Thus, the 200–300 pen grade is the softest grade and the 40–50 pen grade is the hardest grade. Typical asphalt binders used in the United States are 65–70 pen grade and 85–100 pen grade.

The penetration requirements of different penetrating grading of binder are listed in Table 4.1.

FIGURE 4.3 Asphalt penetration testing.

Penetration grade	Penetration at 77°F (25°C)	
	Minimum penetration	Maximum penetration
40–50	40	50
60–70	60	70
85–100	85	100
120–150	120	150
200–300	200	300

TABLE 4.1 Penetration Grading for Asphalt Binder

AASHTO M 226	ASTM D 3381	Absolute viscosity at 60°C (140°F) (poises)
Grading Based on Original Asphalt (AC)		
AC-2.5	AC-2.5	250 ± 50
AC-5	AC-5	500 ± 100
AC-10	AC-10	1,000 ± 200
AC-20	AC-20	2,000 ± 400
AC-30	AC-30	3,000 ± 600
AC-40	AC-40	4,000 ± 800
Grading Based on Aged Residue (AR)		
AR-10	AR-1000	1,000 ± 200
AR-20	AR-2000	2,000 ± 400
AR-40	AR-4000	4,000 ± 800
AR-80	AR-8000	8,000 ± 1,600
AR-160	AR-16000	16,000 ± 3,200

TABLE 4.2 Standard Viscosity Graded Binder

4.3.2 Viscosity Grading

Viscosity grading is a better grading system than penetration grading, but it does not test asphalt binder rheology at low temperatures. This grading system is based on the absolute viscosity of virgin binder and rolling thin-film oven (RTFO) test aged binder. RTFO test simulates the effects of short-term aging during mixing and compaction of the asphalt mixture by heating the asphalt binder film in an oven at 163°C (325°F) for 5 h. Absolute viscosity or simply viscosity is defined as the resistance to flow of a fluid. The absolute viscosity test is discussed later in this chapter.

Grading on virgin binder is denoted by two letters (AC) and a number. AC means asphalt cement and the numerical value indicates viscosity at 60°C (140°F) in hundreds of poises. Both ASTM D 3381 and AASHTO M 226 use this procedure. Table 4.2 lists the AC grading for virgin binder where the AC grades are listed in hundreds of poises (cm-g-s = dyne-s/cm^2). The less is the number of poises, the lower is the viscosity and thus the easier is the flow of a substance. Thus, AC-5 [viscosity is 500 ± 100 poise at 60°C (140°F)] is less viscous than AC-40 [viscosity is 4,000 ± 800 poise at 60°C (140°F)].

Grading on RTFO aged binder is denoted by the two-letter, AR grading and a number. AR means asphalt residue and the numerical value indicates viscosity at 60°C (140°F) in hundreds of poises by AASHTO M 226 (ASTM D 3381 uses the viscosity in poise) as listed in Table 4.2. For example, according to AASHTO M 226, AR-40 means the RTFO aged binder has an absolute viscosity of 4,000 poise. ASTM D 3381 expresses this binder as AR-4000. Common asphalt binders used in the United States are AC-10, AC-20, AC-30, AR-4000, and AR-8000.

4.3.3 Performance Grading (PG)

As part of the Superpave research effort, the PG system was developed to more accurately and fully characterize asphalt binders for use in asphalt pavements. This method tests asphalt binder at an upper performance temperature and at a lower performance temperature. This method also tests virgin asphalt, RTFO aged asphalt residue, and pressure-aging vessel (PAV) aged asphalt residue. The PAV provides simulated long-term aged asphalt binder for in-service aging over a period of 7 to 10 years. The basic PAV procedure takes RTFO aged asphalt binder specimens, places them in stainless steel pans, and then ages them for 20 h in a heated vessel pressurized to 305 psi (2.10 MPa) at 90°C or 100°C.

The PG system is based on the idea that the properties of an asphalt binder should be related to the conditions under which it is used. For asphalt binders, this involves expected climatic conditions as well as aging considerations. Therefore, the PG system uses a common battery of tests (as the older penetration and viscosity grading systems do) but specifies that an asphalt binder must pass these tests at specific temperatures that are dependent upon the specific climatic conditions in the area of intended use. Therefore, a binder used in Colorado would be different than a binder used in, say, New York.

Superpave performance grading is reported using two numbers: the first being the average 7-day maximum pavement temperature in degrees Celsius and the second being the single-day minimum pavement design temperature likely to be experienced in degrees Celsius. Thus, a PG 64-16 is intended for use where the average 7-day maximum pavement temperature is 64°C and the expected single-day minimum pavement temperature is –16°C. Notice that these numbers are pavement temperatures and not air temperatures. The PG system requirement is listed in Table 4.3.

Example 4.1: PG Binder Selection In a region, the average 7-day maximum pavement design temperature is 50°C, and the minimum pavement design temperature is –12°C. Determine the recommended asphalt binder for this region.

Solution

From Table 4.3, for the average 7-day maximum pavement design temperature less than 52°C, PG 52 is recommended. Then, for the minimum pavement design temperature –12°C, the second number is –16 (as $-12 > -16$). Therefore, the PG 52-16 binder is recommended.

Example 4.2: Test Temperature For the asphalt binder PG 64-28, determine the recommended design temperature for direct tension test of binder.

Performance Grading (PG)	PG 52							PG 58					PG 64				
	10	-16	-22	-28	-34	-40	-46	-16	-22	-28	-34	-40	-16	-22	-28	-34	-40
Average 7-Day Maximum Pavement Design Temperature, °C	<52							<58					<64				
Minimum Pavement Design Temperature, °C	>-10	>-16	>-22	>-28	>-34	>-40	>-46	>-16	>-22	>-28	>-34	>-40	>-16	>-22	>-28	>-34	>-40
Original Binder																	
Flash Point Temp, T48: Minimum °C	230																
Viscosity, ASTM D 4402; Maximum, 3 Pa-s (3,000 cP), Test Temp, °C	135																
Dynamic Shear, TP5: G*/sinδ, Minimum, 1.00 kPa Test Temperature @ 10 rad/sec, °C	52							58					64				
Rolling Thin Film Oven (T 240) or Thin Film Oven (T 179) Residue																	
Mass Loss, Maximum, %	1.00																
Dynamic Shear, TP5: G*/sinδ, Minimum, 2.20 kPa Test Temp @ 10 rad/sec, °C	52							58					64				

Pressure Aging Vessel (PAV) Residue																	
PAV Aging Temperature, °C	90							100					100				
Dynamic Shear, TP5: $G^*/\sin\delta$, Minimum, 5,000 kPa Test Temp @ 10 rad/sec, °C	25	22	19	16	13	10	7	25	22	19	16	13	28	25	22	19	16
Physical Hardening	Report																
Creep Stiffness, TP1: S, Maximum, 300 MPa m-value, Minimum, 0.300 Test Temp, @ 60 sec, °C	0	-6	-12	-18	-24	-30	-36	-6	-12	-18	-24	-30	-6	-12	-18	-24	-30
Direct Tension, TP3: Failure Strain, Minimum, 1.0% Test Temp @ 1.0 mm/min, °C	0	-6	-12	-18	-24	-30	-36	-6	-12	-18	-24	-30	-36	-12	-18	-24	-30

Source: Adapted from FHWA (2017). Superpave Fundamentals Reference Manual. NHI Course #131053. Washington, DC: Federal Highway Administration (FHWA).

Table 4.3 Performance Grading (PG) Binder System

Solution

From Table 4.3, the recommended design temperature for PG 64-28 binder for direct tension test is –18°C.

The following tests are required for the PG system.

Determining Temperature

The pavement surface temperatures for the pavement site can be determined using the LTPPBind 3.0/3.1 software. The software is available at the FHWA website.

Rolling Thin-Film Oven Test

The rolling thin-film oven (RTFO) procedure following the AASHTO T 240 and ASTM D 28723 provides simulated short-term aged asphalt binder for physical property testing. This test simulates the aging during mixing and placement. It also provides a quantitative measure of the volatiles lost during the aging process. This test takes virgin (unaged) asphalt binder specimens in cylindrical glass bottles and places these bottles in a rotating carriage within an oven (Fig. 4.4). The carriage rotates within the oven while the 325°F (163°C) temperature ages the specimens for 85 min.

Pressure Aging Vessel (PAV)

The pressure aging vessel (PAV) provides simulated long-term aged asphalt binder for physical property testing. This test simulates the aging that occurs during in-service life. The asphalt binder is exposed to heat and pressure to simulate in-service aging over a span of 7 to 10 years. The basic PAV procedure takes RTFO aged asphalt binder specimens, places them in stainless steel pans, and then ages them for 20 h in a heated vessel pressurized to 305 psi (2.10 MPa or 20.7 atmospheres) at 90°C or 100°C as shown

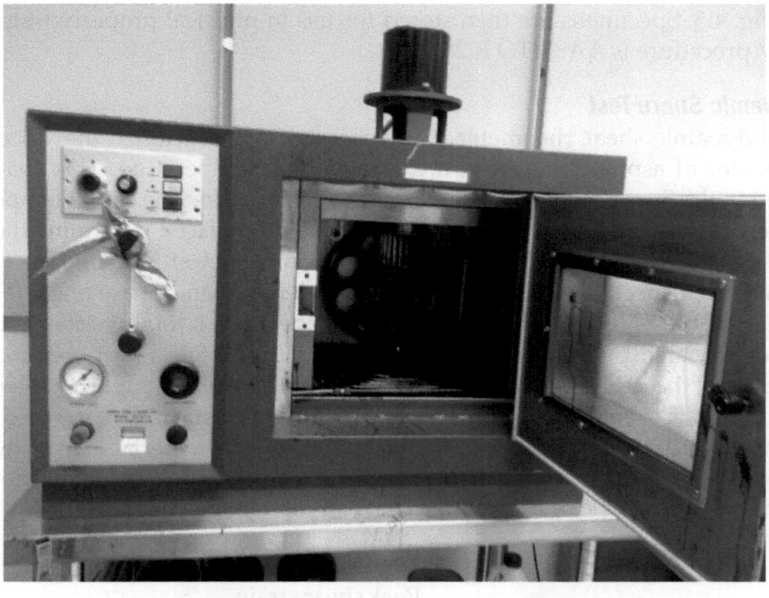

FIGURE 4.4 Rolling thin-film oven test apparatus.

FIGURE 4.5 Pressure aging vessel test apparatus.

in Fig. 4.5. Specimens are then stored for use in physical property tests. The standard PAV procedure is AASHTO R 28.

Dynamic Share Test

The dynamic shear rheometer (DSR) is used to characterize the viscous and elastic behavior of asphalt binders at medium to high temperature. This characterization is used in the Superpave PG asphalt binder specification. The actual temperatures anticipated in the area where the asphalt binder will be placed determine the test temperatures used. The standard dynamic shear rheometer test is AASHTO T 315. The DSR measures a specimen's complex shear modulus (G^*) and phase angle (δ). The complex shear modulus (G^*) can be considered the specimen's total resistance to deformation when repeatedly sheared, while the phase angle (δ) is the lag between the applied shear stress and the resulting shear strain. The larger is the phase angle (δ), the more viscous is the material. A zero-degree phase angle means an elastic material, whereas a 90-degree phase angle means a pure viscous material. Asphalt material is in between these two values. The shear modulus (G^*) is defined mathematically as the ratio of peak shear stress (τ_o) and the peak recoverable shear strain (γ_o), which is presented by the equation:

$$G^* = \frac{\text{Peak shear stress}}{\text{Peak shear strain}} = \frac{\tau_0}{\gamma_0} \tag{4.1}$$

FIGURE 4.6 Dynamic shear rheometer testing by one of the authors.

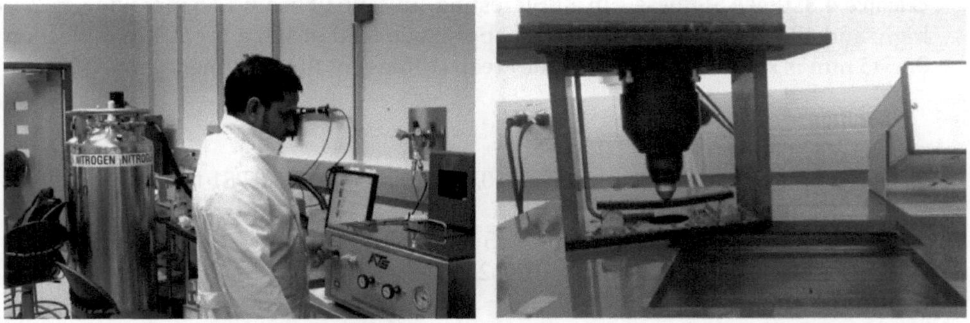

FIGURE 4.7 Bending beam rheometer testing by one of the authors.

A small specimen of asphalt binder is sandwiched between two plates shown in Fig. 4.6. The test specimen is kept at near constant temperature as desired. The top plate oscillates at 10 rad/s (1.59 Hz) in a sinusoidal waveform while the equipment measures the maximum applied shear stress, the resulting maximum shear strain, and the time lag between them. The software then automatically calculates the complex modulus (G^*) and phase angle (δ).

Bending Beam Rheometer (BBR) Creep Stiffness

Creep stiffness means the stiffness or modulus measured using creep (sustained) loading. The BBR test provides a measure of low-temperature creep stiffness and relaxation properties of asphalt binders. These parameters give an indication of an asphalt binder's ability to resist low-temperature cracking.

According to the ASTM D 6648 or AASHTO T 313 standard, an asphalt beam of 4.0 in. (102 mm) long, 0.5 in. (12.5 mm) wide, and 0.25 in. (6.25 mm) high is prepared by pouring the heated binder in a mold. After cooling, the beam is then kept in the test bath for an hour. Then, a load of 100 g (980 mN) is applied at the center of the beam for a total of 240 s, as shown in Fig. 4.7. The deflection of the beam is recorded during this loading period. Using the classical strength of materials equation for a center-point loaded beam, the stiffness after 60 s of loading is calculated as:

$$S(t) = \frac{PL^3}{4bh^3\delta(t)} \tag{4.2}$$

where $S(t)$ = Creep stiffness at time t, the standard is 60 s
P = Applied constant load, the standard is 100 g (980 mN)
L = Beam span, the standard is 4.0 in. (102 mm)
b = Beam width, the standard is 0.5 in. (12.5 mm)
h = Beam height, the standard is 0.25 in. (6.25 mm)
$\delta(t)$ = Deflection at time $t = 60$ s

The m-value is the slope of "Log$S$ versus Log-time" curve at 60 s of loading. The m-value indicates the rate of change of stiffness with loading time. The PG binder specification requires the m-value to be equal or greater than 0.30 at 60 s of loading (McGennis et al., 1994).

Example 4.3: Creep Stiffness In a BBR testing, an asphalt beam 12.5 mm wide, 6.25 mm high, and 102 mm long is used. After applying a load of 100 g for 60 s, a deflection of 0.9545 mm is recorded. Calculate the creep stiffness of the specimen.

Solution

$S(t)$ = Creep stiffness at 60 s = ?
P = Applied constant load, 100 g = 0.1 kg (9.81 m/s²) = 0.981 N
L = Beam span = 102 mm = 0.102 m
b = Beam width = 12.5 mm = 0.0125 m
h = Beam height = 6.25 mm = 0.00625 m
$\delta(t)$ = Deflection 60 s = 0.9545 mm = 0.0009545 m

$$\text{Creep stiffness, } S(t) = \frac{PL^3}{4bh^3\delta(t)} = \frac{0.981\text{N}(0.102\text{ m})^3}{4(0.0125\text{ m})(0.00625\text{ m})^3(0.0009545\text{ m})}$$

$$= 89{,}347{,}732\text{ Pa} = 89.3\text{ MPa}$$

Answer *The creep stiffness is 89.3 MPa.*

Direct Tension Test (DTT)

The direct tension test (DTT), which measures the tensile strength of asphalt binder at a critical cracking temperature, is conducted following the AASHTO T 314 test protocol at different temperatures (commonly 10°C higher than the low temperature grade of the binder). The effective length of the specimen is slightly more than 1.5 in. (38 mm) and the effective area of the cross section of the specimen is about 1.5 in. (38 mm) square. The specimen is pulled at a constant strain rate of 3% per minute. The variations of stress and strain are recorded. The DTT setup is shown in Fig. 4.8.

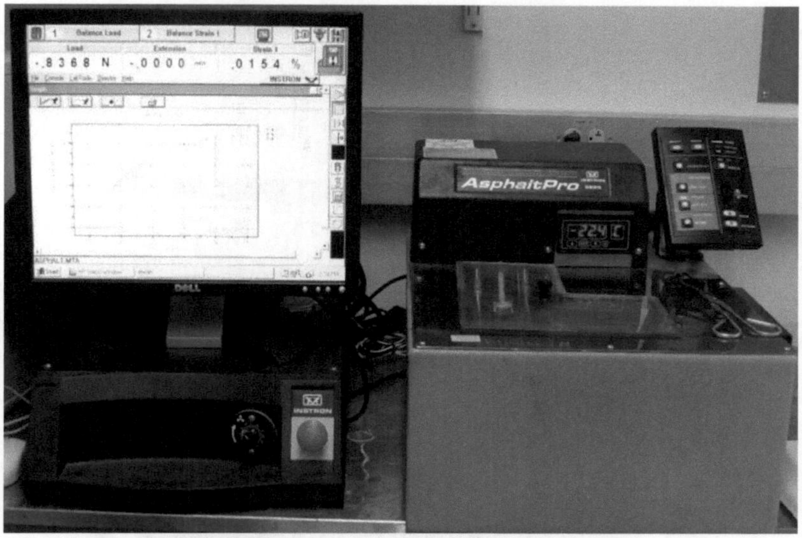

FIGURE 4.8 Direct tension test setup.

4.4 Other Tests on Asphalt Binder

To characterize asphalt binder, any or more of the following tests are recommended.

4.4.1 Absolute Viscosity

The basic absolute viscosity test (ASTM D 2171 and AASHTO T 202) measures the time it takes for a fixed volume of asphalt binder to be drawn up through a capillary tube, as shown in Fig. 4.9. Asphalt drawing is carried out by means of closely controlled vacuum suction conditions at 60°C. This temperature is selected because it is approximately the maximum asphalt pavement surface temperature during summer in most areas.

The viscometer is a U-tube with a reservoir where the asphalt is introduced and a section with a calibrated diameter and timing marks. Vacuum is applied at one end and the time during which the asphalt flows between two timing marks on the viscometer is measured. The flow time in seconds is multiplied by the calibration factor of the viscometer in order to obtain absolute viscosity in poises. Although absolute viscosity is an improvement over the penetration test, it still only measures viscosity at one temperature and thus does not fully characterize an asphalt binder's consistency over the expected range of construction and service conditions.

4.4.2 Kinematic Viscosity

The kinematic viscosity of a liquid is the absolute (or dynamic) viscosity divided by the liquid's density at the temperature of measurement. The 135°C (275°F) measurement temperature is selected to simulate the mixing and compaction temperatures commonly used during asphalt pavement construction. The basic kinematic viscosity test (ASTM D 2170 and AASHTO T 201) measures the time it takes for a fixed volume of asphalt binder to flow through a Zeitfuchs Cross-Arm Viscometer under closely controlled conditions of head and temperature. The kinematic viscosity in centistoke is obtained by multiplying the time taken by the calibration factor of the viscometer provided by

Figure 4.9 Capillary viscometer. *(Courtesy of Raysky Scientific Instruments, Labfreez Group, Guangdong, China. Used with Permission.)*

the manufacturer (Brown et al., 2009). Absolute viscosity can be obtained from this kinematic viscosity by multiplying it by the density of asphalt binder as follows:

$$\textit{Absolute viscosity (poises)} = \textit{Kinematic viscosity (stokes)} \times \textit{Specific gravity} \tag{4.3}$$

Example 4.4: Absolute Viscosity An asphalt binder has a kinematic viscosity of 1,200 centistoke. If its specific gravity is 0.98, determine its absolute viscosity in poise.

Solution

1 stock = 100 centistoke

Therefore, 1,200 centistoke = 12 stokes

$$\begin{aligned}\textit{Absolute viscosity (poises)} &= \textit{Kinematic viscosity (stokes)} \times \textit{Specific gravity}\\ &= 12 \times 0.98\\ &= 11.76 \text{ poise}\end{aligned}$$

Answer *The absolute viscosity is 11.76 poise.*

4.4.3 Brookfield Viscosity

The rotational viscometer (RV) or Brookfield viscometer is used to determine the viscosity of asphalt binders in the high-temperature range of manufacturing and

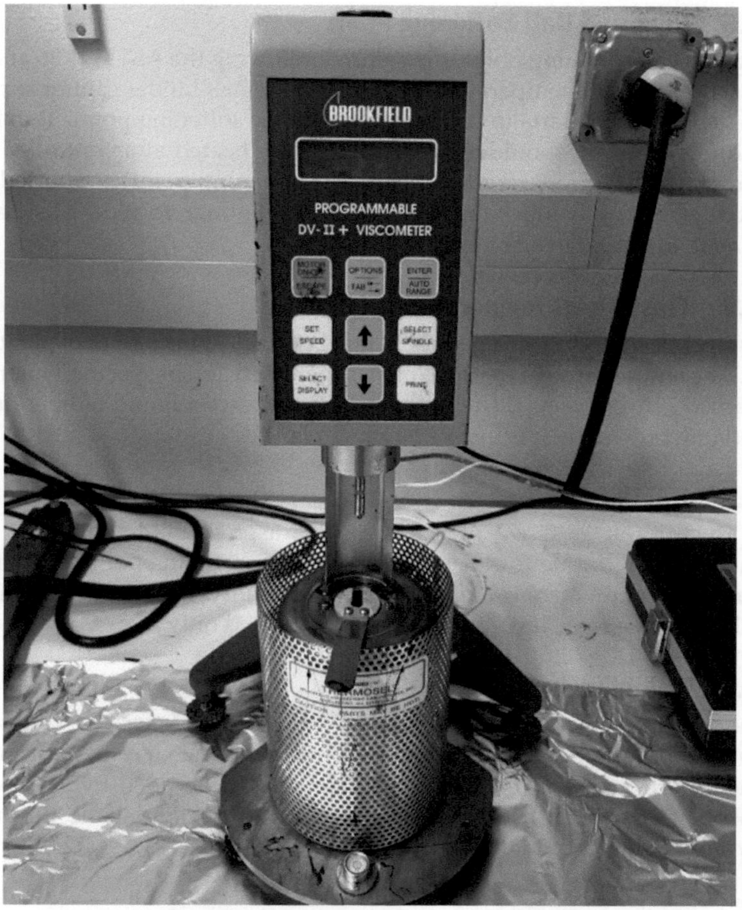

FIGURE 4.10 Rotational (Brookfield) viscometer. (*Photo courtesy of Dr. Mehedi Hasan.*)

construction. The RV test can be conducted at various temperatures, but since manufacturing and construction temperatures are similar regardless of the environment, the test for Superpave PG asphalt binder specification is always conducted at 275°F (135°C). About 11 g of asphalt binder is poured into the chamber. The test measures the torque required to maintain a cylindrical spindle's rotational speed (20 rpm) while submerged in the asphalt binder (Fig. 4.10). This torque is then converted to a viscosity and displayed automatically by the RV in the unit of centipoise (1,000 centipoise = 1 Pa·s). The standard test methods are ASTM D 4402 and AASHTO T 316.

4.4.4 Specific Gravity

Since the specific gravity of the asphalt binder varies with temperature, specific gravity tests are useful in making volume corrections based on temperature. The specific gravity at 15.6°C (60°F) is commonly used when buying/selling asphalt cements. A typical specific gravity for asphalt is around 1.03. The standard test method is AASHTO T 228.

4.4.5 Ring and Ball Softening Point

Ring and ball softening point is measured following the ASTM D 36 and AASHTO T 53 test standards. The temperature at which an asphalt binder cannot support the weight of a steel ball and starts to flow is known as the softening point. Two horizontal disks of bitumen, cast in shouldered brass rings, are heated at a controlled rate in a liquid bath while each supports a steel ball. The softening point is reported as the mean of the temperatures at which the two disks soften enough to allow each ball, enveloped in bitumen, to fall 1.0 in. (25 mm).

4.4.6 Flash Point Temperature

A typical flash point test involves heating a small specimen of asphalt binder in a test cup. The temperature of the specimen is increased and at specified intervals a test flame is passed across the cup. The flash point is the lowest liquid temperature at which the test flame causes the specimen's vapors to ignite.

4.4.7 Ductility

The ductility at 25°C (77°F) test measures asphalt binder ductility by stretching a standard-sized briquette of asphalt binder to its breaking point. The stretched distance at break in centimeters is then reported as ductility. Like the penetration test, this test has limited use since it is empirical and conducted only at 25°C (77°F).

4.4.8 Solubility in Trichloroethylene

Asphalt cement, as used for asphalt paving, should consist of almost pure bitumen. Impurities are not active constituents of cement and may be detrimental to asphalt cement performance. Mineral impurities can be quantified by dissolving a specimen of asphalt cement in trichloroethylene or 1,1,1 trichloroethane through a filter mat. Anything remaining on the mat is considered an impurity.

4.5 Asphalt Mixtures

Asphalt mixture is broadly of three types:

1. Hot-mix asphalt (HMA)
2. Warm-mix asphalt (WMA)
3. Cold-mix asphalt

4.5.1 Hot-Mix Asphalt (HMA)

Asphalt mixture is commonly known as hot-mix asphalt (HMA), although it is one kind of AC mixture. HMA is an AC mixture that is produced by heating the mixture at a certain level of temperature, commonly 285 to 325°F (140–160°C). It should be noted that AC mixture can also be produced by heating up or even at ambient air temperature without heating.

Dense-Graded Mix

A dense-graded mix is a well-graded HMA intended for typical use. When properly designed and constructed, a dense-graded mix is relatively impermeable. Dense-graded

FIGURE 4.11 Comparison of open-graded and dense-graded AC.

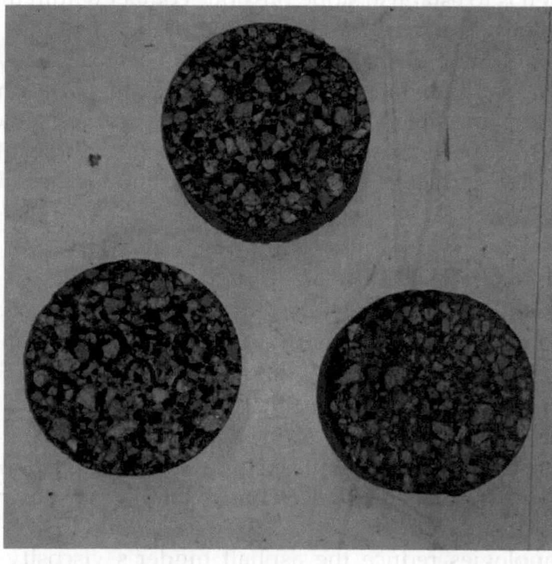

FIGURE 4.12 OGFC specimens.

mixes are generally referred to by their nominal maximum aggregate size and can further be classified as either fine-graded or coarse-graded. Fine-graded mixes have more fines and sand-sized particles than coarse-graded mixes. It is suitable for all pavement layers and for all traffic conditions and works well for structural, friction, leveling, and patching needs. Figure 4.11 shows a pavement with both dense-graded and open-graded mixes.

Open-Graded Friction Course (OGFC)

Open-graded mix (Fig. 4.12) uses only crushed stone (or gravel) and a small percentage of manufactured sands with about 15% air voids. It is used for surface courses only to provide good friction and drain-infiltrated water laterally. Thus, it reduces tire splash/spray in wet weather and typically results in smoother surfaces than dense-graded HMA. It also results in smoother surfaces than dense-graded HMA and reduces the thermal cracking in asphalt pavement (Islam et al., 2018b).

Asphalt-Treated Permeable Bases (ATPBs)

Asphalt-treated permeable base has less stringent specifications than OGFC since it is used only under dense-graded HMA, SMA, or Portland cement concrete for drainage. ATPB is used as a drainage layer below dense-graded HMA, SMA, or PCC.

Sand-Asphalt Mix

Sand-asphalt mix is a dense-graded mix of asphalt and sand with nominal maximum aggregate size less than 3/8 in. (9.5 mm).

Stone Matrix Asphalt (SMA)

Stone matrix asphalt, sometimes called stone mastic asphalt, is a gap-graded HMA originally developed in Europe to maximize rutting resistance and durability. The goal of the mix design is to establish stone-on-stone contact within the mixture. Since aggregates do not deform as much as asphalt binder under load, this stone-on-stone contact greatly reduces rutting. SMA is generally more expensive than a typical dense-graded HMA because it requires more durable aggregates, higher asphalt content, and modified asphalt binder and fibers. It is used to improve rut resistance and durability. SMA is almost exclusively used for surface courses on high-volume interstate pavements and roads. Gap-graded aggregate, modified asphalt binder, fiber filler, etc. are commonly used to produce this.

4.5.2 Warm-Mix Asphalt (WMA)

Warm-mix asphalt technology allows the producers of asphalt pavement material to lower the temperatures at which the material is mixed and placed on the road. Reductions of 30 to 120°F (17–67°C) have been documented. Such drastic reductions have the obvious advantages of reducing fuel consumption and decreasing greenhouse gas production. Fuel consumption during WMA manufacturing is typically reduced by 20%. In addition, engineering benefits include better compaction on the road, the ability to haul paving mix for longer distances, and extending the paving season by being able to pave at lower temperatures.

WMA technologies reduce the asphalt binder's viscosity (thickness) so asphalt aggregates can be coated at lower temperatures. The key is the addition of additives (water-based, organic, chemical, or hybrid) to the asphalt mix (Bonaquist, 2011). The additives allow the asphalt binders and asphalt aggregates to be mixed at the lower temperatures. Reducing the viscosity also makes the mixture easier to manipulate and compact at the lower temperature.

4.5.3 Cold-Mix Asphalt (CMA)

Cold-mix asphalt concrete is formed by emulsifying asphalt with (essentially) soap in water before mixing with the aggregate. While in its emulsified state the asphalt is less viscous, and the mixture is easy to work and compact. The emulsion will break after enough water evaporates and the cold mix will, ideally, take on the properties of HMA. Cold mix is commonly used as a patching material and on lesser trafficked service roads.

Figure 4.13 Cold-mix asphalt touched by one of the authors.

Typically, it is based on two types of processing location: central plant processed and cold-in-place recycling. In central plant processing, milled asphalts are transported to a plant, screened, and emulsifying agents are mixed. The produced mix is transported back to the site, placed, and compacted. The cold-in-place recycling is produced on site. The milled asphalts are screened in a large truck on site, mixed, placed, and compacted with emulsifying agents. Compacted cold mixes look similar to the conventional mixes. A cold-mix asphalt ready to be paved is shown in Fig. 4.13. Proper usages of cold-mix asphalt is very competitive with the conventional mixes when used for low-traffic roads (Islam et al., 2018a).

4.6 Recycled Asphalt Materials

4.6.1 Reclaimed Asphalt Pavement (RAP)

Reclaimed or recycled asphalt pavement is the milling of asphalt surface layer from old pavement containing aggregates and asphalt binder. By milling the old pavement, the RAP is obtained as shown in Fig. 4.14, screened, and mix with the new aggregates. Nowadays, up to 40% by weight of the whole mixture is being used in the United States (Hasan et al., 2018; Islam et al., 2014). As RAP has some amount of aged-stiff binder, the binder grade of the combined binder stiffens. Therefore, a revised mix design is sought out if the mix has considerable amount of RAP.

The amount of RAP used in asphalt mixtures was 66.7 million tons in 2011, a 19% increase over 2009 (56 million tons) and about a 7% increase over 2010 (62.1 million tons). Assuming 5% liquid asphalt in RAP, this represents approximately 3.6 million tons of virgin asphalt binder conserved, or about 12% of the total binder used in 2011. Looking at 2011 U.S. data, approximately 87 million tons of RAP that was milled from

Figure 4.14 Collection of RAP from an interstate pavement in New Mexico.

existing pavements was run through asphalt mixing plants that year, with approximately 74 million tons of the 81 million tons of RAP (92%) recycled into new AC materials. For the years 2009 through 2011, RAP that was not recycled into AC was used for aggregate base (less than 10% annually) and cold mix (less than 3% annually) and less than 0.1% was landfilled (Hansen and Copeland, 2013).

RAP has been used to replace virgin materials in dense AC by up to 50%. However, where mixture performance is most critical, such as in asphalt surface layers, the level of replacement is often lower. Based on their risk assessment, many agencies set limits on how much RAP can be used for different applications. In general, replacement at up to 15% is considered to have minimal effects on properties. Most state highway agencies allow up to 15% or 30% replacement for structural layers, and some also allow those amounts for surface layers. The average RAP content in AC mixtures in the United States in 2009–2010 was about 13% for U.S. Department of Transportation (DOT) mixtures, 15% for other agency mixtures, and 18% for commercial and residential paving mixtures (Hansen and Newcomb, 2011).

4.6.2 Reclaimed Asphalt Shingles (RAS)

Reclaimed asphalt shingles are collected from roof tear-offs and reused to the pavement and in many cases may improve the quality. Shingles can contain between 20% and 36% asphalt. This asphalt can be used to bind aggregates like the conventional asphalt. Shingle wastes either from the manufacturer or roof tear-off can be used to save virgin asphalt and avoid shingle landfills. The literature reports that RAS has increased resistance to rutting, reduced cracking and requires less compaction effort (Roque et al. 2018).

4.6.3 Rubberized Asphalt Concrete (RAC)

Rubberized asphalt concrete, also known as asphalt rubber or just rubberized asphalt, is noise-reducing pavement material that consists of regular AC mixed with crumb rubber from recycled tires. Approximately 2.4 million tires are recycled every year as asphalt rubber and are expected to grow (Dower et al., 1985). RAC is made by blending ground-up recycled tires with asphalt to produce a binder that is then mixed with conventional aggregate materials. This mix is then placed and compacted into a road surface. RAC is a cost-effective, sustainable, safe, and environmentally friendly alternative to traditional road paving materials. The performances of RAC is very similar to conventional asphalt materials but it needs production machinery.

4.6.4 Reclaimed Asphalt Pavement (RAP) in Base and Subgrade

In addition to the asphalt mix, the usages of RAP in aggregate base or subbase are also becoming popular nowadays (Hasan et al., 2018; Islam et al., 2014; Tarefder and Islam, 2015). There are different ways RAP can be used in base and subbase layers. One approach is plant processing where friction is transported, crushed, and screened to a central plant. The better quality RAP is used with the new asphalt mix production as discussed earlier. The inferior RAP is then added with the virgin base or subgrade materials. It improves the base and subgrade strength and saves the RAP from being dumps. An example of RAP-mixed base course used in an interstate highway in New Mexico is shown in Fig. 4.15. About 50% of RAP is mixed with virgin aggregates to produce this base layer. In this interstate highway, RAP is also mixed with subgrade to produce a subbase layer.

FIGURE 4.15 RAP-mixed base course in an interstate pavement in New Mexico.

4.7 Surface Treatment Materials

Surface treatment materials are used to provide routine maintenance or repair works. Some of the popular surface treatment materials are introduced here.

4.7.1 Fog Seal

Fog seal is a spray application of a special asphalt emulsion (a thin liquid oil) to an existing asphalt pavement surface (Fig. 4.16). It seals narrow cracks and narrow pores, restores surface color of the pavement, and preserves the underlying pavement structure.

Fog seal contains globules of paving asphalt, water, an emulsifying agent or surfactant, and sometimes a rejuvenator. Soap is a common form of a surfactant. The surfactant helps to remove the dirt and suspend the dirt particles in the wash water while washing clothes or dishes. Similarly, in asphalt emulsions, the surfactant keeps the paving asphalt globules in suspension until it is applied to the pavement surface when the water in the asphalt emulsion starts to evaporate. A rejuvenator is an asphalt additive that when applied to the existing pavement will slightly soften the pavement it is applied to creating a better bond.

4.7.2 Slurry Seal

Slurry seal is the mixture of water, asphalt emulsion, well-graded fine aggregate, and some additives. It is applied in a thin layer of 3- to 6-mm layers (Fig. 4.17) to seal the

FIGURE 4.16 Fog seal. (*Courtesy of Public Works of LA County.*)

Figure 4.17 Slurry seal. (*Courtesy of Public Works of LA County.*)

pavement and prevent some minor distresses such as raveling, minor/narrow cracks, crack joints, mat tearing, etc. However, it provides no structural strength. A slurry seal is similar to a fog seal except the slurry seal has aggregates as part of the mixture. This combined mixture of the emulsion and aggregates represents slurry. Polymer is commonly added to the asphalt emulsion to provide better mixture properties.

4.7.3 Chip Seal

Chip seal is a rapid setting emulsion sprayed onto the pavement followed by rolling in the high-quality, washed, crushed, and single-sized aggregate typically 9.5 or 6.7 mm (Fig. 4.18). A chip seal is named after the chips or the small crushed rock on the surface. A chip seal seals the narrow cracks, helps bind the cracked pavement together, provides a wearing surface, and prevents reflective cracking. Asphalt emulsions used in chip seal applications contain globules of paving asphalt, water, an emulsifying agent or surfactant, polymer, and sometimes a rejuvenator.

Soap is a common form of a surfactant. In washing clothes or dishes, the surfactant helps remove the dirt and suspend the dirt particles in the wash water. Similarly, in asphalt emulsions, the surfactant keeps the paving asphalt globules in suspension until it is applied to the pavement surface when the water in the asphalt emulsion starts to evaporate. The chips (small crushed rocks) are immediately applied after the asphalt emulsion is applied to the pavement surface. The polymer in the asphalt emulsion is a hardener that increases adhesion to the crushed rock and surface of the pavement. A rejuvenator is an asphalt or additive that when applied to the existing pavement will slightly soften the pavement it is applied to creating a better bond.

4.7.4 Microsurfacing

Microsurfacing is a polymer-modified emulsion mix of aggregates, mineral fillers, water, and additives. It uses a 100% crushed, high-quality aggregate that passes through 9.5-mm sieve. Microsurfacing is similar to slurry seal. It consists of the application of a mixture of water, asphalt emulsion, aggregate (very small crushed rock), and chemical additives to an existing AC pavement surface as shown in Fig. 4.19. Polymer is commonly added to the asphalt emulsion to provide better mixture properties. The major difference between slurry seal and microsurfacing is in how they break or harden. Slurry

FIGURE 4.18 Chip seal. (*Courtesy of Public Works of LA County.*)

FIGURE 4.19 Microsurfacing. (*Courtesy of Public Works of LA County.*)

relies on evaporation of the water in the asphalt emulsion. The asphalt emulsion used in microsurfacing contains chemical additives which allow it to break without relying on the sun or heat for evaporation to occur. Microsurfacing is therefore an application that hardens faster than slurry seals and can be used when conditions do not permit the

Figure 4.20 Scrub seal. (*Courtesy of Public Works of LA County.*)

effective placement of slurry seals. It is effective in treating rutting, moderate distress, and narrow crack width.

4.7.5 Scrub Seal

Scrub seal (Fig. 4.20) is an application that is very close to a chip seal treatment where asphalt emulsion and crushed rock are placed on an asphalt pavement surface. The only difference is that the asphalt emulsion is applied to the surface of the road through a series of brooms placed at different angles. These brooms guide the asphalt emulsion into the pavement distresses to ensure sealing the road. These series of brooms, known as a scrub broom, give the treatment its title, scrub seal. A scrub seal provides an excellent treatment opportunity to treat a heavily distressed road cost-effectively.

4.7.6 Cape Seal

Cape seal is an application of a chip or scrub seal followed by the application of slurry seal or microsurfacing at a later date. The chip or scrub seal is used to seal and bind the cracks in the existing pavement. The slurry seal or microsurfacing serves to improve the chip retention and smoothness of the driving surface.

4.7.7 Coats

Prime coat is a low-viscosity liquid asphalt or emulsified asphalt used to seal a granular surface prior to the placement of surface treatment. Tack coat is an emulsified asphalt-diluted slow setting that is applied to existing pavement surfaces when stated. The placing of tack coats ensures good bonding between layers.

4.8 Characterization of New Asphalt Mixtures

4.8.1 Dynamic Modulus

Dynamic modulus ($|E^*|$) is the primary material property required for pavement design. It dictates the potentiality of deformation upon loading. Dynamic modulus test is conducted by applying sinusoidal loads at different frequencies uniaxial on a cylindrical AC specimen. The dynamic modulus is defined mathematically as the ratio of peak dynamic stress (σ_o) and the peak recoverable axial strain (ε_o):

$$|E^*| = \frac{\text{Peak stress}}{\text{Peak strain}} = \frac{\sigma_o}{\varepsilon_o} \tag{4.4}$$

where $|E^*|$ = Dynamic modulus
σ_o = Peak dynamic stress, applied by loading frame
ε_o = Peak recoverable axial strain, measured upon loading

Thus, the dynamic modulus is defined mathematically as the ratio of σ_o and ε_o. The dynamic modulus is similar to Young's modulus, the difference being that the load applied in dynamic (sinusoidal) and the resulting strain is also dynamic. In Young's modulus test studied in the strength of material class, a monotonic increasing load is applied (say, 10, 20, 30, etc.).

To prepare test specimen, cylindrical specimens of 6 in. (150 mm) in diameter and about 7 in. (170 mm) in height are compacted using the gyratory compactor following the AASHTO T 312 test standard. The compacted specimens are then cored and sawed to diameter of 4 in. (100 mm) and height of 6 in. (150 mm), as shown in Fig. 4.21. Air voids of the finished specimens should be the within the design limits, commonly 4% to 6%.

The dynamic modulus test is conducted according to the AASHTO T 342 standard protocol. The test setup is shown in Fig. 4.22. The linear variable displacement transducer (LVDT) attached to the specimen measures the axial deformation of the specimen upon loading. The strain is calculated by dividing this deformation by the length of the gage (LVDT length). Stress is calculated by dividing the applied load by the

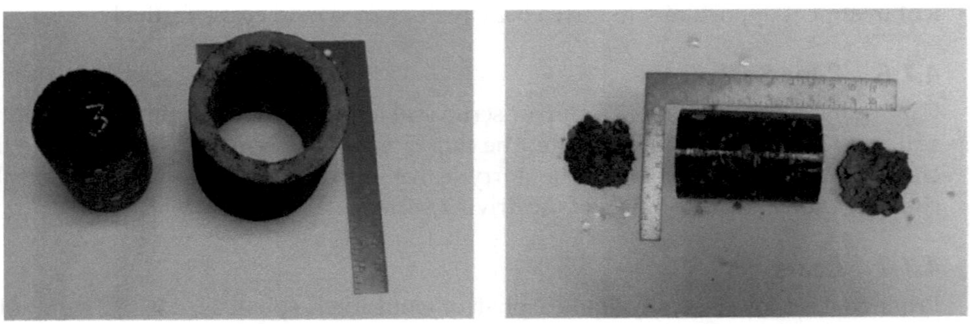

(a) Cored specimen (b) Edge cut specimen

FIGURE 4.21 Specimen preparation for dynamic modulus testing.

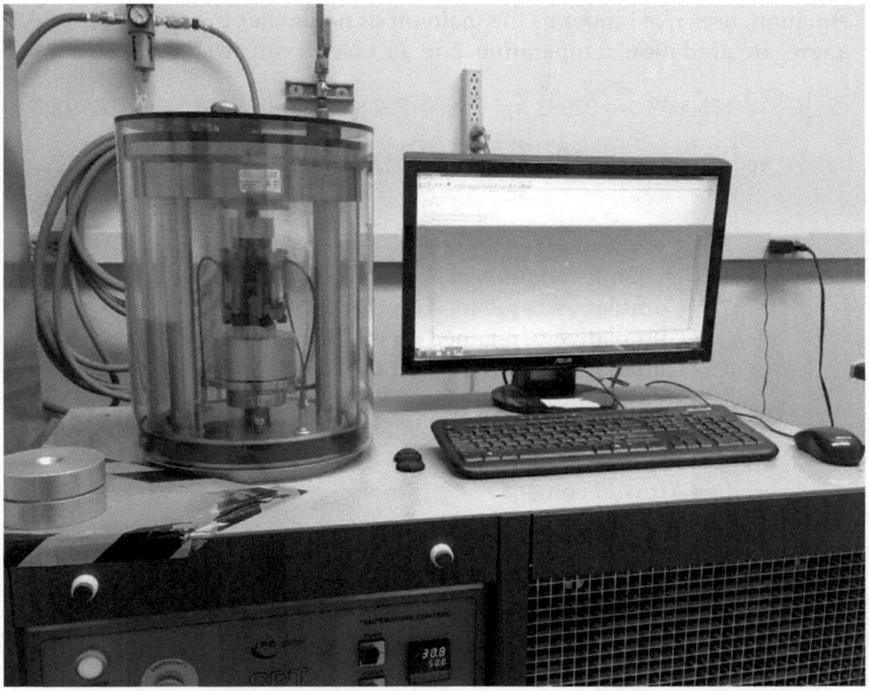

FIGURE 4.22 Dynamic modulus test setup.

cross-sectional area of specimen. The ratio of the peak stress and the peak strain is the dynamic modulus. The specimens are tested at five different temperatures [14, 40, 70, 100, and 130°F (–10, 4.4, 21.1, 37.8, and 54.4°C)], and six loading frequencies (25, 10, 5, 1, 0.5, and 0.1 Hz) applying sinusoidal loading, $\sigma_o \sin(\omega t)$, where σ_o = peak dynamic stress, ω = angular frequency = $2\pi f$, f = frequency of loading, and t = time of loading.

Upon applying loading, sinusoidal strain $\varepsilon_o \sin(\omega t + \phi)$ develops in the materials with some phase lag as asphalt is a viscoelastic material. Here, ε_o = peak dynamic strain and ϕ = phase lag.

After the test, the dynamic modulus values at different temperatures are converted into values of a reference temperature using the time-temperature superposition principle (also called the Arrhenius equation). Then, the converted frequency from the temperature is called the reduced frequency loading. Then, the master curve of dynamic modulus value is plotted. This concept of developing master curve is commonly studied in a higher-level course.

If the test cannot be possible, two empirical equations can be used to determine the $|E^*|$. These are detailed below.

Viscosity-Based Equation

It is the primary $|E^*|$ prediction model in the recently developed AASHTOWare pavement ME design software. The viscosity-based model, presented in the following equation, uses η of binder as the main input parameter to capture the effect of binders, aggregate gradation, temperature, and air void. It can be shown as:

$$\log\left|E^*\right| = 3.750063 + 0.02932\,\rho_{200} - 0.001767\,(\rho_{200})^2 - 0.002841\,\rho_4 - 0.058097\,V_a$$
$$-0.802208\left(\frac{V_{beff}}{V_{beff}+V_a}\right) + \frac{3.871977 - 0.0021\,\rho_4 + 0.003958\,\rho_{38} - 0.000017(\rho_{38})^2 + 0.00547\,\rho_{34}}{1+e^{(-0.603313-0.313351\log(f_r)-0.393532\log(\eta))}} \tag{4.5}$$

where $|E^*|$ = Dynamic modulus of asphalt concrete, psi
ρ_{34} = Cumulative % retained on the ¾ in. sieve
ρ_{38} = Cumulative % retained on the ⅜ in. sieve
ρ_4 = Cumulative % retained on the No. 4 sieve
ρ_{200} = Passing through the No. 200 sieve, %
η = Viscosity of binder at the temperature of interest, 10^6 poise
V_{beff} = Effective binder content, % by volume
V_a = Air void content, %
f_r = Reduced frequency, Hz

Shear Modulus (G)-Based Equation*

The shear modulus (G*)–based predictive equation is expressed as:

$$\log E^* = -0.349 + 0.754\left(\left|G_b^*\right|^{-0.0052}\right) \times \begin{pmatrix} 6.65 - 0.032\rho_{200} + 0.0027(\rho_{200})^2 + 0.011\rho_4 \\ -0.0001(\rho_4)^2 + 0.006\rho_{38} - 0.00014(\rho_{38})^2 - 0.08V_a \\ -1.06\left(\dfrac{V_{beff}}{V_a+V_{beff}}\right) \end{pmatrix}$$
$$+\frac{2.56 + 0.03V_a + 0.71\left(\dfrac{V_{beff}}{V_a+V_{beff}}\right) + 0.012\rho_{38} - 0.0001(\rho_{38})^2 - 0.01\rho_{34}}{1+e^{(-0.7814-0.5785\log\left|G_b^*\right|+0.8834\log\delta_b)}} \tag{4.6}$$

where E^* = Dynamic modulus of asphalt concrete, psi
ρ_{200} = Aggregates (by weight of the total aggregates) passing through No. 200 sieve, %
ρ_{34} = Cumulative % retained on the ¾ in. sieve
ρ_{38} = Cumulative % retained on the ⅜ in. sieve
ρ_4 = Cumulative % retained on the No. 4 sieve
V_{beff} = Effective binder content, % by volume
V_a = Air void content, %
δ_b = Phase angle of binder associated with $|G^*|$, degree
$|G^*|$ = Dynamic shear modulus of binder, psi

Example 4.5: Dynamic Modulus In a dynamic modulus test of AC specimen, a sinusoidal stress of $390\sin(\omega t)$ is applied and a sinusoidal strain of $0.001\sin(\omega t + 80)$ is obtained; where the stress is in psi, ω is the angular frequency, and t is the time. What is the dynamic modulus of the asphalt specimen?

Solution

$$|E^*| = \frac{\text{Peak stress}}{\text{Peak strain}} = \frac{\sigma_0}{\varepsilon_0} = \frac{390 \text{ psi}}{0.001} = 390{,}000 \text{ psi}$$

Answer *The dynamic modulus of the asphalt specimen is 390 ksi.*

4.8.2 Indirect Tensile Strength Test

Indirect tensile strength (ITS) is an important input parameter in the transverse cracking model used in the AASHTOWare pavement ME design guide. Due to decrease in temperature, AC contracts and tensile stress develops in the asphalt mixture. Once the developed tensile stress exceeds the tensile capacity of asphalt mixture, thermal cracks develop. Testing an asphalt specimen in tension similar to a steel rod is quite impossible. Thus, tensile test is conducted with ease in indirect mode and is called the ITS. It is a measure of resistance capacity to low-temperature cracking of asphalt pavement (Islam et al., 2015b). This value is usually calculated in the laboratory using indirect tensile (IDT) specimens and used in the design guide as an input parameter.

To prepare specimen, cylindrical specimens of 7-in. height and 6-in. diameter are prepared as a first step, using a Superpave gyratory compactor following AASHTO T 312 test standard. The specimens are then cut into thin circular specimens of 4-in. diameter and 2-in. thickness. The ITS of asphalt specimens at different temperatures are determined using an IDT test device in an environmental chamber, as shown in Fig. 4.23. Uniform compressive load of 50 mm/min was applied until failure. The peak load was recorded to determine the ITS of the asphalt specimens. The ITS is calculated using the equation:

$$\text{ITS} = \frac{2P}{\pi D t} \tag{4.7}$$

Prepared specimens　　Test setup　　Few failed specimens

FIGURE 4.23 Indirect tensile strength test.

where ITS = Indirect tensile strength, psi
P = Peak load required to crack the specimen diagonally, lb
D = Diameter of the specimen, in.
t = Length or thickness of the specimen, in.

Example 4.6: Indirect Tensile Strength Test In an ITS test, the peak load required to fail a specimen of 4.2-in. diameter and 2.25-in. thickness is 3,200 lb. Calculate the ITS of the specimen.

Solution

Given:
Peak load, $P = 3{,}200$ lb
Diameter, $D = 4.2$ in.
Thickness, $t = 2.25$ in.

$$\text{Indirect tensile strength, ITS} = \frac{2P}{\pi Dt} = \frac{2(3{,}200\ \text{lb})}{\pi(4.2\ \text{in.})(2.25\ \text{in.})} = 216\ \text{psi}$$

Answer *The indirect tensile strength is 216 psi.*

4.8.3 Fatigue Endurance Limit (FEL)

Fatigue endurance limit is an input parameter in the AASHTOWare pavement ME design software. Fatigue damage occurs in asphalt pavement due to repeated tensile strain at the bottom of AC under traffic loading (Islam, 2015). If the developed strain value is below the threshold value (called FEL), no fatigue damage occurs. Beam fatigue test is conducted according to the AASHTO T 321 standards. To prepare specimen, as a first step, beam slabs of 18 × 6 × 3 in. (450 × 150 × 75 mm) are prepared as shown in Fig. 4.24 and then each slab is cut into two beams of 15 × 2.5 × 2 in. (375 × 63 × 50 mm) using a laboratory saw.

The specimen is clamped as shown in Fig. 4.25 and loading is done using a sinusoidal waveform at a frequency of 5 or 10 Hz at a fixed temperature of 68°F (20°C). Test is conducted at different strain levels and different fatigue lives are obtained. The material is considered failed when the stiffness ratio decreases by 50%, as shown in Fig. 4.25. This is because soon after decreasing the stiffness by 50%, microcracks form into macrocracks and the stiffness drops dramatically. Stiffness ratio is the ratio of the stiffness at current cycle of loading to the stiffness at the initial cycle of loading.

(a) Compaction of mixture (b) Cutting the slabs (c) Prepared specimens

FIGURE 4.24 Beam specimen preparation.

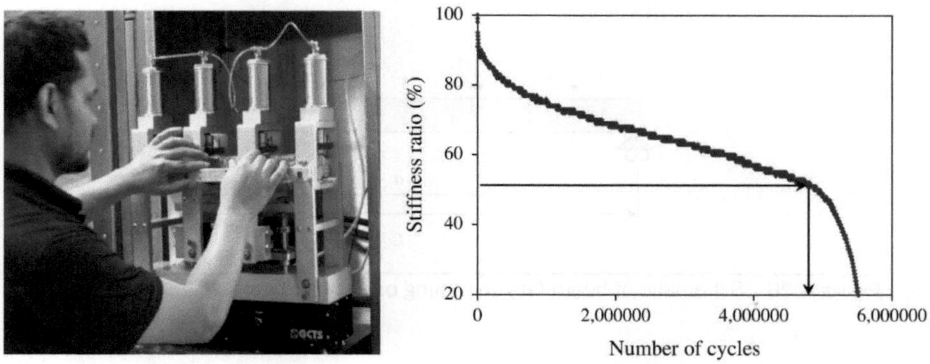

(a) Test setup (b) Test results

FIGURE 4.25 Beam fatigue test.

Specimen's flexural stiffness (E) is calculated using the resulting maximum stress (σ_o) and the applied maximum strain (ε_o) data are recorded from each cycle, as depicted in Eq. (4.8).

$$E = \frac{\sigma_o}{\varepsilon_o} \tag{4.8}$$

The maximum strain and stress in the specimen are calculated using Eq. (4.9) and Eq. (4.10), respectively. These equations can be derived using the classical mechanics of materials principle.

$$\sigma_o = \frac{PL}{bh^2} \tag{4.9}$$

$$\varepsilon_o = \frac{12h\delta}{3L^2 - 4a^2} \tag{4.10}$$

where ε_o = Maximum applied strain
σ_o = Maximum developed stress
P = Load applied by actuator at time t
b = Average specimen width (commonly 2.5 in.)
h = Average specimen height (commonly 2.0 in.)
δ = Deflection at center of beam at time t
a = Distance between inside clamps (commonly $L/3$) (shown in Fig. 4.26)
L = Distance between outside clamps (shown in Fig. 4.26)

Then, a curve (shown in Fig. 4.27) is drawn considering applied strain (ε) and fatigue life (N). From the regression model, the strain value (ε) is calculated at which the fatigue life (N) is 50 million. This strain value is referred to as FEL. The reason for choosing 50 million is that if a specimen can withstand 50 million load cycles in the laboratory, it is expected to be able to withstand more than that in field, which can be considered perpetual.

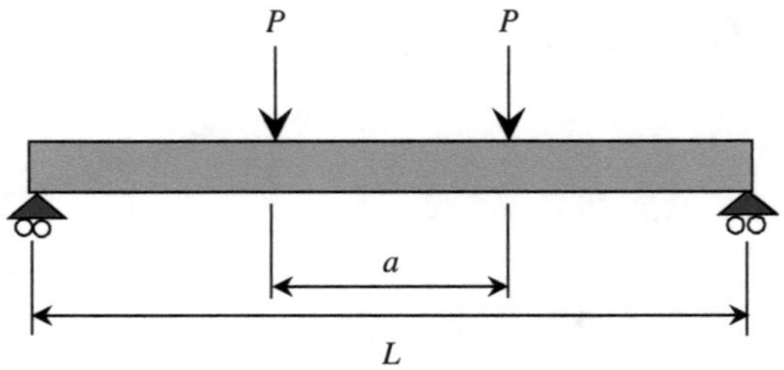

Figure 4.26 Schematic of beam fatigue testing on asphalt beam.

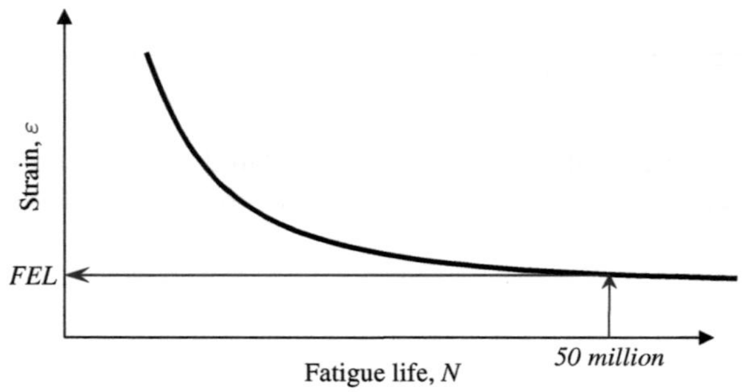

Figure 4.27 Determining FEL from ε-N curve.

Example 4.7: Determining the FEL In a fatigue test of asphalt specimen, the data shown in Fig. 4.28 are obtained:

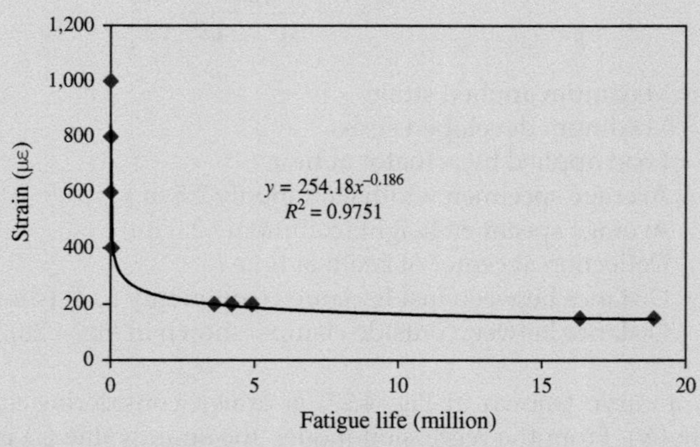

Figure 4.28 Fatigue test results for an asphalt mix.

Determine the fatigue endurance limit (FEL) if the perpetual pavement takes 50 million cycles of load.

Solution

Obtained regression curve, $y = 254.18x^{-0.186}$

If $x = 50$ million, then $y = 254.18x^{-0.186} = 254.18(50)^{-0.186} = 123\ \mu\varepsilon$

Answer *The fatigue endurance limit is 123 microstrains.*

4.8.4 Creep Compliance Test

Creep compliance is defined as the time-dependent strain divided by the applied stress. This test is conducted following the AASHTO T 322-07 test protocol on compacted cylindrical HMA specimen with a diameter of 6 in. (150 mm) and a thickness of 1.5 to 2 in. (38–50 mm). A static load is imposed along a diametric axis of the temperature-controlled specimen for 100 seconds, as shown in Fig. 4.29. During the loading period, vertical and horizontal deformations are measured on the two parallel faces of the specimen using four extensometers. Using these displacements (or strains) and the applied stress, creep compliance can be calculated.

4.8.5 Poisson's Ratio Test

Poisson's ratio of AC can be determined by laboratory test or by using the AASHTOWare pavement ME design default relationship. Poisson's ratio is the ratio of the proportional decrease in a lateral measurement to the proportional increase in length in a specimen of material that is elastically stretched. It is an important property necessary to determine the lateral deformation of a material when it is loaded. Very often a flat value of 0.35 is assumed, although Poisson's ratio varies with loading frequency and temperature

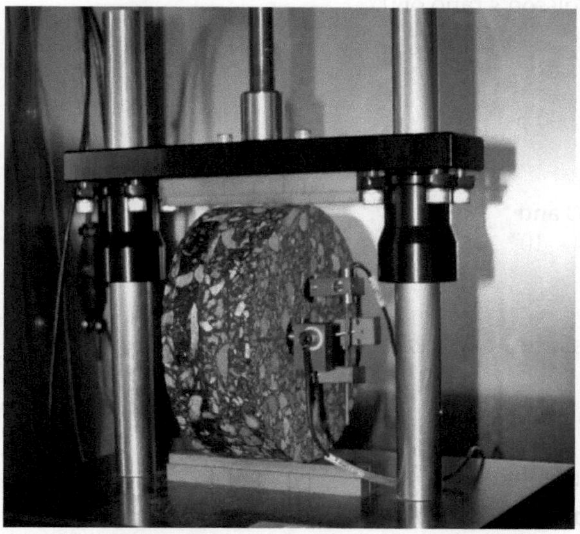

Figure 4.29 Creep compliance test setup.

Temperature (°F)	Dense-graded HMA	Open-graded HMA
Less than 0	0.15	0.35
0–40	0.20	0.35
41–70	0.25	0.40
71–100	0.35	0.40
101–130	0.45	0.45
More than 130	0.48	0.45

Source: From AASHTO (2015). Mechanistic-Empirical Pavement Design Guide: A Manual of Practice, Washington, DC: American Association of State Highway and Transportation Officials. Table 10-3. Used with permission.

Table 4.4 Common Poisson's Ratio of HMA

(Islam et al., 2015a). In the AASHTOWare pavement ME design software, μ-value is predicted from the $|E^*|$ (in psi) of AC using the equation:

$$\mu = 0.15 + \frac{0.35}{1 + e^{a+b(E^*)}} \tag{4.11}$$

where the coefficients a and b are –1.63 and 3.84×10^{-6} respectively, and E^* is the dynamic modulus in psi.

Equation (4.11) is recommended for new HMA mixes. For existing, age-hardened HMA layers, use the typical values listed in Table 4.4.

Example 4.8: Poisson's Ratio The dynamic modulus of an asphalt mixture is 500 ksi. Calculate Poisson's ratio of it.

Solution

Given:

$E = 500{,}000$ psi

Known:

$a = -1.63$ and

$b = 3.84 \times 10^{-6}$

$$\text{Poisson's ratio, } \mu = 0.15 + \frac{0.35}{1 + e^{a+b(E^*)}} = 0.15 + \frac{0.35}{1 + e^{-1.63+3.84x10^{-6}(500{,}000\text{ psi})}} = 0.299$$

Answer *The Poisson's ratio is 0.30.*

4.8.6 Miscellaneous Tests

Surface Shortwave Absorptivity

The surface shortwave absorptivity of pavement depends on pavement's composition, color, and texture. It directly correlates with the amount of solar energy absorbed by pavement surface. The lower the value, the lower the solar energy is absorbed. The AASHTOWare pavement ME design default value is 0.85. No test standard is still available.

Thermal Conductivity

It is defined as the rate at which heat passes through a specified material, expressed as the amount of heat that flows per unit time through a unit area with a temperature gradient of one degree per unit distance. It can be measured using the ASTM E 1952 test standard. If test facility is not available, typical values for HMA ranging from 0.44 to 0.81 Btu/(ft)(h)(°F) can be used. The AASHTOWare pavement ME design default value set in program is 0.67 Btu/(ft)(h)(°F).

Heat Capacity

The heat capacity of a substance is the amount of heat required to change its temperature by one degree and has units of energy per degree. It can be measured using the ASTM D 2766 test standard. If test facility is not available, typical values for HMA ranging from 0.22 to 0.40 Btu/(lb)(°F) can be used. The AASHTOWare pavement ME design default value set in program is 0.23 BTU/lb·°F.

Coefficient of Thermal Contraction (CTC)

It can be defined as the fractional decrease in dimensions (x, y, z direction) due to unit change in temperature. There is no standard test standard yet but the AASHTOWare default value may be used. The AASHTOWare pavement ME design software calculates CTC internally using the mix volumetric properties such as proportions of aggregates. For example, the linear CTC for low temperature range is determined using the equation:

$$L_{\text{mix}} = \frac{VMA \times B_{ac} + V_{agg} \times B_{agg}}{3V_{\text{total}}} \tag{4.12}$$

where L_{mix} = Linear CTC of the AC mix (typically, 2.2 to 3.4 × 10^{-5} per degree Celsius)
B_{ac} = Volumetric CTC of the asphalt cement (binder) in the solid state (typically, 3.5 to 4.3 × 10^{-4} per degree Celsius)
B_{agg} = Volumetric CTC of the aggregate
VMA = % volume of voids in mineral aggregates
V_{agg} = % volume of aggregate in the mixture (typically, 21 to 37 × 10^{-6} per degree Celsius)
V_{total} = 100%

Effective Asphalt Content by Volume

It is defined as the total asphalt binder content of the HMA less the portion of asphalt binder that is lost by absorption into the aggregate. AASHTO T 308 test protocol is to be used to determine the effective asphalt content by volume.

Air Voids

Air voids refer to the total volume of the small pockets of air between the coated aggregate particles throughout a compacted paving mixture, expressed as a percent of the bulk volume of the compacted paving mixture. The amount of air voids in a mixture is extremely important and closely related to stability and durability. AASHTO T 166 test protocol is to be used to determine the air void. The air void can be determined using the equation:

$$\text{Air void} = \left(1 - \frac{G_{mb}}{G_{mm}}\right) \times 100 \quad (4.13)$$

where

G_{mm} = Theoretical maximum specific gravity determined using the AASHTO T 209 test protocol

G_{mb} = Bulk specific gravity determined using the AASHTO T 166 test protocol

Aggregate Specific Gravity

Specific gravity is a measure of a material's density (mass per unit volume) as compared to the density of water at 73.4°F (23°C). Therefore, by definition, water at a temperature of 73.4°F (23°C) has a specific gravity of 1. Specific gravities of fine and coarse aggregate can be determined using the AASHTO T 84 and T 85 test protocols.

Unit Weight

Unit weight means the weight of HMA per unit bulk volume. The AASHTO T 166 test protocol is to be used to determine the bulk specific gravity (G_{mb}) of the material which is to be multiplied by the water unit weight to determine the unit weight of the material.

Voids Filled with Asphalt (VFA)

The portion of the voids in the mineral aggregate that contains asphalt binder is called VFA. This represents the volume of the effective asphalt content. It can also be described as the percent of the volume of the void in mineral aggregate (VMA) that is filled with asphalt cement. VFA is inversely related to air voids: as air voids decrease, the VFA increases. The AASHTO T 209 test protocol is to be used to determine VFA.

4.9 Characterization of Existing Asphalt Mixtures

The falling weight deflectometer (FWD) test is carried out according to the test standards AASHTO T 256 and ASTM D 5858. It is a nondestructive test to measure the pavement surface deflection upon applying predefined load. In this test, an impulse load is applied by a circular steel plate on the pavement surface using a mechanical device. The pavement surface deflects downwards in response to the load applied, creating a deflection basin. Sensors (geophones) installed at different offsets from loading point measure these vertical deflections. The FWD device can accommodate seven to nine sensors for the measurement of vertical deflections. Geophones are used at different radial offsets from the loading plate such as 0 in. (0 mm), 8 in. (200 mm), 12 in. (300 mm), 1.5 ft (450 mm), 2 ft (600 mm), 3 ft (900 mm), 5 ft (1.5 m), etc. The sensor at 0-mm distance means the surface deflection at the loading point. The FWD testing and the corresponding surface deflection schematic are shown in Fig. 4.30. The magnitudes of the load are typically varied at several load levels such as 6 kip (27 kN), 9 kip (40 kN), 12 kip (53 kN), and 16 kip (71 kN).

Backcalculation of the layer moduli is the interpretation of the state of the pavement strength from the FWD test data. Therefore, it also involves some layer properties of the pavement to carry out the analysis. The layer properties cover the layer number and thicknesses, initially assumed modulus of elasticity and Poisson's ratio of each layer

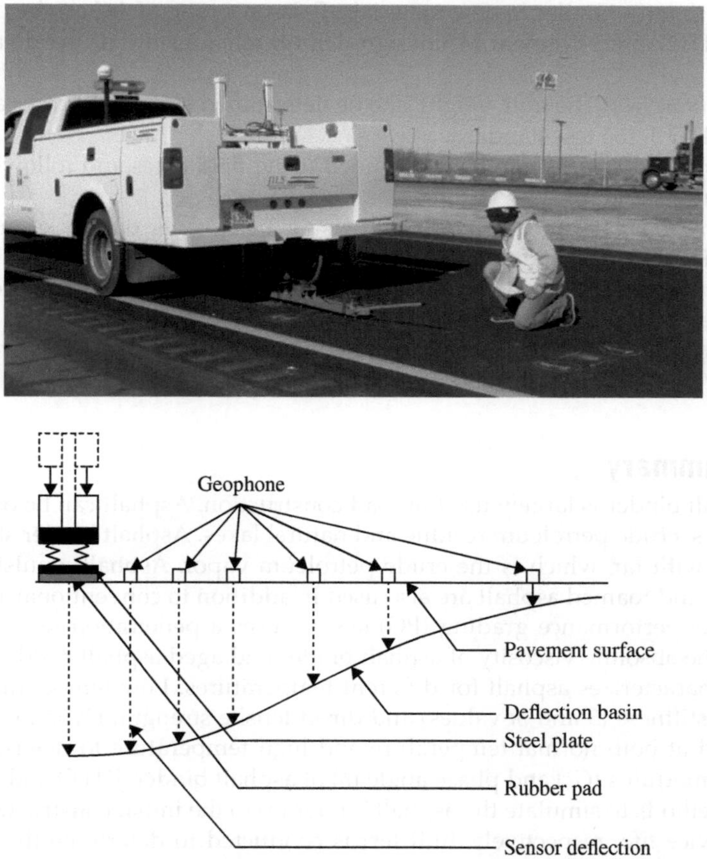

FIGURE 4.30 FWD testing operation.

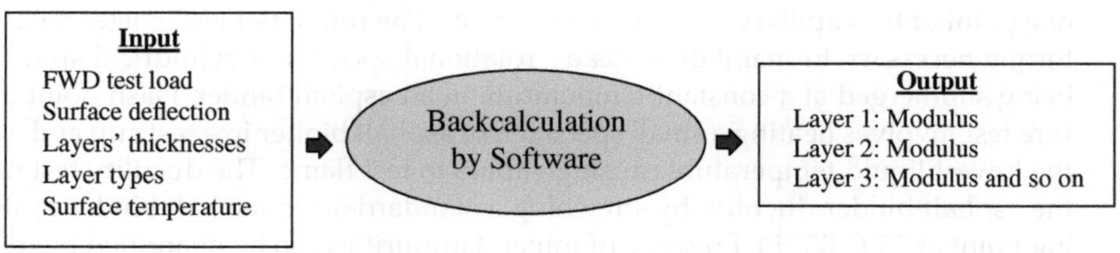

FIGURE 4.31 Backcalculation procedure using the FWD data.

material, and pavement surface temperature. The reverse process of determining the layer moduli from the FWD data as well as the pavement layer properties are the basic tasks of the backcalculation of moduli. The backcalculation procedure is described in the flow chart shown in Fig. 4.31. This layer modulus determined from the FWD data is termed as *backcalculated modulus*. It is mentionable that very often "modulus" is termed as "stiffness" in the current study; this is very common in pavement research.

Poisson's ratio. Similar to new materials, Poisson's ratio of AC can be obtained from the AASHTOWare pavement ME design default relationship, or the default value can be used.

Unit weight. The unit weight is to be determined using field cores and following the AASHTO T 166 test standard.

Asphalt content. It can be determined using field cores and following the AASHTO T 164 test standard.

Gradation. It can be determined using field cores or blocks and following the AASHTO T 27 test standard.

Air voids. It can be determined using field cores and following the AASHTO T 209 test standard.

Asphalt recovery. It can be determined using field cores and following the AASHTO T 164/T 170/T 319 test standards.

4.10 Summary

Asphalt binder is largely used in road construction. Asphalt can be obtained from two sources: crude petroleum residue and natural lakes. Asphalt binder should not be confused with tar, which is the crude petroleum vapor. Asphalt emulsions, asphalt cutbacks, and foamed asphalt are also used in addition to conventional asphalt.

The performance grading (PG) system uses a penetration test. Viscosity grading uses the absolute viscosity of asphalt binder and aged asphalt resides at 60°C. PG system characterizes asphalt for different temperatures. Low-temperature properties are creep stiffness (S and m values) and direct tensile strength. Dynamic shear test is conducted at both normal temperature and high temperature to determine the complex shear modulus (G^*) and phase angle (δ) of asphalt binder. RTFO and PAV are two testing methods to simulate the asphalt binder up to the initial construction and long term of service life, respectively. BBR test is conducted to determine the low-temperature creep stiffness (S-value) and the m-parameter. The DTT measures the asphalt binder's tensile strength at a critical cracking temperature by pulling the specimen under an alcoholic bath at a constant strain rate. Absolute viscosity is determined by using the capillary viscometer measuring the time required by an asphalt specimen to flow from one point of the capillary tube to another point. The rotational viscometer measures the torque necessary to maintain a steady rotational speed of a cylindrical spindle while being submerged at a constant temperature in an asphalt binder. Flash point temperature test involves heating a small specimen of asphalt binder in a test cup and measures the lowest liquid temperature causing vapors to test flame. The ductility test measures the asphalt binder ductility by stretching a standard-sized asphalt binder to its breaking point at 25°C (77°F). Presence of mineral impurities can be quantified by dissolving asphalt binder in trichloroethylene or 1,1,1 trichloroethane.

Asphalt mixture is broadly of three types: HMA, WMA, and cold-mix asphalt. HMA is produced by heating the aggregates and binder at high temperatures. HMA can be dense-graded mixture, OGFC, sand asphalt mix, and SMA. WMA is heated lower than that of HMA with proper mixing of additives. Cold-mix AC is produced at normal temperature by emulsifying the asphalt in water. Recycled asphalt materials such as RAP, RAS, and RAC are now commonly used.

Dynamic modulus is the most important material property required for pavement design. It is determined by applying uniaxial cyclic load on a cylindrical specimen. ITS is determined by compressing a thin cylindrical specimen diametrically so that indirect tension is developed along its diameter. Fatigue endurance limit is the applied strain level at which there is no fatigue damage. Creep compliance is defined as the time-dependent strain divided by the applied stress.

In AC, several modifiers are used to produce desirable properties of AC. For example, fillers such as crushed fines, lime, cement, fly ash, etc. can be used to increase stability, improve the asphalt cement-aggregate bond, reduce optimum binder content, and so on.

4.11 Fundamentals of Engineering (FE) Exam–Style Questions

FE4.1 Asphalt binder is a:

A. Crude petroleum vapor
B. Crude petroleum residue
C. Bituminous coal destructive distillation
D. All of the above

Solution B

Asphalt binder is crude petroleum residue; the other two options for true for tars.

FE4.2 In a region, the average 7-day maximum pavement design temperature is 65°C, and the minimum pavement design temperature is –23°C. The recommended asphalt binder for this region is most appropriately:

A. PG 64-22
B. PG 70-22
C. PG 70-28
D. PG 76-34

Solution C

PG 70-28 binder is recommended; PG 76-34 is over designed.

FE4.3 Sand asphalt mix is a mixture of:

A. Sand and asphalt
B. Stone and asphalt
C. Course aggregate, asphalt, and some sand
D. Fines and asphalt

Solution A

Sand asphalt mix is a dense-graded mix with nominal maximum aggregate size less than 3/8 in. (9.5 mm).

FE4.4 Creep compliance is defined as the:

A. Time-dependent strain divided by the applied stress
B. Strain divided by the time-dependent applied stress
C. Strain divided by the applied stress
D. Time-dependent strain divided by the time-dependent applied stress

Solution A

Creep compliance is defined as the time-dependent strain divided by the applied constant stress. A constant stress is applied, and the resulting time-dependent strain is measured with time.

4.12 Practice Problems

4.1 What are the differences between tar and asphalt binder?

4.2 Which grading system of asphalt binder does best characterize the asphalt binder?

4.3 The average 7-day maximum pavement design temperature of a pavement is 51°C and the minimum pavement design temperature is –18°C. Determine the recommended asphalt binder.

4.4 For the asphalt binder, PG 64-22, determine the recommended design temperature for dynamic shear test of binder.

4.5 In an ITS test, the peak load required to fail a specimen of 6-in. diameter and 2-in. thickness is 3,500 lb. Determine the tensile strength of the material.

4.6 Determine the dynamic modulus of an AC for the following conditions:

$\rho_{34} = 21$
$\rho_{38} = 46$
$\rho_{4} = 65$
$\rho_{200} = 4.5$
$\eta = 2.3 \times 10^6$ poise
$V_{beff} = 3.9\%$
$V_a = 4.5\%$
$f_r = 10$ Hz

4.7 Determine the dynamic modulus of an AC for the following conditions:

$\rho200 = 3.6\%$;	$\rho34 = 21\%$
$\rho38 = 46\%$;	$\rho4 = 86\%$
$V_{beff} = 4.1\%$;	$V_a = 5.9\%$
$\delta_b = 30$ (degree);	$\lvert G^* \rvert = 15$ psi

4.8 The dynamic modulus of an AC specimen is 13,000 MPa. Determine Poisson's ratio of it.

4.9 Poisson's ratio of an asphalt specimen is calculated to be 0.39. Determine how much a 4-in.-diameter and 6-in.-high cylindrical asphalt specimen deforms upon applying a load of 800 lb.

CHAPTER 5

Portland Cement Concrete

5.1 Background

Portland cement concrete (PCC), shown in Fig. 5.1 and produced using portland cement, is one of the most versatile construction materials available in the world. The low cost and widespread availability of portland cement and aggregates make PCC the lowest-cost material widely used over the last century. In rigid pavement, the surface layer is constructed with PCC, reinforced or unreinforced. The reinforcement is provided by structural steel rebar. PCC is also used for some high structures such as curbs, footpaths, roadside rigid barriers, and retaining walls. This chapter discusses the knowledge of PCC required for rigid pavement design.

FIGURE 5.1 Pouring portland cement concrete on a highway.

5.2 PCC Characterizations

5.2.1 Elastic Modulus and Poisson's Ratio

The modulus of elasticity is the slope of the stress-strain curve at the initial phase or until the linear elastic region (proportional limit). The ratio of lateral to longitudinal strain within linear elastic region is known as Poisson's ratio. The modulus of elasticity and Poisson's ratio are used in sizing of reinforced and unreinforced structural members, establishing the quantity of reinforcement, and computing developed stress-strain. The elastic modulus (modulus of elasticity) and Poisson's ratio of PCC are determined using the ASTM C 469 standard using cylindrical specimens, as shown in Fig. 5.2. This test method provides the stress-strain curve until failure. The stress-strain curves of concretes of different strengths show that the curve becomes nonlinear after about 30% to 40% of the ultimate stress.

Poisson's ratio (μ) can be defined as the ratio of lateral strain (ε_t) and axial strain (ε_a). It can be expressed as:

$$\mu = -\frac{\text{Lateral strain}}{\text{Axial strain}} = -\frac{\varepsilon_t}{\varepsilon_a} \tag{5.1}$$

Figure 5.2 Modulus of elasticity and Poisson's ratio of concrete testing.

where Lateral strain, $\varepsilon_t = \frac{\Delta D}{D_o} = \frac{\Delta r}{r_o} = \frac{\Delta w}{w_o} = \frac{\Delta t}{t_o}$

Axial strain, $\varepsilon_a = \frac{\Delta L}{L_o}$

L_o = Original length of member
ΔL = Change in length (final length minus the original length)
r_o = Original radius
Δr = Change in radius (final radius minus the original radius)
w_o = Original width
Δw = Change in width (final width minus the original width)
t_o = Original thickness
Δt = Change in thickness (final thickness minus the original thickness)

Typical elastic modulus values to be used for Levels 2 or 3 analysis of pavement for existing intact and fractured PCC of rigid pavements are provided in Table 5.1.

Poisson's ratio for new PCC of rigid pavements typically ranges between 0.11 and 0.21. Values between 0.15 and 0.18 are typically considered for rigid pavement design. See Table 5.2 for typical (default) Poisson's ratios for PCC of rigid pavements for Levels 2 or 3 analysis of pavement using the AASHTOWare pavement ME design software.

Pavement type	Elastic modulus (ksi)	
Existing intact PCC	Adequate	3,000–4,000
	Marginal	1,000–3,000
	Inadequate	300–1,000
Existing fractured PCC	Crack and seat or break and seat	150–1,000
	Rubblized	50–150

Source: From AASHTO (2015). Mechanistic-Empirical Pavement Design Guide: A Manual of Practice. Washington, DC: American Association of State Highway and Transportation Officials. Table 10-5. Used with permission.

Table 5.1 Elastic Modulus Values of Existing Intact and Fractured PCC (Levels 2 or 3)

PCC materials	Poisson's ratio
PCC slabs (new or existing)	0.20
Fractured slab–crack/seat	0.20
Fractured slab–break/seat	0.20
Fractured slab–rubblized	0.30

Source: From AASHTO (2015). Mechanistic-Empirical Pavement Design Guide: A Manual of Practice. Washington, DC: American Association of State Highway and Transportation Officials. Table 10-5. Used with permission.

Table 5.2 Default Poisson's Ratios of PCC Materials (Levels 2 or 3)

Note that default values used in the AASHTOWare pavement ME design software may change as the software undergoes continuous improvement. Default values listed in this textbook are the most updated ones.

Example 5.1: Modulus of Elasticity and Poisson's Ratio A compressive force of 15 kip is applied on a 6-in.-diameter and 12-in.-high long cylindrical concrete specimen. After applying the load, the diameter of the rod increases to 6.0002 in. and the length decreases to 11.9998 in. Assuming no permanent deformation occurs in that material, calculate the modulus of elasticity and Poisson's ratio.

Solution

Axial stress, $\sigma = \dfrac{\text{Load}}{\text{Area}} = \dfrac{15{,}000\text{ lb}}{\dfrac{\pi}{4}(6\text{ in.})^2} = 530.5\text{ psi}$

Axial strain, $\varepsilon_a = \dfrac{\Delta L}{L_o} = \dfrac{11.998 - 12}{12} = -0.000167$ (negative sign is commonly omitted)

Lateral/transverse strain, $\varepsilon_t = \dfrac{\Delta D}{D_o} = \dfrac{6.0002 - 6}{6} = 0.000033$

Elasticity, $E = \dfrac{\sigma}{\varepsilon_a} = \dfrac{530.5\text{ psi}}{0.000167} = 3{,}176{,}600\text{ psi}$

Poisson's ratio, $\mu = \dfrac{\varepsilon_t}{\varepsilon_a} = \dfrac{0.000033}{0.000167} = 0.198$

Answers *The modulus of elasticity is 3,176,600 psi and Poisson's ratio is 0.198.*

5.2.2 Flexural Strength

Modulus of rupture (R) or the flexural strength of concrete is the bending strength capacity of a PCC slab. When a PCC slab is loaded by traffic, the slab bends and bending stress develops similar to a beam/slab. This is why the flexural strength is an important input for rigid pavement design especially for predicting the amount of transverse cracking in a pavement PCC slab. The flexural strength is determined by applying a third-point loading on a beam specimen following the AASHTO 97 test protocol. The specimen is placed on its side in the machine in such a manner that a minimum of 1 in. (25 mm) of the beam extends outside the support rollers as shown in Fig. 5.3. A load of between 3% and 6% of the expected ultimate load is applied. The beam specimen is subjected to an increasing load until failure, with the extreme fiber stress between 125 psi/min (860 kPa/min) and 175 psi/min (1,200 kPa/min).

a. If the fracture initiates in the tension surface within the middle-third of the span length, the modulus of rupture is calculated as follows:

$$R = \frac{Mc}{I} = \frac{\left(\dfrac{PL}{6}\right)\dfrac{d}{2}}{\dfrac{bd^3}{12}} = \frac{PL}{bd^2} \tag{5.2}$$

FIGURE 5.3 Concrete modulus of rupture testing applying a third-point loading.

where R = Modulus of rupture, psi or MPa
P = Maximum applied load indicated by the testing machine, lb or N
L = Span length, in. or mm
b = Average width of specimen at the fracture, in. or mm
d = Average depth of specimen at the fracture, in. or mm

b. If the fracture occurs in the tension surface outside of the middle-third of the span length by not more than 5% of the span length, the modulus of rupture is calculated as follows:

$$R = \frac{3Pa}{bd^2} \tag{5.3}$$

where a = Average distance between line of fracture and the nearest support measured on the tension surface of the beam, in. or mm

c. If the fracture occurs in the tension surface outside of the middle-third of the span length by more than 5% of the span length, the result of the test is discarded.

Example 5.2: Flexural Strength A 6 × 6 × 18-in. concrete beam is being tested applying the third-point loading for flexure as shown in Fig. 5.4. The beam fails when the load value reaches 6.0 kip. The failure is within the middle-third of the beam. Calculate the flexural strength of the beam.

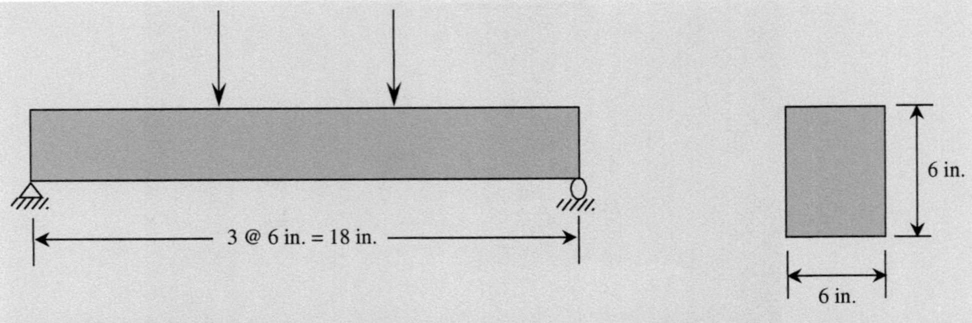

Figure 5.4 Modulus of rupture testing applying a third-point loading for Example 5.2.

Solution

Given:

Width, $b = 6$ in.

Depth, $d = 6$ in.

Peak load, $P = 6{,}000$ lb

Modulus of rupture, $R = \frac{PL}{bd^2} = \frac{(6{,}000\ \text{lb})(18\ \text{in.})}{(6\ \text{in.})(6\ \text{in.})^2} = 500$ psi

Answer *The flexural strength of the beam is 500 psi.*

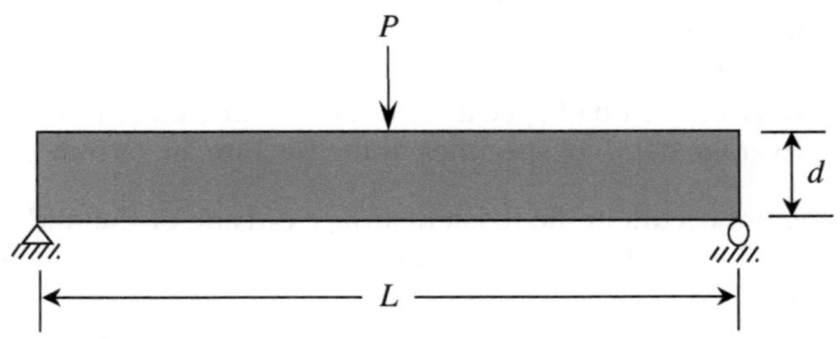

Figure 5.5 Concrete modulus of rupture testing applying a center-point loading.

Flexural strength can also be determined using the center-point concentrated loading as shown in Fig. 5.5 following the ASTM C 293 test protocol. If a center-point loading is applied on the concrete beam, the flexural strength of the concrete beam is determined using the following equation:

$$R = \frac{3}{2}\frac{PL}{bd^2}$$

where R = Modulus of rupture, psi or MPa
P = Maximum applied load indicated by the testing machine, lb or N
L = Span length, in. or mm
b = Average width of specimen at the fracture, in. or mm
d = Average depth of specimen at the fracture, in. or mm

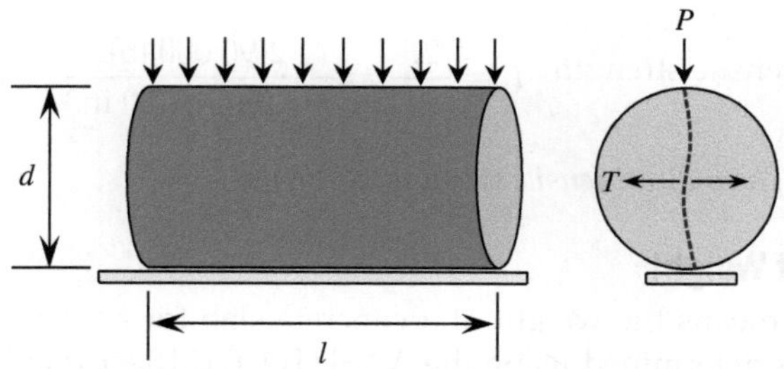

FIGURE 5.6 Concrete splitting tension testing.

5.2.3 Indirect Tensile Strength

It is about impossible to test a concrete specimen in direct tension. An indirect method of determining tensile strength is used where a diametric compressive load is then applied along the length of the cylindrical specimen as shown in Fig. 5.6. Tensile strength is typically used as a PCC performance measure for pavements because it best simulates tensile stresses at the bottom of the PCC surface course as it is subjected to loading. These stresses are typically the controlling structural design stresses. PCC tensile strength is important in pavement applications, although PCC is not nearly as strong in tension as it is in compression.

Indirect tensile strength test (also known as splitting tension test) is conducted using the AASHTO T 198 and ASTM C 496 test standards. A splitting tension test uses a standard 6-in. (150-mm) diameter, 12 in. (300 mm) long test cylinder laid on its side. A diametric compressive load is then applied along the length of the cylinder until it fails. The cylinder typically fails due to horizontal tension and not vertical compression because PCC is much weaker in tension than compression.

The indirect tensile strength or splitting tensile strength or fracture strength T is calculated as:

$$T = \frac{2P}{\pi l D} \tag{5.4}$$

where T = Indirect or splitting or fracture tensile strength, psi or MPa
P = Peak force needed to crack the specimen diagonally, lb or N
D = Diameter of the specimen, in. or mm
l = Length of the specimen, in. or mm

Example 5.3: Tensile Strength Test In a splitting tensile strength test, the peak load required to fail a specimen of 6.0-in. diameter and 12-in. length is 90,000 lb. Calculate the indirect or splitting tensile strength of the specimen.

Solution

Given:
Peak load, $P = 90{,}000$ lb
Diameter, $D = 6$ in.
Length, $l = 12$ in.

Indirect tensile strength, $T = \frac{2P}{\pi l D} = \frac{2(90{,}000 \text{ lb})}{\pi(6.0 \text{ in.})(12.0 \text{ in.})} = 796 \text{ psi}$

Answer *The indirect tensile strength is 796 psi.*

5.2.4 Unit Weight

Unit weight means the weight of a concrete slab for a unit volume. The unit weight of concrete is determined using the AASHTO T 121 test standard. A cylindrical mold, shown in Fig. 5.7, is filled up with freshly mixed concrete in three equal layers. Each layer is compacted 25 times using a standard tamping rod. The surface of the fill is leveled using a cutting plate. The weight of the material filled is divided by the interior volume of the mold to calculate the unit weight. Agency historical data or typical range for normal weight concrete of 140 to 160 lb/ft^3 (22–25 kN/m^3) can be used.

5.2.5 Air Content

Air space (voids) inside a concrete slab is required if there is a probability that concrete may be exposed to freeze-thaw. The air content of concrete can be determined using the AASHTO T 152, which determines the air content of freshly mixed concrete from observation of the change in volume of concrete with a change in pressure.

Figure 5.7 Unit weight testing of concrete.

5.2.6 Other Properties

Some other minor properties required in the AASHTOWare pavement ME design program are discussed in this section.

Coefficient of Thermal Expansion

The coefficient of thermal expansion can be determined using the AASHTO TP 60 test protocol. Some common values are listed in Table 5.3.

Surface Shortwave Absorptivity

The short-wave radiation absorptivity indicates the fraction of the total (i.e., long- and short-wave) solar radiation incident on the pavement surface which is absorbed. This value is used to calculate the surface layer temperature variation within a day. The AASHTOWare pavement ME design default value of 0.85 is to be used unless the test standard is available. This value was used in the global calibration process.

Thermal Conductivity

Thermal conductivity measures the heat flux flowing through a material at unit temperature gradient. This value is used to calculate the surface layer temperature variation within a day. ASTM E 1952 test standard can be used to determine the thermal conductivity. However, this value is not measured very often. Rather, typical values of the thermal conductivity for PCC are used which range from 0.2 to 2.0 BTU/(ft)(h)(°F). The default value set in the AASHTOWare pavement ME design software is 1.25 BTU/(ft)(h)(°F).

Heat Capacity

Heat capacity or thermal capacity is defined as the amount of heat to be supplied to a given mass of a material to produce a unit change in its temperature. This value is used

Aggregate type	Coefficient of thermal expansion (10^{-6} per °F)
Andesite	5.3
Basalt	5.2
Diabase	4.6
Gabbro	5.3
Granite	5.8
Schist	5.6
Chert	6.6
Dolomite	5.8
Limestone	5.4
Quartzite	6.2
Sandstone	6.1
Expanded shale	5.7

Source: From AASHTO (2015). Mechanistic-Empirical Pavement Design Guide: A Manual of Practice. Washington, DC: American Association of State Highway and Transportation Officials. Table 10-5. Used with permission.

TABLE 5.3 Coefficients of Thermal Expansions of PCC Materials (Levels 2 or 3)

to calculate the surface layer temperature variation within a day. ASTM D 2766 test standard can be used to determine the heat capacity. Typical heat capacity values for PCC range from 0.10 to 0.50 BTU/(lb)(°F). The default value set in the AASHTOWare pavement ME design software is 0.28 BTU/(lb)(°F).

PCC Zero-Stress Temperature

Zero-stress temperature (also called PCC set temperature) is considered the temperature at which concrete has no residual stress; below that temperature concrete shrinks and above that temperature concrete expands. There is no national test protocol available to determine the PCC zero-stress temperature. Zero-stress temperature, *Tz*, can be input directly or can be estimated from monthly ambient temperature and cement content using the equations:

$$Tz = (\text{CC} \times 0.59328 \times H \times 0.5 \times 1{,}000 \times 1.8/(1.1 \times 2{,}400) + \text{MMT})\ or \tag{5.5}$$

$$Tz = (0.20225 \times \text{CC} \times H + \text{MMT}) \tag{5.6}$$

where T_z = Zero-stress temperature (allowable range: 70 to 212°F)
CC = Cementitious content, lb/yd^3
$H = -0.0787 + 0.007 \times \text{MMT} - 0.00003 \times \text{MMT}^2$
MMT = Mean monthly temperature for month of construction, °F

The zero-stress temperatures for different mean monthly temperatures and different cement contents in the PCC mix design are presented in Table 5.4.

Example 5.4: Zero-Stress Temperature A concrete pavement is to be constructed in April whose mean monthly temperature is 60°F. The concrete slab uses the cement with a proportion of 500 lb/yd^3. Calculate the zero-stress temperature of the produced concrete slab in the pavement.

Mean monthly temperature (°F)	Cement content (lb/yd^3)			
	400	500	600	700
40	52	56	59	62
50	66	70	74	78
60	79	84	88	93
70	91	97	102	107
80	103	109	115	121
90	115	121	127	134
100	126	132	139	145

Source: From AASHTO (2015). Mechanistic-Empirical Pavement Design Guide: A Manual of Practice. Washington, DC: American Association of State Highway and Transportation Officials. Table 10-5. Used with permission.

TABLE 5.4 Zero-Stress Temperatures (°F) for Different Mean Monthly Temperatures (Levels 2 or 3)

Solution

Equation to be used: $T_z = (0.20225 \times CC \times H + MMT)$

$$T_z = \text{Zero-stress temperature} = ?$$
$$CC = \text{Cementitious content} = 500\ \text{lb/yd}^3$$
$$H = -0.0787 + 0.007 \times MMT - 0.00003 * MMT^2$$
$$= -0.0787 + 0.007 \times 60 - 0.00003 \times 60^2$$
$$= 0.2333$$

Then $T_z = (0.20225 \times CC \times H + MMT)$
$$= 0.20225 \times 500 \times 0.23333 + 60$$
$$= 83.6$$

Answer *The zero-stress temperature is 84°F.*

Cement Type

Cement type is an input while designing rigid pavement, as cement type plays a role in strength, hydration rate, heat generation, setting, etc. It is selected based on actual or expected cement source.

Cementitious Material Content

Cement content is an input while designing rigid pavement and is selected based on actual or expected cement source.

Water-to-Cement Ratio

Water to cement ratio is an input while designing rigid pavement and is selected based on actual or expected cement source.

Aggregate Type

Cement content is an input while designing rigid pavement and is selected based on actual or expected cement source.

Curing Method

Curing method is an input while designing rigid pavement and is selected based on actual or expected cement source.

Ultimate Shrinkage

The probable ultimate shrinkage is an input while designing rigid pavement. Testing the ultimate shrinkage is not practical and thus the prediction equation used in AASHTOWare pavement ME design can be used.

Reversible Shrinkage

The amount of reversible shrinkage is considered 50% of the ultimate shrinkage as the default value in the AASHTOWare pavement ME design software.

Time to Develop 50% of Ultimate Shrinkage

The AASHTOWare pavement ME design software uses the default value of 35 days for the time to develop 50% of ultimate shrinkage.

The summary of different Level 1 properties of PCC and the test protocols for new overlays and existing PCC when subject to a bonded PCC overlay are listed in Table 5.5.

Measured property	Test	Estimate	Recommended test protocol and/or data source
Elastic modulus	x		ASTM C 469
Poisson's ratio	x		ASTM C 469
Flexural strength	x		AASHTO T 97
Indirect tensile strength	x		AASHTO T 198
Unit weight	x		AASHTO T 121 M/T 121
Air content	x		AASHTO T 152 or T 196 M/T 196
Coefficient of thermal expansion	x		AASHTO TP 60
Surface shortwave absorptivity		x	Use AASHTOWare pavement ME design default value
Thermal conductivity	x		ASTM E 1952
Heat capacity	x		ASTM D 2766
Zero-stress temperature		x	Use AASHTOWare pavement ME design default value
Cement type		x	Select based on actual or expected cement source
Cementitious material content		x	Select based on expected concrete mix design
Water-to-cement ratio		x	Select based on expected concrete mix design
Aggregate type		x	Select based on expected aggregate source
Curing method		x	Select based on agency recommendations and practices
Ultimate shrinkage		x	Estimate using prediction equation in the AASHTOWare pavement ME design
Reversible shrinkage		x	Estimate using agency historical data or select AASHTOWare pavement ME design defaults
Time to develop 50% of ultimate shrinkage		x	Estimate using agency historical data or select AASHTOWare pavement ME design defaults

Source: From AASHTO (2015). Mechanistic-Empirical Pavement Design Guide: A Manual of Practice. Washington, DC: American Association of State Highway and Transportation Officials. Table 10-4. Used with permission.

Table 5.5 Summary of Different Level 1 Properties of PCC and Test Protocols for New Overlays and Existing PCC When Subject to a Bonded PCC Overlay

Measured property	Test	Estimate	Recommended test protocol and/or data source
Elastic modulus	x		ASTM C 469 (extracted cores) AASHTO T 256 (non-destructive deflection testing)
Poisson's ratio	x		ASTM C 469 (extracted cores)
Flexural strength	x		AASHTO T 97 (extracted cores)
Unit weight	x		AASHTO T 121 M/T 121
Surface shortwave absorptivity		x	Use AASHTOWare pavement ME design default value
Thermal conductivity	x		ASTM E 1952 (extracted cores)
Heat capacity	x		ASTM D 2766 (extracted cores)

Source: From AASHTO (2015). Mechanistic-Empirical Pavement Design Guide: A Manual of Practice. Washington, DC: American Association of State Highway and Transportation Officials. Table 10-4. Used with permission.

TABLE 5.6 Summary of Different Level 1 Properties of PCC and the Test Protocols for Existing Intact and Fractured PCC

The summary of different Level 1 properties of PCC and the test protocols for existing intact and fractured PCC are listed in Table 5.6.

5.3 Chemically Stabilized PCC Materials

The summary of different chemically stabilized PCC and the test protocols are listed in Tables 5.7 and 5.8.

The recommended Level 2 and 3 input parameters of different chemically stabilized PCC and the test protocols are summarized in Tables 5.9 and 5.10.

Example 5.5: Modulus of Elasticity A lean concrete specimen failed at a compressive force of 50 kip is applied on a 6-in.-diameter and 12-in.-high long cylinder. Assuming no permanent deformation occurs in that material, calculate the modulus of elasticity.

Solution

$$\text{Failure stress, } f_c' = \frac{\text{Load}}{\text{Area}} = \frac{50{,}000 \text{ lb}}{\frac{\pi}{4}(6 \text{ in.})^2} = 1{,}768 \text{ psi}$$

$$\text{Elasticity, } E = 57{,}000\sqrt{f_c'} = 57{,}000\sqrt{1{,}768 \text{ psi}} = 2{,}396{,}700 \text{ psi}$$

Answer *The modulus of elasticity is 2,396,700 psi.*

	Measured property	Test	Estimate	Recommended test protocol and/or data source
Lean concrete cement-treated aggregate	Elastic modulus	x		ASTM C 469
	Flexural strength	x		AASHTO T 97
Lime cement-fly ash	Elastic modulus		x	No test protocols available. Estimate using Levels 2 and 3
	Flexural strength	x		AASHTO T 97
Soil cement	Elastic modulus		x	No test protocols available. Estimate using Levels 2 and 3
	Flexural strength	x		ASTM D 1635
Limestabilized soil	Resilient modulus	x		AASHTO T 307
	Flexural strength		x	No test protocols available. Estimate using Levels 2 and 3
All	Unit weight		x	No testing required. Estimate using Levels 2 and 3
	Poisson's ratio		x	No testing required. Estimate using Levels 2 and 3
	Thermal conductivity	x		ASTM E 1952
	Heat capacity	x		ASTM D 2766
	Surface short-wave absorptivity		x	No test protocols available. Estimate using Levels 2 and 3

Source: From AASHTO (2015). Mechanistic-Empirical Pavement Design Guide: A Manual of Practice. Washington, DC: American Association of State Highway and Transportation Officials. Table 10-4. Used with permission.

Table 5.7 Chemically Stabilized Materials Level 1 Input Requirements and Test Protocols for New Chemically Stabilized Materials

5.4 Summary

Elastic modulus, Poisson's ratio, flexural strength, and indirect tensile strength are the primary properties of PCC used in rigid pavement design. Elastic modulus and Poisson's ratio are determined by applying uniaxial compression load on cylindrical specimens. The peak load is used to calculate the compressive strength and the linear-elastic region of the stress-strain is used to calculate Poisson's ratio. Flexural strength is

Measured property	Test	Estimate	Recommended test protocol and/or data source
Calculated modulus from FWD deflection basins	x		AASHTO T 256 & ASTM D 5858
Flexural strength		x	Estimate using Levels 2 and 3
Unit weight		x	Estimate using Levels 2 and 3
Poisson's ratio		x	Estimate using Levels 2 and 3
Thermal conductivity	x		ASTM E 1952 (cores)
Heat capacity	x		ASTM D 2766 (cores)
Surface shortwave absorptivity		x	Estimate using Levels 2 and 3

Source: From AASHTO (2015). Mechanistic-Empirical Pavement Design Guide: A Manual of Practice. Washington, DC: American Association of State Highway and Transportation Officials. Table 10-4. Used with permission.

Table 5.8 Chemically Stabilized Materials Level 1 Input Requirements and Test Protocols for All Existing Chemically Stabilized Materials

Material	Relationship for modulus	Test method	Common values (psi)
Lean concrete	$E = 57{,}000(f_C')^{0.5}$	AASHTO T 22	2,000,000
Cement-treated aggregate	$E = 57{,}000(f_C')^{0.5}$	AASHTO T 22	1,000,000
Open-graded cement-stabilized aggregate	Use input Level 3	None	750,000
Lime-cement-fly ash	$E = 500 + q_u$	ASTM C 593	1,500,000
Soil cement	$E = 1{,}200q_u$	ASTM D 1633	500,000
Lime-stabilized soil	$M_R = 124(q_u) + 9{,}980$	ASTM D 5102	45,000

E = Elastic modulus; psi; M_R = Resilient modulus, psi; q_u or f_C' = unconfined compressive strength (psi).

Source: From AASHTO (2015). Mechanistic-Empirical Pavement Design Guide: A Manual of Practice. Washington, DC: American Association of State Highway and Transportation Officials. Table 10-4. Used with permission.

Table 5.9 Recommended Input Levels 2 and 3 Elastic (E)/Resilient Modulus (M_R) Values for Chemically Stabilized Material

measured by applying the third-point loading in a beam specimen loading until failure. The peak load and the beam's dimensions are used to calculate the maximum bending stress (flexural strength). Indirect tensile strength is determined by splitting test where load is applied diagonally on a cylindrical specimen and the peak load is used to

Property	Materials	Common values
Flexural strength	Use 20% of the compressive strength of lab specimens or extracted cores as an estimate of the flexural strength for all chemically stabilized materials. OR	
	Chemically stabilized material placed under flexible pavement (base)	750 psi
	Chemically stabilized material used as subbase, select material, or subgrade under flexible pavement	250 psi
Poisson's ratio	Lean concrete and cement-stabilized aggregate	0.1 o 0.2
	Soil cement	0.15 to 0.35
	Lime-fly ash materials	0.1 to 0.15
	Lime-stabilized soil	0.15 to 0.2
Unit weight	Use default AASHTOWare pavement ME design values of 150 pcf	
Thermal conductivity	Use default AASHTOWare pavement ME design values of 1.25 BTU/h·ft·°F	
Heat capacity	Use default AASHTOWare pavement ME design values of 0.28 BTU/lb·°F	

Source: From AASHTO (2015). Mechanistic-Empirical Pavement Design Guide: A Manual of Practice. Washington, DC: American Association of State Highway and Transportation Officials. Table 10-4. Used with permission.

TABLE 5.10 Recommended Input Levels 2 and 3 Properties for Chemically Stabilized Material

calculate the tensile strength of the specimen. Some other minor properties such as coefficient of thermal expansion and contraction, surface shortwave absorptivity, and heat capacity are commonly not measured in laboratory or field. These values are estimated or the AASHTOWare pavement ME design default values are commonly used.

5.5 Fundamentals of Engineering (FE) Exam-Style Questions

FE5.1 In the flexural testing of concrete using the third-point loading, the test result is discarded if

A. The failure occurs inside the middle-third
B. The failure occurs outside the middle-third but no more than 5% of the span
C. The failure occurs outside the middle-third but more than 5% of the span
D. Never unless mesh occurs

Solution C

In flexural testing of concrete using the third-point loading, the test result is discarded if the fracture occurs in the tension surface outside of the middle-third of the span length but more than 5% of the span length.

FE5.2 Modulus of elasticity is defined as the
A. Capacity of deflection resistance
B. Slope of stress versus strain curve until elastic region
C. Initial slope of the stress-strain curve
D. Energy required before the failure

Solution C

Modulus of elasticity is defined as the initial slope of stress versus strain curve or slope at any point up to the proportional limit.

FE5.3 A concrete specimen has the diameter of 6.0 in. The modulus of elasticity of concrete is 3,000 ksi and Poisson's ratio is 0.30. If the specimen is subjected to an axial compression, the lateral strain of 0.0000366 occurs. The increase in the diameter is most nearly:
A. 2.196 μ in.
B. 220 μ in.
C. 3.66 μ in.
D. 2.38 nano in.

Solution B

$$\varepsilon_t = \frac{\Delta D}{D_o}$$

$$\Delta D = \varepsilon_t(D_o) = 0.0000366(6.0 \text{ in.}) = 2.196 \times 10^{-4} \text{ in.} = 219.6\ \mu \text{ in.} \approx 220 \text{ in.}$$

5.6 Practice Problems

5.1 A compressive force of 12 kip is applied on a 6-in.-diameter and 12-in.-long cylinder. After applying the load, the diameter of the rod increases by 0.00016 in. and the length decreases by 0.0016 in. Assuming no permanent deformation occurs in the material, calculate the modulus of elasticity.

5.2 A center-point loading flexural test was performed on a concrete beam of 8-in. span. The beam cross-section was 2 in. wide and 1.95 in. deep. The beam failed at the load of 420 lb. Calculate the flexural strength of the concrete beam.

5.3 A center-point loading flexural test was performed on a concrete beam of 18-in. span. The beam cross-section was 6-in. square. The beam failed at the load of 7,500 lb. Calculate the flexural strength of the concrete beam.

5.4 A third-point loading flexural test was performed on a concrete beam of 18-in. span. The beam cross-section was 6-in. square. The beam shows a crack at 5.5 in. from the end of the beam at the load of 9,000 lb. Calculate the flexural strength of the concrete beam.

5.5 In a splitting tensile strength test, the peak load required to fail a specimen of 5.5-in. diameter and 10-in. thickness is 75,000 lb. Calculate the tensile strength of the specimen.

5.6 A concrete pavement is to be constructed in April whose mean monthly temperature is 90°F. The concrete slab uses the cement with a proportion of 600 lb/yd^3. Calculate the zero-stress temperature of the produced concrete slab in the pavement.

5.7 A soil-cement specimen failed at a compressive force of 10 kip is applied on a 6-in.-diameter and 12-in.-long cylinder. Assuming no permanent deformation occurs in that material, calculate the modulus of elasticity.

5.8 A lime-stabilized soil specimen failed at a compressive force of 1.5 kip is applied on a 2.5-in.-diameter and 5-in.-long cylinder. Assuming no permanent deformation occurs in that material, calculate the modulus of elasticity.

CHAPTER 6

Traffic Analysis for Pavement Design

6.1 Background

Traffic is considered the only load for flexible and rigid pavements design. This is why accurate determination of traffic loading and its distribution throughout the lanes at different times of a year is critical when analyzing and designing the pavement. Larger vehicles or heavy trucks are more important for pavement design than passenger cars since larger vehicles or heavy trucks are likely to cause more damage in pavement layers than passenger cars. A large truck is shown in Fig. 6.1.

6.2 Fundamentals of Traffic Analysis

6.2.1 Tire Imprint Areas

Tire loads are the fundamental loads at the actual tire-pavement contact areas. The load on a wheel is transferred to pavements through tires. Accurate shapes of the

FIGURE 6.1 A truck on a highway.

contact areas between tires and pavements are not known. Three shapes are commonly assumed (Fig. 6.2):

1. Circular area
2. Composite area
3. Rectangular area

For most pavement analyses, it is assumed that the tire load is uniformly applied over a circular area. Tire inflation pressure and contact pressure are also generally assumed to be equal. This is not valid, but it is suitable for approximations. The pressure on the surface of the pavement can be calculated by dividing the load of the tire by circular contact area, as depicted in Eq. (6.1). Then, the radius of contact can be calculated using Eq. (6.2).

$$p = \frac{\text{Load}}{\text{Area}} = \frac{P}{\pi a^2} \tag{6.1}$$

$$a = \sqrt{\frac{P}{p\pi}} \tag{6.2}$$

where a = Radius of tire contact, in. or m
P = Total load on tire, lb or kN
p = Tire inflation pressure, psi or kPa

There are other forms of tire area: composite and rectangular (Fig. 6.3). The width of the shapes represents the tire width. For composite area, there are two semi-circles

FIGURE 6.2 Imprint area of a tire.

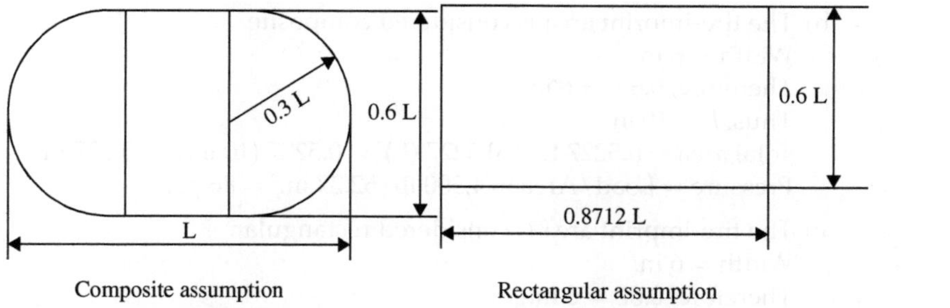

Figure 6.3 Assumed tire contact areas.

at the ends and an interior rectangle. The length of the whole composite is considered "L." The width of the composite shape is 0.6*L*, which equals the tire width. Therefore, the radius of the semi-circle is 0.3*L*. The length of the interior rectangular area is then $L - 2\,(0.3L) = 0.4L$. Therefore, the total area of the composite area can be calculated as:

$$\begin{aligned}\text{Composite area} &= \text{Two semi-circles} + \text{One rectangle}\\ &= 2\,(½ \times \pi \times (0.3L)^2) + (0.4L \times 0.6L)\\ &= 0.2828L^2 + 0.24L^2\\ &= 0.5227L^2\end{aligned} \tag{6.3}$$

For the rectangular tire-imprint area, the width of the composite shape is 0.6*L*, which equals the tire width. The length of the rectangular area is 0.8712*L*. Therefore, the total rectangular area can be calculated as:

$$\begin{aligned}\text{Rectangular area} &= \text{Length} \times \text{Width}\\ &= 0.8712L \times 0.6L\\ &= 0.5227L^2\end{aligned} \tag{6.4}$$

From Eqs. (6.3) and (6.4), both the composite and rectangular tire-imprint areas assume an equal amount of areas with different shapes.

Example 6.1: Pressure on the Pavement The total load on a single tire is 4.5 kip. The tire width is 6 in. Determine the exerted pressure on the pavement surface if:

a) The tire-imprint area is considered circular.

b) The tire-imprint area is considered composite.

c) The tire-imprint area is considered rectangular.

Solution

a) The tire-imprint area is considered circular.

Load, $P = 4{,}500$ lb

Radius, $a = 6$ in./2 = 3 in.

$$\text{Applied pressure, } p = \frac{P}{\pi a^2} = \frac{4{,}500 \text{ lb}}{\pi (3 \text{ in.})^2} = 159 \text{ psi}$$

b) The tire-imprint area is considered composite.
Width = 6 in.
Therefore, 0.6 L = 6 in.
Thus, L = 10 in.
Total area = $0.5227\ L^2 = 0.5227\ (L)^2 = 0.5227\ (10\ \text{in.})^2 = 52.27\ \text{in.}^2$
Pressure = Load/Area = 4,500 lb/52.27 in.^{2} = 86 psi

c) The tire-imprint area is considered rectangular.
Width = 6 in.
Therefore, $0.6L$ = 6 in.
Thus, L = 10 in.
Total area = $0.5227L^2 = 0.5227(L)^2 = 0.5227(10\ \text{in.})^2 = 52.27\ \text{in.}^2$
Pressure = Load/Area = 4,500 lb/52.27 in.^{2} = 86 psi

Answers *The applied pressures at circular, composite, and rectangular area are 159, 86, and 86 psi, respectively.*

This example shows that the circular assumption of the tire-imprint area is very conservative, which calculates the applied pressure higher than that by other tire-imprint areas. The other two assumptions, composite and rectangular, yield the equal pressure. Still though, the circular assumption is popular being conservative. Continuous research is ongoing in this topic to evaluate the effects of different tire-imprint areas on the structural responses of pavement.

6.2.2 Axle Types

While the tire contact pressure and tire-imprint area are of crucial concern for pavement performance, the number and spacing of contact points per vehicle are also important. As tire loads get closer, their areas of impact on the pavement begin to overlap, and the damage to the pavement varies. Axle and tire arrangements are therefore also important. Figure 6.4 shows different axle types commonly available in highways. These can be described as follows:

- A single axle, single tire means there is a single tire at the ends of the single axle (total of two tires in the axle).
- A single axle, dual tires has double tires at the axle ends (total of four tires in the axle).
- A tandem axle has four tires in two rows at the ends of the axle (total of eight tires in the axle).
- A tridem axle has three rows of tires with a total of six tires at ends of the axle (total of 12 tires in the axle).
- A quad axle has four rows of tires with a total of eight tires at ends of the axle (total of 16 tires in the axle).

6.2.3 Counting Traffic and Measuring Axle Load

Counting traffic and measuring axle load are important for pavement design, as these parameters are for design input. Automatic traffic recorders (ATR) with different

Front: single axle, single tire
Rare: single axle, dual tires

Tandem axle

Front: single axle, single tire
Rare: tridem axle

Quad axle

Figure 6.4 Different types of axles.

technologies such as inductive loop (change in magnetic field), piezoelectric sensors (signal voltage or current), road tube (air pulse), and infrared/radar detectors can constantly count the number of vehicles (traffic volumes) passing through a location. The inductive loop is often used with piezoelectric sensors for axle detection for obtaining comprehensive data about traffic number, length, and speed. Through telemetry, data can be transmitted from the field to a central location. The above sensors do not, however, calculate the wheel load and lateral wheel load distribution.

Weigh-in-motion (WIM) device is a popular device for counting the number of vehicles passing through a pavement section by capturing and recording number of axles, axle spacing, weight, and classification of vehicles according to the Federal Highway Administration (FHWA) vehicle classification (FHWA 2013). FHWA vehicle classification is discussed in Sec. 6.2.4. However, a WIM device does not measure the wheel wander. A WIM device installed on I-40 near the city of Albuquerque in New Mexico is shown in Fig. 6.5. It can record the wheel distribution, vehicle classification, axle types, loads on different wheels, etc. The WIM sensor records the traffic data into two special format files (C-files and W-files). Descriptions of each file format are given in Table 6.1.

FIGURE 6.5 Weigh-in-motion device.

Record type	File extension	Data type
"C" variant	"*.CLA" (C-file)	Classification, hourly distribution, axle configuration, etc.
"W" variant	"*.WGT" (W-file)	Load distribution in wheels

TABLE 6.1 Description of Record Types

The C-files contain the classification data, and W-files contain the weight data. The classification data consist of a total number of vehicles in each route and direction during each hour of the day and the distribution of these vehicles as per the FHWA vehicle classification. Taking into account the number of axles, spacing, and weight, the weight file stores the data of each vehicle. A WIM device does not calculate the wheels' lateral distribution.

Wheel centerline and lateral distribution can be measured with axle sensing strips (Fig. 6.6). The axle sensing strips are installed and their configuration is accurately measured. Data acquisition systems record the timing of the tires that hit the sensors. Then using the trigonometry, the position of the tires can be calculated as the spacing of sensors and their time gaps are known.

6.2.4 FHWA Vehicle Classifications

FHWA classifies vehicle into 13 classes, Class 1 to Class 13. Class 1 comprises the motorbikes and Class 13 has the seven or more axle multitrailer trucks. Other vehicles are also classified in between these classes. Their definitions are listed in Table 6.2 and the vehicles are shown in Fig. 6.7.

Figure 6.6 Axle sensing strips.

Class	Description
0	Unclassified vehicles that do not fit into any other classification. Vehicles that do not activate the system sensors are also unclassified
1	Motorcycles: All two- or three-wheeled motorized vehicles. This category includes motorcycles, motor scooters, mopeds, and all three-wheel motorcycles
2	Passenger cars: All sedans, coupes, and station wagons manufactured primarily for the purpose of carrying passengers
3	Other two-axle, four-tire single units: Included in this classification are pickups, vans, campers, and ambulance units, one of which is a tractor or straight truck power unit
4	Buses: All vehicles manufactured as traditional passenger-carrying buses with two axles and six tires or three or more axles
5	Two-axle, single-unit trucks: All vehicles on a single frame, including trucks and camping and recreation
6	Three-axle, single-unit trucks: All vehicles on a single frame, including trucks and camping and recreational vehicles
7	Four or more axle, single-unit trucks: All vehicles on a single frame with four or more axles
8	Four or less axle, single-trailer trucks: All vehicles with four or less axles consisting of two units, one of which is a tractor or straight truck power unit
9	Five-axle, single-trailer trucks: All five-axle vehicles consisting of two units, one of which is a tractor or straight truck power unit
10	Six or more axle, single-trailer trucks: All vehicles with six or more axles consisting of two units, one of which is a tractor or straight truck power unit
11	Five or less axle, multitrailer trucks: All vehicles with five or less axles consisting of three or more units, one of which is a tractor or straight truck power unit
12	Six-axle, multitrailer trucks: All six-axle vehicles consisting of three or more units, one of which is a tractor or straight truck power unit
13	Seven or more axle, multitrailer trucks: All vehicles with seven or more axles consisting of three or more units, one of which is a tractor or straight truck power unit

Table 6.2 Vehicle Classification

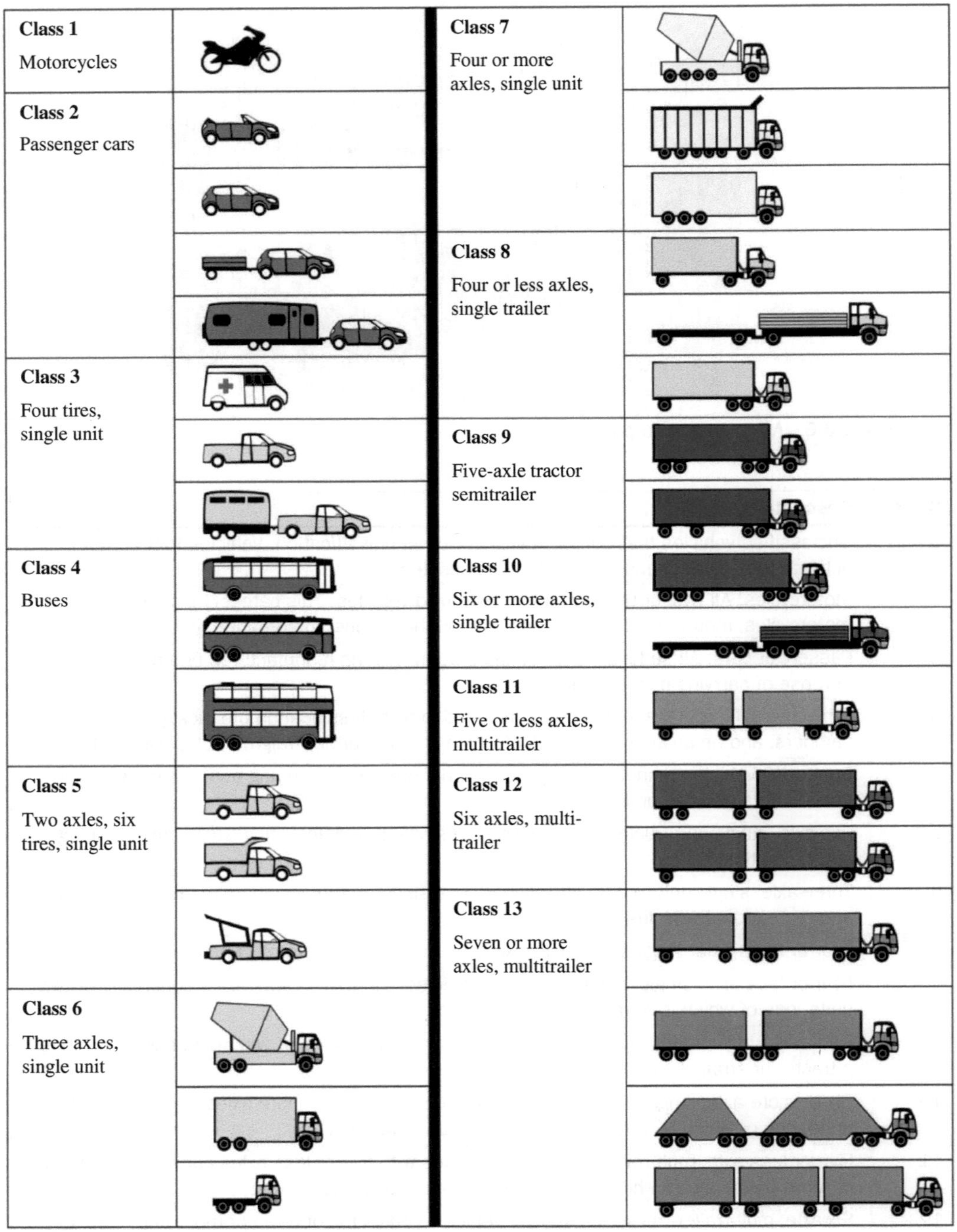

FIGURE 6.7 FHWA vehicle classification.

6.3 Traffic Analysis for the AASHTO 1993 Design

6.3.1 Equivalent Single-Axle Load (ESAL)

AASHTO 1993 pavement design procedure uses the equivalent single-axle load (ESAL) system to count the total number of traffic at the site concerned. The ESAL system is also used in Superpave mix design. The level of compaction in the Superpave mix design is directly associated with the total number of ESAL in the proposed pavement site. Therefore, the study of ESAL is very important.

The ESAL establishes a relationship between the reference 18,000-lb single axle with dual tires and different axle types carrying different loads, as shown in Fig. 6.8. ESAL systems used for pavement design include, in effect, a cumulative traffic load summary statistic (over the design period) representing a mixed stream of axle types and loads converted into an equivalent number of 18,000-lb single-axle loads totaled over that design period.

Note that when passenger cars are converted into a large truck commonly seen in Interstates, thousands of cars are found to be equivalent as a single large truck. Unless the weights in cars and trucks are known, the exact number of cars equivalent to a large truck cannot be determined. However, it is roughly 15 to 25 thousands, meaning 15 to 25 thousand passenger cars produce the equivalent damage of a large truck in a single pass over a pavement.

6.3.2 Equivalent Axle Load Factor (EALF)

An equivalent axle load factor defines the damage per pass to a pavement by the axle in question relative to the damage per pass of a standard axle load, usually the 18,000-lb single-axle load. The design is based on the total number of passes of the standard axle load during the design period. The EALF can be determined using the equation:

$$\text{EALF} = \frac{W_{t18}}{W_{tx}} \tag{6.5}$$

where W_{tx} = Number of x-axle load application at the end of time t

W_{t18} = Number of 18,000-lb single-axle load applications to time t

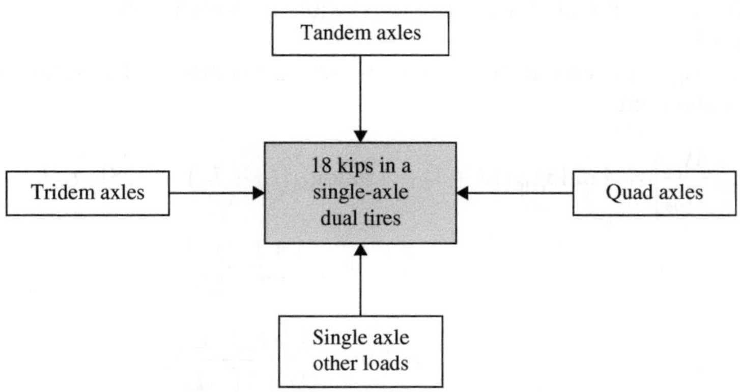

Figure 6.8 Converting different load groups into a standard load.

EALF calculation is quite cumbersome as the likely amount of damage caused by an axle is not readily known. It requires load value, configuration of loading, pavement serviceability requirements, etc.

In the AASHTO design process, the terminal serviceability index, p_t, and the rigidity of the pavement structure (denoted by structural number, SN, for a flexible pavement and slab thickness, D, for a rigid pavement) are utilized, along with the specific axle/load information to determine a load equivalency factor. The serviceability index is a ride quality rating that required a panel of observers to actually ride in an automobile over the pavement in question. The scale varies from 0 to 5 (0 being the no drivable condition and 5 being the most comfortable condition). A new asphalt pavement commonly has an initial serviceability index of 4.2. It is common to use the terminal PSI value of 2.5 for major highways and 2.0 for others; this value can be considered less than 2 for less important roads.

For flexible pavement, the following equations are used to determine the EALF of flexible pavement:

$$\log_{10}\left[\frac{W_{tx}}{W_{t18}}\right] = 4.79\log_{10}(18+1) - 4.79\log_{10}(L_x + L_2) + 4.33\log_{10} L_2 + \frac{G_t}{\beta_x} - \frac{G_t}{\beta_{18}} \tag{6.6}$$

$$G_t = \log_{10}\left[\frac{4.2 - p_t}{4.2 - 1.5}\right] \tag{6.7}$$

$$\beta_x = 0.40 + \frac{0.081(L_x + L_2)^{3.23}}{(\mathrm{SN}+1)^{5.19} L_2^{3.23}} \tag{6.8}$$

where L_x = Load on one single-axle or one tandem-axle or one tridem-axle set (kip)
L_2 = Axle code (1 for single axle, 2 for tandem axle, 3 for tridem axle)
SN = Structural number
p_t = Terminal serviceability (2.5 for major highway, 2 or less than 2 is possible for minor highway)
β_{18} = Value of β_x when L_x is equal to 18 and L_2 is equal to 1
W_{tx} = Number of x-axle load applications for a given traffic to time t
W_{t18} = Number of 18-kip single axle load application to time t

The load equivalency factors for flexible pavement (SN = 5, p_t = 2.5) are listed in Table 6.3.

For rigid pavement, the following equations are used to determine the EALF of rigid pavement.

$$\log_{10}\left[\frac{W_{tx}}{W_{t18}}\right] = 4.62\log_{10}(18+1) - 4.62\log_{10}(L_x + L_2) + 3.28\log_{10} L_2 + \frac{G_t}{\beta_x} - \frac{G_t}{\beta_{18}} \tag{6.9}$$

$$G_t = \log_{10}\left[\frac{4.2 - p_t}{4.2 - 1.5}\right] \tag{6.10}$$

$$\beta_x = 1.0 + \frac{3.63(L_x + L_2)^{5.2}}{(D+1)^{8.46} L_2^{3.52}} \tag{6.11}$$

where D = slab thickness.

Gross axle load		Load equivalency factors		Gross axle load		Load equivalency factors	
kN	lb	Single axles	Tandem axles	kN	lb	Single axles	Tandem axles
4.45	1,000	0.00002		187.0	42,000	25.64	2.51
8.9	2,000	0.00018		195.7	44,000	31.00	3.00
17.8	4,000	0.00209		200.0	45,000	34.00	3.27
22.25	5,000	0.00500		204.5	46,000	37.24	3.55
26.7	6,000	0.01043		213.5	48,000	44.50	4.17
35.6	8,000	0.0343		222.4	50,000	52.88	4.86
44.5	10,000	0.0877	0.00688	231.3	52,000		5.63
53.4	12,000	0.189	0.0144	240.2	54,000		6.47
62.3	14,000	0.360	0.0270	244.6	55,000		6.93
66.7	15,000	0.478	0.0360	249.0	56,000		7.41
71.2	16,000	0.623	0.472	258.0	58,000		8.45
80.0	18,000	1.000	0.0773	267.0	60,000		9.59
89.0	20,000	1.51	0.1206	275.8	62,000		10.84
97.8	22,000	2.18	0.180	284.5	64,000		12.22
106.8	24,000	3.03	0.260	289.0	65,000		12.96
111.2	25,000	3.53	0.308	293.5	66,000		13.73
115.6	26,000	4.09	0.364	302.5	68,000		15.38
124.5	38,000	5.39	0.495	311.5	70,000		17.19
133.5	30,000	6.97	0.658	320.0	72,000		19.16
142.3	32,000	8.88	0.857	329.0	74,000		21.32
151.2	34,000	11.18	1.095	333.5	75,000		22.47
155.7	35,000	12.50	1.23	338.0	76,000		23.66
160.0	36,000	13.93	1.38	347.0	78,000		26.22
169.0	38,000	17.20	1.70	356.0	80,000		28.99
178.0	40,000	21.08	2.08				

Note: kN converted to lb are within 0.1% of lb shown.

Table 6.3 Load Equivalency Factors for Flexible Pavement (SN = 5, $p_t = 2.5$)

Table 6.3 lists EALFs for a flexible pavement with SN = 5, and $p_t = 2.6$ for single, tandem, and tridem axles. Single axle means a single axle with dual tires on each end, a total of four tires. For other conditions, the EALFs can be found in AASHTO (1993), Appendix D. To clarify, the ESAL factor for 40-kip tandem axle is 2.08. It means this axle produces 2.08 times of damage compared to 18-kip single axle. In other words, an 18-kip single axle is required to pass 2.08 times to produce the damage that is produced by single pass of 40-kip tandem axle. The effect (in terms of ESAL) of single tire (carrying the same load as dual tires) is 10% higher. This means the EALF listed in Table 6.3 for a single axle is to be increased by 10% if the single axle has a single tire on each end.

Example 6.2: ESAL Comparison A common full-size car has two single axles with 2-kip load on each. A Class 9 truck has two tandem axles and one single axle. The tandem axles carry 48-kip load on each axle. The single axle of the Class 9 truck carries 14 kip. If SN = 5, p_t = 2.5, how many full-sized cars produce the equivalent damage of a Class 9 truck?

Solution

From Table 6.3:

ESAL of a 2-kip single axle = 0.00018

Therefore, ESAL of one full-size car = 0.00018 × 2 = 0.00036

ESAL of one 48-kip tandem axle = 4.17

ESAL of a 14-kip single axle = 0.36

Therefore, total ESAL of one Class 9 truck = 2 × 4.17 + 0.36 = 8.7

Number of cars = 8.7/0.00036 = 24,167

Answer *A total of 24,167 full-size cars produce the equivalent damage of a Class 9 truck.*

Example 6.3: ESAL for Flexible Pavement Consider a vehicle with two single axles, each with dual tires. The total load on each axle is 10,000 lb. Consider a p_t of 2.5 and SN of 4. Calculate the ESAL for the vehicle considering flexible pavement.

Solution

Given:

Axle load, L_x = 10 kip

Terminal serviceability, p_t = 2.5

Structural number, SN = 4

L_2 = 1 for single axle

Known: $$\log_{10}\left[\frac{W_{tx}}{W_{t18}}\right] = 4.79\log_{10}(18+1) - 4.79\log_{10}(L_x + L_2) + 4.33\log_{10} L_2 + \frac{G_t}{\beta_x} - \frac{G_t}{\beta_{18}}$$

$$G_t = \log_{10}\left[\frac{4.2 - p_t}{4.2 - 1.5}\right] = \log_{10}\left[\frac{4.2 - 2.5}{4.2 - 1.5}\right] = -0.2009$$

$$\beta_x = 0.4 + \frac{0.081(L_x + L_2)^{3.23}}{(\text{SN}+1)^{5.19} L_2^{3.23}} = 0.4 + \frac{0.081(10+1)^{3.23}}{(4+1)^{5.19} 1^{3.23}} = 0.4441$$

$$\beta_{18} = 0.4 + \frac{0.081(L_x + L_2)^{3.23}}{(\text{SN}+1)^{5.19} L_2^{3.23}} = 0.4 + \frac{0.081(18+1)^{3.23}}{(4+1)^{5.19} 1^{3.23}} = 0.6578$$

Therefore,

$$\log_{10}\left[\frac{W_{tx}}{W_{t18}}\right] = 4.79\log_{10}(19) - 4.79\log_{10}(10+1) + 4.33\log_{10}1 + \frac{-0.2009}{0.441} - \frac{0.2009}{0.6578}$$

$$\log_{10}\left[\frac{W_{tx}}{W_{t18}}\right] = 0.9900$$

$$\frac{W_{tx}}{W_{t18}} = 9.7724$$

$$\frac{W_{t18}}{W_{tx}} = 0.1023$$

EALF = 0.1023 for a single tire
ESAL = 0.1023 × 2 axles = 0.2046

Answer *The ESAL for the vehicle is 0.2046.*

Note: Increase this ESAL of 0.2046 by 10% if the single axle has a single tire on each end.

Example 6.4: ESAL for Rigid Pavement Consider a vehicle with two single axles, each with dual tires. The total load on each axle is 10,000 lb. Consider a p_t of 2.5 and slab thickness of 10 in. Calculate the total ESAL for the vehicle considering rigid pavement.

Solution
Given:
Axle load, $L_x = 10$ kip
Terminal serviceability, $p_t = 2.5$
Slab thickness, $D = 10$ in.
$L_2 = 1$ for single axle

Known: $$\log_{10}\left[\frac{W_{tx}}{W_{t18}}\right] = 4.62\log_{10}(18+1) - 4.62\log_{10}(L_x + L_2) + 3.28\log_{10}$$

$$L_2 + \frac{G_t}{\beta_x} - \frac{G_t}{\beta_{18}}$$

$$G_t = \log_{10}\left[\frac{4.2 - p_t}{4.2 - 1.5}\right] = \log_{10}\left[\frac{4.2 - 2.5}{4.2 - 1.5}\right] = -0.2009$$

$$\beta_x = 1.0 + \frac{3.63(L_x + L_2)^{5.2}}{(D+1)^{8.46}L_2^{3.52}} = 1.0 + \frac{3.63(10+1)^{5.2}}{(10+1)^{8.46}1^{3.52}} = 1.0015$$

$$\beta_{18} = 1.0 + \frac{3.63(L_x + L_2)^{5.2}}{(D+1)^{8.46}L_2^{3.52}} = 1.0 + \frac{3.63(18+1)^{5.2}}{(10+1)^{8.46}1^{3.52}} = 1.0251$$

$$\log_{10}\left[\frac{W_{tx}}{W_{t18}}\right] = 4.62\log_{10}(19) - 4.62\log_{10}(10+1) + 3.28\log_{10}1 + \frac{-0.2009}{1.0015} - \frac{-0.2009}{1.0251}$$

$$\log_{10}\left[\frac{W_{tx}}{W_{t18}}\right] = 1.0920$$

$$\frac{W_{t18}}{W_{tx}} = 0.0809$$

$$\text{EALF} = 0.0809$$

$$\text{ESAL} = 0.0809 \times 2 \text{ axles} = 0.1618$$

Answer *The total ESAL for the vehicle is 0.1618.*

6.3.3 Calculation of Projected Design ESAL

The basic steps of calculating the ESAL can be summarized as:

Step 1. Count the initial two-way traffic of different types.

Step 2. Determine the design traffic based on growth factor for the design life.

Step 3. Determine equivalent axle load factor (EALF or simply, F) based on structural number (SN) and terminal serviceability (p_t).

Step 4. Convert traffic into design ESAL and sum up to calculate the total ESAL as follows:

$$\text{Total ESAL} = \sum_{i=1}^{m} F_i n_i \tag{6.12}$$

where m is the number of axle groups, F_i is the EALF for the i-th-axle load group, and n_i is the number of passes of the i-th-axle load group during the design period.

Step 5. Apply the distribution factor as appropriate. The design ESAL in the design lane can be expressed as follows:

$$\text{Design lane ESAL} = D_D(D_L)(\text{Total ESAL}) \tag{6.13}$$

where D_D = Directional distribution factor accounting the distribution of ESAL in the design direction (commonly 50%)

D_L = Lane distribution factor accounting the percentage of traffic in design lane

The lane distribution factor can be determined, as shown in Table 6.4.

The growth factor is listed in Table 6.5. It is calculated using the equation:

$$\text{Factor} = \frac{(1+g)^n - 1}{g} \tag{6.14}$$

where $g = \dfrac{\text{Growth rate}}{100}$ (If the annual growth rate is zero, the growth factor is equal during the analysis period.)

n = Analysis period (Consider the half of the service life; example: for 20-year service life, $n = 20/2 = 10$ years to consider the average traffic.)

No. of lanes in each direction	% of ESAL in the design lane
1	100
2	80–100
3	60–80
4	50–75

Source: AASHTO (1993). AASHTO Guide for Design of Pavement Structures. page II-9. Washington, DC: American Association of State Highway and Transportation Officials. Used with permission.

TABLE 6.4 Lane Distribution Factor

Analysis period years (*n*)	Annual growth rate, *g* (%)							
	0	2	4	5	6	7	8	10
1	1.0	1.0	1.0	1.0	1.0	1.0	1.0	1.0
2	2.0	2.02	2.04	2.05	2.06	2.07	2.08	2.10
3	3.0	3.06	3.12	3.15	3.18	3.21	3.25	3.31
4	4.0	4.12	4.25	4.31	4.37	4.44	4.51	4.64
5	5.0	5.20	5.42	5.53	5.64	5.75	5.87	6.11
6	6.0	6.31	6.63	6.80	6.98	7.15	7.34	7.72
7	7.0	7.43	7.90	8.14	8.39	8.65	8.92	9.49
8	8.0	8.58	9.21	9.55	9.90	10.26	10.64	11.44
9	9.0	9.75	10.58	11.03	11.49	11.98	12.49	13.58
10	10.0	10.95	12.01	12.58	13.18	13.82	14.49	15.94
11	11.0	12.17	13.49	14.21	14.97	15.78	16.65	18.53
12	12.0	13.41	15.03	15.92	16.87	17.89	18.98	21.38
13	13.0	14.68	16.63	17.71	18.88	20.14	21.50	24.52
14	14.0	15.97	18.29	19.16	21.01	22.55	24.21	27.97
15	15.0	17.29	20.02	21.58	23.28	25.13	27.15	31.77
16	16.0	18.64	21.82	23.66	25.67	27.89	30.32	35.95
17	17.0	20.01	23.70	25.84	28.21	30.84	33.75	40.55
18	18.0	21.41	25.65	28.13	30.91	34.00	37.45	45.60
19	19.0	22.84	27.67	30.54	33.76	37.38	41.45	51.16
20	20.0	24.30	29.78	33.06	36.79	41.00	45.76	57.28
25	25.0	32.03	41.65	47.73	54.86	63.25	73.11	98.35
30	30.0	40.57	56.08	66.44	79.06	94.46	113.28	164.49
35	35.0	49.99	73.65	90.32	111.43	138.24	172.32	271.02

Note: The above growth factors multiplied by the first-year traffic estimates provide the total volume of traffic expected during the analysis period.

TABLE 6.5 Traffic Growth Factors

Example 6.5: Determination of Design ESAL A two-way roadway (as shown in Fig. 6.9) has two lanes each way with a two-way annual average daily truck traffic (AADTT) of 5,250 in 2017 and 5,460 in 2018. The EALF of the traffic can be assumed as listed in Table 6.6.

Volume of traffic	EALF
50% of Traffic (say, single axles)	0.25
30% of Traffic (say, tandem axles)	1.50
20% of Traffic (say, tridem axles)	2.38

Table 6.6 Traffic Data for Example 6.5

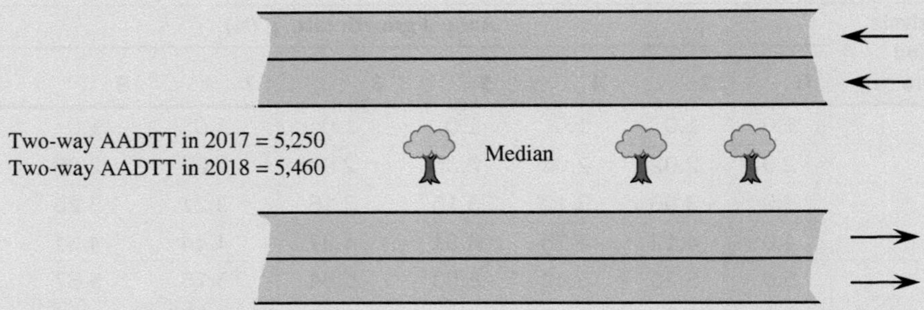

Figure 6.9 Planview of the pavement for Example 6.5.

The pavement is to be designed for a design life of 20 years starting from 2018. Determine the design ESAL in the driving lane.

Solution

Step 1. Count the initial two-way traffic of different types.

Given, two-way AADTT at the initial year = 5,460 (in 2018)

Step 2. Determine the design traffic based on growth factor for the design life.

AADTT in 2017 = 5,250
AADTT in 2018 = 5,460

Growth in 1 year = 5,460 − 5,250 = 210
Growth rate = (210/5,250) × 100 = 4.0%

From Table 6.5, traffic growth factor at mid-life (10 years) for the growth rate of 4.0% = 12.01.

Total design traffic = 5,460 × 12.01 = 65,575

Step 3. Determine EALF.

Given data are listed in Table 6.6.

Step 4. Convert traffic into design ESAL and sum up to calculate the total ESAL as shown in Table 6.7.

$$\text{Total ESAL} = \sum_{i=1}^{m} F_i n_i$$

Volume of traffic, n_i	EALF, F_i	ESAL, $F_i\,n_i$
50% of 65,575 = 32,788	0.25	8,197
30% of 65,575 = 19,672	1.50	29,508
20% of 65,575 = 10,915	2.38	25,978
Total ESAL $\left(\sum_{i=1}^{m} F_i n_i\right)$		67,213

Table 6.7 Traffic Data Analysis for Example 6.5

Step 5. Calculate the design ESAL in the design lane.

$D_D = 50\%$ in each direction (commonly assumed)
$D_L = 80$–100% from Table 6.4 for two-lane road. Take the average value, i.e., 90%.
Design lane ESAL $= D_D(D_L)(\text{Total ESAL}) = (50\%)(90\%)(67{,}213) = 30{,}246$

Answer *The design ESAL is 30,246 each way in the driving lane.*

6.4 Traffic Analysis for the AASHTOWare Design

6.4.1 Traffic Data Summary

The AASHTOWare pavement ME design software uses four design inputs for pavement design and analysis, including traffic, climate material properties, and section properties. Traffic is one of the most critical inputs required for pavement analysis; it represents the magnitude and frequency of the load applied to a pavement. Four types of traffic inputs are provided in the AASHTOWare pavement ME design software for structural design of pavement for any level (Level 1, Level 2, and Level 3) of analysis. These inputs are listed below:

Type 1: Traffic volume–base year information
- a) Two-way AADTT
- b) Number of lanes in the design direction
- c) Percent trucks in design direction
- d) Percent trucks in design lane
- e) Vehicle operational speed

Type 2: Traffic volume adjustment factors
- a) Monthly adjustment
- b) Vehicle class distribution
- c) Hourly truck distribution
- d) Traffic growth factors

Type 3: Axle load distribution factors

Type 4: General traffic inputs

a) Number of axles per truck

b) Axle configuration

c) Wheel base

To remind you, AASHTOWare pavement ME design considers only the two-way AADTT. Truck traffic means the vehicles from Class 4 to Class 13. The first three classes (motorbikes, passenger cars, and four-tire single-unit vehicles) are considered here as they produce negligible amount of damage to pavement materials and structures. However, the first three classes are very critical in designing the number of lanes, signaling times, and other highway infrastructures.

Hasan et al. (2016a) and Tarefder and Islam (2015) have shown that the nationally determined traffic data (Level 3) are not close to the measured traffic data in New Mexico. Studies conducted in other states revealed the same outputs (Haider et al., 2010; Ishak et al., 2010; Papagiannakis et al., 2006; Romanoschi et al., 2011; Tran and Hall 2007a, 2007b). In addition, it was also observed that the predictions of performances were dependent on the accurate determination of traffic classification and weight distribution. Therefore, it is always better to use the site-specific data.

6.4.2 Developing Traffic Data

Traffic input data to be used in the AASHTOWare pavement ME design software can be developed using the measured traffic data onsite or national data can be used depending on the level of analysis. Level 1 data for a pavement site can be determined by direct measurement of the traffic volume, axle load, etc. using WIM data, automatic vehicle classification (AVC), and so on. Level 2 data are determined from regional/statewide WIM data, AVC, and so on. Cluster analysis is the best way to develop the Level 2 data (Hasan et al., 2016b). Level 3 data are determined from national/statewide WIM data, AVC, and so on. This data can be determined by cluster analysis also.

6.5 Details of Traffic Inputs for the AASHTOWare Design

6.5.1 Type 1: Traffic Volume–Base Year Information

The following input parameters are considered site-specific and need to be obtained from the traffic or planning department.

Initial Two-Way Annual Average Daily Truck Traffic

Two-way annual average daily truck traffic data (the trucks are known as FHWA's Class 4 to 13 vehicles) are required for the opening year condition. These data are used as the base for future growth projection.

Number of Lanes in the Design Direction

The number of lanes in the design direction can be obtained based on the design site.

Percent Trucks in Design Direction

This value represents the percentage of trucks using the roadway in both directions in the design direction relative to all trucks. This value may be estimated from AVC data or manual vehicle count data. Unless a road has an unbalanced truck movement, the percentage of truck traffic in the design direction is 50%.

Percent Trucks in Design Lane

Percent trucks in design lane means the percentage of total truck traffic that runs through the design lane, typically the outside lane (driving lane) in a multilane highway (more than one lane in each travel direction). This is because most of the traffic runs through the driving lane. Usually, the percentage of truck in the design lane is determined by calculating the percentage of truck traffic in the design lane in one direction relative to all truck traffic. However, the definition used in the pavement AASHTOWare ME design is slightly different; it is defined by the primary truck class for the roadway. The primary truck class represents the truck class with the majority of applications using the roadway. In other words, the percentage of trucks in the design lane is estimated for each truck class, and the predominant truck class is used to estimate this value. The percent of trucks in the design lane may be estimated from AVC data or manual vehicle count data.

Vehicle Operational Speed

Vehicle operational speed can be obtained from the speed limit of the design site. The default operational speed value in pavement ME design is 60 mph. Truck speed has a definite impact on the predicted dynamic modulus of asphalt concrete and, thus, distresses. Lower speeds result in higher incremental damage values calculated by the AASHTOWare pavement ME design (more fatigue cracking and deeper ruts or faulting). The posted speed limit has been used in all calibration efforts. As such, it is suggested that the posted truck speed limit be used to evaluate trial designs, unless the pavement is located in special low-speed areas such as a steep upgrade and bus stop.

6.5.2 Type 2: Traffic Volume Adjustment Factors

Monthly Adjustment

The truck monthly adjustment factor (MAF) reflects truck travel patterns throughout the year. There are 10 categories of trucks (FHWA vehicle Class 4 to 13) that can generate 10 different temporal patterns over a 12-month period. Mathematically, the monthly adjustment factor for a given vehicle class and a given month is obtained by dividing the average monthly average daily truck traffic (MADTT) for the month by the summation of all the 12-month MADTTs and then multiplied by 12. There are total of 120 MAFs [10 vehicle classes × 12 months = 120 individual MAF]. The default MAF value is 1.0 in the AASHTOWare pavement ME design software. However, the MAF deviates a lot from the default value in actual pavements. For example, the MAFs for I-40 at milepost 141, near the city of Albuquerque, New Mexico, calculated in 2014 are listed in Table 6.8. It has been observed that the MAF varies from 0.05 to 6.66 with an average value of 1.02 and standard deviation of 0.78.

Vehicle Class Distribution

Truck class distribution (TCD) refers to AADTT distribution among the 10 vehicle types (Class 4 to Class 13). The percentage data are computed directly from the vehicle

MADTT	Class 4	Class 5	Class 6	Class 7	Class 8	Class 9	Class 10	Class 11	Class 12	Class 13
January	0.50	0.71	0.99	0.45	0.65	0.99	0.10	0.96	6.66	0.51
February	0.47	0.66	0.97	0.67	0.61	0.95	0.11	0.89	6.13	0.68
March	1.09	1.04	1.05	1.05	1.05	1.05	1.24	1.05	0.10	1.30
April	1.07	0.97	1.01	0.81	1.12	0.99	1.27	1.05	0.10	1.03
May	1.12	1.07	1.02	1.23	1.14	1.01	1.23	1.02	0.12	1.05
June	1.17	1.15	0.96	1.41	1.22	0.98	1.17	0.98	0.15	1.31
July	1.13	1.19	1.02	1.27	1.13	0.99	1.17	1.02	0.13	1.13
August	1.15	1.13	1.02	1.56	1.07	1.02	1.18	1.05	0.18	1.10
September	1.18	1.00	1.00	1.48	1.08	1.01	1.17	1.01	0.18	1.17
October	1.31	1.05	1.07	1.21	1.29	1.10	1.30	1.16	0.14	1.13
November	0.94	0.97	0.98	0.45	0.86	1.00	1.14	0.94	0.05	0.72
December	0.87	1.07	0.90	0.41	0.76	0.91	0.94	0.88	0.05	0.86
Total	12	12	12	12	12	12	12	12	12	12

TABLE 6.8 Monthly Adjustment Factor for a Section on Interstate 40

classification data. The default traffic class distribution data used in the AASHTOWare pavement ME design software are listed in Table 6.9.

The TCD in actual pavements may vary a lot. For example, TCDs calculated in two pavement sites near the city of Albuquerque in New Mexico in 2014 is shown in Table 6.10.

Hourly Truck Distribution (HTD)

Hourly truck distribution refers to the percentage of hourly AADTT among a 24-hour period starting at midnight. There are 24 HTDs in 24 hours of a day. The AASHTOWare pavement ME design default values are shown in Fig. 6.10. It is possible to estimate the hourly distribution factors from WIM data, AVC, or manual truck traffic counts. Average default values were determined from an analysis of the LTPP WIM data. Hourly distribution factors are only required for the analysis of rigid pavements, which keys hourly truck volume to temperature gradients through the PCC slab. The flexible pavement analysis bases all computations related to temperature on a monthly basis.

Traffic Growth Factors

Truck traffic growth is difficult to estimate accurately as there are many sites and socio-economic factors that are difficult to predict over 20 years. The traffic and/or planning departments within an agency may be consulted to estimate the increase in truck traffic over time. The AASHTOWare pavement ME design has the capability to use different growth rates for different truck classes, but assumes that the growth rate is independent over time; in other words, the rate of increase remains the same throughout the analysis period. Truck class–dependent growth rates have a significant effect on the predicted pavement output and can be calculated with as much information as possible about the commodities being transported within and through the location of the project.

Truck traffic classification group and description		Truck class distribution (%)									
		4	5	6	7	8	9	10	11	12	13
1	Major single-trailer truck route (type I)	1.3	8.5	2.8	0.3	7.6	74.0	1.2	3.4	0.6	0.3
2	Major single-trailer truck route (type II)	2.4	14.1	4.5	0.7	7.9	66.3	1.4	2.2	0.3	0.2
3	Major single-trailer truck route (type I)	0.9	11.6	3.6	0.2	6.7	62.0	4.8	2.6	1.4	6.2
4	Major single-trailer truck route (type III)	2.4	22.7	5.7	1.4	8.1	55.5	1.7	2.2	0.2	0.4
5	Major single and multitrailer truck route (type II)	0.9	14.2	3.5	0.6	6.9	54.0	5.0	2.7	1.2	11.0
6	Intermediate light and single-trailer truck route (type I)	2.8	31.0	7.3	0.8	9.3	44.8	2.3	1.0	0.4	0.3
7	Major mixed truck route (type I)	1.0	23.5	1.2	0.5	10.2	42.2	5.8	2.6	1.3	8.4
8	Major multitrailer truck route (type I)	1.7	19.3	4.6	0.9	6.7	44.8	6.0	2.6	1.6	11.8
9	Intermediate light and single-trailer truck route (type II)	3.3	34.0	11.7	1.6	9.9	36.2	1.0	1.8	0.2	0.3
10	Major mixed truck route (type II)	0.8	30.8	6.9	0.1	7.8	37.5	3.7	1.2	4.5	6.7
11	Major multitrailer truck route (type II)	1.8	24.6	7.6	0.5	5.0	31.3	9.8	0.8	3.3	15.3
12	Intermediate light and single-trailer truck route (type III)	3.9	40.8	11.7	1.5	12.2	5.0	2.7	0.6	0.3	1.3
13	Major mixed truck route (type III)	0.8	33.6	6.2	0.1	7.9	26.0	10.5	1.4	3.2	10.3
14	Major light truck route (type I)	2.9	56.9	10.4	3.7	9.2	15.3	0.6	0.3	0.4	0.3
15	Major light truck route (type II)	1.8	56.5	8.5	1.8	6.2	14.1	5.4	0.0	0.0	5.7
16	Major light and multitrailer truck route	1.3	48.4	10.8	1.9	6.7	13.4	4.3	0.5	0.1	12.6
17	Major bus route	36.2	14.6	13.4	0.5	14.6	17.8	0.5	0.8	0.1	1.5

Source: AASHTO (2015). *Mechanistic-Empirical Pavement Design Guide: A Manual of Practice. Washington, DC: American Association of State Highway and Transportation Officials. Table 8-3. Used with permission.*

Table 6.9 Default Truck Classifications in the AASHTOWare Pavement ME Design Software

Class	Distribution (%)	
	I-40	I-25
Class 4	1.9	3.6
Class 5	14.7	57.2
Class 6	1.3	2.7
Class 7	0.1	0
Class 8	4.7	8.9
Class 9	72.1	24.4
Class 10	2.4	1.1
Class 11	2.2	2.0
Class 12	0.5	0
Class 13	0.1	0.1
Total	100	100

TABLE 6.10 Truck Classification Calculated on Two Interstate Highways in New Mexico in 2014

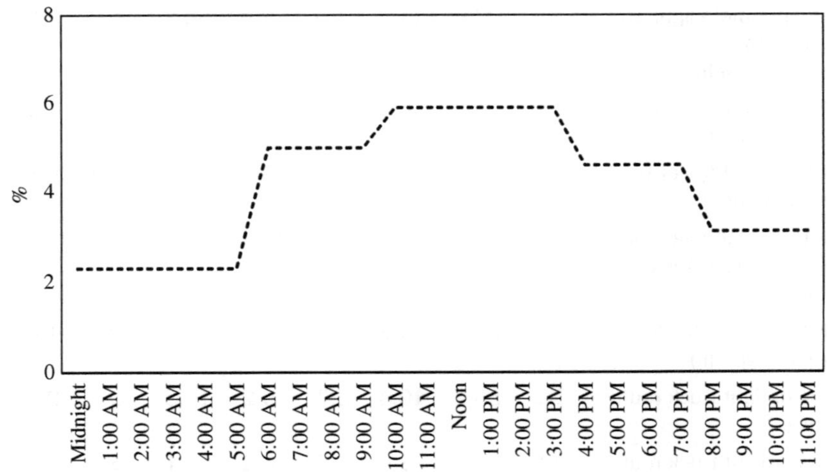

FIGURE 6.10 Default hourly distribution.

6.5.3 Type 3: Axle Load Distribution Factors

FHWA Class 4 to Class 13 vehicles can have a variety of axle configurations, including single axle, tandem axle, tridem axle, and quad axle. For a given vehicle class and axle configuration, axle weight varies depending on vehicle load. Axle load distribution factor (ALDF) is to capture that information in terms of distributions of vehicles based on axle weight under a given vehicle class and axle configuration for a given month.

This is one of the most demanding data sets. Mathematically, the ALDF is the percentage of a given axle load among all axle loads under a given vehicle axle configuration.

1. *Single axle:* There are 39 axle weight groups for single-axle configuration vehicles. The axle weight group ranges from 3,000 lb to 41,000 lb with increments of 1,000 lb.
2. *Tandem axle:* For tandem-axle vehicles, the axle weight group starts at 6,000 lb and ends at 82,000 lb with increments of 2,000 lb.
3. *Tridem axle:* For tridem-axle vehicles, the axle weight group starts at 12,000 lb and ends at 102,000 lb with increments of 3,000 lb.
4. *Quad axle:* Similar to tridem-axle vehicles, for quad-axle vehicles the axle weight group starts at 12,000 lb and ends at 102,000 lb with increments of 3,000 lb.

A part of axle load spectra calculated in two pavement sites on I-40 and I-25 near the city of Albuquerque in New Mexico in January 2014 is shown in Tables 6.11 and 6.12.

6.5.4 Type 4: General Traffic Inputs

Number of Axles per Truck

The number of axles per vehicle class for a given axle configuration is an annual average number of axles per vehicle category (per vehicle class and vehicle axle configuration). Tables 6.13 and 6.14 list the measured number of axles per truck calculated in two pavement sites near the city of Albuquerque in New Mexico in 2014.

Axle Configuration

Several types of input are required to specify the axle configuration, such as axle spacing, axle width, mean wheel location, traffic wander, and lane width. Axle spacing is the distance between two consecutive tandem, tridem, and quad axles. The distance between the two outside edges of an axle is defined as axle width. The mean wheel location is the distance of the centerline of the wheel from the outer edge of the lane. It is measured to be 27 in. (675 mm) on I-40, which is not close to the AASHTOWare pavement ME design software's default value of 36 in. (3 ft) (Islam et al., 2014). However, the mean wheel location does not affect the structural response or performance of pavement. The wheel wander value of 8 in. is suggested for narrow lanes unless the user has measured the value. For wider lanes, the wheel wander is recommended to be 12 in. The spacing of the axles is recorded in the WIM database. These values have been found to be relatively constant for the standard truck classes. The values used in all calibration efforts are listed below and suggested for use, unless the predominant truck class has a different axle configuration.

- Tandem-axle spacing, 51.6 in.
- Tridem-axle spacing, 49.2 in.
- Quad-axle spacing, 49.2 in.

Wheel Base

The distance between the steering and the first axle of a tractor or a heavy single unit is used to classify the truck as short, medium, or long vehicle. The recommended distances are 12, 15, and 18 ft for short, medium, and long axle spacing, respectively.

Class	3,000 lb	4,000 lb	5,000 lb	6,000 lb	7,000 lb	8,000 lb	9,000 lb	1,0000 lb	1,1000 lb	1,2000 lb	1,3000 lb	1,4000 lb	1,5000 lb	1,6000 lb	1,7000 lb	1,8000 lb
4	0.085	0	0.021	0.191	0.234	10.08	12.19	13.13	8.701	13.15	8.935	10.38	6.611	3.582	4.265	2.58
5	16.01	40.47	21.06	7.804	2.887	3.102	1.997	1.612	0.931	1.02	0.656	0.623	0.489	0.327	0.274	0.21
6	0.279	0.685	1.294	2.207	1.877	2.41	4.415	12.38	16.92	22.48	12.73	11.03	6.48	2.41	1.522	0.837
7	0	4.347	10.86	2.173	2.173	0	2.173	6.521	6.521	13.04	10.86	8.695	6.521	0	2.173	0
8	0.332	17.64	27.08	14.6	6.768	6.729	6.113	6.65	3.468	2.973	1.666	1.582	1.362	0.906	1.092	0.596
9	0.133	0.503	0.667	0.926	0.915	1.222	1.327	4.45	16.65	47.74	16.1	2.966	1.353	1.254	1.942	1.347
10	0.43	0.43	0	0.573	1.004	2.152	3.873	10.18	22.23	33.14	12.48	7.604	2.152	1.578	1.147	0.286
11	0	0.302	1.247	3.082	2.656	3.712	6.232	20.07	19.9	7.585	4.855	6.727	6.578	4.725	4.935	3.051
12	0	0.051	0.382	1.558	2.204	2.932	3.895	9.503	22.6	33.87	10.85	6.805	2.829	0.999	0.83	0.374
13	0	0	0.84	3.361	10.92	23.52	3.361	6.042	6.722	10.08	3.361	8.403	8.403	6.722	3.361	0.84

TABLE 6.11 Axle Load Spectra for Single Axle on I-40 in January 2014 (Small Part)

Class	3,000 lb	4,000 lb	5,000 lb	6,000 lb	7,000 lb	8,000 lb	9,000 lb	1,0000 lb	1,1000 lb	1,2000 lb	1,3000 lb	1,4000 lb	1,5000 lb	1,6000 lb	1,7000 lb	1,8000 lb
4	0	0	0	0.056	0.126	16.16	12.8	13.02	9.38	11.38	8.365	8.153	6.474	3.667	3.343	2.144
5	6.809	36.25	26.33	16.26	3.399	3.306	2.412	2.147	1.15	1.084	0.667	0.713	0.543	0.359	0.372	0.25
6	0.092	0.092	0.169	0.292	0.799	2.829	6.643	14.25	13.6	18.59	10.97	10.97	8.688	4.305	3.813	1.722
7	0	0	0	3.076	0	1.538	1.538	3.076	0	3.076	3.076	9.23	9.23	3.076	4.615	3.076
8	0.068	6.177	20.15	16.88	6.572	7.173	6.54	10.6	7.23	6.33	3.04	2.31	1.516	0.947	1.165	0.879
9	0.518	3.251	2.536	1.849	0.913	1.509	2.133	8.542	14.42	26.22	16.04	8.846	2.096	0.977	1.705	1.994
10	0	0	0	0.246	0	0.246	2.091	8.856	17.46	26.07	14.02	13.03	8.856	3.936	2.46	1.23
11	0	3.482	4.334	3.796	2.651	4.225	6.1	13.54	8.799	8.212	3.019	3.244	3.366	3.107	4.266	6.132
12	0	0.152	0.305	2.772	6.061	8.621	7.07	12.02	14.95	19.91	11.36	7.527	4.501	2.187	1.373	0.737
13	0	1.03	3.092	1.03	1.03	3.092	4.123	9.278	9.278	9.278	10.3	9.278	14.43	6.185	9.278	4.123

TABLE 6.12 Axle Load Spectra for Single Axle on I-25 in January 2014 (Small Part)

Class	Single axle	Tandem axle	Tridem axle	Quad axle
Class 4	1.70	0.30	0	0
Class 5	2.00	0	0	0
Class 6	1.00	1.00	0	0
Class 7	0.30	1.50	0.30	0
Class 8	2.20	0.80	0	0
Class 9	1.10	1.90	0	0
Class 10	1.00	1.00	1.00	0
Class 11	3.00	0.90	0.10	0
Class 12	1.90	1.60	0.30	0
Class 13	0.80	1.40	0.10	0.10

Table 6.13 Number of Axle per Truck Measured on I-40

Class	Single axle	Tandem axle	Tridem axle	Quad axle
Class 4	1.72	0.28	0	0
Class 5	2.00	0	0	0
Class 6	1.00	1.00	0	0
Class 7	0.54	0.93	0.54	0
Class 8	2.14	0.86	0	0
Class 9	1.34	1.82	0.01	0
Class 10	1.01	1.01	0.99	0
Class 11	3.83	0.54	0.03	0
Class 12	2.35	1.81	0.01	0
Class 13	0.62	1.05	0.21	0.10

Table 6.14 Number of Axle per Truck Measured on I-25

The default percentages of short, medium, and long vehicles are 17%, 22%, and 61%, respectively (AASHTO, 2015).

Readers are reminded again that continuous researches and improvements are going on in pavement engineering. The AASHTOWare pavement ME design software's default value may change over time.

Example 6.6: Truck Class Distribution The TCD of a pavement site is provided in Table 6.15.

Determine the percentage of Class 13 truck on that pavement site.

Solution

Sum of trucks' percentage except Class 13 = 8.1% + 12.1% + 3.3% + 2.1% + 4.8% + 60.7% + 2.4% + 2.1% + 0.5% = 96.1%

Percentage of Class 13 = 100% − 96.1% = 3.9%

Answer *The percentage of Class 13 trucks on the site is 3.9%.*

Class	Percentage
Class 4	8.1
Class 5	12.1
Class 6	3.3
Class 7	2.1
Class 8	4.8
Class 9	60.7
Class 10	2.4
Class 11	2.1
Class 12	0.5

Table 6.15 Truck Class Distribution for Example 6.6

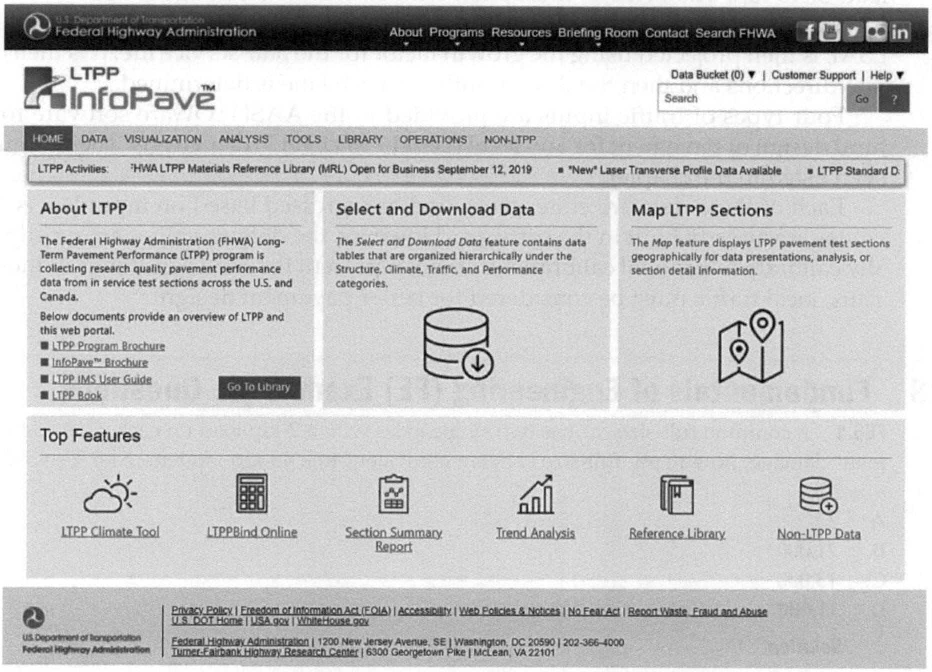

Figure 6.11 LTPP InfoPave homepage (accessed on October 6, 2019).

6.6 Traffic Data Source

A very good source of traffic data for different sites in any state can be the long-term pavement performance (LTPP) database (http://www.infopave.com), as shown in Fig. 6.11. Most of the inputs required for the AASHTOWare pavement ME design software are available at this website. However, there are still no axle load spectra available in the LTPP database that can be used for the software.

6.7 Summary

Accurate determination of traffic loading and its distribution throughout the lanes is important while analyzing and designing pavement as traffic is considered the only load for both flexible and rigid pavement designs. The load on a wheel is transferred through the tire to the pavement. The contact between the tire and the pavement can have three shapes: circular, composite, and rectangular. Circular assumption is very conservative, and the other two assumptions, composite and rectangular, yield the equal pressure. Axles are of five types: single axle, single tire; single axle, dual tires; tandem axle; tridem axle; and quad axle.

FHWA classifies vehicles into 13 classes, Class 1 to Class 13. Class 1 comprises motorbikes and Class 13 refers to the seven or more axle multitrailer trucks. There are other vehicles in between these classes. AASHTO 1993 does not use this classification of FHWA vehicles, but rather uses the ESAL system to count the total number of vehicles at the site concerned. Different types of axles and their loads are converted into a standard axle. The standard axle is the single axle with dual tires on each end with the axle load of 18 kip. The EALF is multiplied with each axle to convert it into the ESAL. The EALF is dependent on the axle type, axle load, SN, and the terminal serviceability. The ESAL is then projected using the growth factor for the half-service life. It is then divided into directions and then the design traffic on each lane is determined.

Four types of traffic inputs are provided in the AASHTOWare software for structural design of pavement for any level (Level 1 to Level 3) of analysis. These inputs have been listed in the chapter.

Each of these data are either measured or regressed based on input levels. Default values are already built in the software. However, the default values are mostly nationally calibrated. National calibrations do not represent the local conditions in most cases. Thus, local traffic must be considered for better pavement design.

6.8 Fundamentals of Engineering (FE) Exam-Style Questions

FE6.1 A common full-size car has two single axles with a 2-kip load on each. Considering pavement damage, how many full-size cars are equivalent to a 48-kip tandem axle? Use SN = 5 and $p_t = 2.5$.

A. 32
B. 21,000
C. 13,000
D. 11,600

Solution D

ESAL of a 2-kip single axle = 0.00018
One full-size car = 0.00018 × 2 = 0.00036
ESAL of a 48-kip tandem axle = 4.17
Number of cars = 4.17/0.00036 = 11,583 = 11,600 (approx.)

FE6.2 In a construction site, a construction vehicle with a 24-kip single axle and a 44-kip tandem axle travels 30 times per day from the plant to the job site. If the construction period is 6 months (180 days) with no rest days, the total ESAL of that vehicle is most nearly:

A. 108,000
B. 54,000
C. 33,000
D. 5.3×10^6

Solution C

ESAL of the vehicle = 3.03 + 3 = 6.03 per travel
No of travel = 6 months × 30 days/month × 30 times a day = 5,400 travels
Total ESAL = 5,400 travels × 6.03 ESAL per travel = 32,562 ≈ 33,000

FE6.3 Which of the following is the standard axle used while calculating the ESAL?

A. Single axle, single tire with the axle load of 18 kip
B. Single axle, dual tires with the axle load of 18 kip
C. Dual axle, single tire with the axle load of 18 kip
D. Tandem axle, single tire with the axle load of 18 kip

Solution B

6.9 Practice Problems

6.1 The total load on a single tire is 3 kip. The tire width is 6 in. If the tire imprint is considered circular, calculate the exerted pressure on the pavement surface.

6.2 The total load on a single tire is 4 kip. The tire width is 6 in. If the tire imprint is considered rectangular, calculate the exerted pressure on the pavement surface.

6.3 The total load on a tandem axle is 32 kip. Consider a p_t of 2.2 and SN of 6. Calculate the EALF for the vehicle considering flexible pavement.

6.4 The total load on a tridem axle is 44 kip. Consider a p_t of 2.5 and SN of 4. Calculate the EALF for the vehicle considering flexible pavement.

6.5 The AADTT of a two-way (two-lane each way) highway is 10,000. The directional distribution may vary between 45% and 55% depending on the season. The lane distribution may vary between 60% and 75%. The driving lane always carries more loads compared to the passing lane. The growth factor for the pavement is 3.5%. Based on the above information, calculate the design AADTT for the driving lane if the pavement service life is 20 years.

6.6 The monthly adjustment factor (MAF) of class vehicle for a pavement site is provided in Table P6.6.

Month	Class 4
February	0.37
March	1.19
April	1.17
May	1.02
June	1.07
July	1.03
August	1.25
September	1.28
October	1.30
November	0.90
December	0.82

Table P6.6 MAF Data for Prob. 6.6

Calculate the MAF for the month of January.

6.7 Why the traffic input data in the AASHTOWare are more robust compared to that in the AASHTO 1993 pavement design procedure?

6.8 Define the short, medium, and long vehicle as per AASHTOWare pavement design procedure. What are the default volumes of short, medium, and long vehicles in the AASHTOWare pavement design procedure?

6.9 What is the default monthly adjustment factor for traffic in the AASHTOWare pavement design procedure?

6.10 Why is standard deviation of wheel wander important in pavement design?

CHAPTER 7

Flexible Pavement Design by AASHTO 1993

7.1 Background

This chapter discusses the AASHTO 1993 method for flexible pavement design. Chapter 1 discusses the history of this design method. The method is developed empirically based on the road test in Illinois during 1956–1960. Several equations are primarily developed based on regression analysis of the test findings. Flexible pavement design using these equations is discussed in this chapter.

7.2 AASHTO 1993 Design Equation

The basic flexible pavement design equation based on AASHTO 1993 method is as follows:

$$\log W_{18} = -Z_R S_o + 9.36\log(\text{SN}+1) - 0.2 + \frac{\log\left[\dfrac{4.2-p_t}{4.2-1.5}\right]}{0.4+\dfrac{1094}{(\text{SN}+1)^{5.19}}} + 2.32\log M_R - 8.07 \tag{7.1}$$

where W_{18} = Number of 18-kip single-axle load repetitions in time t (or the total ESAL)

SN = Structural number representing the composite stiffness of pavement systems considering all layers

M_R = Effective roadbed soil resilient modulus, psi

p_t = Terminal serviceability (commonly used as 2.5 for major highways and 2.0 for others; this value can be considered less than 2 for less important highways). Very often, $(4.2 - p_t)$ is written as ΔPSI, called decrease in present serviceability index (PSI).

Z_R = Normal deviate for a given reliability, R. Consider the positive sign of Z_R. In other words, the term $Z_R S_o$ is to be negative in Eq. (7.1).

S_o = Standard deviation that accounts for both chance variations in the traffic prediction and performance prediction for a given ESAL. The AASHO road test obtained S_o of 0.35 for flexible pavement.

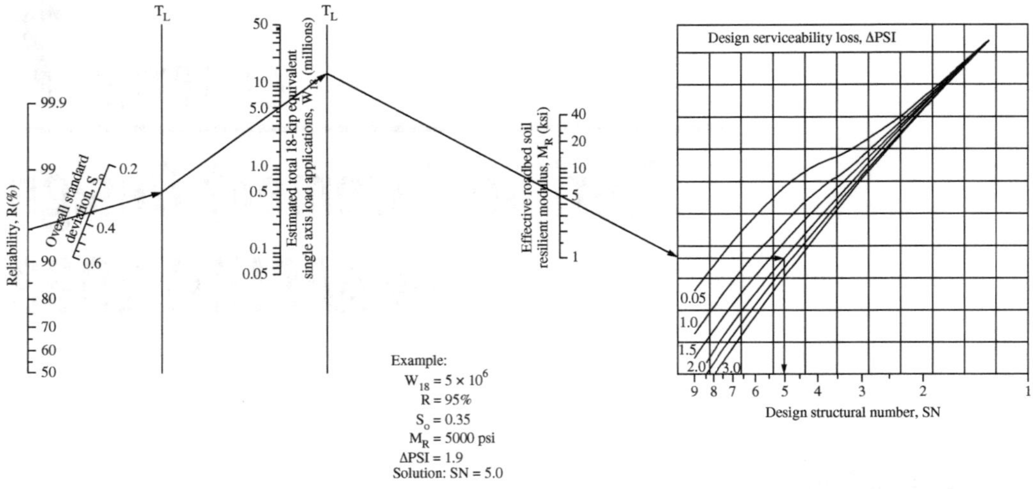

FIGURE 7.1 AASHTO 1993 design nomograph. *(From AASHTO (1993). AASHTO Guide for Design of Pavement Structures. AASHTO Guide for Design of Pavement Structures.. Washington, DC: American Association of State Highway and Transportation Officials. Figure 3-1. Used with permission.)*

The left side of Eq. (7.1) is the demanded pavement strength by the traffic loading. The right side represents the materials capacity of the trial section considering the roadbed soil's modulus and desired serviceability. All the parameters except the SN are the input in Eq. (7.1) and must be measured or estimated before starting the design process. The use of Eq. (7.1) requires the selection of reliability levels in terms of Z_R and S_o, initial and terminal PSI or change in PSI, and determination of resilient modulus, M_R. Then the required SN is calculated using Eq. (7.1) or using a nomograph shown in Fig. 7.1. The nomograph is read from left to right.

The following sections explain the various parameters of Eq. (7.1) in details.

7.3 Load Repetitions (W_{18})

The number of 18-kip single-axle load repetitions W_{18} in time t (or total ESAL) is calculated based on the traffic count, applying the distribution, growth factor, etc., which are discussed in Chapter 6. The basic steps of calculation of W_{18} in time t (or total ESAL) can be summarized as follows:

Step 1. Count the initial two-way traffic of different types.

Step 2. Determine the design traffic based on growth factor for the design life.

Step 3. Determine equivalent axle load factor (EALF) based on structural number (SN) and terminal serviceability (p_t).

Step 4. Convert traffic into design ESAL and sum up to calculate the total ESAL.

Step 5. Apply the directional distribution factor and the lane distribution factor to find out the design ESAL in the design lane.

7.4 Structural Number

7.4.1 Definition

Structural number (SN) is a number that represents the strength of the pavement to produce tolerable vertical stress on the top of the subgrade soil. In other words, the SN is a measure of the relative ability of the material to function as a structural component of a pavement. The SN also includes the performance of the drainage quality of the soil/aggregate layers. The SN requirement is determined based on the expected traffic volume, available material strength, desired service condition, and reliability level. It can be calculated using the equation:

$$SN = a_1 D_1 + a_2 D_2 m_2 + a_3 D_3 m_3 + \cdots \tag{7.2}$$

where a_i = Layer structural coefficient (depends on layer modulus)
D_i = Layer depth, in.
m_i = Layer drainage coefficient (increases with increase in drainage quality)

Each layer is assigned a layer coefficient to convert thicknesses of different layers into a single number, SN. The layer coefficient is primarily dependent on the elastic modulus of the materials. However, other factors such as layer thickness, underlying support, and position in the pavement structure also affect the layer coefficients. Commonly used values of layer coefficients are listed in Table 7.1.

The SN value also includes the drainage coefficients which range from 0.4 (slow-draining, saturated layers) to 1.4 (fast-draining layers that do not get saturated). The details of the layer coefficients and drainage coefficients are discussed here.

Materials	a_i
HMA	0.44
Crushed stone base	0.14
Stabilized base	0.30–0.40
Crushed stone subbase	0.11

Table 7.1 Layer Coefficients

7.4.2 Layer Coefficient of Asphalt Layer

For asphalt layer, the layer coefficient is determined using its resilient modulus or elastic modulus measured at 68°F. Figure 7.2 can be used to determine the layer coefficient of asphalt material. Caution is used to determine the layer coefficients for modulus value above 450,000 psi as shown by the dashed portion of the curve. This is because stiffer asphalt is more prone to cracking due to low temperature and fatigue loading, although good for permanent deformation.

7.4.3 Layer Coefficient of Base Layer

Layer coefficient of base layer (a_2) can be determined in different ways. First, Fig. 7.3 can be used if any of the four laboratory parameters [modulus, Texas Triaxial, California bearing ratio (CBR), or *R*-value] is known.

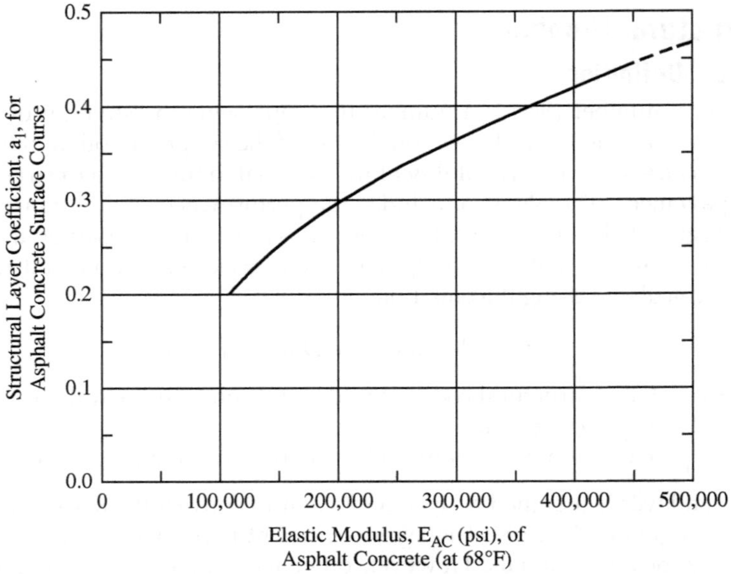

FIGURE 7.2 Layer coefficient of asphalt material. *(From AASHTO (1993). AASHTO Guide for Design of Pavement Structures. Washington, DC: American Association of State Highway and Transportation Officials. Figure 2-5. Used with permission.)*

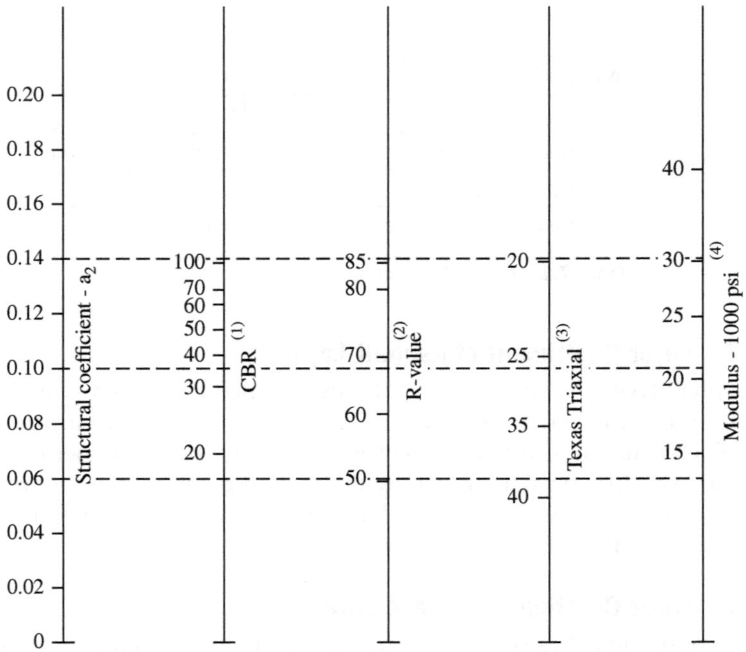

FIGURE 7.3 Layer coefficient of base material. *(From AASHTO (1993). AASHTO Guide for Design of Pavement Structures Washington, DC: American Association of State Highway and Transportation Officials. Figure 2-6. Used with permission.)*

Asphalt concrete thickness (in.)	Roadbed soil resilient modulus (psi)		
	3,000	7,500	15,000
Less than 2	20	25	30
2–4	10	15	20
4–6	5	10	15
Greater than 6	5	5	5

Source: From AASHTO (1993). AASHTO Guide for Design of Pavement Structures. Washington, DC: American Association of State Highway and Transportation Officials. Page II-20. Used with permission.

TABLE 7.2 Stress State for Base Layer

Moisture condition	K_1	K_2
Dry	6,000–10,000	0.5–0.7
Damp	4,000–6,000	0.5–0.7
Wet	2,000–4,000	0.5–0.7

Source: From AASHTO (1993). AASHTO Guide for Design of Pavement Structures. Washington, DC: American Association of State Highway and Transportation Officials. Page II- 20. Used with permission.

TABLE 7.3 Typical K_1 and K_2 Values for Base Layer

Layer coefficient of base layer (a_2) can also be determined using the equation:

$$a_2 = 0.249(\log E_2) - 0.977 \tag{7.3}$$

where E_2 = resilient modulus of base layer and is expressed as:

$$E_2 = K_1\theta^{K_2} \tag{7.4}$$

where θ = Stress state = sum of principal stresses = $\sigma_1 + \sigma_2 + \sigma_3$ (psi)
K_1 and K_2 = Regression constants, which are the functions of material type

Typical values of stress state (θ) for base layer are listed in Table 7.2.

Typical K_1 and K_2 values of base materials are listed in Table 7.3.

Example 7.1: Layer Coefficient of Base Layer In an asphalt pavement, the thickness of asphalt layer is 5.0 in. The roadbed soil resilient modulus is 7,500 psi. If the base layer is dry, determine the layer coefficient of the base layer.

Solution

Layer coefficient of base layer (a_2) can be determined using the following equation:

$$a_2 = 0.249(\log E_2) - 0.977$$

where $E_2 = K_1\theta^{K_2}$

From Table 7.2, for asphalt thickness of 5 in. and M_R of 7,500, θ = 10 psi.

From Table 7.3, for dry base, K_1 = 8,000 and K_2 = 0.6 (the average numbers).

Then, $E_2 = K_1\theta^{K_2} = 8{,}000(10 \text{ psi})^{0.6} = 31{,}849$ psi

$$a_2 = 0.249(\log E_2) - 0.977$$
$$= 0.249(\log 31{,}849) - 0.977 = 0.14$$

Answer *The layer coefficient of the base layer is 0.14.*

7.4.4 Layer Coefficient of Subbase Layer

Coefficient of subbase layer (a_3) can be determined using Fig. 7.4 if its modulus, Texas triaxial, CBR, or *R*-value is known.

Layer coefficient of subbase layer (a_3) can be determined using the equation:

$$a_3 = 0.227(\log E_3) - 0.839 \quad (7.5)$$

The resilient modulus (E_3) of subbase can be determined in the laboratory or the AASHTO-recommended value can be used. The E_3 is expressed as:

$$E_3 = K_1\theta^{K_2} \quad (7.6)$$

where θ = stress state = sum of principal stresses = $\sigma_1 + \sigma_2 + \sigma_3$ (psi)

K_1 and K_2 = Regression constants, which are the functions of material type

Typical values of stress state (θ) for subbase layer are listed in Table 7.4.

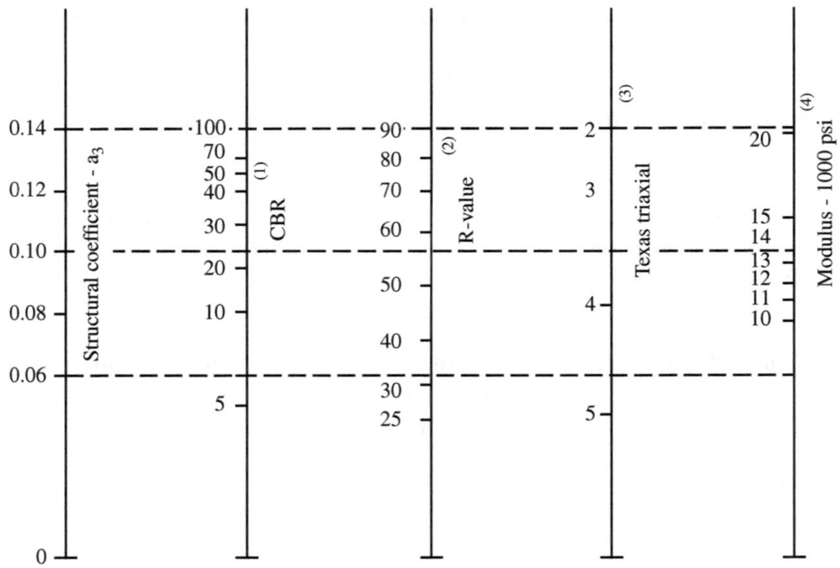

FIGURE 7.4 Layer coefficient of subbase material. (*From AASHTO (1993). AASHTO Guide for Design of Pavement Structures. Washington, DC: American Association of State Highway and Transportation Officials. Figure 2-7. Used with permission.*)

Asphalt concrete thickness (in.)	Stress state (psi)
Less than 2	10.0
2–4	7.5
Greater than 4	5.0

Source: From AASHTO (1993). AASHTO Guide for Design of Pavement Structures. Washington, DC: American Association of State Highway and Transportation Officials. Page II- 22. Used with permission.

TABLE 7.4 Stress State for Subbase Layer

Moisture condition	K_1	K_2
Dry	6,000–8,000	0.4–0.6
Damp	4,000–6,000	0.4–0.6
Wet	1,500–4,000	0.4–0.6

Source: From AASHTO (1993). AASHTO Guide for Design of Pavement Structures. Washington, DC: American Association of State Highway and Transportation Officials. Page II–20. Used with permission.

TABLE 7.5 Typical K_1 and K_2 Values for Subbase Layer

Typical K_1 and K_2 values of subbase materials are listed in Table 7.5.

Example 7.2: Layer Coefficient of Subbase Layer The K_1 and K_2 of a subbase layer are 5,000 and 0.5, respectively. The expected stress state at the subbase is 5 psi. Determine the layer coefficient of the subbase layer.

Solution

Known conditions:

$$a_3 = 0.227(\log E_3) - 0.839$$
$$E_3 = K_1\theta^{K_2}$$

Now, $E_3 = 5000(5\text{ psi})^{0.5} = 11{,}180\text{ psi}$

Then, $a_3 = 0.227(\log 11{,}180) - 0.839$

Therefore, $a_3 = 0.08$

Answer *The layer coefficient of the subbase layer 0.08.*

7.4.5 Drainage Coefficients

The recommended drainage coefficients of base and subbase can be determined using Table 7.6. It lists the recommended drainage coefficients based on the quality of drainage and the percent of time during the year the pavement structure would normally

Quality of drainage	Water removed within	Percent of time pavement structure is exposed to moisture levels approaching saturation			
		Less than 1%	1–5%	5–25%	Greater than 25%
Excellent	2 hours	1.40–1.35	1.35–1.30	1.30–1.20	1.20
Good	1 day	1.35–1.25	1.25–1.15	1.15–1.00	1.00
Fair	1 week	1.25–1.15	1.15–1.05	1.00–0.80	0.80
Poor	1 month	1.15–1.05	1.05–0.80	0.80–0.60	0.60
Very poor	(water will not drain)	1.05–0.95	0.95–0.75	0.75–0.40	0.40

Source: From AASHTO (1993). AASHTO Guide for Design of Pavement Structures. Washington, DC: American Association of State Highway and Transportation Officials. Table 2-4. Used with permission.

TABLE 7.6 Recommended Drainage Coefficients of Base and Subbase

be exposed to moisture levels approaching saturation, which depends on the annual precipitation and prevailing drainage conditions. AASHTO (1993) does not explain the drainage performance of different measures. It is the responsibility of the design engineer to find the quality of drainage under an adopted condition.

7.5 Effective Roadbed Soil Resilient Modulus

The soil resilient or elastic modulus (M_R) is moisture dependent; the moisture of subgrade soil changes month to month due to rainfall variation in ground water table. The M_R of soil thus changes month to month and the damage in pavement also changes month to month. However, only a single M_R is used in the design equation. This is why a single M_R is established and termed as "effective roadbed soil resilient modulus." It is the combined effect of all the seasonal modulus values. The process can be described as follows:

Step 1. A year is divided into a number of periods (say, 12 months or four seasons) such that moisture condition in each period is uniform.

Step 2. The roadbed soil modulus is determined using the soil-moisture condition, or by nondestructive testing, or by laboratory resilient modulus testing.

Step 3. The relative damage (u_f) during each month is calculated using Eq. (7.7), where M_R is in psi:

$$u_f = 1.18 \times 10^8 (M_R)^{-2.32} \tag{7.7}$$

Step 4. The relative damage is summed up for the entire year as $\sum_{i=1}^{n} u_{f,i}$.

Step 5. The average relative damage is then determined as the sum of the relative damage over the number of period (n), i.e., $\overline{u_f} = \frac{\sum_{i=1}^{n} u_{f,i}}{n}$.

Step 6. The M_R can be determined using the equation:

$$\overline{u_f} = 1.18 \times 10^8 (M_R)^{-2.32} \tag{7.8}$$

Now let us see Example 7.3 to clarify this process.

Example 7.3: Effective Roadbed Soil Modulus The roadbed soil modulus of a pavement site is listed in Table 7.7. Determine the effective roadbed soil modulus of the pavement.

Period	Soil Modulus, psi
January	24,500
February	17,650
March	12,170
April	8,430
May	5,230
June	3,570
July	4,220
August	6,650
September	9,670
October	11,330
November	18,270
December	21,300

Table 7.7 Roadbed Soil Modulus of a Pavement for Example 7.3

Solution

Steps 1 and 2 details are provided.

Step 3. Calculate the relative damage during each month using the following equation:

$$u_f = 1.18 \times 10^8 (M_R)^{-2.32}$$

$$u_{f(\text{January})} = 1.18 \times 10^8 (24{,}500 \text{ psi})^{-2.32} = 0.01$$

$$u_{f(\text{February})} = 1.18 \times 10^8 (17{,}650 \text{ psi})^{-2.32} = 0.02$$

$$u_{f(\text{March})} = 1.18 \times 10^8 (12{,}170 \text{ psi})^{-2.32} = 0.04$$

$$u_{f(\text{April})} = 1.18 \times 10^8 (8{,}430 \text{ psi})^{-2.32} = 0.09$$

$$u_{f(\text{May})} = 1.18 \times 10^8 (5{,}230 \text{ psi})^{-2.32} = 0.28$$

$$u_{f(\text{June})} = 1.18 \times 10^8 (3{,}570 \text{ psi})^{-2.32} = 0.68$$

$$u_{f(\text{July})} = 1.18 \times 10^8 (4{,}220\ \text{psi})^{-2.32} = 0.46$$

$$u_{f(\text{August})} = 1.18 \times 10^8 (6{,}650\ \text{psi})^{-2.32} = 0.16$$

$$u_{f(\text{September})} = 1.18 \times 10^8 (9{,}670\ \text{psi})^{-2.32} = 0.07$$

$$u_{f(\text{October})} = 1.18 \times 10^8 (11{,}330\ \text{psi})^{-2.32} = 0.05$$

$$u_{f(\text{November})} = 1.18 \times 10^8 (18{,}270\ \text{psi})^{-2.32} = 0.02$$

$$u_{f(\text{December})} = 1.18 \times 10^8 (21{,}300\ \text{psi})^{-2.32} = 0.01$$

Step 4. The relative damage is summed up as $\sum u_f = 0.01 + 0.02 + \ldots.. = 1.87$.

Step 5. The average relative damage is then determined as $\overline{u_f} = \frac{\sum u_f}{n} = \frac{3.80}{12} = 0.16$.
$n = 12$ (12 months)

Step 6. The roadbed soil modulus M_R can be determined using the equation:

$$\overline{u_f} = 1.18 \times 10^8 (M_R)^{-2.32}$$
$$0.16 = 1.18 \times 10^8 (M_R)^{-2.32}$$
$$M_R = 6{,}642\ \text{psi}$$

Answer *The effective roadbed soil modulus of the pavement is 6,640 psi (as the data is given in nearest 10).*

7.6 Terminal Serviceability

The serviceability index or the present serviceability index (PSI) is based on the original AASHO Road Test_Present Serviceability Rating (PSR). Basically, it is a ride quality rating that required a panel of observers to actually ride in an automobile over the pavement in question. The scale varies from 0 to 5 (0 being the no drivable condition and 5 being the most comfortable condition). The terminal (or failure) serviceability can be determined using the rating conditions listed below:

- If 12% of the people state unacceptable, the terminal serviceability is 3.0.
- If 55% of the people state unacceptable, the terminal serviceability is 2.5.
- If 85% of the people state unacceptable, the terminal serviceability is 2.0.

PSI is similar to PSR but it is based on measured physical roughness. The scale of 0 to 5 is used with 5 being very good and 0 being very poor. A new asphalt pavement commonly has an initial PSI value of 4.2. It is common to use the terminal PSI value of 2.5 for major highways and 2.0 for others; this value can be considered less than 2 for less important roads.

Functional classification	Recommended levels of reliability	
	Urban	Rural
Interstate and other freeways	85–99.9	80–99.9
Principal arterials	80–99	75–95
Collectors	80–95	75–95
Local	50–80	50–80

Source: From AASHTO (1993). AASHTO Guide for Design of Pavement Structures. Washington, DC: American Association of State Highway and Transportation Officials. Page II–9. Used with permission.

TABLE 7.8 Recommended Level of Reliability

7.7 Reliability

Reliability means incorporating some degree of certainty into the design process to ensure that the various design alternatives will last the analysis period. The AASHTO 1993–recommended levels of reliability are listed in Table 7.8.

The standard normal deviates (Z_R) for various levels of reliability are listed in Table 7.9.

Reliability, R (%)	Standard normal deviates (Z_R)
50	0
60	0.253
70	0.524
80	0.841
90	1.282
95	1.645
98	2.054
99	2.327
99.90	3.090
99.99	3.750

TABLE 7.9 Standard Normal Deviate for Different Reliability Values

7.8 Selection of Layers' Thicknesses

The design equation discussed earlier in this chapter provides design SN for the pavement. Once design SN is obtained, layer combination must be selected. Note that there is no unique solution for pavement design. A set of combinations is possible, such as increases in asphalt layer and thin base layer or vice versa. The final design should be selected based on cost analysis. For example, if crushed stone is available very nearby, then thicker base layer may be cost-effective. On the other hand, if crushed

Minimum thickness (in.)		
Traffic, ESAL	Asphalt concrete	Aggregate base
Less than 50,000	1.0 (or surface treatment)	4
50,001–150,000	2.0	4
150,001–500,000	2.5	4
500,001–2,000,000	3.0	6
2,000,001–7,000,000	3.5	6
Greater than 7,000,000	4.0	6

Source: From AASHTO (1993). AASHTO Guide for Design of Pavement Structures. Washington, DC: American Association of State Highway and Transportation Officials. Page II–35. Used with permission.

TABLE 7.10 Recommended Minimum Thickness Values

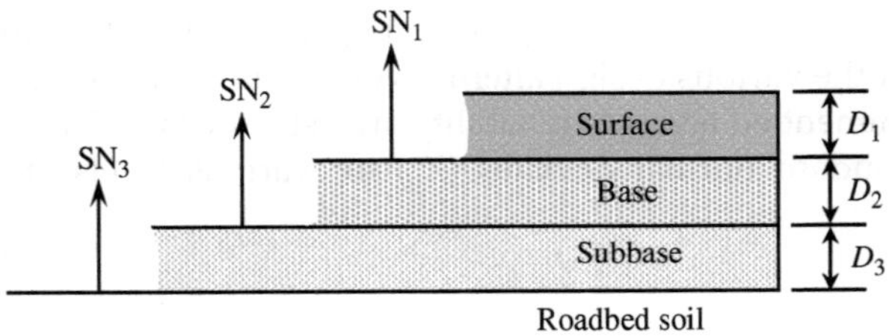

FIGURE 7.5 Thickness design for layered systems.

stone is needed to be brought in from location far away costing a huge transportation cost, then a thin base layer may be cost-effective. However, this is to remind you that there is a minimum thickness recommendation considering the level of traffic, which is listed in Table 7.10.

The thickness design procedure starts from the top layer. The steps for the thickness design are explained using Fig. 7.5.

Step 1. Determine the structural number SN_1 required to protect the second layer by assuming $E_2 = M_R$ in the design equation. Then, the first layer thickness can be computed as:

$$D_1 \geq \frac{SN_1}{a_1} \tag{7.9}$$

Round off D_1 to the nearest 0.5 in.

Step 2. Determine the structural number SN_2 required to protect the third layer (subbase) by assuming $E_3 = M_R$ in the design equation. Then, the second layer thickness can be computed as:

$$D_2 \geq \frac{SN_2 - a_1 D_1}{a_2 m_2} \tag{7.10}$$

Step 3. Determine the structural number SN_3 required to protect the fourth layer (roadbed soil) by considering the roadbed soil resilient modulus M_R in the design equation. Then, the third layer thickness can be computed as:

$$D_3 \geq \frac{SN_3 - a_1 D_1 - a_2 D_2 m_2}{a_3 m_3} \tag{7.11}$$

Now let us see Example 7.4 for better understanding of the design process.

Example 7.4: Thickness Design A pavement section with the layer coefficients and moduli of surface, base, and subbase layers is shown in Fig. 7.6. The predicted design ESAL is 13 million and the standard deviation is 0.35. The expected terminal serviceability for this pavement is 2.5. For these conditions, determine the layers' thicknesses at 99% reliability. Round off the thicknesses to 0.50 in.

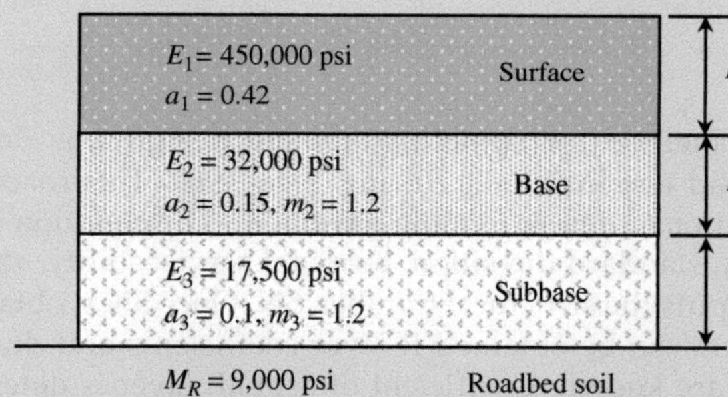

Figure 7.6 Pavement layers for Example 7.4.

Solution

Step 1. Assume, $E_2 = M_R$

Using the nomograph presented in Fig. 7.1, $SN_1 = 3.0$.

Now,

$$\begin{aligned} SN &= a_1 D_1 \\ SN_1 &= 3 \\ 3 &= 0.42 D_1 \\ D_1 &= 7.1 \text{ in.} \end{aligned}$$

Therefore, $D_1 \approx 7.5$ in. [We cannot provide thicknesses less than the required (7.1 in.).]

Step 2. Assume, $E_3 = M_R$

Using the nomograph presented in Fig. 7.1, $SN_2 = 4.0$.

$$\begin{aligned} SN &= a_1 D_1 + a_2 m_2 D_2 \\ SN_2 &= 4 \\ 4 &= 0.42(7.5) + 0.15(1.2) D_2 \\ D_2 &= 4.72 \text{ in.} \end{aligned}$$

From Table 7.10, the minimum base thickness is 6.0 in. for ESAL greater than 7,000,000.

Therefore, $D_2 = 6.0$ in.

Step 3. Assume $M_R = M_R$

Using the nomograph presented in Fig. 7.1, $SN_3 = 5.0$.

$$\begin{aligned} SN &= a_1D_1 + a_2m_2D_2 + a_3m_3D_3 \\ SN_3 &= 5 \\ 5 &= 0.42(7.5) + 0.15(1.2)(6.0) + 0.1(1.2)D_3 \\ D_3 &= 6.42 \text{ in.} \end{aligned}$$

Therefore, $D_3 = 6.5$ in.

Answers *The thicknesses of surface, base, and subbase layers are 7.5, 6.0, and 6.5 in., respectively.*

7.9 Summary

AASHTO 1993 method for flexible pavement design was developed empirically based on the road test in Illinois during 1956–1960. This road test led to development of an equation to design flexible pavement. The equation takes inputs of traffic ESAL, reliability, standard deviation, terminal serviceability, and roadbed soil modulus. It then outputs the structural number, SN. Once SN is obtained, the thicknesses of different layers can be obtained if layer coefficients and drainage coefficients of different layers are known. Coefficient of asphalt layer is determined based on the elastic or resilient modulus of asphalt at 68°F. Coefficients of base and subbase layers can be obtained if any of the four laboratory parameters (modulus, Texas triaxial, CBR, or *R*-value) is known. There are several equations available to calculate layers' coefficients as well. The recommended drainage coefficients of base and subbase can be determined based on the quality of drainage and the percent of time during the year the pavement structure would normally be exposed to moisture levels approaching saturation.

Only a single M_R is used in the design equation. However, the M_R of soil thus changes month to month and the damage in pavement also changes month to month. This is why a single M_R is established, which is known as "effective roadbed soil resilient modulus." It is the combined effect of all the seasonal modulus values. Reliability is also an important consideration to increase the confidence level in the analysis.

7.10 Fundamentals of Engineering (FE) Exam-Style Questions

FE7.1 The layer coefficients of different layers of a pavement system are shown in Fig. P7.1. The required structural number of this pavement system is 6.0. The thickness (in.) of the asphalt layer is most nearly:

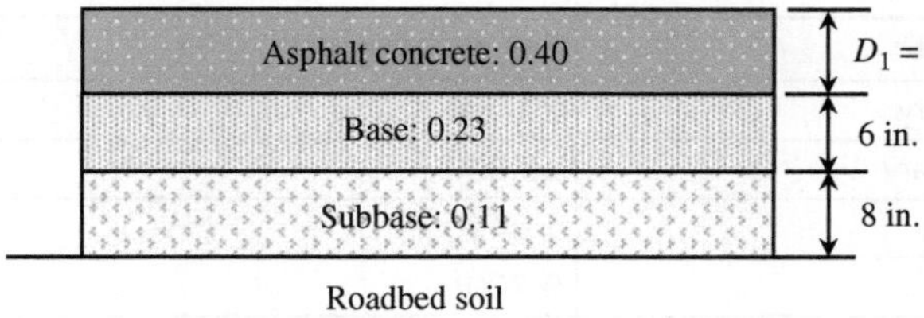

FIGURE P7.1 Layer coefficients of a pavement system for FE Prob. 7.1.

A. 9.0
B. 9.5
C. 10.5
D. 11.0

Solution B

Known: $\text{SN} = a_1D_1 + a_2D_2m_2 + a_3D_3m_3$

As moisture coefficients are not mentioned, let us skip m_i.

Therefore, $6.0 = a_1D_1 + a_2D_2 + a_3D_3$

$$6.0 = 0.4D_1 + 0.23(6.0) + 0.11(8.0)$$

Thus, $D_1 = 9.35$ in. ≈ 9.5 in.
You cannot provide 9.0 in. of asphalt layer as requirement is 9.35 in.

FE7.2 The resilient modulus of aggregate/soil layer is dependent on (select all that apply):
A. Moisture content of the materials
B. Applied stress on the materials
C. Temperature of the materials
D. All of the above

Solution A, B

Temperature affects the modulus of asphalt materials, not the aggregates/soils. Modulus of aggregates/soils is severely affected by moisture content and applied stress.

7.11 Practice Problems

7.1 The K_1 and K_2 of a subbase layer are 5,000 and 0.6, respectively. The expected stress state at subbase is 6 psi. Determine the layer coefficient of the subbase layer.

7.2 The roadbed soil moduli of a pavement site for 12-month are listed in Table P7.2. Calculate the effective roadbed soil modulus of the pavement.

Period	Soil modulus, psi
January	23,500
February	16,300
March	11,400
April	9,700
May	6,950
June	3,650
July	2,150
August	4,150
September	7,950
October	9,950
November	15,650
December	19,650

TABLE P7.2 Roadbed Soil Moduli of a Pavement Site for Prob. 7.2

7.3 A pavement section with the layer coefficients and moduli of surface, base, and subbase layers is shown in Fig. P7.3. The predicted design ESAL is 35 million and the standard deviation is 0.35. The expected terminal serviceability for this pavement is 2.0. For these conditions, determine the layers' thicknesses at 95% reliability.

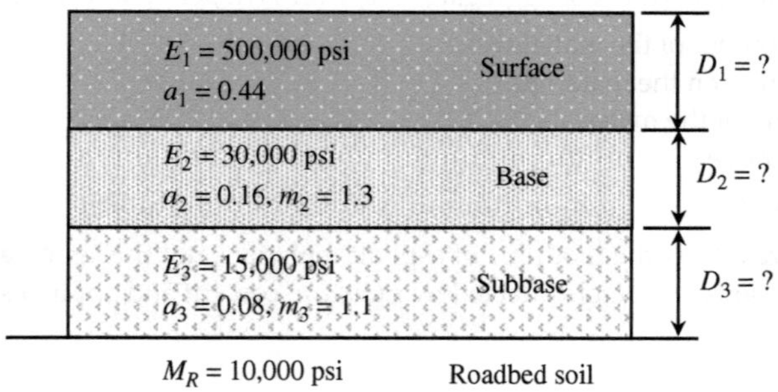

FIGURE P7.3 Pavement layers for Prob. 7.3.

7.4 Given that $R = 90\%$, $SN = 7$, $S_o = 0.35$, $M_R = 7{,}500$ psi, $\Delta PSI = 2.0$, determine the value of W_{18}.

7.5 Given that $R = 95\%$, $W_{18} = 12$ million, $S_o = 0.35$, $M_R = 5{,}000$ psi, $\Delta PSI = 1.7$, determine the structural number required for the pavement.

CHAPTER 8
Distresses in Flexible Pavement

8.1 Background

It is impossible to find people who have not seen any distress (cracking or deformation) in the pavement. Pavement is designed based on the consideration that it will fail (show cracking or deformation) after few years of service. The pavement is then repaired regularly or periodically to monitor the level of damage or to provide the expected quality of service. When any or more of the distress levels exceed the threshold (limiting) value, the pavement is considered failed and rehabilitation is recommended. This chapter discusses different types of distresses, their causes, and possible measures to control them. Almost all types of distresses allow moisture to penetrate into the pavement and deteriorate the pavement structure and subsequently decrease the pavement serviceability and quality service life.

8.2 Major Distresses

Four major distresses are considered while analyzing and designing flexible pavement. These are alligator cracking, top-down cracking, rutting, and transverse cracking. These distresses are calculated using the AASHTOWare pavement ME design software using some mechanistic-empirical equations (discussed in the next chapter) (AASHTO, 2015). If the distresses are greater than the predefined threshold values, then the trial section is considered adequate. An optimum section is finally selected after several software simulations. The four major distresses are discussed in this section.

8.2.1 Alligator Cracking

Definition. Series of interconnected cracks looking similar to a chicken wire or back of an alligator, as shown in Fig. 8.1, are defined as alligator cracking. This crack is initiated at the bottom of asphalt layer due to the damage caused by the developed tensile strain induced by the repeated wheel load. With increase in severity, the cracking propagates to the surface. This is why this crack is very often called "bottom-up fatigue crack" or simply "fatigue crack." However, some agencies use "fatigue crack" for both bottom-up fatigue and top-down fatigue (discussed later) cracks. Therefore, care should be taken while using the term "fatigue crack." The details of the cracking mechanism are discussed in Chap. 9.

Figure 8.1 Alligator or bottom-up fatigue cracking.

Causes. Crack initiates at the bottom of asphalt layer due to repeated tensile strain caused by wheel loading (AASHTO, 2015). It propagates upward and appears at the surface as several short, longitudinal, or transverse cracks in the wheel path and interconnects laterally with continued repeated (fatigue) loadings forming many-sided, sharp-angled pieces that develop a pattern which resembles the skin of an alligator or the shape of chicken wire. This crack occurs because of thin pavement or pavement with weak base/subgrade, excessive loading, or combination of these. Other factors such as poor drainage accelerate the progress of this crack.

Treatments. Full-depth rehabilitation is the best option to get rid of large cracked areas. As alligator cracking starts at the bottom of asphalt layer and propagates upward, replacing a part of the asphalt layer or overlay does not eliminate this cracking. However, replacing a part of the asphalt layer or overlay be used for low-severity alligator cracking to defer the high-severity cracking for few years.

8.2.2 Top-Down Longitudinal Cracking

Definition. Longitudinal cracks, near the wheel paths, parallel to the centerline of the roadway are called top-down longitudinal cracks or top-down cracks or longitudinal cracks. Figure 8.2 shows the photos of top-down longitudinal cracks.

Causes. Crack initiates at the surface of asphalt layer due to repeated tensile and/or shear strain caused by wheel loading (AASHTO, 2015). It can also develop along the construction joint between adjacent passes of the paver or at a location that corresponds to the center of the paver.

Treatments. A thin layer of asphalt may be replaced if the cracks do not penetrate to full depth of asphalt layer. Otherwise, full-depth rehabilitation may be required. Two major options are available to mitigate or avoid top-down cracking. The first one is improved heavy vehicle loadings control and the second is improved asphalt surface courses.

8.2.3 Rutting

Definition. Rutting is the longitudinal surface depression in the wheel path resulting from permanent deformation in each pavement layer along the wheel path (AASHTO, 2015). Ruts are particularly evident after a rain when they are filled with water. The rut depth is the vertical difference in elevation between the transverse profile of the surface

Figure 8.2 Longitudinal cracking along wheel path.

Severe mix rutting　　　　Mix rutting

Figure 8.3 Rutting in asphalt pavements.

and a wire-line across the lane width, as shown in Fig. 8.3. As the rut depth increases, the amount of water ponded along the wheel paths increases. It causes a reduction of surface friction and increase of hydroplaning.

Causes. Rutting occurs in three basic ways:

- Wearing of asphalt surface
- Vertical deformation of layers (also called structural rutting)
- Shear deformation of asphalt layer (also called instability rutting)

Wearing of asphalt surface is the progressive loss of material from the roadway surface due to poor quality asphalt mixture, studded tire and inadequate compaction during construction.

Structural rutting is the sum of permanent vertical deformations (densification or deformable materials) of all pavement layers due to repeated traffic loads. It occurs due to less stiff materials and low thickness of layers. Improper compaction or mix design such as high asphalt content, excessive fines, and insufficient amount of angular aggregate particles are responsible for rutting.

Shear rutting is the lateral/shear displacement of asphalt materials under wheel loading. The hot-mix asphalt (HMA) is displaced downward, laterally, and then upward to form ridges on either side of the wheel paths. Unstable asphalt layer or too less air void leads to shear rutting.

Treatments. Small amount of rutting, less than ⅓ in. (8.5 mm) deep, can generally be left untreated. If the depth is more than ¾ in. (19 mm), milling out certain depth and replacing it with better rutting-resistant asphalt may be the right option (ARRA, 2015). If the rutting is due to the base/subgrade layer, then full-depth reclamation is the only option.

8.2.4 Transverse Cracking

Definition. Transverse top-down surface cracking (also known as transverse cracking or thermal cracking) of asphalt pavement is one of the important distresses especially observed in cold regions. Thermal crack consists of two types of cracking, namely, the thermal fatigue cracking and the low-temperature cracking. Repetitive thermal cycles may cause thermal fatigue cracking even though the amplitude of the temperature cycle is not so severe. Single cooling due to extreme cold temperature or high rate of cooling may cause low-temperature cracking (AASHTO, 2015). Low-temperature cracking occurs when the induced tensile stress exceeds the tensile strength of the asphalt concrete.

A transverse crack, also called a thermal crack as it is not load related, runs perpendicular to the pavement's centerline or lay down direction, as shown in Fig. 8.4. This is caused due to single-day extreme cooling of pavement or cyclic cooling-heating loading.

It allows moisture infiltration, and causes an increased roughness.

Causes. Transverse cracking occurs due to shrinkage of the HMA surface as a result of low temperatures or asphalt binder hardening. Proper selection of asphalt binder considering the possible single-day extreme low temperature may avoid the transverse crack.

Treatments. Based on the crack-depth penetrated, a thin- or the full-depth replacement may be required.

Large transverse crack | Medium transverse crack

Figure 8.4 Transverse cracking in asphalt pavements.

8.3 Minor Distresses

The above-discussed four major distresses are considered while designing the pavement. There are some minor distresses which are considered for asphalt-mix design and maintenance and rehabilitation operation.

8.3.1 Stripping

Definition. Stripping is a moisture-related damage—the loss of bond between aggregates and asphalt binder beginning at the bottom of the HMA layer and progressing upward. It is very difficult to identify stripping as it starts at the bottom of asphalt layer due to the presence of moisture and when it reaches the surface it accompanies other forms of distress such as rutting, shoving/corrugations, raveling, or cracking. Coring is the only option to identify stripping. Figure 8.5 shows two candidate pavements with stripping. Full-depth coring is required to confirm these are stripping.

Causes. Infiltration of moisture and poor drainage of moisture are the primary causes of stripping.

Treatments. Repairs in both asphalt layer and underneath base/subgrade are required. Thus, full-depth reclamation is the only remedy for stripping. Stripping may be avoided with proper drainage system and by sealing cracks regularly.

8.3.2 Raveling

Definition. Raveling is the loss of bond between aggregates and binder starting from the surface and propagating downward, as shown in Fig. 8.6. It causes loose debris on the pavement, increased roughness, loss of skid resistance, and so on.

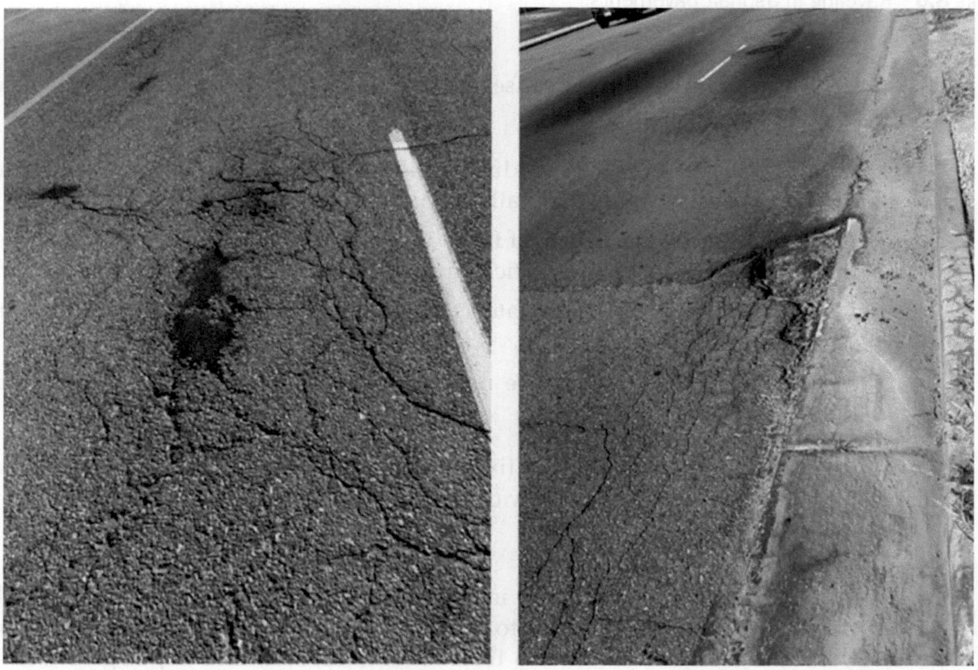

Figure 8.5 Pavements of potential stripping. (*Photos by Carl Brown.*)

Figure 8.6 Raveling in asphalt pavements.

Causes. There may be several causes for asphalt raveling (loss of bond between aggregate and binder):

- *Dusty aggregate.* Presence of dust in aggregate particles allows the asphalt binder to bond with the dust rather than the aggregate.
- *Less fines.* If there is shortage of fine particles, the asphalt binder is only able to bind the remaining coarse particles at their relatively few contact points.
- *Inadequate compaction.* Adequate compaction is required to allow bonding between aggregate and binder.
- *Mechanical action.* Studded tire may break the bond between aggregate and binder.

Treatments. Small raveling may be fixed by patching, micro-surfacing, or chip seal. Large raveled areas require the overlay or surface replacement.

8.3.3 Potholes

Definition. Bowl-shaped depressions in the pavement surface penetrating all the way through the asphalt layer down to the base course are called potholes. They are generally circular, with diameter less than 30 in. (750 mm) and depth greater than

Figure 8.7 Potholes in asphalt pavement.

Figure 8.8 Bleeding in asphalt pavements.

3/8 in. (10 mm), and have sharp edges, and vertical sides are near the top, as shown in Fig. 8.7 (ARRA, 2015). Serious vehicular damage can result from driving across potholes at higher speeds.

Causes. Potholes occur due to poor-quality asphalt mixture (low asphalt binder contents, soft aggregates, etc.) or construction deficiencies (low compaction, mix segregation, etc.), or structural problems (insufficient pavement thickness, reduced base support, etc.), and environmental conditions (freeze-thaw cycles, etc.). Sometimes, potholes are the result of alligator cracking. As a vehicle runs over severe alligator cracking, the interconnected cracks create small chunks and they take off.

Treatments. Potholes are fixed by patching or filling. It consists of removing the loose materials, cleaning the potholes, applying some tack coat (prime coat if base is exposed), applying the mix (hot or cold), and compacting.

8.3.4 Bleeding

Definition. Bleeding is a flushing film of asphalt binder on the pavement surface, looking shiny and glass-like reflecting surface (as shown in Fig. 8.8) that can become quite sticky. Bleeding areas have less skid resistance, i.e., they are slippery especially in wet weather. Bleeding is not a reversible process. This means asphalt binder accumulates on the pavement surface over time and does not go down.

Causes. Bleeding occurs when asphalt binder fills the aggregate voids during hot weather and then expands onto the pavement surface (Brown et al., 2009). This can be caused by one or a combination of the following:

FIGURE 8.9 Block cracking in asphalt pavement.

- Improper mix such as high asphalt binder, too soft asphalt binder, low air voids, etc.
- Construction deficiencies (excessive tack or prime coat, etc.)
- Environmental conditions (excessively hot weather, etc.)
- Traffic effects (high traffic volumes, etc.)

Treatments. Minor bleeding may be corrected by applying coarse, hot sand or grit to blow up the excess asphalt binder. Major bleeding can be corrected by removing a thin layer and resurfacing.

8.3.5 Block Cracking

Definition. Block cracking happens when interconnected cracks divide the pavement up into rectangular pieces. The blocks may range in size from 12 in. by 12 in. (300 mm by 300 mm) to 10 ft by 10 ft (3 m by 3 m), as shown in Fig. 8.9.

Causes. Block cracking is associated with shrinkage of the asphalt layer or the base/subgrade materials and hardening of asphalt binder.

Treatments. Crack seal may prevent the low-severity cracks. High-severity cracks may be removed and replaced with an overlay.

8.3.6 Reflection Cracking

Definition. Reflective cracks (Fig. 8.10) are cracks in asphalt overlays reflecting the existing cracks in the underlying pavement before the placement of the overlay. These cracks are larger than block cracking and can be longitudinal, transverse, or random depending on the preexisting cracks in the underlying layers.

Causes. Reflection cracking occurs due to stress created by the horizontal and vertical movement of the underlying pavement structure following the fracture mechanics rule.

Treatments. Linear and localized cracks can be sealed. Replacing the layer is the solution if cracks are extensive and nonlinear. Use of reinforced asphalt (with grids) may be a way to avoid reflection cracking.

8.3.7 Depression

Definition. Depressions are localized lower elevations of pavement surface, as shown in Fig. 8.11, due to subgrade settlement or frost heave. They are readily noticeable after

FIGURE 8.10 Reflection cracking in asphalt pavement.

FIGURE 8.11 Depressions in asphalt pavements.

rain fills it with water. It causes an increased vehicle hydroplaning and accelerates the bottom-up fatigue damage process.

Causes. Depression occurs due to frost heave or failure of subgrade (settlement) resulting from inadequate compaction during construction.

Figure 8.12 Corrugation in asphalt pavement at signalized intersection.

Treatments. Depressions of less than ¾ in. (19 mm) may be micro-surfaced. Otherwise, they should be repaired by digging out and replacing the area of poor subgrade.

8.3.8 Corrugation and Shoving

Definition. Corrugation and shoving is somewhat a ripple (corrugation) or an abrupt wave (shoving) on the pavement surface at points where traffic starts and stops (corrugation) or areas where asphalt abuts a rigid object (shoving), as shown in Fig. 8.12. This is a form of plastic movement of asphalt materials perpendicular to the traffic direction and does not accompany cracking.

Causes. Corrugation and shoving is usually caused by traffic action (starting and stopping) on less stable and less shear resistant materials, which is because of the following reasons:

- Improper mix design such as high asphalt content, soft binder, excessive fines, mix contamination, poor mix design, poor HMA manufacturing, lack of aeration of liquid asphalt emulsions, etc.
- Possibly excessive moisture in the subgrade sometimes leading to cracking

Treatments. Repairing of corrugation and shoving is carried out following one of the two strategies:

- Small corrugation or shoving can be removed and patched with well-designed and properly produced asphalt; better to avoid cut-back asphalt.
- Large corrugated or shoved areas are indicative of general asphalt layer failure. Removing the damaged pavement and overlaying may be an appropriate solution.

8.3.9 Slippage Cracking

Definition. During braking or turning vehicles, pavement surface slides forming crescent or half-moon-shaped cracks with two ends pointed into the direction of traffic known as slippage cracks. Figure 8.13 shows two slippages cracks; the left photo shows clearly that the two ends are pointed toward the direction of traffic.

Figure 8.13 Slippage cracking in asphalt pavements.

Figure 8.14 Microcracking in asphalt pavements for rolling cold mat.

Causes. Slippage crack occurs in asphalt overlays with lack of tack coat or poor bond between the old surface and the new overlay. It also occurs due to excessive deflection of the underlying pavement structure under the compaction equipment or haul trucks. As the roller or truck passes, the surface of the pavement gets tension, forming the slippage crack.

Treatments. Removal and replacement of the affected area is the only solution of this distress.

8.3.10 Microcracking

Definition. Microcracking is usually 1 to 3 in. (25–75 mm) long and 1 to 3 in. (25–75 mm) apart that appear on the mat surface during compaction of cold mat, as shown in Fig. 8.14. Microcracking is usually shallow, small, and not so detrimental to pavement performance.

Causes. There are three major causes of microcracking:

- Excessive pavement deflection during compaction creates surface tension in mat, causing tiny fissures or cracks on the surface.
- Tender mixes tend to shove when compacted and create surface tension in mat, causing tiny fissures or cracks on the surface.
- Compacting cool mat shows microcracks as cool mat is stiff.

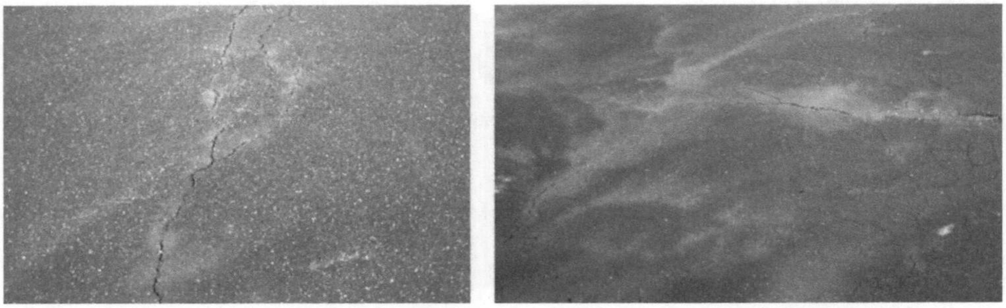

FIGURE 8.15 Pumping in asphalt pavements (water dried, leaving the mud).

Treatments. The existing microcracks can be treated by applying micro-surfacing, chip seal, or a thin overlay. For excessive microcracks, a thin surface may be removed and micro-surfaced or overlaid. To avoid microcracking, the causes discussed previously should be avoided (ARRA, 2015).

8.3.11 Water Bleeding and Pumping

Definition. Water bleeding means seeping water out of joints or cracks or through an asphalt layer. If seeping consists of water and some fine materials, it is termed "pumping," as shown in Fig. 8.15. It causes a decreased skid resistance, an indication of high pavement porosity (water bleeding) and decreased structural support (pumping).

Causes. Causes of water bleeding and pumping may include the following:

- Porous pavement due to inadequate compaction
- High water table
- Poor drainage

Treatments. If the problem is a high water table or poor drainage, subgrade drainage should be improved. If the problem is a porous mix (in the case of water bleeding), a fog seal or slurry seal may be applied to limit water infiltration.

8.3.12 Polished Aggregate

Definition. Polished aggregates are those that have lost their original surface texture and angularity due to abrasion action by tires, as shown in Fig. 8.16. Polished aggregates have a decrease in skid resistance, especially when they are wet.

Causes. As a pavement ages the protruding rough, angular particles become polished due to repeated traffic as the aggregates are less abrasive resistant. The process accelerates if the aggregate is susceptible to abrasion or subject to excessive studded tire wear.

Treatments. A skid-resistant slurry seal or overlay can be applied to remove the polished aggregate.

8.3.13 Mat Tearing

Definition. Mat tearing (Fig. 8.17) is defined as the surface dragging defects (torn or streaked areas) that occur in the asphalt surface during the laydown and compaction operations. Torn or streaked areas may have higher air voids and are susceptible to

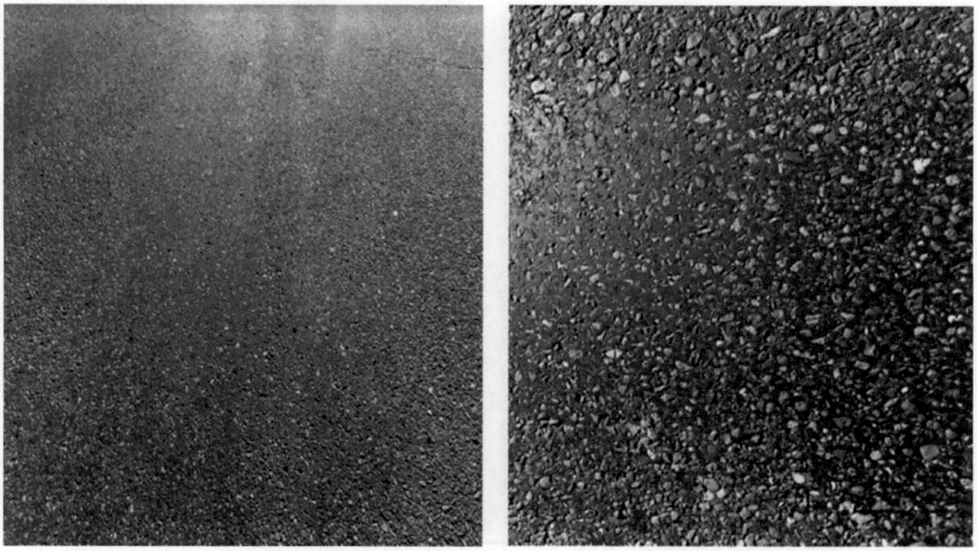

FIGURE 8.16 Polished aggregates in asphalt pavements. (*Photos by Caleb Johnson.*)

FIGURE 8.17 Mat tearing in asphalt pavement. (*Photo by Chris Meeks.*)

decreased stiffness, reduced fatigue life, accelerated aging/decreased durability rutting, raveling, and moisture damage.

Causes. There are several causes of mat tearing depending upon the type of mat tearing or streaking. Some major causes are listed below:

- Warped or worn screed plates have the increased likelihood of dragging aggregate particles along the mat surface for larger friction and have more potential defects that can grab aggregates.
- A screed plate tends to tear the mat at the beginning of paving operations if it is not preheated. As the hot asphalt heats the screed, this type of tearing will generally disappear.
- If screed extension is not properly installed for its crown and elevation, it could create a small streak at the point of transition from screed to extension.

FIGURE 8.18 Nonuniform texture in asphalt pavements.

- High paver speed can stimulate small aggregate particle disturbances to the point where the result becomes a visible streak.
- Stacked aggregates caught between the screed and the surface being paved may roll or be dragged forward.

Treatments. Chip seal or thin overlay may be a solution for existing mat tearing. To avoid the mat tearing, the above-discussed causes are to be avoided.

8.3.14 Nonuniform Texture

Definition. Nonuniform texture is the difference in appearance at the pavement surface as shown in Fig. 8.18. Separation of coarse particles and fine particles means the properties, strength, and durability of these separated coarse and fine particles are different. This separation may substantially reduce pavement life.

Causes. There are many potential causes for nonuniform mat texture. Many times, the same mechanisms can cause mat tearing and nonuniform texture. Therefore, an investigation of mat tearing causes is also warranted. The causes listed below are typical of the nonuniform texture:

- Segregation is one of the main causes of nonuniform texture.
- Temperature differentials may not allow proper mixing of aggregates.
- If the mixture is delivered at low temperatures, it is an indication that the mix might not have adequate time for compaction, which results in segregated mix.
- Handwork behind the paver screed generally involves depositing asphalt material on top of the mat in either thin layers or individual aggregate particles and therefore detrimental to pavement appearance.

Treatments. Small-scale nonuniform mat texture may be left untreated. Chip seal or thin overlay may be a solution for large-scale nonuniform mat texture (ARRA, 2015). To avoid the nonuniform mat texture, the above-discussed causes are to be avoided.

8.3.15 Miscellaneous Distresses

Four major cracks have been discussed in Sec. 8.2 and 14 minor cracks have been discussed in this section. Additionally, there are some cracks which are not so common but exist. Some of them are discussed here.

Edge cracks. Edge cracks are parallel to roadway and occur within 1 to 2 ft (300–600 mm) of the outer edge of the asphalt pavement with no or partially paved shoulders. This cracking is initially caused by lack of lateral support from the shoulder or base weakening due to poor drainage and/or frost action.

Lane/shoulder drop-off. Lane/shoulder drop-off is a difference in elevation between the pavement edge and the roadway shoulder due to shoulder erosion, shoulder settlement, or the buildup of the roadway through asphalt overlays without adjustment of the shoulder level.

Base/subgrade problems. Poor drainage or inadequate compaction of base/subgrade causes some major cracks, as discussed earlier. Some other forms of distresses such as swells, bumps, and sags also occur due to base/subgrade problems.

- *Swells* are characterized by upward displacement of the pavement surface due to frost heaving or swelling soils in a long, gradual wave that is more than 10 ft (3 m) in length. *Bumps* are usually less than 10 ft (3 m).
- *Sags* are abrupt downward displacement of the pavement surface initiated with loss or settlement of the underlying base/subgrade. Sags accompanied by cracking are known as dipping or cupping.

Joint separation. Joints are required between two sequences of construction. Sometimes joints are provided between lane and shoulder. The widening of joints may turn into distresses if it is not sealed and moisture infiltrates through it. Figure 8.19 shows an example of an extreme consequence of a joint separation.

FIGURE 8.19 Extreme joint separation.

Figure 8.20 Short waves in asphalt pavement. (*Photo by Jon Alfonso.*)

Surface waves. Different degrees of surface waves, such as washboarding (3–4 in. apart), short wave (1–3 ft apart), and long waves (4–100 ft apart), may occur due to excessive roller speeds, low vibratory frequencies, excessive vibratory amplitudes, sharp breaks, fluctuating screeds, etc. Improper mix design for which stiffness is extremely sensitive to temperature or tender mixes may show surface waves. Figure 8.20 shows an example of short waves.

8.4 Summary

This chapter discusses the distresses that are frequently seen in asphalt pavements. Depending on the material, climate, composite layers, and local practices, other kinds of distresses are also possible. Some pavements may show a combination of two or more distresses. For example, a pavement with alligator cracking might have rutting or other kinds of distress as well. There may be multiple causes of these distresses. For example, alligator cracking may occur because of the following:

- Thin asphalt layer, or
- Heavy load on thick asphalt layer, or
- Weak base/subgrade with thick asphalt layer with normal loading
- Combination of two or more causes

As an example, Fig. 8.21 shows a pavement with alligator cracking, edge cracking, and rutting at the same time. The probable causes are base/subgrade deficiencies and repeated loading.

Regarding the treatment option, every highway agency is unique, and they have their own innovative strategy to repair these distresses. The application of treatment also depends on local materials, climate, budget, etc. The above-discussed treatment prescriptions are general in nature. Based on local needs, an optimum solution is sought out.

In addition to general repair strategies, some repair and preventive treatments are summarized in Table 8.1.

FIGURE 8.21 Cracking and rutting due to base/subbase deficiencies.

Distress	Preventive treatments	Repair treatments
Alligator cracking	Surface/fog seal surface patch	Full-depth repair
Longitudinal cracking	Crack sealing	Partial-depth repair
Reflective cracking	Seal cracks Saw and seal cuts above joints	Full-depth repair
Block cracking	Seal cracks and chip seal	Chip seal
Depression	None	Leveling course Mill surface
Rutting	None	Leveling course Mill surface
Raveling	Rejuvenating seal	Chip seal/surface seal
Potholes	Crack sealing Surface patch	Full- or partial-depth repair
Polished aggregate	Seal coat	Slurry seal, chip seal, open-graded friction course
Slippage cracks	Better mix design and compaction	Slurry seal, chip seal, open-graded friction course
Joint cracks	Better construction practice	Seal
Corrugation	Better mix design and compaction	Thin overlay
Bleeding	Better mix design	Chip seal

TABLE 8.1 Candidate Repair and Preventive Treatments for Flexible Pavements

8.5 Fundamentals of Engineering (FE) Exam-Style Questions

FE8.1 Alligator cracking occurs due to:

A. Repeated wheel-induced surface tension
B. Repeated horizontal strain developed at the bottom of asphalt layer
C. Permanent deformation under the wheel
D. Loss of bond between aggregate and binder by moisture damage

Solution B

Repeated wheel-induced surface tension—top-down longitudinal cracking
Permanent deformation under the wheel—rutting
Loss of bond between aggregate and binder by moisture damage—stripping or raveling

FE8.2 Which of the following statements is not correct for the permanent deformation (rutting) of asphalt pavement?

A. Rutting occurs due to the permanent deformation of any or all layers of pavement
B. Rutting occurs along the wheel paths
C. Stiffer (high modulus) mix is not good against rutting
D. If the rutting is due to the base/subgrade layer, then full-depth reclamation is the only option

Solution C

All statements are correct for rutting except the third one. Stiffer mix is good for rutting resistance. Consider a one-dimensional member, say steel rod. From Hooke's law, Strain = Stress/Modulus. The higher the modulus, the lower the strain and thus, the lower the permanent deformation.

FE8.3 Which of the following statements is correct for the transverse (thermal or low-temperature) cracks of asphalt pavement?

A. Thermal or low-temperature cracks occur due to increase in pavement temperature
B. Thermal or low-temperature cracks start at the bottom of asphalt layer
C. Stiffer (high-modulus) mix is not good against thermal or low-temperature cracks
D. A thin-depth replacement is good enough if the crack depth is small

Solution C

All statements are wrong for the transverse (thermal or low temperature) cracks except the third one.

Stiffer mix is not good for thermal or low-temperature crack resistance. Consider a one-dimensional member, say a steel rod. From Hooke's law, Thermal Stress = Thermal Strain × Modulus. The higher the modulus, the higher the developed stress. When the developed thermal stress exceeds the tensile capacity of the mix, thermal cracks initiate. Thus, stiffer (high-modulus) mix is not good against thermal or low-temperature cracks.

Thermal or low-temperature cracks occur due to a decrease in pavement temperature. With a decrease in temperature, asphalt contracts and tensile stress develops in the materials. As the decrease in temperature is highest on the surface, thermal or low-temperature cracks begin to appear on the surface of asphalt layer. Based on the crack-depth penetrated, a thin- or the full-depth replacement may be required. For full-depth penetration of the crack, full-depth replacement is a must.

8.6 Practice Problems

8.1 Name and define the distresses of flexible pavement which are related to surface defects.

8.2 Name and define the distresses of flexible pavement which are related to deformation of pavement layers.

8.3 Name and define the cracking of flexible pavement which are load-related.

8.4 Name and define the cracking of flexible pavement which are non-load-associated.

8.5 Name and define the distresses of flexible pavement which are caused by base/subgrade deficiencies.

8.6 Name and define the distresses of flexible pavement which are caused by high stiff binder and low stiff binder.

8.7 List the distresses considered while designing flexible pavement using the AASHTOWare pavement ME design software.

CHAPTER 9

Distress Models in Flexible Pavement

9.1 Background

The AASHTOWare pavement ME design software uses several regression equations to determine the distresses with service life of pavement. If any of the distresses exceeds the threshold value within the design life, the trial section is revised and rechecked. This procedure is schematically shown by a flowchart in Fig. 9.1. A trial pavement section is assumed first. The climate and traffic data are provided as input to that trial section. The software then divides the pavement structure, including subgrade, into many sublayers for analysis. The software divides the top 8 ft (2.4 m) of a pavement structure into the maximum of 19 sublayers. The remaining subgrade is treated as the semi-infinite layer. The software then calculates the structural responses at different sublayers of this trial pavement. One of the structural responses is horizontal strains at the bottom of the lowest layer, and at the surface of the top sublayer of asphalt layer. The other structural response is the vertical strain at the middle of each sublayer of the trial pavement. Using the responses, the material damages are calculated for bottom-up alligator cracking and top-down longitudinal cracking. Finally, the field distresses are calculated using some empirical equations. For rutting and transverse cracking, the distresses are calculated directly using the vertical strain values. The process of the AASHTOWare pavement ME design method is presented by a flowchart in Fig. 9.1. In the first stage, the information of the site, materials to be used, desired distress levels, traffic information, climate data, etc. are collected. Most of the data come from the historical information or the agency database. In the second stage, a trial pavement is built using a computer. Distresses and IRI are predicted using the available models in the software for the predefined reliability. All the distresses listed in Fig. 9.1 have been discussed in the previous chapter. Once a feasible design is obtained, it is further analyzed for alternative options in the third stage: constructability, life-cycle cost analysis, etc. Then a final pavement section is proposed to be built in the field.

The iterative process shown in Fig. 9.1 can be summarized as follows:

1. Select a trial pavement.
2. Select appropriate performance indicator criteria for the project.

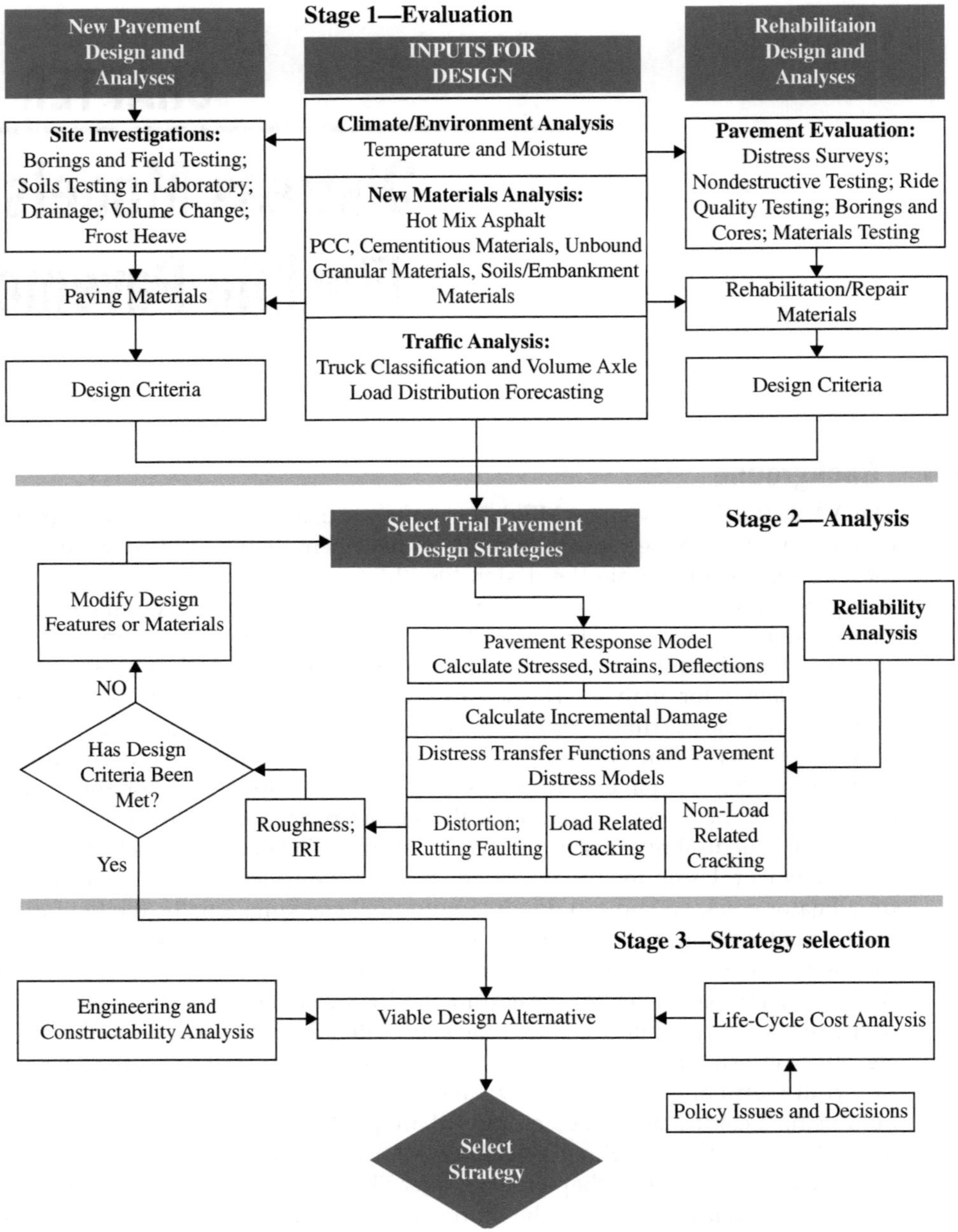

FIGURE 9.1 Conceptual flow chart of the three-stage design/analysis process for the AASHTOWare pavement ME design. (*From AASHTO. (2015). Mechanistic-Empirical Pavement Design Guide: A Manual of Practice. Washington, DC: American Association of State Highway and Transportation Officials. Figure 1-1. Used with permission.*)

3. Select appropriate reliability level for the project.
4. Assemble all inputs for the pavement trial design under consideration.
5. Run the AASHTOWare pavement ME design software: The software calculates changes in layer properties, damage, key distresses, and IRI over the design life. The key steps include the following:

   a. Processing input to obtain monthly values of traffic, seasonal variations of material, and climatic inputs needed in design evaluations for the entire design period.
   b. Computing structural responses (stresses and strains) using multilayer elastic theory or finite element–based pavement response models for each axle type and load and each damage-calculation increment throughout the design period.
   c. Calculating accumulated distress at the end of each analysis period for the entire design period.
   d. Predicting key distresses (rutting, bottom-up/top-down fatigue cracking, and thermal cracking) at the end of each analysis period throughout the design life using calibrated mechanistic-empirical performance models.
   e. Predicting IRI as a function of initial IRI, distresses accumulating over time, and site factors at the end of each analysis increment.

6. Evaluate adequacy of the trial design: The trial design is considered adequate if all the predicted distresses and IRI are equal or less than the threshold values. If any criterion exceeds the threshold value, then a remedy is sought out by altering material types, properties, layer thicknesses, or other design features.

7. Revise the trial design, as needed: If the trial design is inadequate or overdesigned, one must revise the inputs and rerun the program until all performance criteria are satisfactory.

The equations used by the AASHTOWare pavement ME design software to determine the materials damages and field distresses are discussed next.

9.2 Alligator Cracking

Alligator cracking, also called bottom-up fatigue cracking or simply fatigue cracking, is the interconnected cracks caused by fatigue damage of asphalt concrete (AC) under repeated traffic loading. Fatigue damage is defined by the decrease in stiffness of AC under repeated loading. For each cycle of traffic loading, tensile strain develops at the bottom of AC of asphalt pavement as shown in Fig. 9.2*a*. Some localized damages occur in the material at minute-scale due to this tensile strain. After a certain level of damage accumulation, bottom-up fatigue cracking initiates and forms alligator cracking, a pattern resembling the back of an alligator or crocodile as shown in Fig. 9.2*b*. Alligator

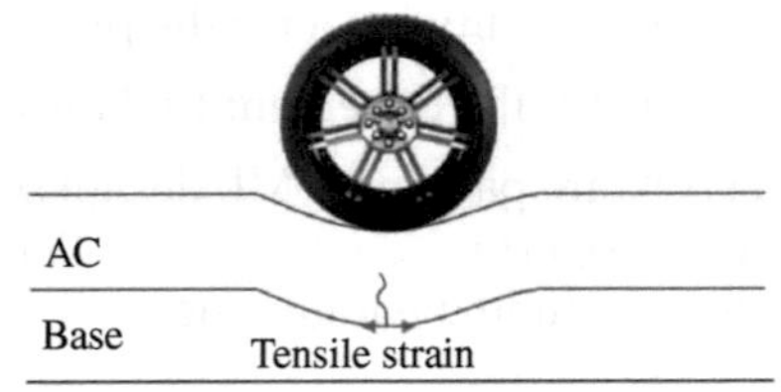

(a) Tensile strain at the bottom of AC due to wheel load

(b) Fatigue cracking when cracks reach the surface

FIGURE 9.2 Bottom-up fatigue (alligator) cracking mechanism.

cracking is calculated as percent of total lane area using the AASHTOWare pavement ME design software.

As the first step, the fatigue damages are determined at the bottom of asphalt layer for the bottom-up cracking. The fatigue damage is then correlated to the fatigue cracking using empirical models. Estimation of fatigue damage is based upon Miner's law, which states that damage is calculated by the equation:

$$DI_{\text{Bottom}} = \sum(\Delta DI_{\text{Bottom}})_{j,m,l,p,T} = \sum\left(\frac{n}{N_{f,\text{Bottom}}}\right)_{j,m,l,p,T} \tag{9.1}$$

where n = Actual number of axle load applications within a specific period
j = Axle-load interval
m = Axle-load type (single, tandem, tridem, quad, or special axle configuration)
l = Truck type using the truck classification groups included in the AASHTOWare pavement ME design software
p = Month
T = Median temperature for the five temperature intervals or quintiles used to subdivide each month, °F

The allowable number of load repetitions for bottom-up cracking ($N_{f,\text{Bottom}}$) is given by the equation:

$$N_{f,\text{Bottom}} = k_{f1}C(C_H)\beta_{f1}\left(\varepsilon_{t,\text{Bottom}}\right)^{k_{f2}\beta_{f2}}\left(E_{\text{HMA}}\right)^{k_{f3}\beta_{f3}} \tag{9.2}$$

where $\varepsilon_{t,\text{Bottom}}$ = Tensile strain at the bottom of AC layer, in./in.
E_{HMA} = Dynamic modulus of AC measured using uniaxial compression testing, psi
k_{f1}, k_{f2}, k_{f3} = Global field calibration parameters
k_{f1} = 0.007566
k_{f2} = –3.9492
k_{f3} = –1.281
$\beta_{f1}, \beta_{f2}, \beta_{f3}$ = Local or mixture-specific field calibration constants; for global calibration effort, these constants are considered 1.0. These coefficients are determined locally for better accuracy.
$C = 10^M$ = Air void constant

$$M = 4.84\left(\frac{V_{be}}{V_a + V_{be}} - 0.69\right)$$

V_{be} = Effective binder content (%)
V_a = Air voids (%)
C_H = Thickness correction factor $= \dfrac{1}{0.000398 + \dfrac{0.003602}{1 + e^{(11.02 - 3.49H_{\text{HMA}})}}}$
H_{HMA} = Total HMA thickness, in.

Defaults values or calibration coefficients used in the AASHTOWare pavement ME design software may change as the software undergoes continuous improvement. Default values or calibration coefficients listed in this textbook are the most updated ones.

Note that AASHTO (2015) uses the term HMA in the equations to represent all types of asphalt layers. However, HMA implies here not only hot-mix asphalt but also all types of ACs such as warm-mix or cold-mix asphalts. This is why it is ideal to use AC instead of HMA. To be consistent with the AASHTO (2015), HMA has been used in the equations.

The area of alligator cracking that is visible at the surface is calculated from the total damage over time using the equation:

$$\text{FC}_{\text{Bottom}} = \left(\frac{1}{60}\right)\left(\frac{C_4}{1 + e^{\left(C_1C_1^* + C_2C_2^*\log(\text{DI}_{\text{Bottom}} \times 100)\right)}}\right) \tag{9.3}$$

where $\text{FC}_{\text{Bottom}}$ = Alligator cracking that is visible at the surface, % of total lane area
$\text{DI}_{\text{Bottom}}$ = Cumulative damage index at the bottom of the AC layer, decimal
$C_{1,2,4}$ = Transfer function regression constants; C_4 = 6,000; C_1 = 1.00; and C_2 = 1.00
$C_1^* = -2C_2^*$

$$C_2^* = -2.40874 - 39.748(1 + H_{\text{HMA}})^{-2.856}$$

The above-mentioned calculation represents the computations at the mean level (i.e., at 50% reliability). For almost all projects, the designer expects a reliability greater than 50% to ensure the performance criteria over the design life. Higher design reliability is desired for project with greater potential and consequences of failure. The AASHTOWare pavement ME design software calculates the reliability of the trial pavement considering the design criteria or threshold values selected by the user. The mean distress or IRI value (50% reliability) is increased by the number of standard error (S_e) that applies to the reliability level selected. For example, a 90% reliability uses a factor of 1.282 times the standard error, and a 95% reliability uses a factor of 1.645. This factor is called the standard normal deviate. Mathematically, the distress at R% reliability can be written as:

$$\text{Distress at } R\% \text{ reliability} = \text{Distress at 50\% reliability} + \text{Standard error} \times \text{Normal deviate at } R\% \text{ reliability}$$

The standard error of predicted alligator cracking of asphalt pavement ($S_{e(\text{Alligator})}$) is calculated using Eq. (9.4). A larger list of the normal deviates can be found in any statistics book. Few normal deviates for different reliability levels are listed in in Table 7.8.

$$S_{e(\text{Alligator})} = 1.13 + \frac{13}{1 + e^{7.57 - 15.5\log(\text{FC}_{\text{Bottom}} + 0.0001)}} \tag{9.4}$$

where $\text{FC}_{\text{Bottom}}$ = Predicted bottom-up alligator cracking based on mean inputs (corresponding to 50% reliability), % of lane area.

Now practice the following examples for better understanding of the bottom-up fatigue cracking mechanism. The first two examples (Examples 9.1 and 9.2) are small parts of the procedure.

Example 9.1: Thickness Correction Factor An asphalt pavement has the surface layer of 4.5 in., base layer of 6.0 in., and subbase layer of 8.0 in. What is the thickness correction factor to be used in the bottom-up fatigue life function of that pavement?

Solution

$$\text{Thickness correction factor, } C_H = \frac{1}{0.000398 + \dfrac{0.003602}{1 + e^{(11.02 - 3.49 H_{\text{HMA}})}}}$$

Given, $H_{\text{HMA}} = 4.5$ in.

$$= \frac{1}{0.000398 + \dfrac{0.003602}{1 + e^{(11.02 - 3.49(4.5\text{ in.}))}}}$$

$$= 252.0766$$

Answer *The thickness correction factor is 252.0766.*

In this example, we do not need the base and subbase thicknesses. The unit of H_{HMA} is in.

Example 9.2: Alligator Cracking An asphalt pavement has the surface layer of 4.5 in., base layer of 6.0 in., and subbase layer of 8.0 in. After 5 years of service, the asphalt layer accumulates a bottom-up fatigue damage index of 4.5%.

a. Determine the amount of bottom-up fatigue cracking as percent of lane area of this pavement at 5-year of service.

b. Determine the amount of bottom-up fatigue cracking as percent of lane area of this pavement at 5-year of service at 90% reliability. The 90% reliability has a standard normal deviate of 1.282.

Solution

Given, $H_{HMA} = 4.5$ in.

$$\text{Known, } FC_{Bottom} = \left(\frac{1}{60}\right)\left(\frac{C_4}{1+e^{(C_1C_1^* + C_2C_2^* \log(DI_{Bottom} \times 100))}}\right)$$

$C_1 = 1.0$
$C_2 = 1.0$

$$C_2^* = -2.40874 - 39.748\left(\frac{1}{1+H_{HMA}}\right)^{2.856}$$

$$= -2.40874 - 39.748\left(\frac{1}{1+4.5}\right)^{2.856} = -2.7141$$

$$C_1^* = -2C_2^* = -2(-2.7141) = 5.5282$$

$$FC_{Bottom} = \left(\frac{1}{60}\right)\left(\frac{6{,}000}{1+e^{(1)(4.9672)+(1)(-2.4836)\log(4.5)}}\right) = 2.52\%$$

The standard error of predicted alligator cracking:

$$S_{e(\text{Alligator})} = 1.13 + \frac{13}{1+e^{7.57-15.5\log(FC_{Bottom}+0.0001)}}$$

$$= 1.13 + \frac{13}{1+e^{7.57-15.5\log(2.52+0.0001)}}$$

$$= 3.8105$$

$$FC_{Bottom} \text{ at 90\% reliability} = 2.52 + 1.282 \times 3.8105 = 7.4050$$

Answer *The amount of bottom-up fatigue cracking (50% reliability) is 2.52% and the amount of bottom-up fatigue cracking at 90% reliability is 7.40%.*

In this example, we do not need the base and subbase thicknesses. In addition, the DI is already given in percent. Therefore, there is no need to multiply DI by 100 in the FC equation. Finally, do not confuse between DI and FC_{Bottom}. DI can be interpreted as the damage in the asphalt material, although actual damage cannot be measured or quantified. FC_{Bottom} is the percent of lane area that is cracked. If the alligator cracking is confined around the wheel path only, the full-lane width is assumed cracked.

Example 9.3: Alligator Cracking An asphalt pavement section has the traffic and response data as shown in Table 9.1.

Axle type	Annual average repetition	Tensile strain produced at the bottom of HMA (yearly average)
Single	500,000	200×10^{-6}
Tandem	80,000	300×10^{-6}
Tridem	10,000	350×10^{-6}
Quad	1,000	370×10^{-6}

Table 9.1 Traffic and Response Data for Example 9.3

The yearly average dynamic modulus of the AC is 500,000 psi.

The pavement section has the following properties:

Air voids of asphalt layer = 5.5%
Effective asphalt content by volume = 13.5%
Thickness of the asphalt layer = 8.0 in.

Calculate the amount of alligator cracking after 10 years, if the above-mentioned information remains the same.

Solution

Coefficients

$$C_H = \frac{1}{0.000398 + \dfrac{0.003602}{1 + e^{(11.02 - 3.49 H_{\text{HMA}})}}}$$

$$C_H = \frac{1}{0.000398 + \dfrac{0.003602}{1 + e^{(11.02 - 3.49(8))}}}$$

$$C_H = 250$$

$$M = 4.84\left(\frac{V_{be}}{V_a + V_{be}} - 0.69\right) = 4.84\left(\frac{13.5}{5.5 + 13.5} - 0.69\right) = 0.099$$

$$C = 10^M = 10^{0.099} = 1.256$$

Fatigue Equation

$$N_{f,\text{Bottom}} = k_{f1}C(C_H)\beta_{f1}(\varepsilon_{t,\text{Bottom}})^{k_{f2}\beta_{f2}}(E_{\text{HMA}})^{k_{f3}\beta_{f3}}$$

$$N_{f,\text{Bottom}} = 0.007566(1.256)(250)(1)\left(\frac{1}{\varepsilon_{t,\text{Bottom}}}\right)^{3.9492(1)}\left(\frac{1}{E_{\text{HMA}}}\right)^{1.281(1)}$$

$$\text{Therefore, } N_{f,\text{ Bottom}} = 2.3757\left(\frac{1}{\varepsilon_{t,\text{ Bottom}}}\right)\left(\frac{1}{E_{HMA}}\right)^{1.281}$$

Calculation of Damage Index (DI) (Table 9.2)

Type	n	$\varepsilon_{t,\text{ Bottom}}$	E_{HMA} (psi)	$N_{f,\text{ Bottom}}$	$\frac{n}{N_{f,\text{ Bottom}}}$
Single	500,000	200×10^{-6}	500,000	48,237,341	0.01037
Tandem	80,000	300×10^{-6}		9,726,661	0.00822
Tridem	10,000	350×10^{-6}		5,291,484	0.00189
Quad	1,000	370×10^{-6}		4,248,824	0.00024
$\text{DI}_{\text{Bottom}} = \sum\left(\frac{n}{N_{f,\text{Bottom}}}\right) =$					0.02072
$\text{DI}_{\text{Bottom}} = \sum\left(\frac{n}{N_{f,\text{Bottom}}}\right) \times 10 \text{ years} =$					0.20715

Table 9.2 Calculation of the Cumulative Damage Index for Example 9.3

Calculation of Alligator Cracking

$$\text{FC}_{\text{Bottom}} = \left(\frac{1}{60}\right)\left(\frac{C_4}{1+e^{(C_1C_1^*+C_2C_2^*\log(\text{DI}_{\text{Bottom}}\times 100))}}\right)$$

$$C_1 = 1.0$$

$$C_2 = 1.0$$

$$C_2^* = -2.40874 - 39.748\left(\frac{1}{1+H_{\text{HMA}}}\right)^{2.856}$$

$$C_2^* = -2.40874 - 39.748\left(\frac{1}{1+8}\right)^{2.856} = -2.4836$$

$$C_1^* = -2C_2^* = -2(-2.4836) = 4.9672$$

$$FC_{\text{Bottom}} = \left(\frac{1}{60}\right)\left(\frac{6{,}000}{1+e^{(1)(4.9672)+(1)(-2.4836)\log(0.20715\times100)}}\right) = 15.5\%$$

Answer *The amount of alligator cracking is 15.5%.*

In this example, the local calibration factors have been used as 1.0, as no local factors are provided. The dynamic modulus of AC is taken as the yearly average value, although it changes every moment with the change in AC temperature and wheel speed of traffic. The strain for each single-axle loading is also not the same as each single-axle loading may not carry the same load and may not follow the same wheel path. The AASHTOWare pavement ME design software calculates the damage index or the cracking for each single loading and cumulates it. It is not possible to calculate the cracking for each single loading for millions of axle loads using calculator. Thus, this example uses the average values of AC dynamic modulus and the strain at the bottom of AC layer.

Example 9.4: Alligator Cracking with Local Coefficients An asphalt pavement section has the traffic and response data as shown in Table 9.3.

Axle type	Annual average repetition	Tensile strain produced at the bottom of HMA (yearly average)
Single	500,000	200×10^{-6}
Tandem	80,000	300×10^{-6}
Tridem	10,000	350×10^{-6}
Quad	1,000	370×10^{-6}

Table 9.3 Traffic and Response Data for Example 9.4

The yearly average dynamic modulus of the AC is 500,000 psi.

The pavement section has the following properties:

Air voids of asphalt layer = 5.5%
Effective asphalt content by volume = 13.5%
Thickness of the asphalt layer = 8.0 in.

The local calibration factors of that pavement are given below:

$\beta_{f1} = 130.3674$
$\beta_{f2} = 1.0$
$\beta_{f3} = 1.217799$

a. Calculate the amount of alligator cracking after 10 years, if the above-mentioned information remains the same.
b. Calculate the amount of alligator cracking after 10 years at 90% reliability, if the above-mentioned information remains the same. The 90% reliability has a standard normal deviate of 1.282.

Solution

Coefficients

$$C_H = \frac{1}{0.000398 + \frac{0.003602}{1 + e^{(11.02 - 3.49H_{HMA})}}}$$

$$C_H = \frac{1}{0.000398 + \frac{0.003602}{1 + e^{(11.02 - 3.49(8))}}}$$

$$C_H = 250$$

$$M = 4.84\left(\frac{V_{be}}{V_a + V_{be}} - 0.69\right) = 4.84\left(\frac{13.5}{5.5 + 13.5} - 0.69\right) = 0.099$$

$$C = 10^M = 10^{0.099} = 1.256$$

Fatigue Equation

$$N_{f,\text{Bottom}} = k_{f1}C(C_H)\beta_{f1}(\varepsilon_{t,\text{Bottom}})^{k_{f2}\beta_{f2}}(E_{HMA})^{k_{f3}\beta_{f3}}$$

$$N_{f,\text{Bottom}} = 0.007566(1.256)(250)(130.3674)\left(\frac{1}{\varepsilon_{t,\text{Bottom}}}\right)^{3.9492(1)}\left(\frac{1}{E_{HMA}}\right)^{1.281(1.217799)}$$

Therefore, $N_{f,\text{Bottom}} = 309.717\left(\frac{1}{\varepsilon_{t,\text{Bottom}}}\right)^{3.9492}\left(\frac{1}{E_{HMA}}\right)^{1.5600}$

Calculation of Damage Index (DI) (Table 9.4)

Type	n	$\varepsilon_{t,\text{Bottom}}$	E_{HMA} (psi)	$N_{f,\text{Bottom}}$	$\frac{n}{N_{f,\text{Bottom}}}$
Single	500,000	200×10^{-6}	500,000	161,770,345	0.00309
Tandem	80,000	300×10^{-6}		32,619,651	0.00245
Tridem	10,000	350×10^{-6}		17,745,697	0.00056
Quad	1,000	370×10^{-6}		14,248,998	0.00007
$DI_{\text{Bottom}} = \sum\left(\frac{n}{N_{f,\text{Bottom}}}\right) =$					0.00618
$DI_{\text{Bottom}} = \sum\left(\frac{n}{N_{f,\text{Bottom}}}\right) \times 10 \text{ years} =$					0.0618

Table 9.4 Calculation of the Cumulative Damage Index for Example 9.4

Calculation of Alligator Cracking

a. Fatigue cracking at 50% reliability

$$FC_{Bottom} = \left(\frac{1}{60}\right)\left(\frac{C_4}{1+e^{(C_1C_1^*+C_2C_2^*\log(DI_{Bottom}\times 100))}}\right)$$

$$C_1 = 1.0$$

$$C_2 = 1.0$$

$$C_2^* = -2.40874 - 39.748\left(\frac{1}{1+H_{HMA}}\right)^{2.856}$$

$$C_2^* = -2.40874 - 39.748\left(\frac{1}{1+8}\right)^{2.856} = -2.4836$$

$$C_1^* = -2C_2^* = -2(-2.4836) = 4.9672$$

$$FC_{Bottom} = \left(\frac{1}{60}\right)\left(\frac{6{,}000}{1+e^{(1)(4.9672)+(1)(-2.4836)\log(0.0618\times 100)}}\right) = 4.7283\%$$

b. Fatigue cracking at 90% reliability

The standard error of predicted alligator cracking:

$$S_{e(\text{Alligator})} = 1.13 + \frac{13}{1+e^{7.57-15.5\log(FC_{Bottom}+0.0001)}}$$

$$= 1.13 + \frac{13}{1+e^{7.57-15.5\log(4.7283+0.0001)}}$$

$$= 13.44$$

$$FC_{Bottom} \text{ at 90\% reliability} = 4.7283 + 1.282 \times 13.44$$
$$= 21.96$$

Answer *The amount of bottom-up fatigue cracking at 50% reliability is 4.73% and the amount of bottom-up fatigue cracking at 90% reliability is 21.96%.*

In this example, some local calibration factors have been used. These are the values obtained by Colorado Department of Transportation (CDOT, 2019). It is found that after using the local factors, the amount of alligator cracking decreases to 4.7283% from 15.5% (obtained in Example 9.3). Similar to Example 9.3, this example uses the average value of AC dynamic modulus and the strain at the bottom of the AC layer.

9.3 Top-Down Longitudinal Cracking

Top-down longitudinal cracking, very often called as longitudinal cracking or top-down cracking (TDC), appears to be a common mode of HMA pavement distress in at least several states and countries. This crack is caused by the surface tension or shear force or combination of both due to wheel load, as shown in Fig. 9.3. The development of TDC for wheel loading can be described into three stages. The first stage is of a single, short longitudinal crack or cracks appearing just outside the wheel path. With repeated load, the short longitudinal cracking grows longer and wider and new sister cracks develop parallel to the original crack (second stage). In the third stage, the cracking evolves parallel to longitudinal cracking that are connected via short transverse cracking. TDC is calculated as feet per mile in the AASHTOWare pavement ME design software.

Top-down longitudinal cracking in asphalt pavement occurs close to the tire loading at or near the surface of the AC. Under each cycle of wheel loading, some localized damages occur in the material at minute scale due to this tensile strain. After a certain level of damage accumulation, top-down fatigue cracking initiates at the AC surface, propagates downward, and forms the cracking. Similar to alligator cracking, calculation of fatigue damage is based upon Miner's law, which states that damage is given by the equation:

$$DI_{Top} = \sum(\Delta DI_{Top})_{j,m,l,p,T} = \sum\left(\frac{n}{N_{f,Top}}\right)_{j,m,l,p,T} \quad (9.5)$$

where DI_{Top} = Damage index for top-down longitudinal cracking
n = Actual number of axle load applications within a specific period
j = Axle load interval
m = Axle load type (single, tandem, tridem, quad, or special axle configuration)

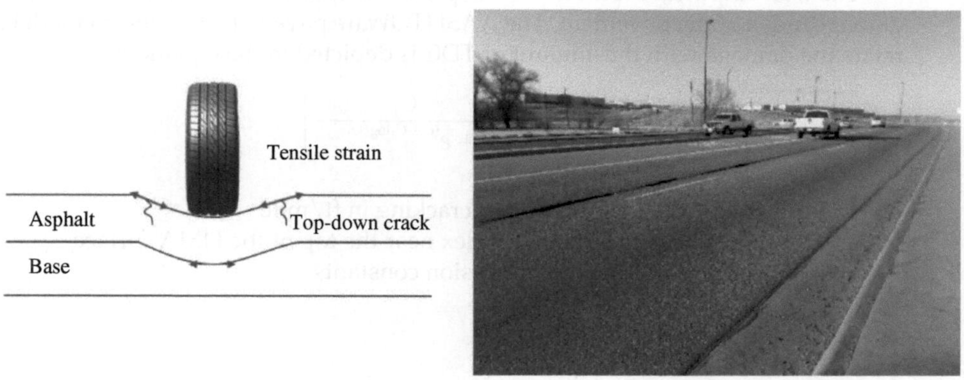

(a) Top-down cracking mechanism (b) Top-down crack in a pavement

Figure 9.3 Top-down longitudinal cracking in asphalt pavement.

l = Truck type using the truck classification groups included in the AASHTOWare pavement ME design software
p = Month
T = Median temperature for the five temperature intervals or quintiles used to subdivide each month, °F

The allowable number of load repetitions for top-down longitudinal cracking ($N_{f,\text{Top}}$) is given by the equation:

$$N_{f,\text{Top}} = k_{f1}C(C_H)\beta_{f1}\left(\varepsilon_{t,\text{Top}}\right)^{k_{f2}\beta_{f2}}\left(E_{\text{HMA}}\right)^{k_{f3}\beta_{f3}} \quad (9.6)$$

$$M = 4.84\left(\frac{V_{be}}{V_a + V_{be}} - 0.69\right) \quad (9.7)$$

where $\varepsilon_{t,\text{Top}}$ = Tensile strain at the surface of AC, in./in.
E_{HMA} = Dynamic modulus of the AC measured in compression, psi
k_{f1}, k_{f2}, k_{f3} = Global field calibration parameters
k_{f1} = 0.007566
k_{f2} = –3.9492
k_{f3} = –1.281
$\beta_{f1}, \beta_{f2}, \beta_{f3}$ = Local or mixture-specific field calibration constants; for global calibration effort, these constants are set to 1.0
$C = 10^M$
V_{be} = Effective binder content, %
V_a = Air voids, %
C_H = Thickness correction factor = $\dfrac{1}{0.01 + \dfrac{12}{1 + e^{(15.676 - 2.8186 H_{\text{HMA}})}}}$
H_{HMA} = Total HMA thickness, in.

Then, an empirical model is used to predict the amount of TDC in ft/mile using the damage index of the pavement. The AASHTOWare pavement ME design model to correlate the damage with the amount of TDC is depicted in the equation:

$$\text{FC}_{\text{Top}} = 10.56\left(\frac{C_4}{1 + e^{(C_1 - C_2\log(\text{DI}_{\text{Top}}))}}\right) \quad (9.8)$$

where FC_{Top} = Top-down longitudinal cracking in ft/mile
DI_{Top} = Cumulative damage index near the top of the HMA surface
$C_{1,2,4}$ = Transfer function regression constants
C_1 = 7.0
C_2 = 3.5
C_4 = 1,000

The standard error of predicted top-down longitudinal cracking of asphalt pavement ($S_{e(\text{Top})}$) can be calculated using the equation:

$$S_{e(\text{Long})} = 200 + \frac{2{,}300}{1 + e^{1.072 - 2.1654\log\left(FC_{\text{Top}} + 0.0001\right)}} \tag{9.9}$$

where FC_{Top} = predicted top-down longitudinal cracking based on mean inputs (corresponding to 50% reliability), ft/mile.

Example 9.5: Top-Down Longitudinal Cracking An asphalt pavement section in a construction zone has the traffic and structural response data in a day, as shown in Table 9.5.

Time	Axle repetition	Average asphalt dynamic modulus (ksi)	Average tensile strain at the surface
12:00 a.m.–7:59 a.m.	1,000	800	35×10^{-6}
8:00 a.m.–12:59 p.m.	8,000	700	32×10^{-6}
1:00 p.m.–5:59 p.m.	6,000	300	35×10^{-6}
6:00 p.m.–11:59 p.m.	4,000	500	39×10^{-6}

Table 9.5 Traffic and Response Data for Example 9.5

The pavement section has the following properties:

Air voids of asphalt layer = 4.5%
Effective asphalt content by volume = 12.2%
Thickness of the asphalt layer = 6.5 in.

a. If the above-mentioned condition sustains for 180 days, determine the amount of top-down longitudinal cracking in that pavement section.
b. If the above-mentioned condition sustains for 180 days, determine the amount of top-down longitudinal cracking in that pavement section at 90% reliability. The standard normal deviate at 90% reliability is 1.282.

Solution

Coefficients

$$C = \frac{1}{0.01 + \dfrac{12}{1 + e^{(15.676 - 2.8186 H_{\text{HMA}})}}} = \frac{1}{0.01 + \dfrac{12}{1 + e^{(15.676 - 2.8186(6.5))}}} = 0.08917$$

$$M = 4.84\left(\frac{V_{be}}{V_a + V_{be}} - 0.69\right) = 4.84\left(\frac{12.2}{4.5 + 12.2} - 0.69\right) = 0.196$$

$$C = 10^M = 10^{0.196} = 1.570$$

Fatigue Equation

$$N_{f,\mathrm{Top}} = k_{f1}C(C_H)\beta_{f1}(\varepsilon_{t,\mathrm{Top}})^{k_{f2}\beta_{f2}}(E_{\mathrm{HMA}})^{k_{f3}\beta_{f3}}$$

$$N_{f,\mathrm{Top}} = 0.007566(1.570)(0.08917)(1)\left(\frac{1}{\varepsilon_{t,\mathrm{Top}}}\right)^{3.9492(1)}\left(\frac{1}{E_{\mathrm{HMA}}}\right)^{1.281(1)}$$

Therefore, $N_{f,\mathrm{Top}} = 0.00106\left(\frac{1}{\varepsilon_{t,\mathrm{Top}}}\right)^{3.9492}\left(\frac{1}{E_{\mathrm{HMA}}}\right)^{1.281}$

Calculation of DI_{Top} (Table 9.6)

Time	n	E_{HMA} (psi)	$\varepsilon_{t,Top}$	$N_{f,Top}$	$\frac{n}{N_{f,Top}}$
12:00 a.m.–7:59 a.m.	1,000	800,000	35×10^{-6}	11,502,938	0.00009
8:00 a.m.–12:59 p.m.	8,000	700,000	32×10^{-6}	19,444,271	0.00041
1:00 p.m.–5:59 p.m.	6,000	300,000	35×10^{-6}	40,408,684	0.00015
6:00 p.m.–11:59 p.m.	4,000	500,000	39×10^{-6}	13,698,961	0.00029
$\mathrm{DI_{Top}} = \sum\left(\frac{n}{N_{f,\mathrm{Top}}}\right) =$					0.00094

Table 9.6 Calculation of the Cumulative Damage Index for Example 9.5

$\mathrm{DI_{Top}}$ after 180 days $= 0.00094 \times 180$ days $= 0.1689$

Calculation of Top-Down Longitudinal Cracking

a. Top-down longitudinal cracking at 50% reliability

$C_1 = 7.0$
$C_2 = 3.5$
$C_4 = 1{,}000$

$$\mathrm{FC_{Top}} = 10.56\left(\frac{C_4}{1+e^{(C_1-C_2\log(\mathrm{DI_{Top}}))}}\right) = 10.56\left(\frac{1{,}000}{1+e^{(7.0-3.5\log(0.1689))}}\right) = 0.6455 \text{ ft/mile}$$

$\mathrm{FC_{Top}}$ at 50% = 0.6455 ft/mile

b. Top-down longitudinal cracking at 90% reliability

The standard error ($S_{e(\mathrm{Top})}$) is listed below:

$$S_{e(\mathrm{Long})} = 200 + \frac{2{,}300}{1+e^{1.072-2.1654\log(\mathrm{FC_{Top}}+0.0001)}}$$

$$= 200 + \frac{2,300}{1 + e^{1.072 - 2.1654 \log(0.6455 + 0.0001)}}$$

$$= 625$$

$$FC_{Top} \text{ at } 90\% = 0.6455 + 1.282 \times 625$$

$$= 802 \text{ ft/mile}$$

Answers *The amount of top-down longitudinal cracking is 0.6 ft/mile at 50% reliability and 802 ft/mile at 90% reliability.*

The calculation of top-down longitudinal cracking is similar to that of alligator cracking with the exception of regression coefficients. The horizontal strain used in this calculation is the surface strain. Finally, the top-down longitudinal cracking obtained in this example is in feet per mile of a pavement. Alligator cracking is expressed as percent of lane. Therefore, do not get confused.

9.4 Rutting

Surface deformation in the form of rutting (Fig. 9.4) is caused by the plastic or the permanent vertical deformation in the AC, unbound layers, and foundation soil. The approach used in the AASHTOWare pavement ME design is based upon calculating incremental distortion or rutting within each sublayer. In other words, rutting is estimated for each sub-season at the mid-depth of each sublayer within the pavement structure. The plastic deformation for a given season is the sum of the plastic vertical deformations within each layer. The model for calculating total permanent deformation uses the plastic vertical strain under specific pavement conditions for the total number of trucks within that condition. Conditions vary from one month to another, so it is necessary to use a special approach called the strain hardening approach to incorporate those plastic vertical strains within each month in a cumulative deformation subsystem. The rate of accumulation of plastic deformation is measured in the laboratory using repeated load permanent deformation triaxial tests for both HMA mixtures and unbound materials. The laboratory-derived relationship is then adjusted to match the rut depth measured on the roadway. Rutting is calculated in inches of permanent deformation in the AASHTOWare pavement ME design software.

FIGURE 9.4 Rutting in pavements.

In summary, the AASHTOWare Pavement ME design software calculates the total rutting of pavement by summing up of ruts at different layers of pavement. To do so, it divides the pavement section into the maximum number of 10 layers or sublayers. The rut at each layer or sublayer is calculated by rutting models and finally summed up. The rutting model used by the AASHTOWare pavement ME design software to determine the rut in HMA layer/sublayer is given by the equation:

$$\Delta_{p(\mathrm{HMA})} = \varepsilon_{P(\mathrm{HMA})} h_{\mathrm{HMA}} = \beta_{1r} k_z \varepsilon_{r(\mathrm{HMA})} 10^{k_{1r}} n^{k_{2r}\beta_{2r}} T^{k_{3r}\beta_{3r}} \tag{9.10}$$

where $\Delta_{p(\mathrm{HMA})}$ = Accumulated permanent or plastic deformation in the HMA layer, in.
$\varepsilon_{P(\mathrm{HMA})}$ = Accumulated permanent strain in the HMA layer, in.
$\varepsilon_{r(\mathrm{HMA})}$ = Resilient or elastic strain calculated at the mid-depth of each layer, in./in.
T = Mix or pavement temperature, °F
h_{HMA} = HMA layer/sublayer thickness, in.
$k_z = (C_1 + C_2 D)0.328196^D$
$C_1 = -0.1039(H_{\mathrm{HMA}})^2 + 2.4868H_{\mathrm{HMA}} - 17.342$
D = Depth below the surface, in.
$C_2 = 0.0172(H_{\mathrm{HMA}})^2 - 1.7331H_{\mathrm{HMA}} + 27.428$
H_{HMA} = Total HMA thickness, in.
n = Number of axle repetitions
k_{1r}, k_{2r}, k_{3r} = Global field calibration parameters; $k_{1r} = -3.35412$, $k_{2r} = 0.4791$, $k_{3r} = 1.5606$
$\beta_{1r}, \beta_{2r}, \beta_{3r}$ = Calibration factors for the asphalt mixtures; for national calibration these constants are set to 1.0

For unbound layers, the rut is calculated using the equation:

$$\Delta_p = \beta_{s1} k_{s1} \varepsilon_v h \left(\frac{\varepsilon_o}{\varepsilon_r}\right) e^{-\left(\frac{\rho}{n}\right)^{\beta}} \tag{9.11}$$

where Δ_p = Plastic deformation for unbound layer/sublayer, in.
n = Number of axle load applications
ε_o = Intercept determined from laboratory-repeated load permanent deformation tests, in./in.
ε_r = Resilient strain imposed in laboratory test to obtain material properties such as ε_o, ε_r, and ρ
ε_v = Average vertical elastic strain in the layer/sublayer
h = Thickness of the unbound layer/sublayer, in.
k_{s1} = Global calibration coefficients; 2.03 for granular materials and 1.35 for fine materials
β_{s1} = Local or mixture field calibration constant; it is set to 1.0 for the global calibration
β = A function of water content
ρ = A function of β and resilient modulus

β can be calculated using the equation:

$$\text{Log}\beta = -0.61119 - 0.017638W_c \tag{9.12}$$

where W_c = water content, %.

ρ can be calculated using the equation:

$$\rho = 10^9\left(\frac{C_o}{1-\left(10^9\right)^\beta}\right)^{1/\beta} \tag{9.13}$$

where $C_o = Ln\left(\frac{a_1 M_r^{b_1}}{a_9 M_r^{b_9}}\right)$

M_r = Resilient modulus of the unbound layer, psi
$a_{1,9}$ = Regression constants: $a_1 = 0.15$ and $a_9 = 20.0$
$b_{1,9}$ = Regression constants: $b_1 = 0$ and $b_9 = 0$

The standard error (S_e) for the total rut depth is the sum of the standard error for the asphalt and unbound layer rut depths and is a function of the average predicted rut depth. Equations of the standard error (standard deviation of the residual errors) for the individual asphalt layer and unbound layers for coarse and fine-grained materials and soils are given below:

$$S_{e(\text{HMA})} = 0.24(\Delta_{\text{HMA}})^{0.8026} + 0.001 \tag{9.14}$$

$$S_{e(\text{Gran})} = 0.1235(\Delta_{\text{Gran}})^{0.5012} + 0.001 \tag{9.15}$$

$$S_{e(\text{Fine})} = 0.1477(\Delta_{\text{Fine}})^{0.6711} + 0.001 \tag{9.16}$$

where $S_{e(\text{HMA})}$ = Standard error of rutting in the asphalt layer, in.
$S_{e(\text{Gran})}$ = Standard error of rutting in the aggregate and coarse-grained layers, in.
$S_{e(\text{Fine})}$ = Standard error of rutting in the fine-grained layers, in.
Δ_{HMA} = Rutting in the asphalt layer, in.
Δ_{Gran} = Rutting in the aggregate and coarse-grained layers, in.
Δ_{Fine} = Rutting in the fine-grained layers, in.

Example 9.6: Rutting in HMA A flexible pavement has 4.5 in. of asphalt layer. While analyzing the asphalt layer is divided into three sublayers of 1.5 in. each. In a certain period, a total of 20,000 axles pass and the average temperatures of the top, middle, and the bottom layers are 90, 85, and 80°F, respectively. The elastic strain at the middle of each layer is 50 microstrains. Calculate the total HMA rut possible during that period.

Solution

Known:

$$\Delta_{P(\text{HMA})} = \beta_{1r} k_z \varepsilon_{r(\text{HMA})} 10^{k_{1r}} n^{k_{2r}\beta_{2r}} T^{k_{3r}\beta_{3r}}$$

$k_{1r} = -3.35412$

$k_{2r} = 0.4791$

$k_{3r} = 1.5606$

$\beta_{1r} = \beta_{2r} = \beta_{3r} = 1.0$ (as no calibration factors are provided)

$$C_1 = -0.1039(H_{\text{HMA}})^2 + 2.4868H_{\text{HMA}} - 17.342$$
$$= -0.1039(4.5\text{ in.})^2 + 2.4868(4.5\text{ in.}) - 17.342$$
$$= -8.2554$$

$$C_2 = 0.0172(H_{\text{HMA}})^2 + 1.7331H_{\text{HMA}} + 27.248$$
$$= 0.0172(4.5)^2 - 1.7331(4.5) + 27.428$$
$$= 19.9774$$

For the Top Layer

D = middle of the top layer = 0.75 in., $T = 90°\text{F}$

$$k_z = (C_1 + C_2D)\,0.328196^D = (-8.2554 + 19.9774(0.75))0.328196^{0.75} = 2.9172$$

$$\Delta_{P(\text{HMA})} = \beta_{1r}k_z\varepsilon_{r(\text{HMA})}10^{k_{1r}}n^{k_{2r}\beta_{2r}}T^{k_{3r}\beta_{3r}}$$
$$= (1.0)(2.9172)(50\times10^{-6})10^{(-3.35412)(1.0)}20{,}000^{(0.4791)(1.0)}90^{(1.5606)(1.0)}$$
$$= 0.0083\text{ in.}$$

For the Middle Layer

D = middle of the middle layer = 2.25 in., $T = 85°\text{F}$

$$k_z = (C_1 + C_2D)\,0.328196^D = (-8.2554 + 19.9774(2.25))0.328196^{2.25} = 2.9915$$

$$\Delta_{P(\text{HMA})} = \beta_{1r}k_z\varepsilon_{r(\text{HMA})}10^{k_{1r}}n^{k_{2r}\beta_{2r}}T^{k_{3r}\beta_{3r}}$$
$$= (1.0)(2.9915)(50\times10^{-6})10^{(-3.35412)(1.0)}20{,}000^{(0.4791)(1.0)}85^{(1.5606)(1.0)}$$
$$= 0.0078\text{ in.}$$

For the Bottom Layer

D = middle of the bottom layer = 3.75 in., $T = 80°\text{F}$

$$k_z = (C_1 + C_2D)\,0.328196^D = (-8.2554 + 19.9774(3.75))0.328196^{3.75} = 1.0218$$

$$\Delta_{P(\text{HMA})} = \beta_{1r}k_z\varepsilon_{r(\text{HMA})}10^{k_{1r}}n^{k_{2r}\beta_{2r}}T^{k_{3r}\beta_{3r}}$$
$$= (1.0)(1.0218)(50\times10^{-6})10^{(-3.35412)(1.0)}20{,}000^{(0.4791)(1.0)}80^{(1.5606)(1.0)}$$
$$= 0.0024\text{ in.}$$

$$\text{Total HMA rut} = \sum\Delta_{P(\text{HMA})} = 0.0083 + 0.0078 + 0.0024 = 0.0186\text{ in.}$$

Answer *The total HMA rut is 0.0186 in.*

9.5 Transverse Cracking

Transverse cracking (also known as thermal cracking) of asphalt pavement, shown in Fig. 9.5, is one of the important distresses especially observed in cold regions. It is not a load-related crack and occurs due to a single extreme, low-temperature or repetitive low-temperature cycles. The AASHTOWare pavement ME design software assumes the fracture mechanics approach to calculate the thermal cracking. Thermal cracking is calculated as feet per mile in the AASHTOWare pavement ME design software.

The amount of crack propagation induced by a given thermal cooling loading cycle is predicted by using the Paris law of crack propagation as depicted in the equation:

$$\Delta C = A(\Delta K)^n \tag{9.17}$$

where ΔC = Change in the crack depth due to a cooling cycle
ΔK = Change in the stress intensity factor due to a cooling cycle

The parameters A, n are the fracture parameters for the asphalt mixture.

$$A = k_t \beta_t 10^{(4.389 - 2.52\log(E_{\text{HMA}}\sigma_m n))} \tag{9.18}$$

Note that the symbol n represents number of axle repetitions in fatigue or rutting models. For transverse cracking model, the n-value is expressed as depicted in the equation:

$$n = 0.8\left(1 + \frac{1}{m}\right) \tag{9.19}$$

where k_t = Coefficient determined by global calibration for each input level (Level 1 = 1.5, Level 2 = 0.5, and Level 3 = 1.5)
E_{HMA} = HMA indirect tensile modulus, psi
σ_m = Mixture tensile strength, psi
m = The m-value derived from the indirect tensile creep compliance curve measured in the laboratory
β_t = Local or mixture calibration factor; the default value is 1.0

FIGURE 9.5 Transverse cracking in pavements.

The stress intensity factor, K, has been incorporated in the AASHTOWare pavement ME design software using a simplified equation developed from theoretical finite element studies, as depicted in the equation:

$$K = \sigma_{\text{tip}}\left[0.45 + 1.99(C_o)^{0.56}\right] \tag{9.20}$$

where σ_{tip} = Far-field stress from pavement response model at depth of crack tip, psi
C_o = Current crack length, ft

The amount of transverse thermal cracking is determined by Eq. (9.21) using an assumed relationship between the probability distribution of the log of the crack depth to HMA-layer thickness ratio and the percent of cracking:

$$\text{TC} = \beta_{t1} N\left[\frac{1}{\sigma_d}\log\left(\frac{C_d}{H_{\text{HMA}}}\right)\right] \tag{9.21}$$

where TC = Observed amount of thermal cracking, ft/mile
β_{t1} = Regression coefficient determined through global calibration, 400
$N[z]$ = Standard normal distribution evaluated at $[z]$
σ_d = Standard deviation of the log of the depth of cracks in the pavement, 0.769 in.
C_d = Crack depth, in.
H_{HMA} = Thickness of HMA layer

The standard error for the transverse cracking prediction equations for the three input levels is depicted in the equations:

$$S_{e(\text{Level 1})} = -0.1468(\text{TC} + 65.027) \tag{9.22}$$

$$S_{e(\text{Level 2})} = -0.2841(\text{TC} + 55.462) \tag{9.23}$$

$$S_{e(\text{Level 3})} = 0.3972(\text{TC} + 20.422) \tag{9.24}$$

Example 9.7: Transverse Tracking A flexible pavement has a 10 in. of asphalt layer. The asphalt mixture has a tensile strength of 150 psi, and tensile modulus of 500,000 psi. The m-value of the mixture derived from the indirect tensile creep compliance curve measured in the laboratory is 0.17. Assume this layer has a preexisting crack with the depth of 1.2 in. and length of 8 ft. Finite element analysis shows the stress at 1.2-in. depth far away from the crack tip is 12 psi. Upon a cooling cycle is applied, the stress increases to 100 psi. Calculate the increase in depth of the crack after a single cooling cycle is applied. Assume Level 1 analysis.

Solution

Tensile modulus, $E_{\text{HMA}} = 500{,}000$
Tensile strength, $\sigma_m = 150$ psi
Far-field stress from pavement response model at depth of crack tip, $\sigma_{\text{tip}} = 100$ psi
Current crack length, $C_o = 8$ ft
Coefficient determined by global calibration, $k_t = 1.5$ for Level 1 analysis
Local or mixture calibration factor, $\beta_t = 1.0$
$m = 0.17$

$$n = 0.8\left(1 + \frac{1}{m}\right) = 0.8\left(1 + \frac{1}{0.17}\right) = 5.506$$

$$\begin{aligned} A &= k_t \beta_t 10^{(4.389 - 2.52\log(E_{\text{HMA}}\sigma_m n))} \\ &= (1.5)(1.0)10^{(4.389 - 2.52\log(500{,}000\times150\times5.506))} \\ &= 7.12\times10^{-18} \end{aligned}$$

$$\Delta K = \sigma_{\text{tip}}\left[0.45 + 1.99(C_o)^{0.56}\right] = (100)\left[0.45 + 1.99(8.0)^{0.56}\right] = 682.65$$

$$\Delta C = A(\Delta K)^n = (7.12\times10^{-18})(682.65)^{5.506} = 0.0287\,\text{ft} = 0.344\ \text{in.}$$

Answer *The increase in depth of the crack is 0.344 in.*

9.6 International Roughness Index

The International Roughness Index (IRI) is the roughness index most commonly obtained from measured mean longitudinal road profiles of each wheel path. IRI is calculated using a quarter-car vehicle math model, whose response is accumulated to yield a roughness index with units of slope (in./mile). Since its introduction in 1986, IRI has become the road roughness index most commonly used worldwide for evaluating and managing road systems (Sayers et al., 1986). The design premise included in the AASHTOWare pavement ME design software for predicting smoothness degradation is that the occurrence of surface distress results in increased roughness (increasing IRI value), or in other words, a reduction in smoothness; see Fig. 9.6. IRI for new HMA pavements and HMA overlays of flexible pavements are calculated using the equation:

$$\text{IRI} = \text{IRI}_o + C_1(\text{RD}) + C_2(\text{FC}_{\text{Total}}) + C_3(\text{TC}) + C_4(\text{SF}) \tag{9.25}$$

where IRI_o = Initial IRI after construction, in./mile
SF = Site factor
FC_{Total} = Area of fatigue cracking (combined alligator, longitudinal, and reflection cracking in the wheel path), % of total lane area. All load-related cracks are combined on an area basis–length of cracks is multiplied by 1 ft to convert length into an area basis
TC = Length of transverse cracking (including the reflection of transverse cracks in existing HMA pavements), ft/mile
RD = Average rut depth, in.
$C_{1,2,3,4}$ = Calibration factors; $C_1 = 40.0$, $C_2 = 0.40$, $C_3 = 0.008$, $C_4 = 0.015$

The site factor (SF) is calculated in accordance with the equation:

$$\text{SF} = \text{Age}^{1.5}\left\{\ln\left[(\text{Precip}+1)(\text{FI}+1)p_{02}\right]\right\} + \left\{\ln\left[(\text{Precip}+1)(\text{PI}+1)p_{200}\right]\right\} \tag{9.26}$$

where Age = Pavement age, year
PI = Plasticity index of the soil
FI = Annual average freezing index, °F days

Smooth pavement Rough pavement

FIGURE 9.6 Roughness in pavement. (*Courtesy of Nasim News.*)

Precip = Annual average precipitation or rainfall, in.
p_{02} = Passing the 0.02-mm sieve, %
p_{200} = Passing the 0.075-mm (No. 200) sieve, %

The above-mentioned IRI equation shows IRI is calculated empirically as a function of pavement distresses, site factors that represent the foundation's shrink/swell and frost heave capabilities, and an estimate of the IRI at the time of construction (the initial IRI). The unit of smoothness calculated by the AASHTOWare pavement ME design software is inches per mile. IRI for HMA overlays of rigid pavements is calculated using the equation:

$$\text{IRI} = \text{IRI}_o + \text{PCC}C_1(\text{RD}) + \text{PCC}C_2(\text{FC}_{\text{Total}}) + \text{PCC}C_3(\text{TC}) + \text{PCC}C_4(\text{SF}) \quad (9.27)$$

where PCC $C_{1,2,3,4}$ = Calibration factors
PCC C_1 = 40.8
PCC C_2 = 0.575
PCC C_3 = 0.0014
PCC C_4 = 0.00825

Equations (9.25) to (9.27) are developed from data collected within the LTPP program and are embedded in the AASHTOWare pavement ME design software to predict the IRI over time for HMA-surfaced pavements.

The standard error of the estimate for new flexible pavements and HMA overlays of flexible and semi-rigid pavements is 18.9 in./mile and for HMA overlays of intact PCC pavements it is 9.6 in./mile. The AASHTOWare pavement ME design software assumes that the standard error for HMA overlays of fractured PCC pavements is the same as for HMA overlays of intact PCC pavements.

Example 9.8: International Roughness Index A flexible pavement site has the following conditions:

Initial IRI = 67

After 12 years of service:

Total fatigue cracking including bottom-up, top-down, and reflection = 8% of lane area
Transverse cracking = 500 ft/mile
Total rutting = 0.17 in.
Plasticity index of subgrade = 18
Freezing index = 50°F days
Annual average precipitation = 0.11 in.
Percent passing the 0.02-mm sieve = 1.2%
Percent passing the 0.075-mm (No. 200) sieve = 3.6%

Determine the IRI after 12 years of service.

Solution

Given data:

Initial IRI, $IRI_o = 67$
Total fatigue cracking, $FC_{Total} = 8\%$
Transverse cracking, TC = 500 ft/mile
Total rutting, RD = 0.17 in.
Plasticity index of subgrade, PI = 18
Freezing index, FI = 50°F days
Annual average precipitation, Precip = 0.11 in.
Percent passing the 0.02-mm sieve, $p_{02} = 1.2\%$
Percent passing the 0.075-mm (No. 200) sieve, $p_{200} = 3.6\%$

Site Factor

$$SF = Age^{1.5}\{\ln[(Precip + 1)(FI + 1)p_{02}]\} + \{\ln[(Precip + 1)(PI + 1)p_{200}]\}$$

$$SF = (12)^{1.5}\{\ln[(0.11 + 1)(50 + 1)(1.2)]\} + \{\ln[(0.11 + 1)(18 + 1)(3.6)]\}$$

$$SF = (12)^{1.5}(4.2185) + 4.3297$$

$$SF = 179.6895$$

International Roughness Index (IRI)

$$IRI = IRI_o + C_1(RD) + C_2(FC_{Total}) + C_3(TC) + C_4(SF)$$

$$IRI = IRI_o + 40(RD) + 0.4(FC_{Total}) + 0.008(TC) + 0.015(SF)$$

$$= 67 + 40(0.17) + 0.4(8.0) + 0.008(500) + 0.015(179.6895)$$
$$= 83.69$$

Answer *The IRI is 83.69 in./mile.*

9.7 Reflective Cracking in HMA Overlay

The AASHTOWare pavement ME design software predicts reflection cracks (Fig. 9.7) in HMA overlays or HMA surfaces of semi-rigid pavements using an empirical equation. The empirical equation is used for estimating the amount of fatigue and thermal cracks from a non-surface layer that has reflected to the surface after a certain period. This empirical equation predicts the percent of area of cracks that propagate through the HMA as a function of time with width of cracks being 1 ft using a sigmoid function, as depicted in the equation:

$$RC = \frac{100}{1 + e^{a(c)+bt(d)}} \tag{9.28}$$

where RC = Reflection cracking (%)
t = Time, year
a, b = Regression fitting parameters defined through calibration process
c, d = User-defined cracking progression parameters

The empirical equation also is used to estimate the reflection of fatigue and thermal cracks from a stabilized layer or existing flexible pavement, as well as from joints and cracks in a rigid pavement. The regression fitting parameters (a and b) are a function of the effective HMA overlay thickness (Heff), the type of existing pavement, and for PCC pavements, load transfer at joints and cracks, as shown below. The user-defined cracking progression parameter d can be used by the user to accelerate or delay the amount of reflection cracks, which also are included in Table 9.7. Non-unity cracking progression parameters (c and d) could be used with caution, after they have been calibrated locally.

$$a = 3.5 + 0.75(H_{eff}) \tag{9.29}$$

$$b = -0.688684 - 3.37302(H_{eff})^{-0.915469} \tag{9.30}$$

FIGURE 9.7 Reflective cracking in pavement.

Effective overlay thickness (H_{eff}), in.	d	
	Delay cracking by 2 years	Accelerate cracking by 2 years
Less than 4.0	0.6	3.0
4.0 to 6.0	0.7	1.7
More than 6.0	0.8	1.4

Source: *From AASHTO. (2015). Mechanistic-Empirical Pavement Design Guide: A Manual of Practice. Washington, DC: American Association of State Highway and Transportation Officials. Table 5-2. Used with permission.*

Table 9.7 Reflective Cracking Fitting and Cracking Progression Parameter, *d*

where For flexible pavement, $H_{eff} = H_{\text{HMA}}$
For rigid pavement, $H_{eff} = H_{\text{HMA}} - 1$ and $H_{eff} = H_{\text{HMA}} - 3$ for good and poor load transfer condition, respectively.
$c = 1.0$

After HMA overlay placement, the underlying bound layers (all HMA layers, asphalt-bound layers, chemically stabilized layers, and PCC layers) undergo load-related damage with continued truck loadings. The continual fatigue damage accumulation of these layers is considered in the AASHTOWare pavement ME design software HMA overlay analysis procedure. For any given month, *m*, the total fatigue damage is estimated by the equation:

$$\text{DI}_m = \sum_{i=1}^{m} \Delta\text{DI}_i \tag{9.31}$$

where DI_m = Damage index for month, *m*
ΔDI_i = Increment of damage index in month, *i*

The minimum recommended HHMA is 2 in. for existing flexible pavements, 3 in. for existing rigid pavement with good load transfer, and 4 in. for existing rigid pavements with poor load transfer.

The area of fatigue damage for the underlying layer at month *m* (CA_m) is given by the equation:

$$\text{CA}_m = \frac{100}{1 + e^{6-(6\text{DI}_m)}} \tag{9.32}$$

For each month *i*, there will be an increment of damage ΔDI_i which will cause an increment of cracking area CA_i to the stabilized layer. To estimate the amount of cracking reflected from the stabilized layer to the surface of the pavement for month *m*, the reflective cracking prediction equation is applied incrementally, in accordance with the equation:

$$\text{TRA}_m = \sum_{i=1}^{m} \text{RC}_t(\Delta\text{CA}_i) \tag{9.33}$$

where TRA_m = Total reflected cracking area for month m
RC_t = % cracking reflected for age t (in years)
ΔCA_i = Increment of fatigue cracking for month i

Example 9.9: Reflective Cracking An asphalt pavement had total HMA thickness of 6 in. The top 3 in. has been milled off and another 4-in. overlay has been applied. The design agency wants to delay the reflective cracking by 2 years. Calculate the expected amount of reflection cracks in HMA overlay after 5 years. Neglect the fatigue damage.

Solution

Effective overlay thickness, $H_{eff} = (6 - 3) + 4 = 7$ in.

$a = 3.5 + 0.75(H_{eff}) = 3.5 + 0.75(7) = 8.75$

$$\begin{aligned} b &= -0.688684 - 3.37302(H_{eff})^{-0.915469} \\ &= -0.688684 - 3.37302(7)^{-0.915469} \\ &= -1.2567 \end{aligned}$$

$c = 1.0$
$d = 0.8$
$t = 5$ years

Amount of reflection cracks, $\text{RC} = \dfrac{100}{1 + e^{ac+btd}}$

$\text{RC} = 2.36\%$

Answer *The expected amount of reflection cracks in HMA overlay is 2.36%.*

9.8 Recommended Design-Performance Criteria

Performance criteria are used to ensure that a pavement design performs satisfactorily over its design life. The designer selects critical limits or threshold values considering the type of pavement such as highways versus local roads. Unless any threshold values are assigned the software selects the default values. Table 9.8 provides values for considerations by highway agencies, realizing that these levels may vary between agencies based on their local conditions.

Performance criteria	Threshold value at end of design life
Alligator cracking (bottom-up fatigue cracking)	Interstate: 10% of lane area Primary: 20% of lane area Secondary: 35% of lane area
Total rutting	Interstate: 0.40 in. Primary: 0.50 in. Others: 0.65 in.
Transverse cracking	Interstate: 500 ft/mile Others: 700 ft/mile
IRI	Interstate: 160 in./mile Others: 200 in./mile

Source: *From AASHTO. (2015). Mechanistic-Empirical Pavement Design Guide: A Manual of Practice. Washington, DC: American Association of State Highway and Transportation Officials. Table 7-1. Used with permission.*

TABLE 9.8 Recommended Threshold Values for Trial Design of HMA Pavement and Overlays

9.9 Reliability

Design reliability (R) is defined as the probability (P) that the predicted distress is less than the critical level over the design period as depicted in the equation:

$$R = P[\text{Distress over design period} < \text{Critical distress level}] \tag{9.34}$$

Design reliability is defined as follows for smoothness (IRI), as depicted in the equation:

$$R = P[\text{IRI over design period} < \text{Critical IRI level}] \tag{9.35}$$

It is possible to incorporate reliability while designing pavement for all pavement types. The designer may select the desired level of reliability for each distress type and smoothness or may choose the default value. The level of design reliability could be based on the general consequence of reaching the threshold value before the design life. More specifically, if 100 projects were designed using the AASHTOWare pavement ME design software and each had a design reliability for fatigue cracking of 95%, five of those projects would show more than the threshold or terminal value of fatigue cracking at the end of the design period. The design reliability could be selected in balance with the performance criteria. For example, the selection of a high-design reliability value (e.g., 99%) and a low-performance criterion (3% alligator cracking) might make it impossible or certainly costly to obtain an adequate design. Table 9.9 provides values that are believed to be in balance with the performance criteria included in Table 9.2 and are suggested for use in design.

Functional classification	Level of reliability	
	Urban	Rural
Interstate/freeways	95	95
Principal arterials	90	85
Collectors	80	75
Local	75	70

Source: *From AASHTO. (2015). Mechanistic-Empirical Pavement Design Guide: A Manual of Practice. Washington, DC: American Association of State Highway and Transportation Officials. Table 7-2. Used with permission.*

Table 9.9 Level of Reliability for Different Functional Classifications of Roadway

9.10 Calibration of Local Calibration Coefficients

The distress models described in this chapter use some local calibration coefficients. For example, the bottom-up fatigue cracking equations use β_{f1}, β_{f2}, β_{f3}, C_1, C_2, and C_4 as local or mixture-specific field calibration constants. For global calibration effort, all these constants are set to 1.0 except $C_4 = 6{,}000$. There are other local calibration factors in other distress models as well. However, the local pavement materials, pavement geometry, climate, and traffic pattern may not reflect the average global conditions. Hence, local calibrations of these calibration coefficients are essential before using them for pavement design. Many design agencies are calibrating these local calibration factors through laboratory testing, field testing, and field investigation. The step-wise procedure to be followed to calibrate the local calibration coefficients can be as follows:

Step 1. Design a pavement using the software with Level 1 analysis using detailed materials properties. The material properties must have been determined using appropriate laboratory or field testing.

Step 2. Construct the pavement and collect the field-distress data with time, say each month (alligator cracking, top-down longitudinal cracking, thermal cracking, rutting, and IRI).

Step 3. Compare the software-predicted distress and the measured distress data.

Step 4. Find out the revised local calibration coefficients to minimize the error of the prediction using regression or appropriate statistical analysis.

The above-mentioned calibration procedure for local calibration coefficients can be presented using a flowchart shown in Fig. 9.8.

Some researchers already performed some studies to find out the local calibration coefficients for their own regions. Some of the obtained local calibration coefficients are listed in Tables 9.10, 9.11, and 9.12. These calibration coefficients are the best obtained values at the time of study. Local calibration is an ongoing process and the coefficients improve with the ongoing data. Thus, these local calibration values may change with time. The authors present these values here to have an idea of how the local calibration coefficients may vary region to region.

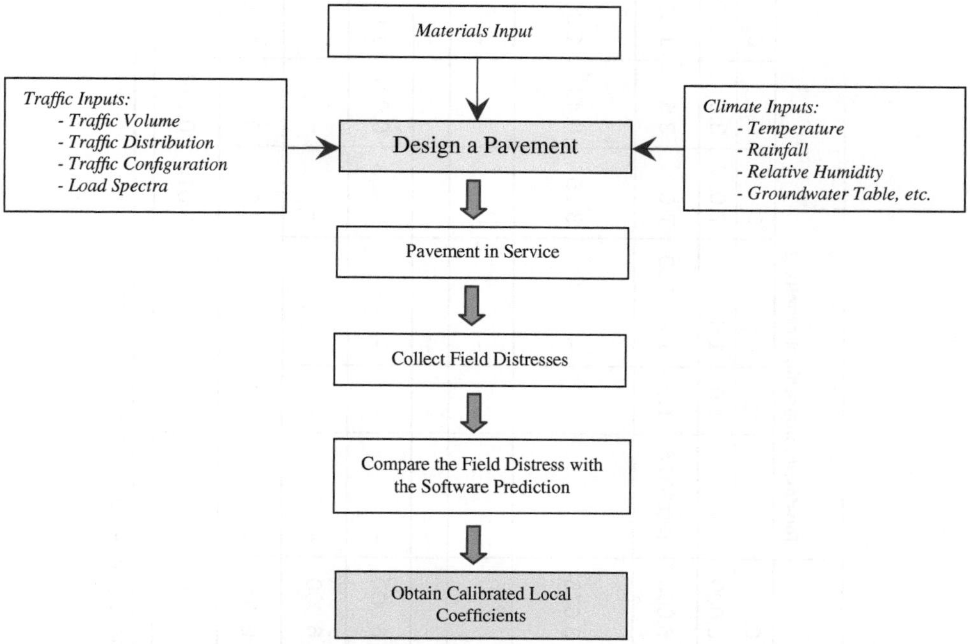

Figure 9.8 Calibration process of local calibration coefficients.

9.11 Summary

This chapter discusses the fundamental distress calculation methods with numerical examples. The newly developed AASHTOWare pavement ME design software uses the above-described approaches to predict the amount of major distresses. The examples discussed in this chapter and the practice problems are short enough to be solved by hand calculations. The software performs numerical analysis to find out the structural responses, and possible distresses for each axle of loading in pavement. More specifically, the software considers each axle, its time of passing, temperature profile at that time, modulus of asphalt at that time, etc. and then calculates the structural responses. From the responses, it calculates the amount of distresses. Then the amount of distresses is summed up to determine the total distresses with the pavement's service life. The calibration procedure for local calibration coefficients is also discussed with some findings from different researchers. The next chapter discusses the design procedure using step-by-step screenshots of the AASHTOWare pavement ME design software.

Reference	Study state	Bottom-up fatigue						Top-down longitudinal cracking					
		β_{f1}	β_{f2}	β_{f3}	C_1	C_2	C_4	β_{f1}	β_{f2}	β_{f3}	C_1	C_2	C_4
Default	–	1.0	1.0	1.0	1.0	1.0	6,000	1.0	1.0	1.0	7.0	3.5	1,000
CDOT (2019)	CO	130.3674	1.0	1.217799	0.021	2.35	6,000	130.3674	1.0	1.217799	7.0	3.5	1,000
Kasperick and Ksaibati (2015)	WY	–	–	–	0.3	0.535	3,675	–	–	–	3.25	0.43	1,300
FHWA (2010)	NC	1.41	–2.82	–6.67	–	–	–	–	–	–	–	–	–
Guo (2013)	AL	1.0	1.0	1.0	–	–	–	–	–	–	–	–	–
Rahman (2014)	OR	–	–	–	0.560	0.225	6,000	–	–	–	1.453	0.097	1,000
Darter et al. (2014)	AZ	249.0087	1.0	1.2334	1.0	4.5	6,000	–	–	–	–	–	–
Smith and Nair (2015)	VA	42.87	1.0	1.0	0.3190	0.3190	6,000	1.0	1.0	1.0	7.0	3.5	1,000
Hall et al. (2011)	AR	–	–	–	0.688	0.294	6,000	–	–	–	3.016	0.216	1,000

TABLE 9.10 Calibrated Coefficients for Bottom-up Fatigue and Top-down Longitudinal Cracking Obtained by Different Researchers

Reference	Study state	AC rutting						Granular subgrade rutting		Fine subgrade rutting	
	–	β_{1r}	β_{2r}	β_{3r}	k_{1r}	k_{2r}	k_{3r}	$-\beta_{s1}$	$-k_{s1}$	$-\beta_{s1}$	$-k_{s1}$
Default	–	1.0	1.0	1.0	–3.35412	0.4791	1.5606	–1.0	–2.03	–1.0	–1.35
CDOT (2019)	CO	7.6742	1.0	1.0	–3.35412	0.3791	1.5606	–0.22	–2.03	–0.37	–1.35
Kasperick and Ksaibati (2015)	WY	1.09	1.0	1.0	–	–	–	–0.95	–	–0.69	–
FHWA (2010)	NC	1.52	4.24	–0.75	–	–	–	–	–	–	–
Guo (2013)	AL	1.0	1.0	1.0	–	–	–	–	–	–	–
Rahman (2014)	OR	1.48	1.0	0.90	–	–	–	0	–	0	–
Darter et al. (2014)	AZ	0.69	1.0	1.0	–3.35412	1.5606	0.4791	–0.14	–2.03	–0.37	–1.35
Smith and Nair (2015)	VA	0.687	1.0	1.0	–3.35412	0.4791	1.5606	–0.153	–2.03	–0.153	–1.35
Hall et al. (2011)	AR	1.20	1.0	0.80	–	–	–	–1.0	–	–0.50	–

Table 9.11 Calibrated Coefficients for Rutting Obtained by Different Researchers

Reference	Study state	IRI (Flexible)				Thermal cracking			
	–	C_1	C_2	C_3	C_4	k_t (Level 1)	k_t (Level 2)	k_t (Level 3)	β_{t1}
Default	–	40	0.040	0.008	0.015	1.5	0.5	1.5	400
CDOT (2019)	CO	50	0.55	0.0111	0.02	6.3	0.5	6.3	
Kasperick and Ksaibati (2015)	WY	27	0.48	0.0015	0.0152	–	–	–	–
FHWA (2010)	NC	–	–	–	–	–	–	–	4,224
Darter et al. (2014)	AZ	1.2281	0.1175	0.008	0.0280	1.5	0.5	1.5	

TABLE 9.12 Calibrated Coefficients for IRI and Thermal Cracking Obtained by Different Researchers

9.12 Fundamentals of Engineering (FE) Exam-Style Questions

FE9.1 Fatigue damage decreases with the increase in the:

A. Modulus of asphalt
B. Tensile strain at the bottom of asphalt
C. Traffic volume
D. None of the above

Solution D

The allowable number of load repetitions for bottom-up cracking ($N_{f,\text{Bottom}}$) is given by:

$$N_{f,\text{Bottom}} = k_{f1}C(C_H)\beta_{f1}(\varepsilon_{t,\text{Bottom}})^{k_{f2}\beta_{f2}}(E_{\text{HMA}})^{k_{f3}\beta_{f3}}$$

where $\varepsilon_{t,\text{Bottom}}$ = Tensile strain at the bottom of AC layer, in./in.
E_{HMA} = Dynamic modulus of AC measured in compression, psi
$k_{f_1}, k_{f_2}, k_{f_3} = 1.0$
$k_{f_1} = 0.007566$
$k_{f_2} = -3.9492$
$k_{f_3} = -1.281$

It can be seen that the power of dynamic modulus ($k_{f_3} = -1.281$) is negative, meaning fatigue life decreases (fatigue damage increases) with the increase in the modulus of asphalt. Similarly, the power tensile strain at the bottom of asphalt ($k_{f_2} = -3.9492$) is negative, meaning fatigue life decreases (fatigue damage increases) with the increase in tensile strain at the bottom of asphalt.

Fatigue damage increases with the increase in the traffic volume according to Miner's law.

FE9.2 Permanent deformation decreases with the increase in the (select all that apply):

A. Modulus of asphalt
B. Temperature of asphalt
C. Traffic volume
D. None of the above

Solution A

The permanent deformation of asphalt is determined using the following equation:

$$\Delta_{p(\text{HMA})} = \varepsilon_{p(\text{HMA})} h_{\text{HMA}} = \beta_{1r} k_z \varepsilon_{r(\text{HMA})} 10^{k_{1r}} n^{k_{2r}\beta_{2r}} T^{k_{3r}\beta_{3r}}$$

- Higher the modulus, lower the strain, and thus permanent deformation decreases with the increase in the modulus of asphalt. Thus, option A is true.
- The power of T is positive in the above equation. This means permanent deformation increases with the increase in the temperature of asphalt.
- The power of n is positive in the above equation. This means permanent deformation increases with the increase in the traffic volume.

Alternative: Applying common sense—higher the modulus, lower the strain or deformation. Higher the temperature, softer the asphalt, and thus produces higher deformation with the increase in the temperature of asphalt. In addition, the higher the load repetitions, the larger the permanent deformation.

9.13 Practice Problems

9.1 A pavement section has accumulated a total of 10% bottom-up fatigue damage. Calculate the fatigue cracking, given that the AC layer has a thickness of 4.5 in.

9.2 The single-axle information of a pavement site is provided in Table P9.2.

Period	Average repetition	Tensile strain produced at the bottom of HMA (period average)	HMA dynamic modulus (psi)
Spring	110,000	200×10^{-6}	1,200,000
Summer	110,000	300×10^{-6}	340,000
Fall	113,000	250×10^{-6}	850,000

TABLE P9.2 Load and Response Data for Prob. 9.2

The pavement section has the following properties:

Air voids of asphalt layer = 5.5%
Effective asphalt content by volume = 12.5%
Thickness of the asphalt layer = 4.0 in.

a. Calculate the amount of alligator cracking produced by the single axle after 10 years, if the above-mentioned information remains the same.
b. Calculate the amount of alligator cracking produced by the single axle after 10 years at 90% reliability, if the above-mentioned information remains the same.

9.3 The single-axle information of a pavement site is provided in Table P9.3.

Period	Average repetition	Tensile strain produced at the bottom of HMA (period average)	HMA dynamic modulus (psi)
Spring	40,000	150×10^{-6}	1,400,000
Summer	19,000	350×10^{-6}	650,000
Fall	11,000	275×10^{-6}	1,050,000

TABLE P9.3 Load and Response Data for Prob. 9.3

The pavement section has the following properties:

Air voids of asphalt layer = 6.5%
Effective asphalt content by volume = 14.5%
Thickness of the asphalt layer = 6.0 in.

The local calibration coefficients C_1, C_2, and C_4 are 0.021, 2.35, and 5,500, respectively.

a. Calculate the amount of alligator cracking produced by the single axle after 10 years, if the above-mentioned information remains the same.
b. Calculate the amount of alligator cracking produced by the single axle after 10 years at 90% reliability, if the above-mentioned information remains the same.

9.4 An asphalt pavement section in a construction zone has the traffic and structural response data in a day, as listed in Table P9.4.

Time	Axle repetition	HMA dynamic modulus (ksi)	Tensile strain at the surface (period average)
12:00 a.m.–7:59 a.m.	12,000	800	150×10^{-6}
8:00 a.m.–12:59 p.m.	16,000	700	170×10^{-6}
1:00 p.m.–5:59 p.m.	13,000	300	185×10^{-6}
6:00 p.m.–11:59 p.m.	9,000	500	190×10^{-6}

TABLE P9.4 Load and Response Data for Prob. 9.4

The pavement section has the following properties:

Air voids of asphalt layer = 6.5%
Effective asphalt content by volume = 14.5%
Thickness of the asphalt layer = 4.5 in.

a. If the above-mentioned condition sustains for 10 months, determine the amount of top-down longitudinal cracking in that pavement section.
b. If the above-mentioned condition sustains for 10 months, determine the amount of top-down longitudinal cracking in that pavement section at 90% reliability.

9.5 An asphalt pavement section in a construction zone has the traffic and structural response data in a day, as listed in Table P9.5.

Time	Annual axle repetition	Annual average HMA dynamic modulus (ksi)	Tensile strain at the surface (period average)
Single axle	120,000	450	65×10^{-6}
Tandem axle	60,000		72×10^{-6}
Tridem axle	30,000		75×10^{-6}
Quad axle	9,000		70×10^{-6}

TABLE P9.5 Load and Response Data for Prob. 9.5

The pavement section has the following properties:

Air voids of asphalt layer = 5.5%
Effective asphalt content by volume = 12.5%
Thickness of the asphalt layer = 5.5 in.

The local calibration coefficients C_1, C_2, and C_4 are 3.5, 0.43, and 1,300, respectively.

a. If the above-mentioned condition sustains for 8 years, determine the amount of top-down longitudinal cracking in that pavement section.
b. If the above-mentioned condition sustains for 8 years, determine the amount of top-down longitudinal cracking in that pavement section at 90% reliability.

9.6 A flexible pavement has 6.0 in. of asphalt layer. While analyzing, the asphalt layer is divided into three sublayers of 2.0 in. each. In a certain period, a total of 50,000 axles pass and the average temperatures of the top, middle, and the bottom layers are 140, 130, and 120°F, respectively. The elastic strain at the middle of each layer is 120 microstrains. Calculate the total HMA rut possible during that period.

9.7 A flexible pavement has 9.0 in. of asphalt layer. While analyzing, the asphalt layer is divided into three sublayers of 3.0 in. each. In a certain period, a total of 350,000 axles pass and the average temperatures of the top, middle, and the bottom layers are 30, 40, and 50°F, respectively. The elastic strain at the middle of each layer is 150 microstrains. Calibration factors for the asphalt mixtures β_{1r}, β_{2r}, and β_{3r} are 1.25, 0.90, and 1.75, respectively. Calculate the total HMA rut possible during this period.

9.8 An unbound aggregate base has a thickness of 6.0 in. In a certain period, a total of 350,000 axles pass and the average conditions of this base layer are listed below:

- Water content = 7.5%
- Resilient modulus = 28,000 psi
- Resilient strain imposed in laboratory testing = 0.001 in./in.
- Intercept determined from laboratory permanent deformation tests = 0.002 in./in.
- Average vertical elastic strain in the layer = 0.0025 in./in.
- Natural water content of the base layer = 7.5%

Calculate the rut possible in this base layer during this period.

9.9 A flexible pavement has 4 in. of asphalt layer. The asphalt mixture has a tensile strength of 180 psi, and tensile modulus of 500,000 psi. The m-value of the mixture derived from the indirect tensile creep compliance curve measured in the laboratory is 0.21. Assume this layer has a preexisting crack with the depth of 0.2 in. and length of 12 ft. Finite element analysis shows the stress at 1.2-in. depth far away from the crack tip is 12 psi. Upon a cooling cycle is applied, the stress increases to 115 psi. Calculate the increased depth of the crack after a single cooling cycle is applied.

9.10 An asphalt pavement has a total HMA thickness of 8 in. The top 3 in. has been milled off and another 4-in. overlay has been applied. The design agency wants to delay the reflective cracking by 2 years. Consider the increment in fatigue damage index for the underlying layer is 40% in 5 years. Calculate the expected total amount of reflection cracks in HMA overlay after 5 years.

9.11 A flexible pavement site has the following conditions:

Initial IRI = 62

After 10 years of service:

Bottom-up fatigue cracking = 3% of lane area
Top-down fatigue cracking = 6.5% of lane area
Transverse cracking = 900 ft/mile
Total rutting = 0.23 in.
Plasticity index of subgrade = 22
Freezing index = 120°F days
Average annual precipitation = 0.85 in.
Percent passing the 0.02-mm sieve = 7.2%
Percent passing the 0.075-mm (No. 200) sieve = 13.6%

Determine the IRI after 10 years of service.

9.12 An asphalt pavement has a total HMA thickness of 6 in. The top 3 in. has been milled off and another 4-in. overlay has been applied. The design agency wants to delay the reflective cracking by 2 years. Consider the increment in fatigue damage index for the underlying layer is 40% in 5 years. Calculate the expected total amount of reflection cracks in HMA overlay after 5 years.

CHAPTER 10

Flexible Pavement Design by AASHTOWare

10.1 Background

This chapter discusses flexible pavement design procedure using the AASHTOWare pavement ME design method. The latest version of the software is available at https://me-design.com/MEDesign/. Many types of flexible pavements such as conventional flexible sections, deep strength sections, full-depth sections, and semi-rigid sections can be analyzed using the AASHTOWare pavement ME design software. In the AASHTO 1993, the old empirical method, all input parameters such as materials properties, traffic, and climate are lightly characterized. For example, the asphalt material is represented by only a layer coefficient. This means the asphalt mix is not differentiated by its dynamic modulus, tensile strength, fatigue performance, binder type, etc. More specifically, all types of asphalt materials are just judged based on a layer coefficient that is derived based on its elastic/resilient modulus. The performance of asphalt materials such as bottom-up fatigue cracking, top-down longitudinal cracking, rutting, and thermal cracking are not evaluated. In the AASHTOWare design method, detailed material characterizations are performed including detailed properties and performance. This is true for base, subbase, and subgrade layers, and climate and traffic data requirements.

10.2 AASHTOWare Design Considerations

10.2.1 Starting the Design

The design using the AASHTOWare pavement ME design software starts with the selection of a trial pavement (geometry and materials). Then, the design requirements such as traffic volume, climatic conditions, threshold performance index, and reliability are assigned. Thereafter, the software predicts the performance over its service life. The prediction is compared with its threshold performance value at its design life. After several trials, an optimum section is selected. The software is capable of designing new or reconstruction of flexible and rigid pavement including overlay and composite sections.

10.2.2 Materials and Layers

The AASHTOWare pavement ME design software provides a wide range of options for asphalt mixtures, aggregate base layers, and subgrades. The designer may not assign more than seven layers to start the analysis. These are three asphalt layers, an unbound aggregate base, a stabilized subgrade or improved embankment, a subgrade layer, and a rigid layer, if present. The number of layers and their materials must be chosen according to the conditions of the site or the proposed conditions.

The number of asphalt layers cannot exceed three in all cases. Thin layers of less than 1.5 in. (37.5 mm) in thickness are to be combined with other layers. Therefore, the minimum asphalt thickness should be at least four times the nominal maximum aggregate size of the asphalt mixture. Thin wearing courses such as seal mix, porous friction course, open-graded friction course, and other similar mixtures could be combined with the next layer below the wearing surface. The low-temperature cracking and load-related top-down (longitudinal) cracking models use the properties of the wearing surface in predicting the length of transverse and longitudinal cracks throughout the asphalt layers. Likewise, the alligator cracking model takes the properties of the lowest asphalt layer and predicts the percentage of the total lane area with alligator cracking. Thus, the designer needs to carefully consider the properties being entered into the AASHTOWare pavement ME design software for the lowest asphalt layer and asphalt wearing surface.

When multiple layers are combined for design, the volumetric properties such as air voids, effective asphalt content, gradation, unit weight, and voids filled with asphalt entered into the software need to be the weighted average values based on the thickness of the layers. A wearing surface greater than 1.5 in. (37.5 mm) in thickness with a different PG asphalt than the underlying asphalt layer needs to be considered as a separate layer. Similarly, a dense-graded asphalt base layer (the lowest asphalt layer) that is more than 3 in. thick could be considered as a separate layer. All other layers could be combined into the intermediate layer, if possible.

10.2.3 Presence of Rigid Layer

A rigid layer is the lower soil layer with the modulus greater than 100,000 psi (689 MPa). A rigid layer may consist of bedrock, weathered bedrock, hardpan, sandstone, shale, or even over-consolidated clays. When a rigid layer is encountered, the software limits the thickness of the last subgrade layer to be 100 in. (2.5 m). The designer may need to use multiple subgrade layers if the depth to bedrock exceeds 100 in. (2.5 m).

10.2.4 Presence of Water Table

Presence of water table is important for determining the variation of base and subgrade moduli due to change in water content. If there is a water table within 5 ft from the surface, a subsurface drainage system is recommended. The depth to a water table that is input into the software is the depth below the final pavement surface. The designer has the option to enter an annual depth to the water table or seasonal water table depths. The average annual depth could be used, unless the designer has historical data to determine the seasonal fluctuations of the water table depth. If a subsurface drainage system is used to lower the water table, the lower depth could be included in the program not the depth measured during the subsurface investigation.

10.2.5 Drainage System

The use of a drainage system to remove surface water infiltration is dependent on the user's standard design practice. The AASHTOWare pavement ME design software recommends that water not be allowed to accumulate within the pavement structure.

Water may significantly weaken aggregate base layers and the subgrade soil and result in stripping of asphalt layers. The AASHTOWare assumes that all water-related problems will be addressed via the materials and construction specifications, and/or inclusion of subsurface drainage features in the design strategy.

10.2.6 Soil Stabilization

Lime and/or lime-fly ash stabilized soils could be considered a separate layer. If these layers are engineered to provide structural support and have enough stabilizer combined with the soil, they must be treated as a structural layer. Under this case, they could be treated as a material that is insensitive to moisture and the resilient modulus or stiffness of these layers can be held constant over time. On the other hand, when a stabilized subgrade is used as a construction platform for compacting other paving layers, only a small amount of lime or lime-fly ash is added and mixed with the soil. Such layers could be treated as unbound soils. However, if these materials are not engineered to provide long-term strength and durability, they could also be considered as an unbound material and possibly combined with the upper granular layer.

10.2.7 Base/Subbase

Unbound aggregate base layers are commonly used in flexible pavement construction unless pavements are of full-depth asphalt. The number of unbound aggregate base layers cannot exceed two when one of those layers is more than 18 in. (0.45 m). Other soil-aggregate layers could be added separately from stone or aggregate base layer if the resilient moduli are remarkably different.

10.2.8 Initial IRI

An initial International Roughness Index (IRI) value is required for each pavement strategy or trial design considered. The initial IRI value could be taken from previous years' construction acceptance records, if available. Not all agencies, however, use IRI in accepting the pavement related to smoothness criteria. Table 10.1 provides some recommendations for those agencies or users that do not use IRI as a basis for accepting the final surface.

10.2.9 Traffic Data

The traffic data required for the AASHTOWare pavement ME design software are discussed in Chapter 6. Four types of traffic inputs are required in the AASHTOWare

	Initial IRI, in./mi.	
Pavement design strategy	**IRI included as an acceptance test**	**IRI excluded from acceptance test**
Conventional flexible pavements	65	80
Deep-strength flexible pavements	60	70
Full-depth asphalt pavements	60	70
Semi-rigid pavements	65	80

Table 10.1 Recommended Initial IRI

pavement ME design software for structural design of pavement for any level (Level 1, Level 2, and Level 3) of analysis. These inputs are listed below:

Type 1: Traffic volume–base year information
- Two-way annual average daily truck traffic (AADTT)
- Number of lanes in the design direction
- Percent trucks in design direction
- Percent trucks in design lane
- Vehicle operational speed

Type 2: Traffic volume adjustment factors
- Monthly adjustment
- Vehicle class distribution
- Hourly truck distribution
- Traffic growth factors

Type 3: Axle load distribution factors

Type 4: General traffic inputs
- Number of axles per truck
- Axle configuration
- Wheel base

Each of these data are measured, calculated, or regressed based on input levels. Nationally calibrated default values are also built in the software.

10.2.10 Climate Data

Climate data include historical air temperature, precipitation, relative humidity, wind speed, longitude, latitude, etc. These data are not input manually. Alternatively, the AASHTOWare pavement ME design software integrated thousands of climate stations and their historical data from the United States and Canadian provinces. A single climate station can be selected if the pavement site is located very close to a site or similar to other areas which have climate stations. Climate data can also be interpolated using two or more climate stations with the help of the AASHTOWare pavement ME design software.

10.2.11 Analysis Procedure

The AASHTOWare method designs asphalt pavement considering the bottom-up fatigue (alligator) cracking, top-down longitudinal cracking, rutting, transverse thermal cracking, and international roughness index. At first, the stress-strains under various traffic loadings are determined for different seasonal conditions using the built-in numerical program in the software. These stress-strains are used to predict different distresses with its service life using empirical distress equations. If all the predicted distress is below the threshold values up to design life, then the trial section is considered adequate. The trial section may be adjusted to determine an optimum section.

10.3 AASHTOWare Input Hierarchy

The AASHTOWare pavement ME design software uses three levels of inputs for most of the parameters of traffic, material, and pavement conditions. This approach offers AASHTOWare pavement ME design users a great deal of flexibility in obtaining project inputs based on their criticality and data availability. The following list defines the input levels:

- Input Level 1 input parameter is measured directly on its site and thus the most accurate.
- Input Level 2 input parameter is the estimated values from correlations or regression equations, or local average values.
- Input Level 3 input parameter is the best-estimated national or regional default values.

10.4 Getting Started with the AASHTOWare Pavement ME Design Software

The interface of the AASHTOWare pavement ME design software is shown in Fig. 10.1*a*. The software is required to be licensed after the installation. After the license is activated the following (Fig. 10.1*a*) interface will be appeared. Once the "OK" button is clicked, the initial window, as shown in Fig. 10.1*b*, will pop up.

In the initial window (Fig. 10.1*b*), a new project can be started using the "New" button, or an existing file can be opened using the "Open" button. If the "New" button is clicked, the window shown in Fig. 10.2 will be popped up asking the following general information:

Design type. When the designer opens a new project, the first step will be to enter information in the General Information area in the top left corner of the Project tab. Any change in inputs in this area will affect the inputs and information in all other parts of the project, as such changes will be updated and visible throughout the project.

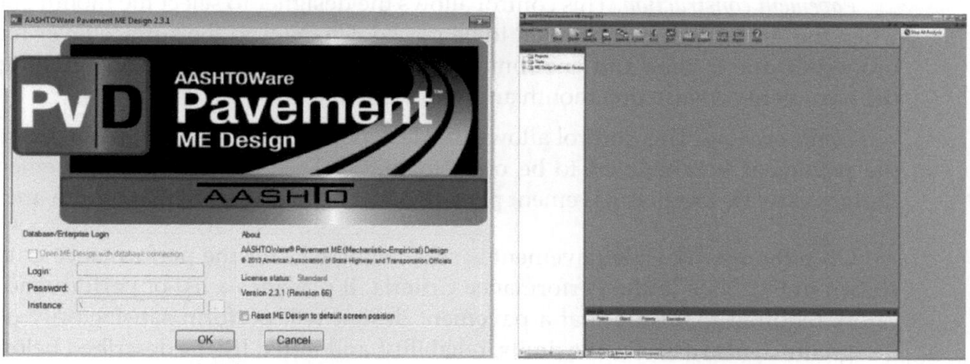

(a) AASHTOWare software (b) Initial window

Figure 10.1 AASHTOWare pavement ME design default window layout.

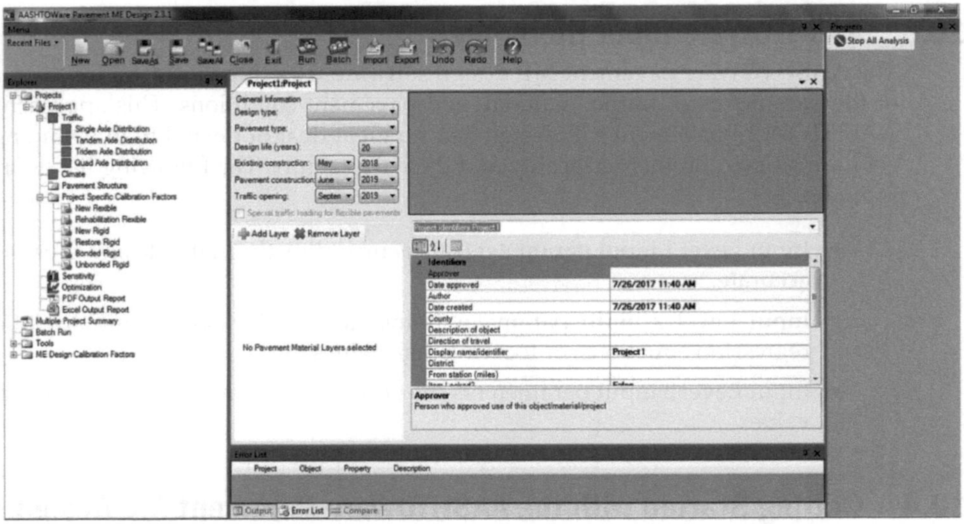

FIGURE 10.2 AASHTOWare window interface of a new project.

Pavement type. Flexible pavement or any other pavement type can be selected.

Design life (years). This control allows the designer to select from a list the period in years from the month and year of completion of construction of the pavement which is expected to perform adequately without significant loss of functional and structural integrity. Pavement performance is predicted over the design life beginning from the month the pavement is open to traffic.

Base construction. This control allows the designer to select the month and year when the base layers and subgrade are scheduled to be completed. Changes to the moisture and temperature profiles of the base and subgrade materials due to time and environmental conditions are considered beginning from the base construction month and year.

Pavement construction. This control allows the designer to select the month and year when the AC surface is scheduled to be placed. Changes to the surface layer material properties due to time and environmental conditions are considered beginning from the pavement construction month and year.

Traffic opening. This control allows the designer to select the month and year when the pavement is scheduled to be open to traffic. The AASHTOWare pavement ME design software predicts pavement performance beginning from that month and year.

Once the new flexible pavement is selected, it asks for the performance criteria as shown in Fig. 10.3. In the Performance Criteria, it provides a list of performance indicators required to ensure that a pavement design will perform satisfactorily over its design life. It also asks for the limits, reliability, and initial IRI, as described below:

Limit. This column allows the designer to define any threshold values of these performance indicators to evaluate the adequacy of a design.

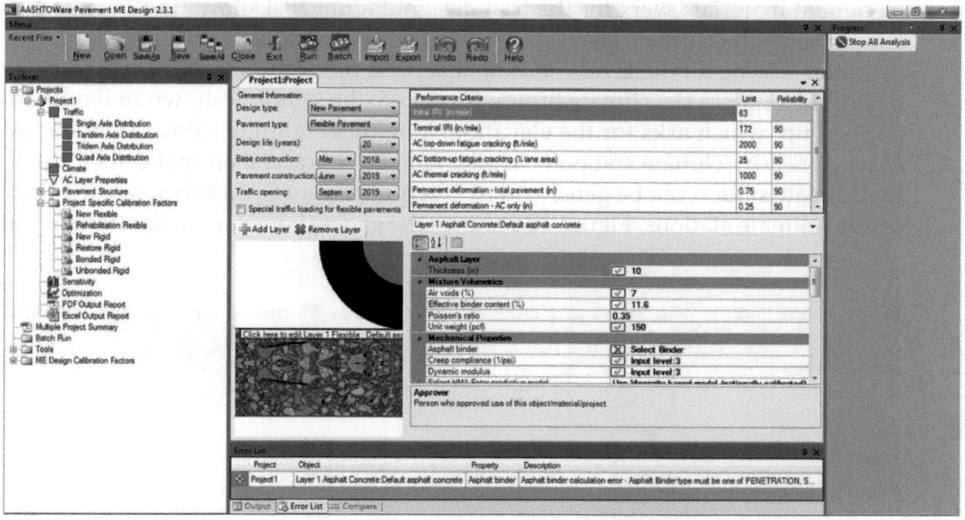

FIGURE 10.3 AASHTOWare window performance criteria and pavement structure.

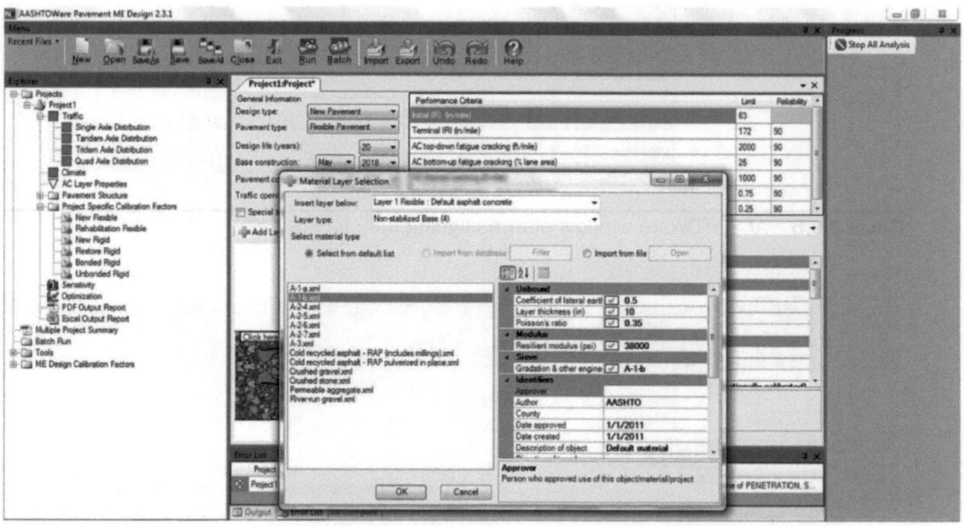

FIGURE 10.4 AASHTOWare window after adding a layer below the surface asphalt layer.

Reliability. This column allows the designer to choose any reliability value at which the predicted distresses and smoothness are to be estimated over the design period.

Initial IRI (in./mile). The limit control allows the designer to define the expected smoothness immediately after new pavement construction is expressed in terms of IRI. The designer can override the default value of 63 in./mile based on local conditions.

In the same window, as shown in Fig. 10.4, materials and layers properties are provided. The controls available here allow the designer to add, remove, and edit pavement material layers for the trial design. Figure 10.4 shows the process of adding a layer below the surface layer.

After adding all the layers and the materials properties, traffic or climate data may be provided. Once the climate button is clicked, the window shown in Fig. 10.5 will be popped up, which asks for the elevation, longitude, latitude, climate station, etc. If an exact or nearby climate station is available in the library, then that can be used. A virtual climate station can be generated by interpolation or extrapolation using the nearby several climate stations. Figure 10.6 shows the generation of a virtual climate station.

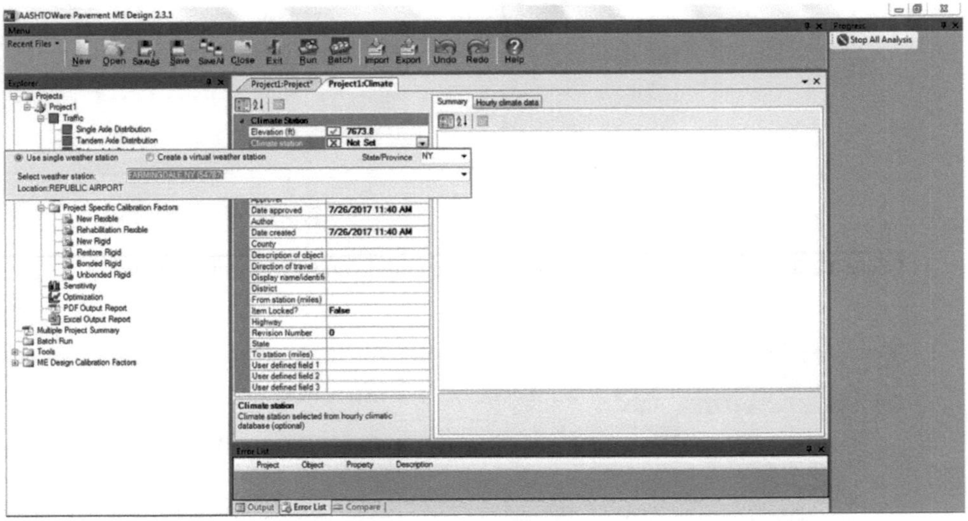

FIGURE 10.5 AASHTOWare window after assigning the climate data.

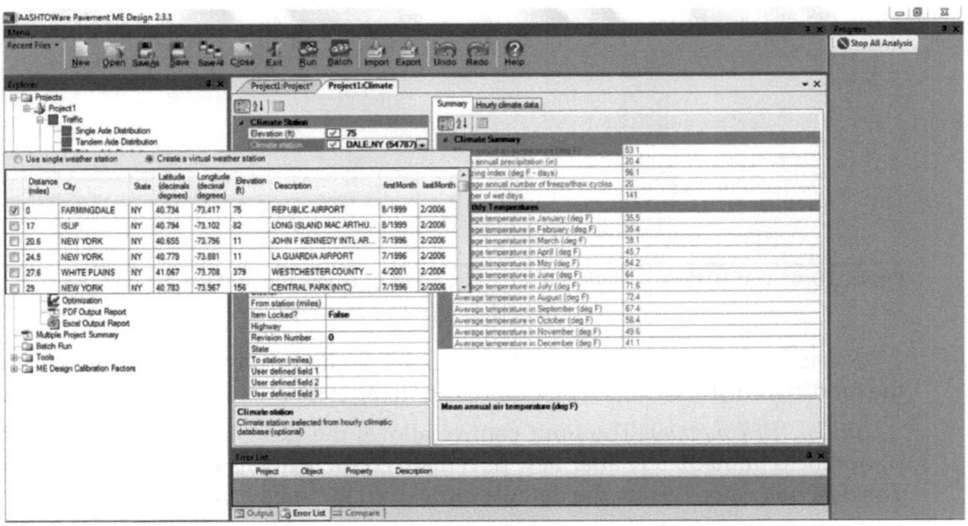

FIGURE 10.6 AASHTOWare window after generating a virtual climate station.

After providing the layers and climate data, the screen looks like the window shown in Fig. 10.7. It shows climate and AC layer properties now in green, whereas the traffic is shown by red. Red color indicates that data are not provided yet or there is an issue in the provided data which must be corrected before running the analysis.

Figure 10.8 shows the window to be popped up once the traffic button is clicked. It asks for the AADTT, number of lanes, design speed, axle configuration, truck class-distribution, monthly distribution, axle per truck, etc. Once these data are provided,

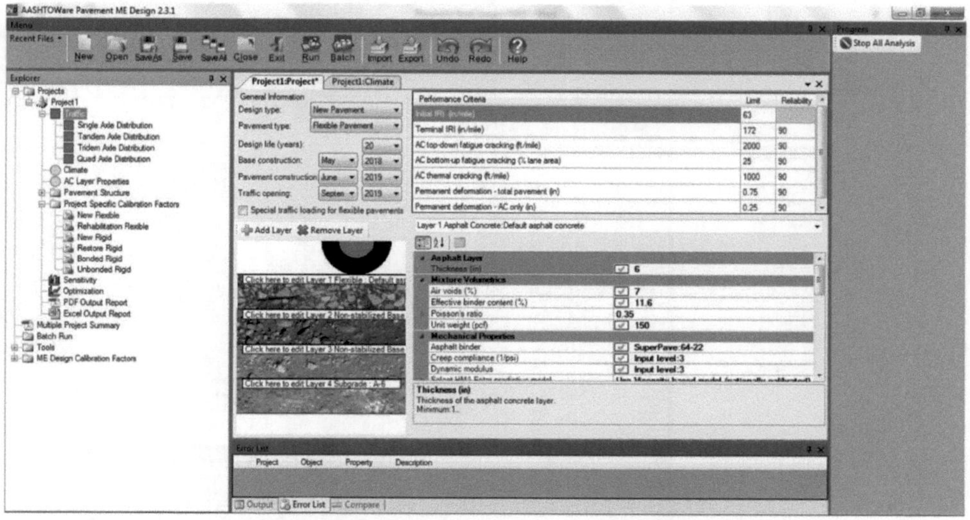

FIGURE 10.7 AASHTOWare window after providing layer and climate data.

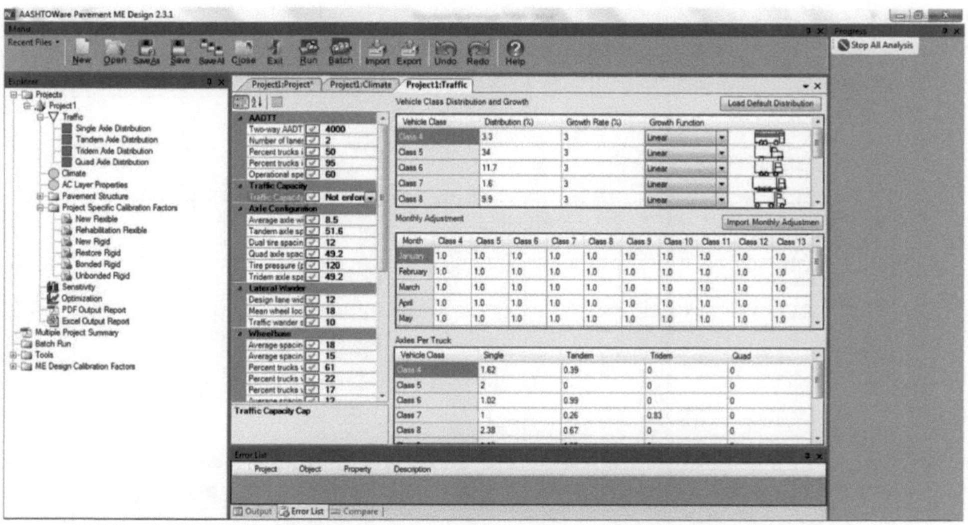

FIGURE 10.8 AASHTOWare window after providing the traffic data.

it can be seen from Fig. 10.8 that traffic is in yellow color, and axle load spectra are in red color.

Figures 10.9 and 10.10 show the input window for axle load spectra for single and quad axles. Following the same process, axle load spectra for tandem and tridem axles are provided. The button color will then turn into green.

The default analysis output file is in the PDF format. If Excel is desired, then it has to be declared before running the execution. Figure 10.11 window pops up if the

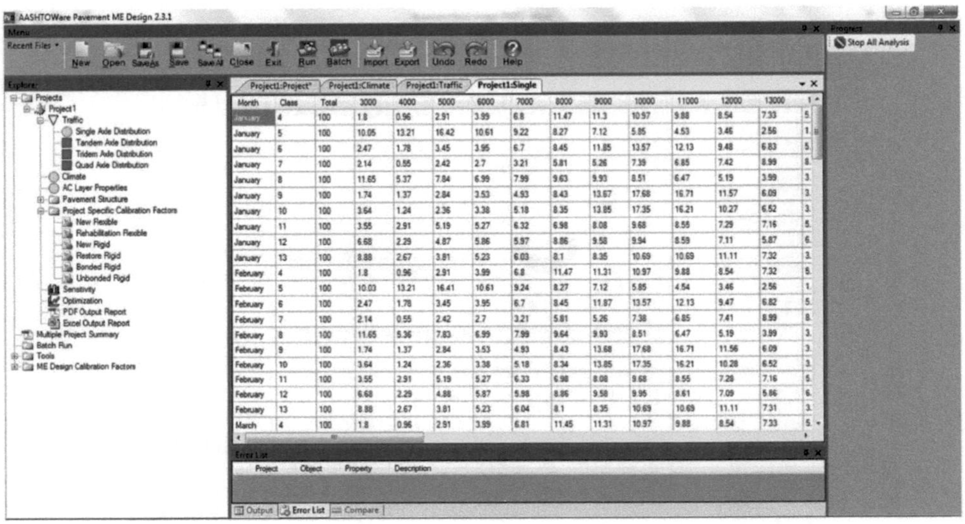

FIGURE 10.9 AASHTOWare window axle load spectra for single axle.

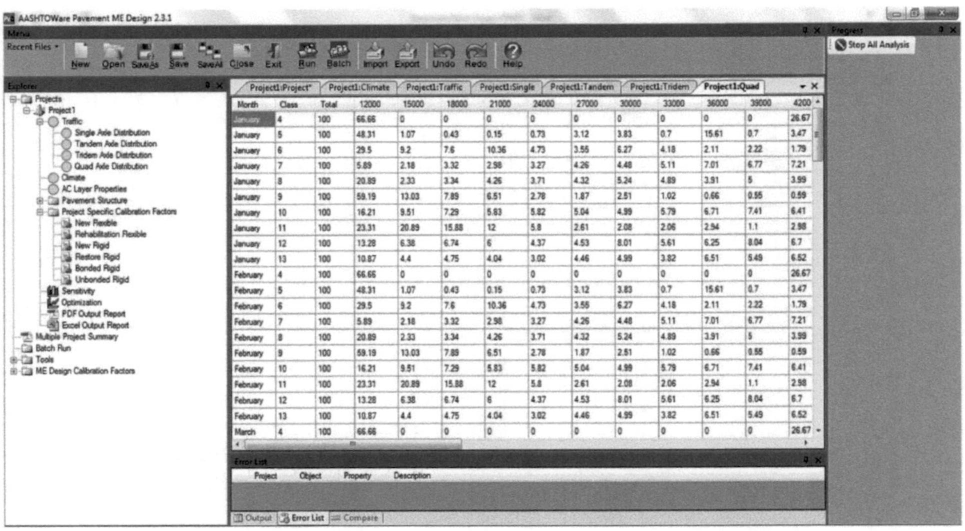

FIGURE 10.10 AASHTOWare window axle load spectra for quad axle.

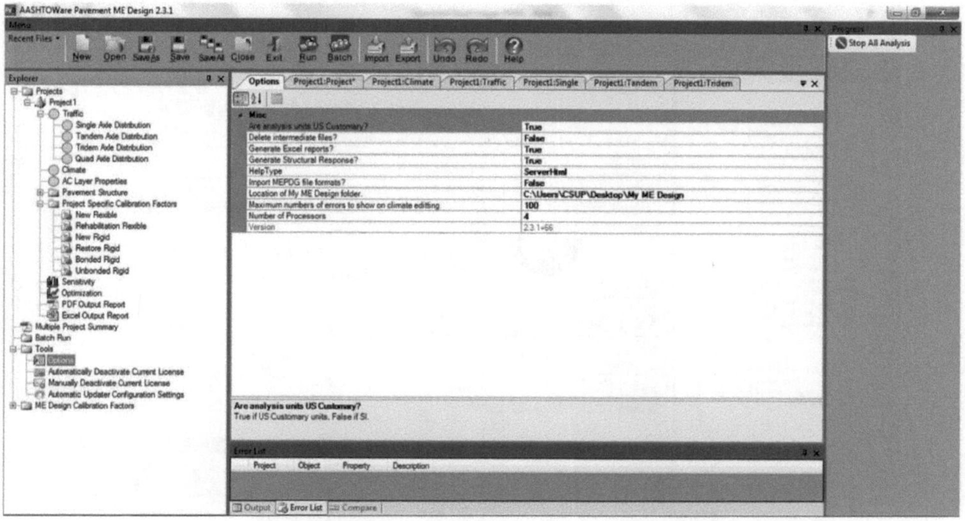

FIGURE 10.11 AASHTOWare window after Excel file request window.

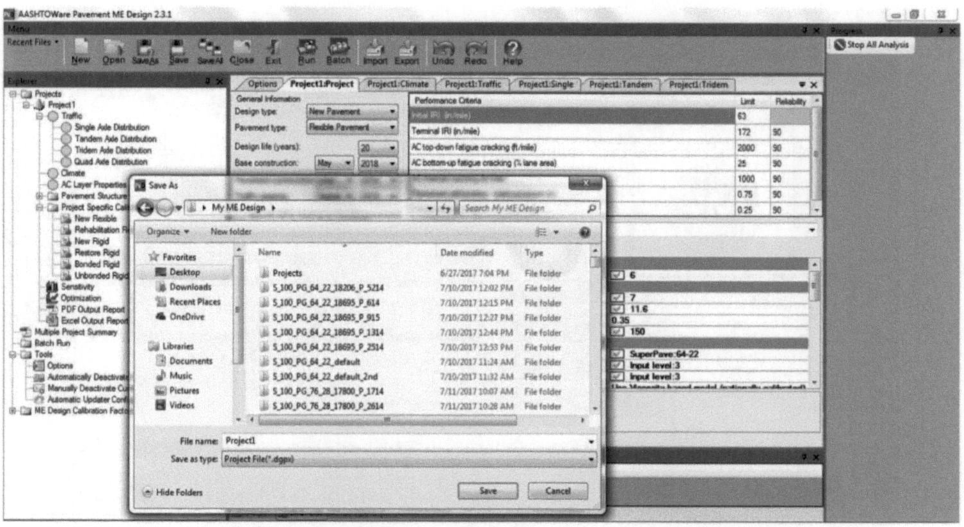

FIGURE 10.12 AASHTOWare window saving option before running.

"Options" button in "Tools" is clicked. In the "Generate Excel reports?" it must be input as "True" to produce the Excel output file.

Once all the inputs are provided, and all the input buttons are green or yellow, the analysis can be executed. Before the execution, the file is to be saved in a convenient place. Desktop is the default destination of the analysis. The saving can be done by clicking "Save" or "Save As" button in the file menu, as shown in Fig. 10.12.

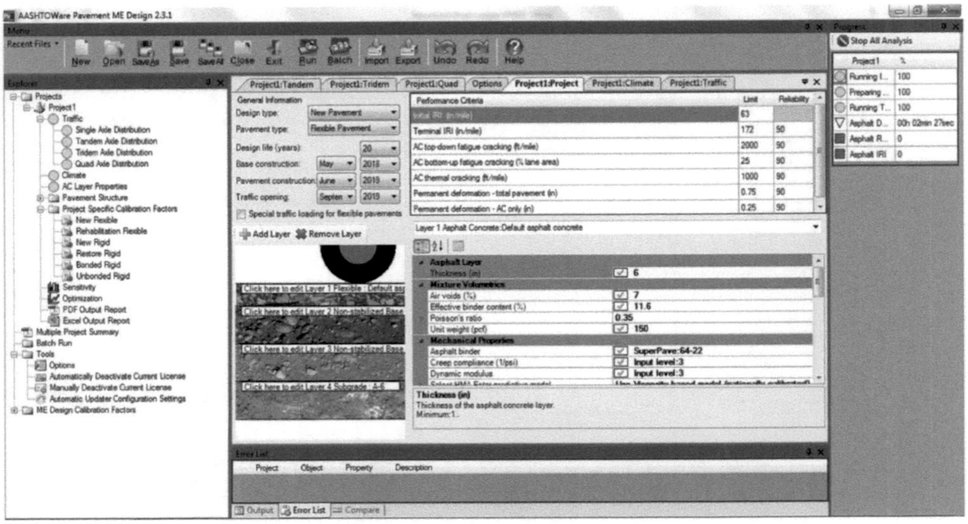

FIGURE 10.13 AASHTOWare window of analysis in progress.

FIGURE 10.14 AASHTOWare window of analysis completed.

The analysis can be executed by pressing the "Run" button in the command toolbar. Figure 10.13 shows an execution process and Fig. 10.14 shows the after-analysis phase of an analysis.

After the execution, the PDF file automatically pops up. A part of the PDF file is shown in Fig. 10.15. The Excel file is typically directed to the analysis folder where the file has been saved. A part of the Excel file is shown in Fig. 10.16.

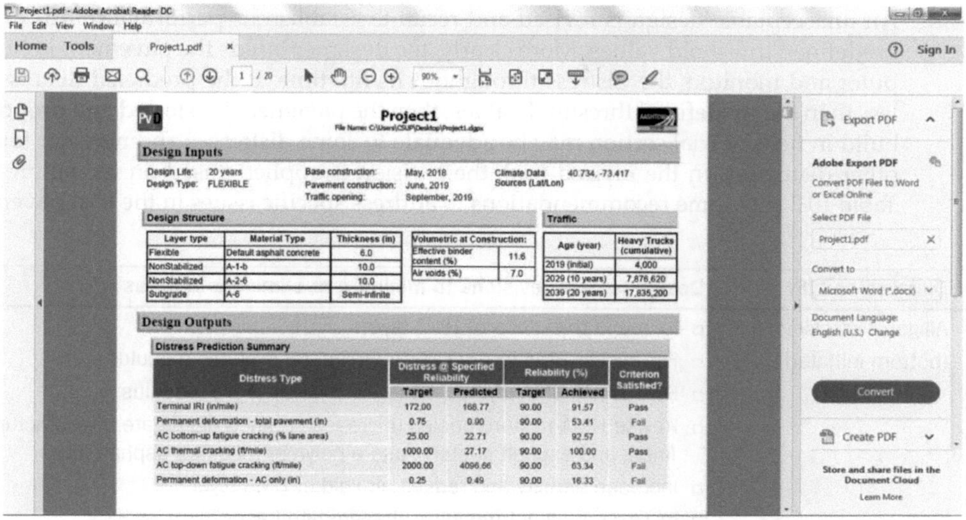

FIGURE 10.15 A part of the generated PDF file.

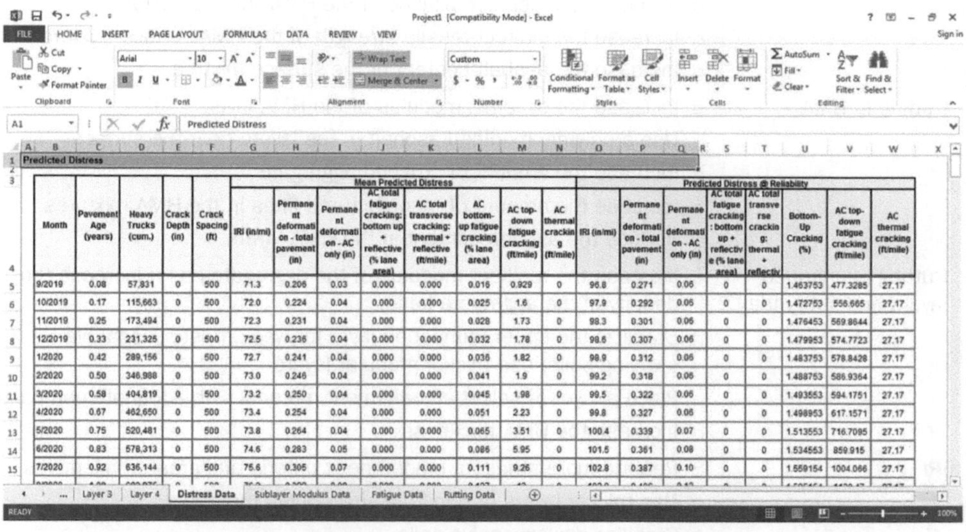

FIGURE 10.16 A part of the generated Excel file.

10.5 Interpretation and Analysis of the Trial Design

As mentioned above, the AASHTOWare pavement ME design software analyzes that trial design over the selected design period. The program outputs the following information: inputs, reliability of design, materials and other properties, and predicted performance using both PDF and Excel. Each of these outputs needs to be examined by the designer to achieve a satisfactory design or less than the predefined threshold values.

An unacceptable design is revised and rerun to establish its performance less than the predefined threshold values. More clearly, the designer builds the pavement in his computer and monitors the distress throughout its life time. If the predicted distresses are less than the predefined threshold values, then the pavement is selected and proposed to build in field. A trial section may be adequate to some distresses and may not for some other distresses. In the revised trial, the revision is applied based on the failure areas. Table 10.2 lists some recommendations to address specific issues in the trial pavement.

Distress and IRI	Design feature revisions to minimize or eliminate distress
Alligator cracking (bottom initiated)	□ Increase thickness of HMA layers □ For thicker HMA layers (>5 in.) increase dynamic modulus □ For thinner HMA layers (<3 in.) reduce dynamic modulus □ Revise HMA mixture design (increase crushed aggregate, manufactured fines, and asphalt content, use a polymer modified asphalt, etc.) □ Increase density and reduce air void of HMA layer □ Increase resilient modulus of aggregate base
Thermal transverse cracking	□ Increase the thickness of the HMA layers □ Use softer asphalt in the surface layer □ Reduce the creep compliance of the HMA-surface mixture □ Increase the indirect tensile strength of the HMA-surface mixture □ Increase the asphalt content of the surface mixture
Rutting in HMA	□ Increase the dynamic modulus of the HMA layers □ Use a polymer-modified asphalt in the layers near the surface □ Increase the amount of crushed aggregate □ Increase the amount of manufactured fines in the HMA mixtures □ Reduce the asphalt content in the HMA layers
Rutting in unbound layers and subgrade	□ Increase the resilient modulus of the aggregate base; increase the density of the aggregate base □ Stabilize the upper foundation layer for weak, frost-susceptible, or swelling soils; use thicker granular layers □ Place a layer of select embankment material with adequate compaction □ Increase the HMA thickness
IRI	□ Require more stringent smoothness criteria and greater incentives □ Use thicker layers of non-frost-susceptible materials □ Stabilize any expansive soils □ Place subsurface drainage system to remove groundwater
Longitudinal fatigue cracking (surface initiated)	□ Reduce the dynamic modulus of the HMA-surface course □ Increase HMA thickness □ Use softer asphalt in the surface layer □ Use a polymer-modified asphalt in the surface layer

Source: From AASHTO (2015). Mechanistic-Empirical Pavement Design Guide: A Manual of Practice. Washington, DC: American Association of State Highway and Transportation Officials. Table 13-3. Used with permission.

Table 10.2 Guidance for Modifying HMA Trial Designs to Satisfy Performance Criteria

10.6 Special Features of the Software

10.6.1 Thickness Optimization

AASHTOWare allows the designer to optimize the thickness of any layer above the foundation (semi-infinite thickness) only available for JPCP design. The designer can only optimize a single layer at a time in 0.5-in. increments. AASHTOWare chooses the minimum thickness value first and determines whether the minimum thickness meets the performance criteria. If the run is successful, the process stops there. Otherwise, the program selects the maximum thickness value for its next run. If both minimum and maximum thicknesses are unsuccessful, the process stops. If the maximum thickness value is successful, the program selects the midpoint thickness value between the maximum and minimum thickness for its third run. Irrespective of the outcome, the program chooses the midpoint between the thickness of the last successful run and the last unsuccessful run for all further runs.

10.6.2 Batch Run

AASHTOWare has the options to run a batch of projects at a time and view the batch report. This is done by selecting the "Load Projects" control on the Batch Run node on the Explorer Pane, as shown in Fig. 10.17, and commanding the "Run Batch Projects."

10.6.3 Structural Response

The AASHTOWare produces the structure response outputs as intermediate files:

1. Vertical compressive strain values are stored in the file "_VertStrain.txt".
2. Horizontal tensile strain values are stored in the file "_TensStrain.txt".

FIGURE 10.17 Batch run procedure.

The information contained in both of the files is as follows:

1. Month
2. Sub-month
3. Sub-season
4. Axle number
5. AC modulus
6. Load location
7. Analysis location
8. Strain (either vertical compression or horizontal tension)

10.6.4 Calibration Factors

The AASHTOWare pavement ME design software allows the designer to specify calibration factors for both the current project and for AASHTOWare as a whole.

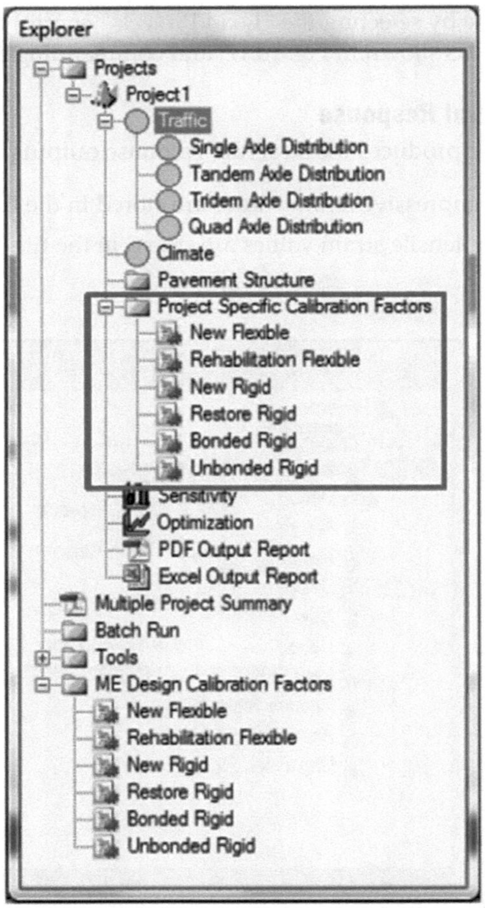

FIGURE 10.18 Calibration factors in Explorer pane.

The variables to be defined depend on the pavement types and distress types. To modify the calibration factors for the current project only, click the plus sign (+) next to Project Specific Calibration Factors under the current project in the Explorer tree. The Project Specific Calibration Factors are shown enclosed by a red rectangle in the Explorer graphic on this page, as shown in Fig. 10.18. Select the type of Project Specific Calibration Factors that matches the currently active project. For Flexible Pavement Design, double-click New Flexible. The designer can also access Project Specific Calibration Factors by selecting New Flexible Pavement-Calibration Settings from the Layer drop-down box.

For further information, readers are referred to the AASHTOWare pavement ME design software manual or AASHTO (2015).

10.7 Summary

This chapter demonstrates the AASHTOWare software, its basic buttons, inputting system, analysis procedure, and the interpretation of the results. This software can be used for different types of flexible pavement such as conventional flexible sections, deep strength sections, full-depth sections, and semi-rigid sections.

The design using AASHTOWare pavement ME design software begins with the selection of a trial pavement geometry and materials to be used. Then, the site parameters such as traffic volume, climate conditions, threshold performance index, reliability, etc. are provided. The software predicts the performance over its service life. The prediction in its design life is compared with its threshold performance value. After several trials, an optimum section is selected. More information is available at AASHTO (2015).

10.8 Fundamentals of Engineering (FE) Exam–Style Questions

FE10.1 How many hierarchical levels of input data does the AASHTOWare pavement ME design software include?

A. 1
B. 2
C. 3
D. 4

Solution C

The AASHTOWare pavement ME design software includes three hierarchical levels of input data:

- Input Level 1 input parameter is measured directly on its site.
- Input Level 2 input parameter is the estimated values from correlations or regression equations, or local average values.
- Input Level 3 input parameter is the best estimated or national average values.

FE10.2 Which of the following distresses does the AASHTOWare pavement ME design software not consider a design criterion?

A. Fatigue/alligator cracking
B. Thermal cracking
C. International Roughness Index
D. Potholes

Solution D

The AASHTOWare pavement ME design software does not consider potholes as a design criterion. In addition to the first three options, the software considers top-down longitudinal cracking and rutting as design criteria.

FE10.3 Which of the following inputs does the AASHTOWare pavement ME design software not include?

A. ESAL
B. Speed limit
C. Climate
D. Lane distribution factor

Solution A

The AASHTOWare pavement ME design software does not consider ESAL. It is the way of inputting traffic volume in the AASHTO 1993 design method.

10.9 Practice Problems

10.1 A new flexible pavement geometry and materials are shown below. The project will start sometime next year.

Climate: Choose
Traffic:

Two-way AADTT = 6,000
No. of lanes in each direction = 3
Design speed = 50 mph
Traffic wander = 8 in.
Lane width = 12 ft

Use the default values for the unmentioned data. Use the AASHTOWare pavement ME design software to analyze the pavement at 90% reliability of the distresses.

After the AASHTOWare pavement ME design software analysis:

a. Compare the bottom-up fatigue cracking after 20 years if the HMA thicknesses are 4, 6, and 8 in. Use a bar chart or line graph.
b. Compare the total rutting after 20 years if the HMA thicknesses are 4, 6, and 8 in. Use a bar chart or line graph.

10.2 A new flexible pavement geometry and materials are shown below. The project will start sometime next year.

Climate: Choose

Traffic:

Two-way AADTT = 8,000
No. of lanes in each direction = 2
Design speed = 70 mph
Traffic wander = 7 in.
Lane width = 12 ft

Use the default values for the unmentioned data. Use the AASHTOWare pavement ME design software to analyze the pavement at 90% reliability of the distresses.

After the AASHTOWare pavement ME design software analysis:

a. Compare the top-down longitudinal cracking after 20 years if the HMA thicknesses are 4, 6, and 8 in. Use a bar chart or line graph.
b. Compare the transverse cracking after 20 years if the HMA thicknesses are 4, 6, and 8 in. Use a bar chart or line graph.

CHAPTER 11

Asphalt Overlay Design by AASHTOWare

11.1 Background

Pavement overlays are used to restore the quality of the surface layer (both HMA and PCC) (such as smoothness, friction, and aesthetics) or add structural strength to the existing pavement (Fig. 11.1). If the existing pavement has distresses such as cracking or rutting or provides insufficient structural support, these defects often reflect even the overlay and cause premature failure of the pavement in the form of cracks and deformations. Pavement cracks should be sealed to maximize service life of overlays. The damaged parts of the existing pavements should be patched or replaced. Overlays not only improve the quality of the underlays but also increase their service life. In the event that the existing pavement failure is caused by inadequate subgrade support or poor subgrade drainage, the existing pavement over the failed area should be removed and the subgrade improved prior to the overlay application.

Before overlaying, a tack coat should be placed on the existing pavement to ensure that the overlay is properly bonded to the existing pavement surface. Proper tack coat

FIGURE 11.1 Overlay in an existing asphalt pavement.

Figure 11.2 Reflective cracking in an asphalt overlaid pavement.

application can be critical to the long-term performance of the pavement. The existing pavement should be made as smooth as possible before being overlaid.

11.2 AASHTOWare Design Method

The AASHTOWare pavement ME design software predicts reflection cracks (Fig. 11.2) in HMA overlays or HMA semi-rigid pavement surfaces using an empirical equation. Empirical equations are used to estimate the amount of fatigue and thermal cracks from a non-surface layer that have been transmitted to the surface after a certain period of time. This empirical equation calculates the percentage of crack area that spreads through the HMA as a function of time using the sigmoid function:

$$RC = \frac{100}{1 + e^{ac+btd}} \tag{11.1}$$

where RC = Percent of reflection cracks with the width of cracks being 1 ft
t = Time, yr
a, b = Regression fitting parameters defined through calibration process
c, d = User-defined cracking progression parameters

Moreover, this empirical equation has not been recalibrated globally. It is still under research. More details of the overlay design have been discussed in Chap. 9.

11.3 Overlay Design Using the AASHTOWare Software

The AASHTOWare pavement ME design software (Fig. 11.3) can help design an asphalt concrete overlay for any type of pavement–flexible, rigid, or composite pavement. First, you need to determine if you need pre-overlay treatment, if any. Pre-overlay treatments may be used to minimize the impact of current pavement problems on the future performance of AC overlays. A combination of milling, full or partial depth repairs, or in-place recycling may include pre-overlay treatments. Not all forms of pre-overlay repairs are specifically considered by AASHTOWare pavement ME design software. Nevertheless, the effect of pre-overlay repairs on the general condition of the existing AC layer and pavement is what should be used in the design of AC overlays for

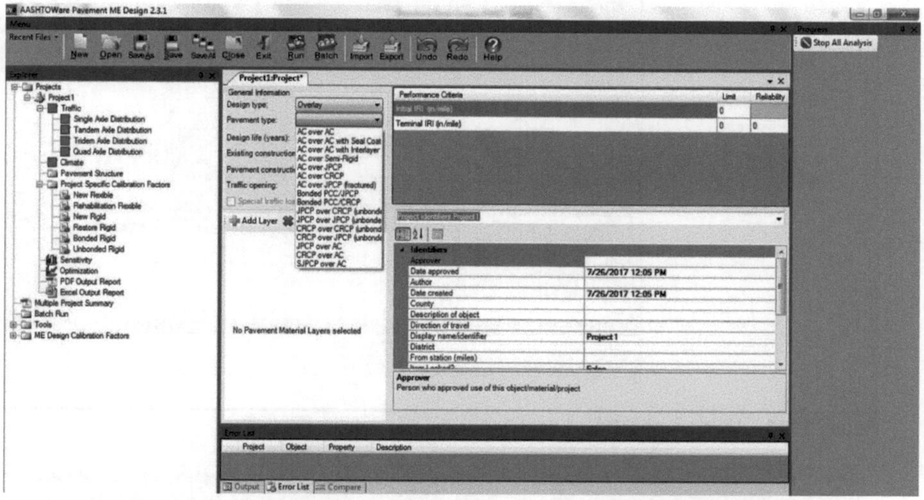

FIGURE 11.3 Overlay design interface.

existing flexible pavements. In either case, an AC overlay of an existing flexible pavement is the resulting analysis.

Once you open a new project, in the top left corner of the Project tab, the first move is to enter information in the General Information field. Inputs in this area have an impact on other inputs and information in all other parts of the project. Changing them once they are entered would change the whole project. The following options allow you to decide your project's basic parameters.

In addition to selecting Design and Pavement type, select values for the following basic design parameters:

- Design life
- Existing construction
- Pavement construction
- Traffic opening

New construction means the month and year in which the construction of the new flexible pavement has been completed. Pavement construction means the month and year in which the AC overlay is expected to be built. Traffic opening means the month and year that the rehabilitated pavement is expected to be open to traffic.

All design inputs are identical to those for a new flexible pavement except for an alternative that asks about the condition/rehabilitation of old layers of asphalt. Figures 11.4 to 11.6 show three input levels of overlay design. The other design inputs such as climate, base, subbase, subgrade, and traffic are similar.

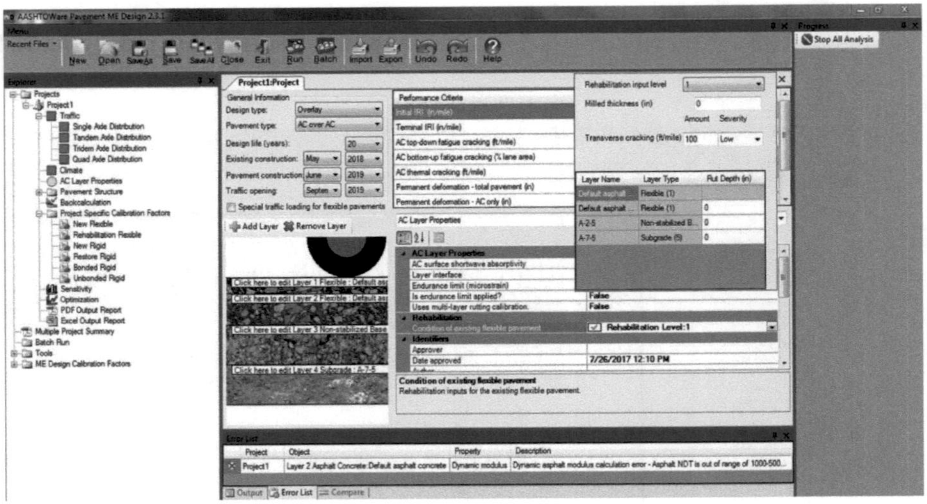

Figure 11.4 Layer properties–Level 1.

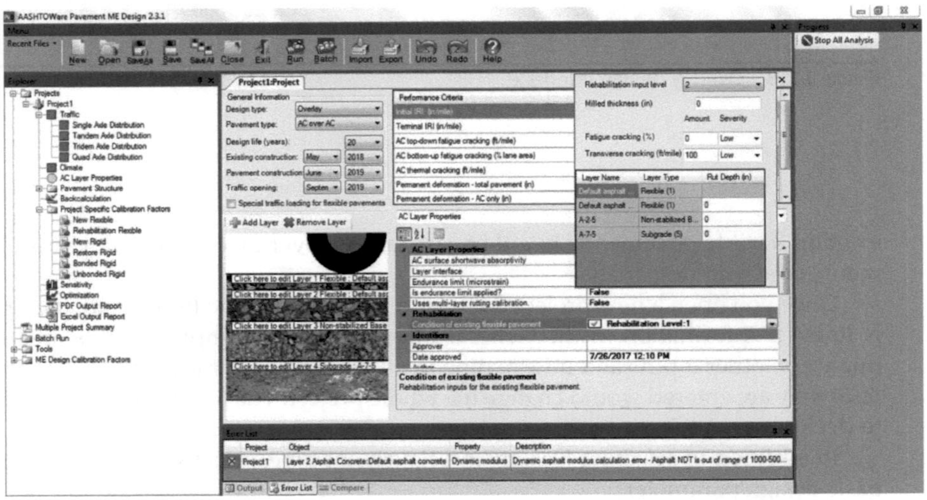

Figure 11.5 Layer properties–Level 2.

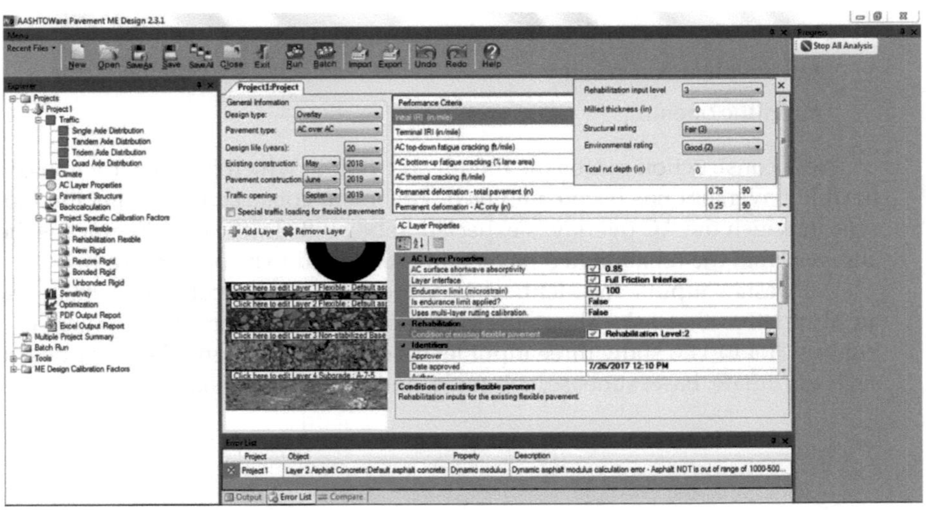

Figure 11.6 Layer properties–Level 3.

11.4 Summary

This chapter deals with the design equation and the overlay design using the AASHTOWare pavement ME design software. The overlay design starts with the compilation of pavement condition data such as current pavement materials, geometry, and distress. Such data are used as software inputs and at the top of the existing pavement a new proposed layer is applied. Therefore, the performance for the design life is estimated based on the climate and traffic level. The quality of current pavement data can be based on different input levels (Level 1, Level 2, or Level 3). You can find more information in AASHTO (2015).

11.5 Fundamentals of Engineering (FE) Exam-Style Questions

FE11.1 Which of the following distresses is considered in the overlay design but not in the conventional flexible pavement design?

A. Block cracking
B. Corrugations
C. Slippage cracking
D. Reflective cracking

Solution D

If the existing pavement is cracked, these cracks often reflect through even the best constructed overlay and are considered with special attention in overlay design.

FE11.2 How many hierarchical levels of input data does the AASHTOWare pavement ME design software include for the overlay design?

A. 1
B. 2
C. 3
D. 4

Solution C

The AASHTOWare pavement ME design software includes three hierarchical levels of input data for the overlay design.

FE11.3 In an overlay design using the AASHTOWare pavement ME design software, the existing asphalt layer in the deteriorated pavement:

A. Should be removed completely
B. Should be inspected to measure/count the distresses
C. Should not be inspected as a new layer is to be placed
D. Should be milled off about half of its existing thickness

Solution B

The current asphalt layer in the damaged pavement should be checked in overlay design using the AASHTOWare pavement ME design software to measure/count the distress. To avoid surface defects such as aged binder, small rutting, small holes, and polished aggregates, a small part of the thickness is milled off. Nevertheless, it is not required to remove half of its current thickness.

11.6 Practice Problems

11.1 Using the AASHTOWare pavement ME design software, complete the following overlay problem.

Figure P11.1 shows an existing flexible pavement with its geometry and materials. A 3.0 in. overlay on the existing 4.0 in. cracked HMA layer is proposed. The paving work starts in April.

FIGURE P11.1 A candidate for flexible pavement for overlay design.

Climate: Choose

Traffic:

AADTT = 4,000
No. of lane in each direction = 2
Design speed = 40 mph
Traffic wander = 12 in.
Lane width = 10 ft
Percentage of Class 9 truck = 40.2%
Percentage of Class 5 vehicle = 30%

Use the default values for the unmentioned data. Use the rehabilitation input Level 3 with 0.5 in. milled thickness, pavement rating–fair.

Deliverables:

a. Draw the total cracking (bottom-up + reflective) with service life of the pavement.
b. Draw the AC top-down fatigue cracking with service life of the pavement.
c. Draw the AC thermal cracking with service life of the pavement.

CHAPTER 12

Rigid Pavement Design by AASHTO 1993

12.1 Background

This chapter describes the thickness design for rigid pavements. The authors do not think that AASHTO (1993) is still in use for rigid pavement design. Thus, this method is very concisely discussed in this chapter. AASHTO (1993) method for rigid pavement design is also based on the road test in Illinois during 1956–1960. Several equations are primarily developed based on regression analysis of the test findings. Rigid pavement designs using these equations are discussed in this chapter.

12.2 AASHTO Thickness Design

The basic equation of AASHTO 1993 design method for designing thickness of rigid pavement is:

$$\log W_{18} = Z_R S_o + 7.35\log(D+1) - 0.06 + \frac{\log\left[\dfrac{\Delta\text{PSI}}{4.5-1.5}\right]}{1+\dfrac{1.624\times 10^{7}}{(D+1)^{8.46}}} + (4.22 - 0.32p_t)\log\left[\frac{S_c' C_d\left[D^{0.75} - 1.132\right]}{215.63J\left[D^{0.75} - \dfrac{18.42}{\left(\dfrac{E_c}{k}\right)^{0.25}}\right]}\right] \tag{12.1}$$

where W_{18} = Number of 18-kip single-axle load repetitions in time t (or total ESAL)
Z_R = Normal deviate for a given reliability, R
S_o = Standard deviation (the AASHO road test obtained S_o of 0.45 for rigid pavement)
ΔPSI = Design serviceability loss
E_c = Concrete elastic modulus
S_c' = Concrete modulus of rupture
J = Load transfer coefficient

C_d = Drainage coefficient
D = Depth of concrete slab
k = Effective modulus of subgrade reaction

All the other parameters except D (depth of concrete slab) are the inputs and D is the output. Solving this equation is cumbersome. This is why AASHTO (1993) developed two nomographs to solve the equation. Segments I and II (Figs. 12.1 and 12.2) of AASHTO (1993) are the nomographs which can be used to solve the Eq. 12.1 to determine the required slab thickness.

Before starting to solve the equation for D, it is required to find out all the input parameters either by laboratory testing or field testing or from historical database or engineer's best judgment for the design of rigid pavement. All the parameters are very briefly discussed in this chapter. For more information, readers are encouraged to read AASHTO (1993). The specific section numbers are given below:

- Estimated future traffic, W_{18} [Sec. 2.1.2 of AASHTO (1993)]
- Desired reliability, R [Sec. 2.1.3 of AASHTO (1993)]
- The overall standard deviation, S_o [Sec. 2.1.3 of AASHTO (1993)]
- Design serviceability loss, $\Delta PSI = p_i - p_t$ [Sec. 2.2.1 of AASHTO (1993)]
- Concrete elastic modulus, E_c [Sec. 2.3.3 of AASHTO (1993)]

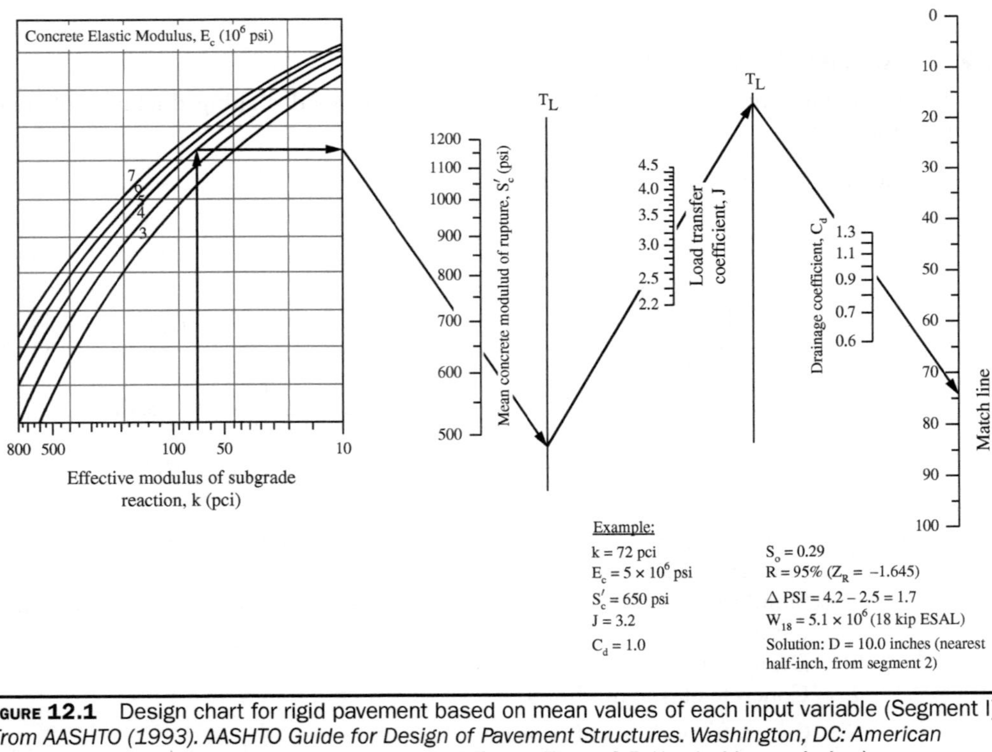

FIGURE 12.1 Design chart for rigid pavement based on mean values of each input variable (Segment I). *(From AASHTO (1993). AASHTO Guide for Design of Pavement Structures. Washington, DC: American Association of State Highway and Transportation Officials. Figure 3-7. Used with permission.)*

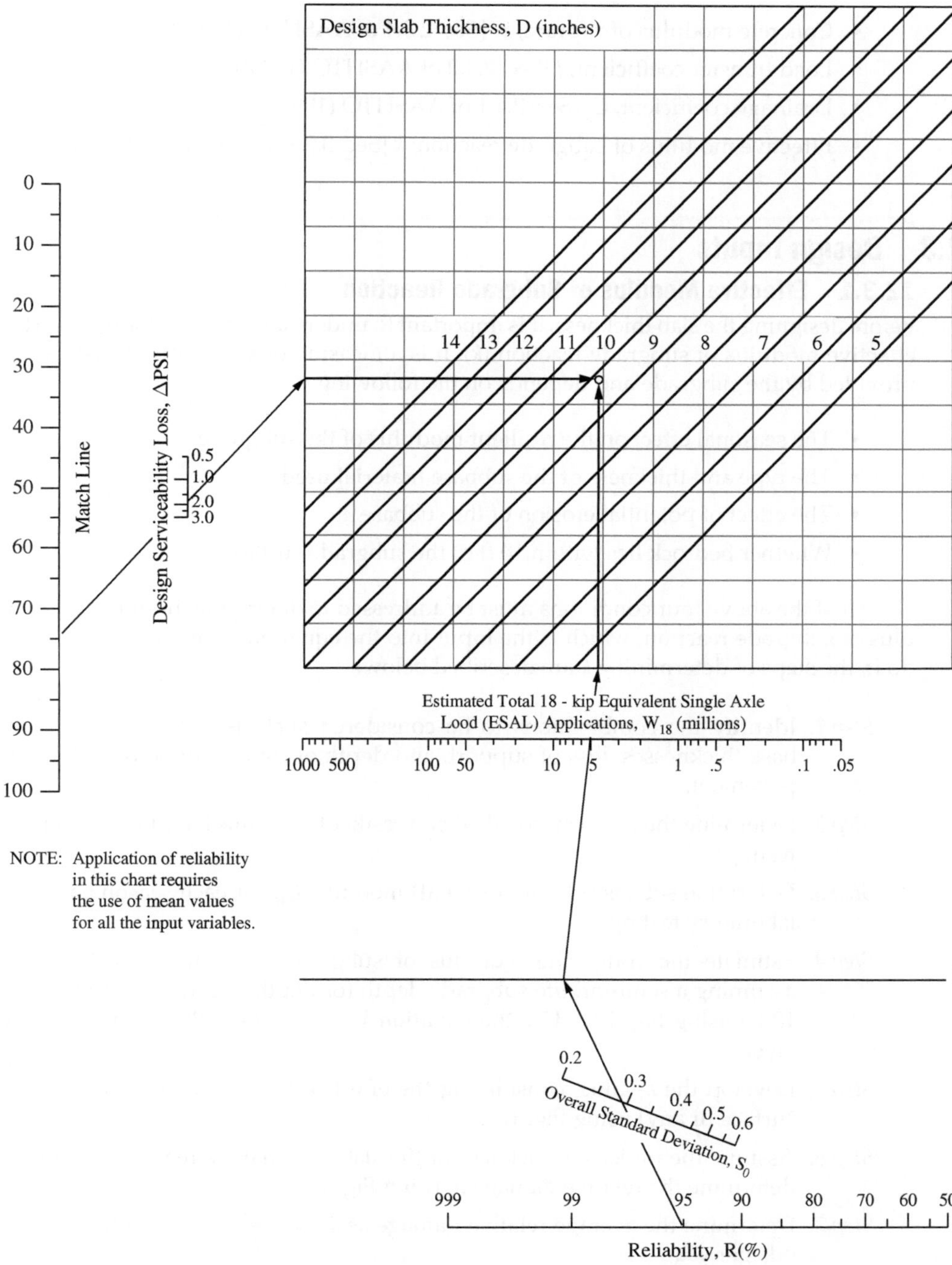

FIGURE 12.2 Design chart for rigid pavement based on mean values of each input variable (Segment II). (*From AASHTO (1993). AASHTO Guide for Design of Pavement Structures. Washington, DC: American Association of State Highway and Transportation Officials. Figure 3-7. Used with permission.*)

- Concrete modulus of rupture, S_c' [Sec. 2.3.4 of AASHTO (1993)]
- Load transfer coefficient, J [Sec. 2.4.2 of AASHTO (1993)]
- Drainage coefficient, C_d [Sec. 2.4.1 of AASHTO (1993)]
- Effective modulus of subgrade reaction, k [Sec. 3.2.1 of AASHTO (1993)]

12.3 Design Inputs

12.3.1 Effective Modulus of Subgrade Reaction

Before designing the slab thickness, it is important to understand the subgrade property–effective modulus of subgrade reaction (k). It is a measure of slab support that can be provided by the subgrade and depends on the following:

- The seasonal effect on the resilient modulus of the subgrade
- The type and thickness of the subbase material used
- The effect of potential erosion of the subbase
- Whether bedrock lies within 10 ft of the subgrade surface

All of the above four conditions must be addressed to determine the effective modulus of subgrade reaction, which is the input into the equation stated into Eq. 12.1. To do it, the steps of determining k are described below:

Step 1. Identify the combinations to be considered such as subbase types, subbase thicknesses, loss of support, and depth of rigid foundation below the pavement.

Step 2. Determine the seasonal roadbed soil resilient modulus based on laboratory testing.

Step 3. Determine subbase elastic (resilient) modulus E_{SB} for each season based on laboratory testing.

Step 4. Estimate the composite modulus of subgrade reaction for each season assuming a semi-infinite subgrade depth (or depth to bedrock greater than 10 ft) using Fig. 12.3. Use the equation $k = M_R/19.4$ if there is no subbase layer.

Step 5. Develop the k-value, considering the effect of a rigid foundation near the surface (if any) using Fig. 12.4.

Step 6. Assume the projected thickness of the slab that may be required, and then determine the relative damage u_r using Fig. 12.5.

Step 7. Determine the average relative damage as $\bar{u}_r = \frac{\sum u_r}{n}$; n = number of periods per year.

Step 8. Determine the k-value for the $\bar{u}_r$ using Fig. 12.5 (reverse process).

Step 9. Adjust the k-value to account for the potential loss of support arising from the subbase erosion using Fig. 12.6.

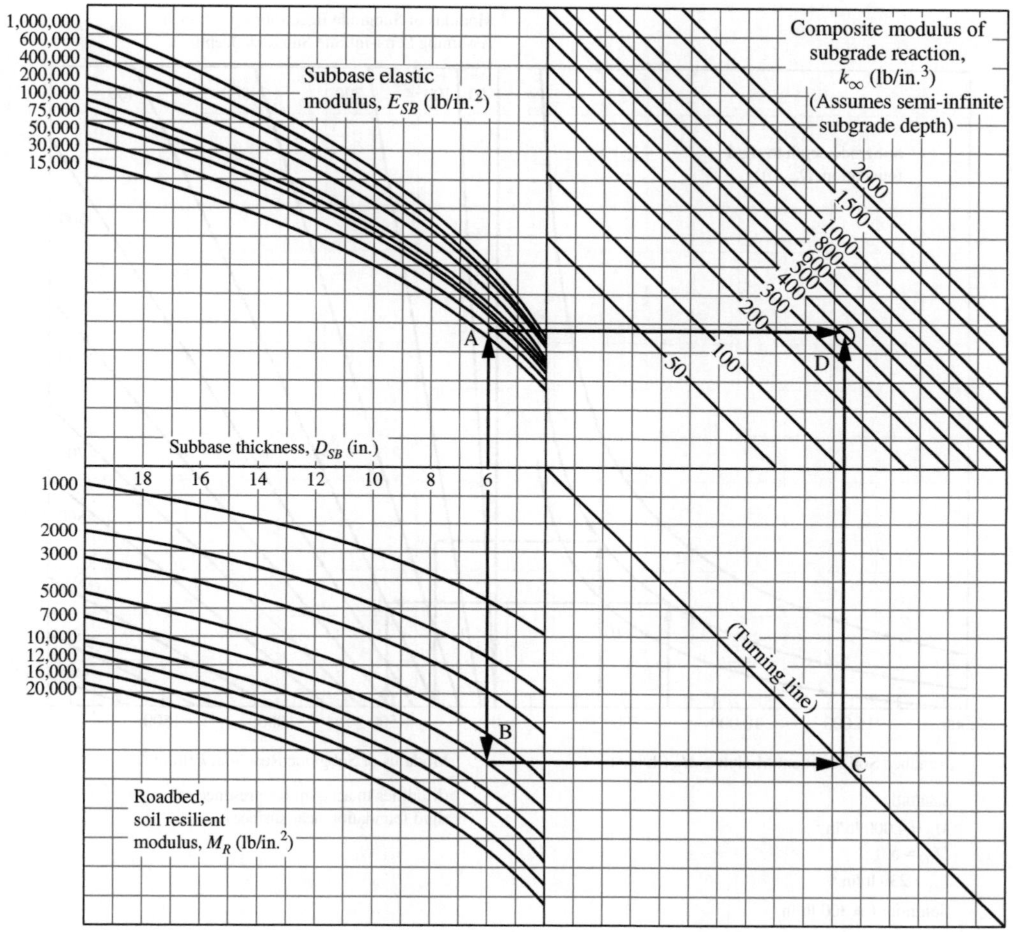

Example:

D_{SB} = 6 in.

E_{SB} = 20,000 lb/in.^{2}

M_R = 7,000 lb/in.^{2}

Solution: k_∞ = 400 lb/in.^{3}

FIGURE 12.3 Chart for estimating composite modulus of subgrade reaction, k_∞, assuming a semi-infinite subgrade depth (or no bedrock within 10 ft below the subgrade). (*From AASHTO (1993). AASHTO Guide for Design of Pavement Structures. Washington, DC: American Association of State Highway and Transportation Officials, Washington, DC. Figure 3-3. Used with permission.*)

If there is no subbase, then k can be determined using the equation:

$$k = M_R/18.8 \tag{12.2}$$

where k is in pci and M_R is in psi.

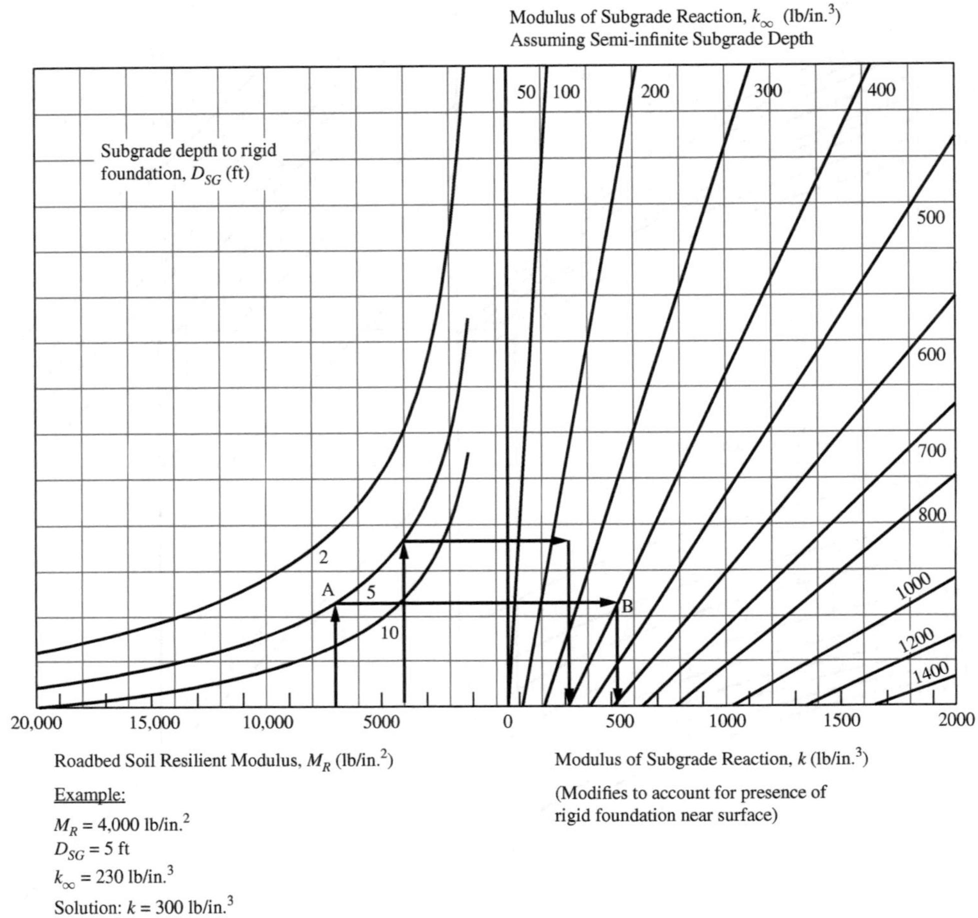

FIGURE 12.4 Chart to modify modulus of subgrade reaction to consider effect of bedrock within 10 ft below the subgrade. *(From AASHTO (1993). AASHTO Guide for Design of Pavement Structures. Washington, DC: American Association of State Highway and Transportation Officials. Figure 3-4. Used with permission.)*

Due to erosion or vertical soil movement, there might be some loss in soil. Figure 12.6 is a scope to account for that loss of subbase support before the thickness design. AASHTO-recommended LS values are listed in Table 12.1.

12.3.2 Concrete Properties

Two concrete properties are required: elastic modulus of concrete and modulus of rupture. Elastic modulus (E_c) can be determined from the compression strength (f_c') testing and then using the following equation:

$$E_c = 57{,}000\left(f_c'\right)^{0.5} \tag{12.3}$$

where E_c = Elastic modulus in psi
f_c' = Concrete compression strength in psi

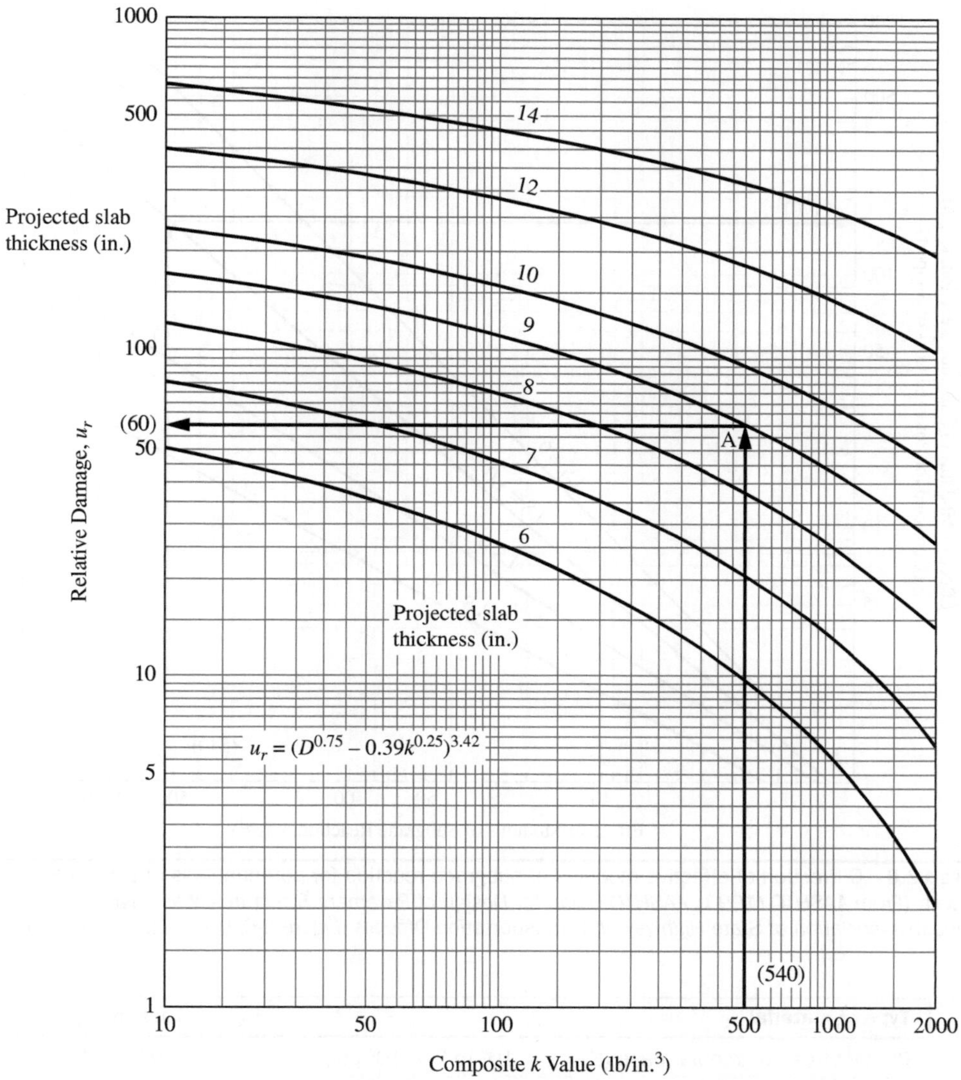

FIGURE 12.5 Chart for estimating relative damage to rigid pavement based on slab thickness and underlying support. *(From AASHTO (1993). AASHTO Guide for Design of Pavement Structures. Washington, DC: American Association of State Highway and Transportation Officials. Figure 3-5. Used with permission.)*

The modulus of rupture at 28 days of the concrete to be used in construction should be determined by conducting a three-point loading test as specified in the AASHTO T 97.

12.3.3 Drainage

The drainage quality of the pavement is considered by introducing a factor (C_d) into the performance equation. This factor depends on the quality of the drainage and the percent of time the pavement structure is exposed to moisture levels approaching saturation. Table 12.2 gives AASHTO-recommended values for C_d.

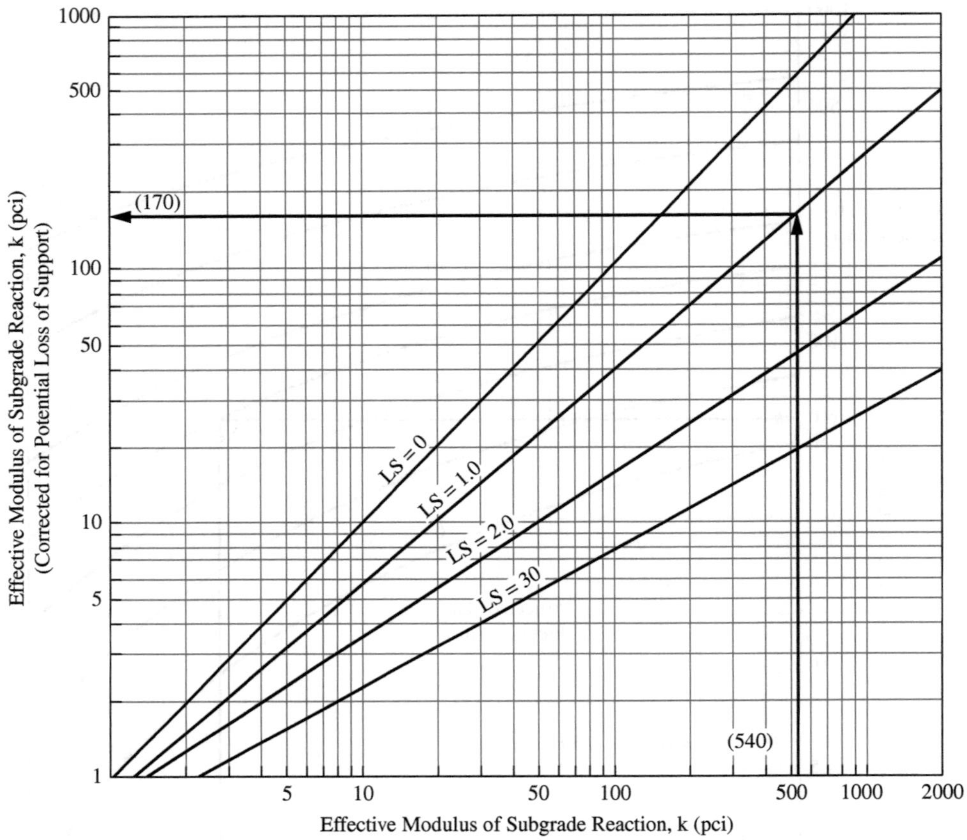

FIGURE 12.6 Correction of effective modulus of subgrade reaction for potential loss of subbase support. (*From AASHTO (1993). AASHTO Guide for Design of Pavement Structures. Washington, DC: American Association of State Highway and Transportation Officials. Figure 3-6. Used with permission.*)

Type of material	Loss of support (LS)
Cement-treated granular base ($E = 1 \times 10^6$ to 2×10^6 psi)	0.0 to 1.0
Cement aggregate mixtures ($E = 500{,}000$ to 1×10^6 psi)	0.0 to 1.0
Asphalt-treated bases ($E = 350{,}000$ to 1×10^6 psi)	0.0 to 1.0
Bituminous-stabilized mixture ($E = 40{,}000$ to $300{,}000$ psi)	0.0 to 1.0
Lime-stabilized materials ($E = 20{,}000$ to $70{,}000$ psi)	1.0 to 3.0
Unbound granular materials ($E = 15{,}000$ to $45{,}000$ psi)	1.0 to 3.0
Fine-grained or natural subgrade materials ($E = 3{,}000$ to $40{,}000$ psi)	2.0 to 3.0

Note: E in this table refers to the general symbol of the resilient modulus.
Source: *From AASHTO (1993). AASHTO Guide for Design of Pavement Structures. Washington, DC: American Association of State Highway and Transportation Officials. Table 2-7. Used with permission.*

TABLE 12.1 AASHTO Recommended LS Values

Quality of drainage	Percent of time pavement structure is exposed to moisture levels approaching saturation			
	<1%	1–5%	5–25%	<25%
Excellent	1.2–1.20	1.20–1.15	1.15–1.10	1.10
Good	1.20–1.15	1.15–1.10	1.10–1.00	1.00
Fair	1.15–1.10	1.10–1.00	1.00–0.90	0.90
Poor	1.10–1.00	1.00–0.90	0.90–0.80	0.80
Very poor	1.00–0.90	0.90–0.80	0.80–0.70	0.70

Source: From AASHTO (1993). AASHTO Guide for Design of Pavement Structures. Washington, DC: American Association of State Highway and Transportation Officials. Table 2-5. Used with permission.

TABLE 12.2 Recommended Values for Drainage Coefficient, C_d, for Rigid Pavements

Type of shoulder	Asphalt		PCC	
Load transfer device	Yes	No	Yes	No
JPCP and JRCP	3.2	3.8–4.4	2.5–3.1	3.6–4.2
CRCP	2.9–3.2	NA	2.3–2.9	NA

Source: From AASHTO (1993). AASHTO Guide for Design of Pavement Structures. Washington, DC: American Association of State Highway and Transportation Officials. Used with permission.

TABLE 12.3 Recommended Values for Load Transfer Coefficient, J, for Rigid Pavements

12.3.4 Load Transfer Coefficient

Load transfer coefficient is an empirically derived number used to account for load transfer efficiency across joints (or cracks in CRCP). The lower the J, the better the load transfer. AASHTO-recommended values for load transfer coefficient, J, for rigid pavements are listed Table 12.3.

12.3.5 Reliability

Reliability considerations for rigid pavement are similar to those for flexible pavement, as presented in Chap. 7. Reliability levels, $R\%$, and the overall standard deviation, S_o, are incorporated directly into the design charts. The recommended reliability values (R) and normal deviate (Z_R) for a given reliability, R, are listed in Chap. 7.

12.3.6 Change in Present Serviceability Index (ΔPSI)

Present serviceability index (PSI) is an index based on measured physical roughness in pavement. The scale of 0 to 5 is used with 5 being very good and 0 being very poor. A new rigid pavement commonly has an initial PSI value of 4.5. It is common to use the terminal PSI value (p_t) of 2.5 for major highways and 2.0 for others; this value can be considered less than 2 for less important roads. Then, $\Delta\text{PSI} = 4.2 - p_t$.

Example 12.1: Determining *k*-Value If Subbase Is Present A rigid pavement has the following conditions:

- 6 in. of unbound granular subbase
- Bedrock located at 5 ft from the bottom of subgrade
- Estimated thickness of concrete slab about 9 in.

The resilient modulus of roadbed soil and subbase at different months of a year is listed in Table 12.4.

Month	Roadbed soil modulus (psi)	Subbase modulus, E_{SB} (psi)
January	7,000	20,000
February	7,000	20,000
March	7,000	20,000
April	4,000	15,000
May	4,000	15,000
June	4,000	15,000
July	4,000	15,000
August	4,000	15,000
September	2,500	15,000
October	2,500	15,000
November	2,500	15,000
December	7,000	20,000

TABLE 12.4 Modulus Data for Example 12.1

Determine the effective modulus of a subgrade reaction.

Solution

Step 1. Identify the combinations to be considered.
Given in the problem.

Step 2. Determine the seasonal roadbed soil resilient modulus based on laboratory testing.
Given in the problem.

Step 3. Determine subbase elastic (resilient) modulus E_{SB} for each season based on laboratory testing.
Given in the problem.

Step 4. Estimate the composite modulus of subgrade reaction for each season assuming a semi-infinite subgrade depth (or depth to bedrock greater than 10 feet) using Fig. 12.3.

Step 5. Develop *k*-value considering the effect of a rigid foundation near the surface (if any) using Fig. 12.4.

Step 6. Assume the projected thickness of the slab that may be required, then determine the relative damage u_r as listed in Table 12.5.

Month	*Step 2.* Roadbed soil modulus (psi)	*Step 3.* Subbase modulus, E_{SB} (psi)	*Step 4.* Composite *k*-value (pci) (Fig. 12.3)	*Step 5.* *k*-value (pci) on rigid foundation (Fig. 12.4)	*Step 6.* Relative damage, u_r (Fig. 12.5)
January	7,000	20,000	410	540	60
February	7,000	20,000	410	540	60
March	7,000	20,000	410	540	60
April	4,000	15,000	230	300	78
May	4,000	15,000	230	300	78
June	4,000	15,000	230	300	78
July	4,000	15,000	230	300	78
August	4,000	15,000	230	300	78
September	2,500	15,000	160	230	86
October	2,500	15,000	160	230	86
November	2,500	15,000	160	230	86
December	7,000	20,000	410	540	60
				Sum of u_r	888
				Step 7. Average, $\bar{u}_r$	888/12 = 74

TABLE 12.5 Relative Damage Calculation for Example 12.1

Step 7. Determine the average relative damage as $\bar{u}_r = \dfrac{\sum u_r}{n}$; n = number of periods per year.

Step 8. Determine the *k*-value for the $\bar{u}_r$ using Fig. 12.5.
Using Fig. 12.5, for $\bar{u}_r$, $k = 340$ pci.

Step 9. Adjust the *k*-value to account the potential loss of support arising from the subbase erosion using Fig. 12.6.

From Table 12.1, for an unbound granular subbase with resilient modulus 15,000 to 45,000 psi, LS = 1.0 – 3.0 (say, 2.0).

From Fig. 12.6, for $k = 340$, LS = 2.0, effective $k = 34$ pci.

Answer *The effective modulus of subgrade reaction is 34 pci.*

Example 12.2: Determining *k*-Value If Subbase Is Not Present A rigid pavement is placed directly on subgrade, with no bedrock existing within 10 ft of the subgrade. The estimated concrete slab is about 7.5 in. There is no possibility of subgrade loss. The resilient modulus of roadbed soil of a year is given in Table 12.6.

Month	Roadbed soil modulus (psi)
January	50,000
February	38,000
March	27,000
April	16,000
May	5,000
June	5,000
July	5,000
August	5,000
September	8,000
October	12,000
November	21,000
December	29,000

Table 12.6 Modulus Data for Example 12.2

Determine the effective modulus of subgrade reaction.

Solution

Step 1. Identify the combinations to be considered.
Given in the problem.

Step 2. Determine the seasonal roadbed soil resilient modulus based on laboratory testing.
Given in the problem.

Step 3. Determine subbase elastic (resilient) modulus E_{SB} for each season based on laboratory testing.
Not applicable as no subbase exists.

Step 4. $k = M_R/19.4$.

Step 5. Develop k-value considering the effect of a rigid foundation near the surface (if any) using Fig. 12.4.
Not applicable as no bedrock exists within 10 ft.

Step 6. Assume the projected thickness of the slab that may be required, and then determine the relative damage u_r as listed in Table 12.7.

Step 7. Determine the average relative damage as $\bar{u}_r = \frac{\sum u_r}{n}$; n = number of periods per year.

Step 8. Determine the k-value for the $\bar{u}_r$ using Fig. 12.5.
Using Fig. 12.5, for $\bar{u}_r$ of 25, $k = 650$ pci.

Step 9. Adjust the k-value to account the potential loss of support arising from the subbase erosion using Fig. 12.6.
LS = 0; no adjustment is required for k-value.

Therefore, effective $k = 650$ pci.

Answer *The effective modulus of subgrade reaction is 650 pci.*

Month	*Step 2.* Roadbed soil modulus (psi)	*Step 3.* Subbase modulus, E_{SB} (psi)	*Step 4.* $k = M_R/19.4$	*Step 5.* k-value (pci) on rigid foundation (Fig. 12.4)	*Step 6.* Relative damage, u_r (Fig. 12.5)
January	50,000	–	2,577	–	6.8
February	38,000	–	1,959	–	9.6
March	27,000	–	1,392	–	13.7
April	16,000	–	825	–	21.2
May	5,000	–	258	–	41.4
June	5,000	–	258	–	41.4
July	5,000	–	258	–	41.4
August	5,000	–	258	–	41.4
September	8,000	–	412	–	32.8
October	12,000	–	619	–	25.8
November	21,000	–	1,082	–	17.1
December	29,000	–	1,495	–	12.8
				Sum of u_r	305
				Step 7. Average, $\bar{u}_r$	$305/12 = 25$

TABLE 12.7 Relative Damage Calculation for Example 12.2

12.4 Thickness Design

Equation (12.1) is used to solve for the concrete thickness, D. All the other parameters except D are the inputs and D is the output. Solving this equation is not so easy. Nomographs (Figs. 12.1 and 12.2) can be used instead to solve for D.

First, Segment I (Fig. 12.1) is used starting to input the k-value and then the E_c. Subsequently, the S_c', J, and C_d are used to find out the match line number. Once the match line number is found out, Segment II (Fig. 12.2) is then used from two sides (left to right and bottom to up). Using the match line number through the ΔPSI, a horizontal line is drawn in the graph area. Then, using the R, S_o, and ESAL, a vertical line is drawn in the graph area. The intersection gives the value of D.

Work on Example 12.3 to clarify your understanding.

Example 12.3: Thickness Design of Concrete Pavement A concrete pavement is to be designed for a four-lane local highway on a subgrade with the following conditions:

- An effective modulus of subgrade reaction = 150 pci
- The accumulated equivalent axle load for the design period = 10 million
- The initial and terminal serviceability indices = 4.5 and 1.5, respectively
- The modulus of rupture of the concrete = 900 lb/in.^{2}
- The modulus of elasticity of the concrete = 4×10^6 lb/in.^{2}
- The standard deviation = 0.25

- Load transfer coefficient = 3.2
- Drainage coefficient = 0.9
- Reliability = 95%

Using the AASHTO 1993 design method, determine a suitable thickness of the concrete slab.

Solution

$W_{18} = 10$ million
$R = 95\%$
$S_o = 0.25$
$\Delta PSI = 4.5 - 1.5 = 3.0$
$E_c = 4 \times 10^6$ lb/in.^{2}
$S'_c = 900$ lb/in.^{2}
$J = 3.2$
$C_d = 0.90$
$k = 150$ pci
$D =$ depth of concrete slab = ?

NOMOGRAPH SOLVES:

$$\log_{10}W_{18} = Z_R*S_o + 7.35*\log_{10}(D+1) - 0.06 + \frac{\log_{10}\left[\frac{\Delta PSI}{4.5-1.5}\right]}{1+\frac{1.624*10^7}{(D+1)^{8.46}}} + (4.22-0.32p_t)*\log_{10}\left[\frac{S'_c*C_d\left[D^{0.75}-1.132\right]}{215.63*J\left[D^{0.75}-\frac{18.42}{\left(\frac{E_c}{k}\right)^{0.25}}\right]}\right]$$

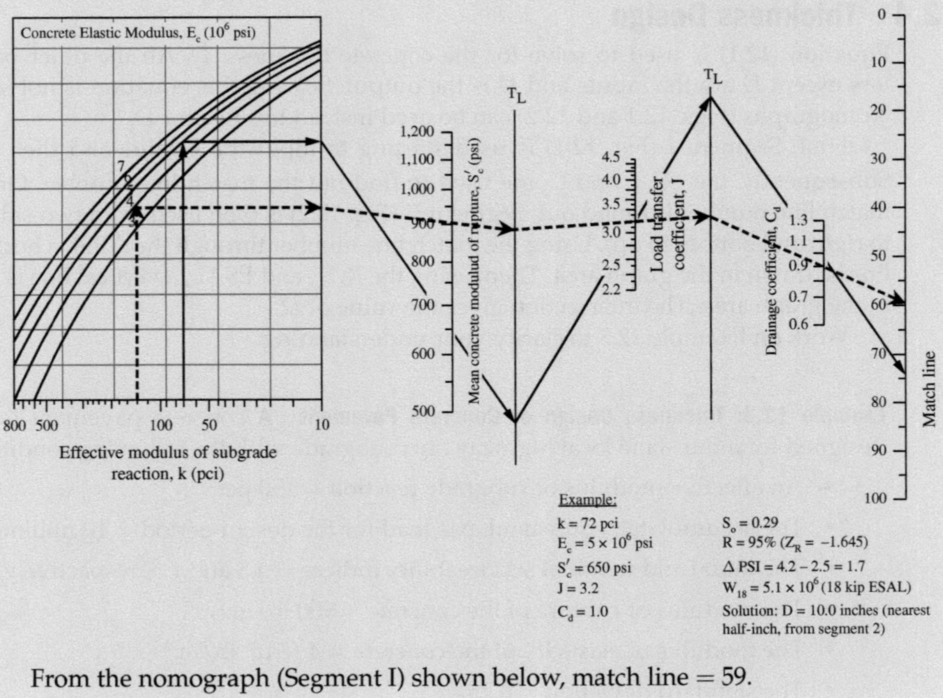

From the nomograph (Segment I) shown below, match line = 59.

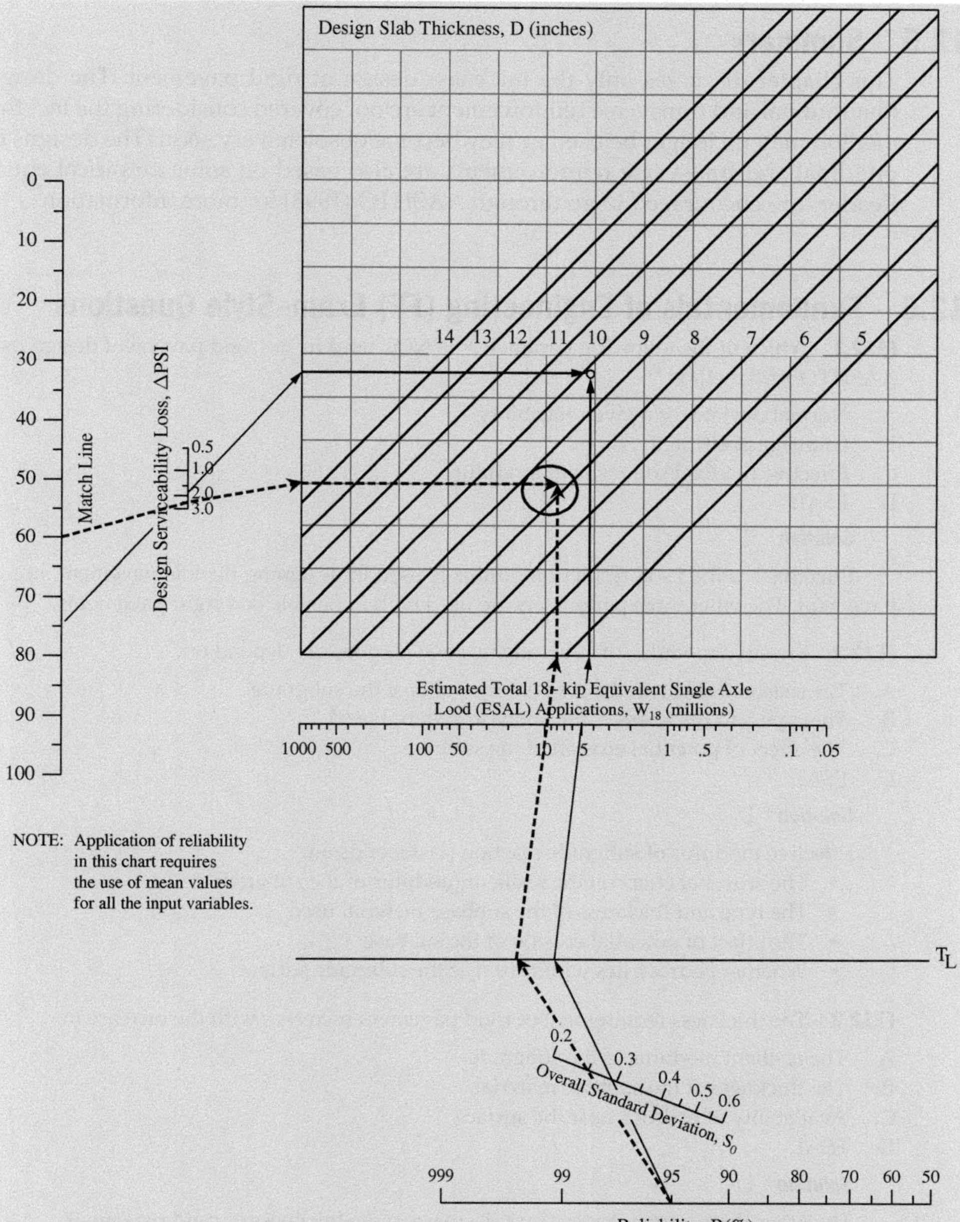

From the nomograph (Segment II) shown, thickness required = 8.3 in. ≈ 8.5 in. to round off

Answer *The thickness of the concrete slab is 8.5 in.*

12.5 Summary

This chapter discusses only the thickness design of rigid pavement. The designs for longitudinal and transverse reinforcement are not covered considering the fact that this method may no longer be used or may become obsolete very soon. The designs of longitudinal and transverse reinforcements are also based on some empirical equations. Readers are encouraged to go through AASHTO (1993) for more information.

12.6 Fundamentals of Engineering (FE) Exam-Style Questions

FE12.1 Which of the following parameters is NOT used in the rigid pavement design using the AASHTO 1993 method?

A. Normal deviate for a given reliability, R
B. Drainage coefficient
C. Effective roadbed soil resilient modulus
D. ESAL

Solution C

Effective roadbed soil resilient modulus is used in designing flexible pavement, not in rigid pavement. The other three parameters are used both in flexible and rigid pavements.

FE12.2 Effective modulus of subgrade reaction (k) does not depend on:

A. The seasonal effect on the resilient modulus of the subgrade
B. The type and thickness of the subbase material used
C. The effect of potential erosion of the subbase
D. ESAL

Solution D

Effective modulus of subgrade reaction (k) depends on:

- The seasonal effect on the resilient modulus of the subgrade
- The type and thickness of the subbase material used
- The effect of potential erosion of the subbase
- Whether bedrock lies within 10 ft of the subgrade surface

FE12.3 The thickness requirement of rigid pavement increases with the increase in:

A. The resilient modulus of the subgrade
B. The thickness of the subbase material
C. Availability of bedrock near the surface
D. ESAL

Solution D

The other three parameters cause a decrease in the thickness of rigid pavement.

12.7 Practice Problems

12.1 A rigid pavement has the following conditions:

- 6 in. of lime-stabilized subbase (consider the average loss of support)
- Bedrock located at 5 ft from the bottom of subgrade
- Estimated thickness of the concrete slab about 9 in.

The resilient modulus of roadbed soil and subbase at different months of a year is listed in Table P12.1.

Month	Roadbed soil modulus (psi)	Subbase modulus, E_{SB} (psi)
January	20,000	50,000
February	20,000	50,000
March	20,000	50,000
April	20,000	50,000
May	20,000	50,000
June	12,000	15,000
July	12,000	15,000
August	12,000	15,000
September	12,000	15,000
October	16,000	35,000
November	16,000	35,000
December	16,000	35,000

TABLE P12.1 Modulus Data for Prob. 12.1

Determine the effective modulus of subgrade reaction.

12.2 A rigid pavement is placed directly on subgrade with no bedrock existing within 10 ft of the subgrade. The estimated concrete slab is about 9.0 in. Assume the subgrade loss factor to be 1.0. The resilient modulus of roadbed soil of a year is listed in Table P12.2.

Month	Roadbed soil modulus (psi)
January	25,000
February	23,000
March	21,000
April	19,000
May	18,000
June	17,000
July	16,000
August	17,000
September	19,000
October	21,000
November	22,000
December	23,000

TABLE P12.2 Modulus Data for Prob. 12.2

Determine the effective modulus of subgrade reaction.

12.3 A concrete pavement is designed for a four-lane local highway on a subgrade with the following conditions:

- An effective modulus of subgrade reaction = 500 pci
- The accumulated equivalent axle load for the design period = 50 million
- The initial and terminal serviceability indices = 4.5 and 2.5, respectively
- The modulus of rupture of the concrete = 1,000 $lb/in.^2$
- The modulus of elasticity of the concrete = 6×10^6 $lb/in.^2$
- The standard deviation = 0.30
- Load transfer coefficient = 3.2
- Drainage coefficient = 0.9
- Reliability = 95%

Using the AASHTO 1993 design method, determine a suitable thickness of the concrete slab.

CHAPTER 13

Distresses in Rigid Pavement

13.1 Background

Rigid pavement suffers from both major and minor distresses like flexible pavement. Major distresses are those distresses whose presence decreases the riding comfort remarkably, caused by load- and temperature-related factors. These distresses are readily controlled by proper structural design of pavement. Minor cracks are essentially maintenance related, and can be controlled by proper mix design. In structural design of pavement, only major distresses are considered.

13.2 Major Distresses

13.2.1 Transverse Slab Cracking in Jointed Plain Concrete Pavement (JPCP)

Transverse cracking occurs in two ways in concrete. First, due to repeated fatigue loading, stress and strain develop at the top and at the bottom of concrete slab. This repeated stress and strain cause fatigue damage in concrete slab. The fatigue damage leads to development of a transverse crack. The cracks may develop at the surface and propagate downward or the opposite. The second type of transverse crack occurs due to thermal expansion and contraction of concrete slab. If the slab length is too long compared to the width of slab, the thermal expansion and contraction cause transverse cracks in the slab, as shown in Fig. 13.1. Slab length should be kept within the limits to avoid transverse cracks.

13.2.2 Transverse Joint Faulting in JPCP

Transverse joint faulting means the difference in elevation (Fig. 13.2) across a joint or crack usually associated with undoweled JPCP. Due to pumping, the underneath material gets off through the joint and one slab gets settled. This is the most common faulting mechanism. Some other reasons are slab settlement, curling, and warping. Diamond grinding may be an appropriate solution if the faulting is small [say, 0.15 in. (4 mm)]. For larger faulting, dowel bar retrofit or reconstruction is required to get rid of faulting.

13.2.3 Punchouts in Continuously Reinforced Concrete Pavement (CRCP)

Small, localized broken pieces of concrete slab are known as punchouts. Punchout occurs when there is loss of load transfer across two closely spaced adjacent cracks,

Figure 13.1 Transverse slab cracking.

Figure 13.2 Transverse joint faulting.

making the concrete between them act as a cantilever beam. When truck axles pass along near the longitudinal edge of the slab between two closely spaced transverse cracks, a high-tensile stress occurs at the top of the slab, some distance from the edge [4 ft (1.2 m)], transversely across the pavement. This increase in stress is very high when there is a loss of load transfer across the transverse cracks or loss of support along the edge of the slab. Repeated loading of heavy axles results in fatigue damage at the top of the slab, which results first in micro-cracks that initiate at the transverse crack and propagate longitudinally across the slab to the other transverse crack resulting in a punchout. Punchout also occurs due to inadequate consolidation, steel corrosion, inadequate amount of steel, wide shrinkage cracks or very close shrinkage cracks, etc. Figure 13.3 shows a punchout due to steel corrosion. Punchouts cause severe roughness and allow moisture infiltration, leading to pumping and spalling. Small punchouts may be patched using wire mesh but full-depth patch is required to get rid of larger punchouts.

Figure 13.3 Concrete punchouts due to steel corrosion.

13.2.4 Smoothness in JPCP and CRCP

Smoothness is the overall measure of surface roughness considering transverse cracks, spalling, joint faulting, etc. Smooth pavement gives good riding comfort, whereas rough surface is not comfortable.

13.3 Selected Minor Distresses

13.3.1 Spalling

After the sequential occurrence of cracking and breaking of joint or crack edges, if some materials fall apart into loose debris, as shown in Fig. 13.4, the resulting fall-off is called spalling. It usually occurs within about 2 ft. (0.6 m) of joint or crack edge. Spalling occurs due to a series of action. Due to infiltration of incompressible materials (say, stone or steel pieces) in to the joint space and subsequent expansion (can also cause blowups), a small part of slab may be disintegrated. Then, the impact of high-speed tires may break more and spall occurs. Extreme freeze-thaw action, weak concrete at a joint caused by inadequate consolidation, misalignment or corroded dowel, heavy traffic loading, etc. are some other causes of spalling. The problem of spalling is roughness, impact on tire, and acceleration for larger defects. Small spalling may be patched but full-depth patch is required to get rid of larger spalling.

13.3.2 Polished Aggregates

If the aggregates extending above the cement paste are not rough or angular or polished due to tire movement, it is called polished aggregates. Figure 13.5 shows a portion of polished aggregate. Repeated traffic and/or low-quality aggregate may be responsible

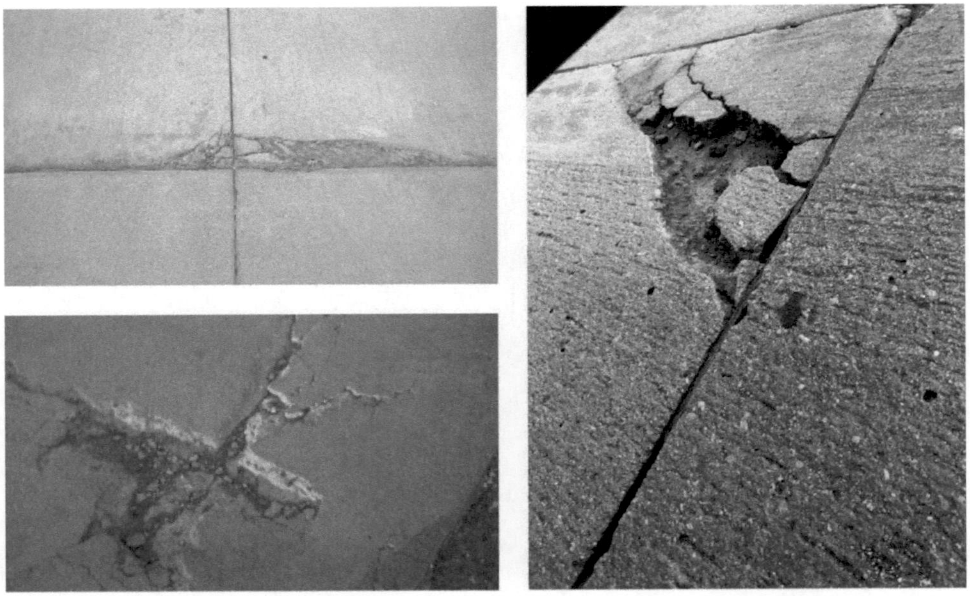

FIGURE 13.4 Spalling in concrete pavement.

FIGURE 13.5 Polished aggregates.

for this distress. The polished concrete suffers from extreme skid resistance problem. Diamond grinding or overlay can be applied to get rid of polished aggregate.

13.3.3 Shrinkage Cracking

Shrinkage cracking means the cracks formed due to insufficient curing during the setting of soft concrete, as shown in Fig. 13.6. The crack width and length depend on the degree of shortage of curing. Shrinkage cracks typically do not extend the entire depth of the slab. Small shrinkage cracks may not be a problem. However, it may widen due to winter shrinkage, application of repeated load, etc. and allow moisture infiltration. Small shrinkage cracks can be sealed but the large crack may require the entire slab to be replaced.

13.3.4 Linear Cracking

Linear cracks are not associated with corner breaks or blowups and extend across the entire slab, as shown in Fig. 13.7. Repeated traffic loading, curling stress, moisture stresses, and loss of support, etc. are some reasons for linear cracking. Like other cracks,

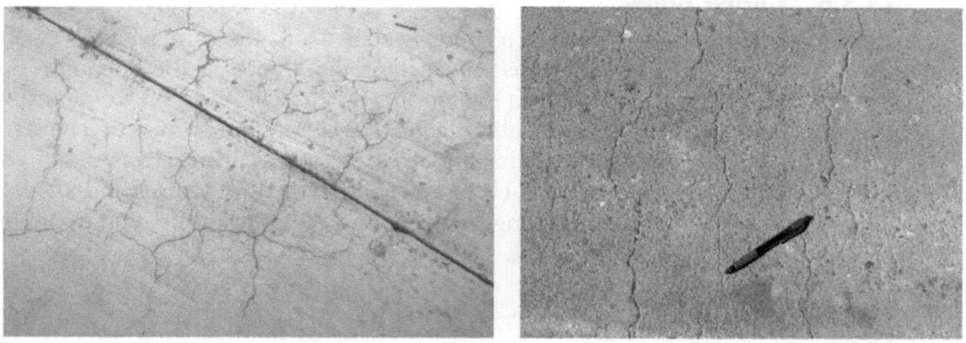

FIGURE 13.6 Shrinkage cracking.

FIGURE 13.7 Linear cracking in concrete pavement.

FIGURE 13.8 Corner break in concrete slabs.

this crack causes an increased roughness, allows moisture infiltration, and leads to pumping, spalling, and ultimate failure. A single narrow, linear crack may be sealed but multiple cracks may require full-depth reclamation.

13.3.5 Corner Break

The corner is the weakest area of a concrete slab. Thus, the corner breaks easily with applied stress. Corner break means the cracking that intersects the slab joints near the corner within about 6 ft (2 m), as shown in Fig. 13.8. In addition to corner stresses caused by load repetitions, some other factors accelerate the process of a corner break such as loss of support, poor load transfer across the joint, curling stresses, and warping stresses. Like other cracks, this crack causes an increased roughness and allows moisture infiltration, leading to pumping, spalling, and ultimate failure. Full-depth patch is the only remedy for a corner break.

13.3.6 Blowup

If there is insufficient room for slab expansion, then a slab may move up due to thermal expansion in the summer. This process is called blowup and is shown in Fig. 13.9. In the winter, slabs contract as temperature decreases. Then, some room may open up that may be filled with dirt or stone. In the following summer, when slabs expand again, high compressive stress may be caused in the slab due to its inability to expand. As a result, the slabs may buckle and shatter to relieve the stresses if these stresses are great enough. Blowup in extreme cases can be a safety issue. Full-depth patch is the only option to get rid of concrete blowups.

13.3.7 Pumping/Water Bleeding

Pumping is the ejection of material from underneath the slab through the joints due to water pressure. Infiltrated water accumulated underneath a concrete slab forces up when the slab deflects under load. If the pumping does not eject any material other than water, it is called water bleeding (Fig. 13.10). Some causes of pumping are high water table, poor drainage, cracks, poor joint seals, etc. This water can do one of the following:

- Move out from underneath the slab through the joints to the pavement surface. This results in a slow removal of underneath material resulting in loss of support.

FIGURE 13.9 Concrete pavement blowups.

FIGURE 13.10 Water bleeding. (*Photo by Chris Meeks.*)

- Migrate underneath an adjacent slab. This type of movement leads to faulting.
- Move around under the same slab if the moisture quantity is less.

Full-depth patch is the only solution to get rid of pumping. To avoid pumping, the following tasks may be carried out:

- Address source of water or cause of poor drainage.
- Install dowel bars to increase load transfer between slabs.
- Fill and stabilize any lost materials.

13.3.8 Other Minor Cracks

Some other minor cracks not mentioned earlier are durability cracking (D-cracking), bar corrosion, alkali–aggregate reaction, etc. (Fig. 13.11). D-cracking is the series of closely spaced, crescent-shaped cracks near a joint, corner, or crack. It is caused by freeze-thaw expansion of the large aggregate within the PCC slab. D-cracking is a general PCC distress and is not unique to pavement PCC. Steel bar corrosion occurs due to moisture infiltration through the cracking of concrete. An alkali–aggregate reaction is the expansive reaction that takes place in PCC between alkali (contained in the cement paste) and elements like silica, etc. within an aggregate. This reaction, which occurs to some extent in most PCC, can result in map or pattern cracking.

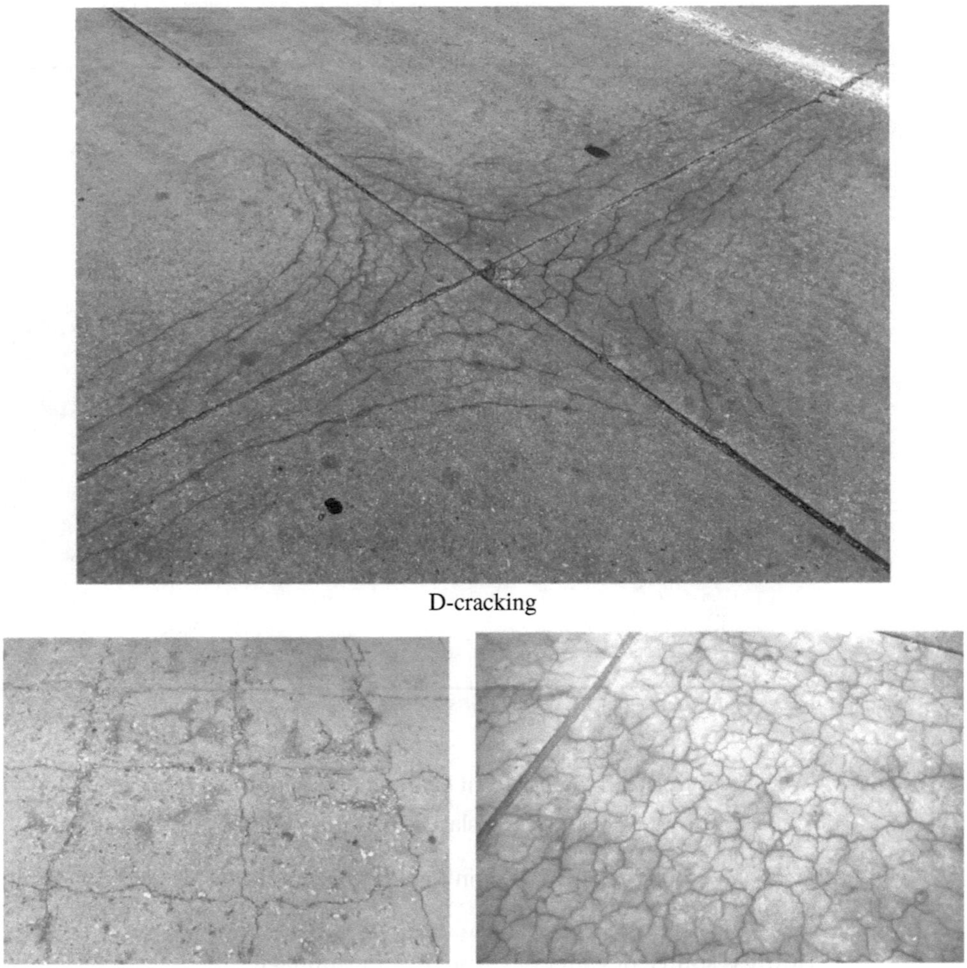

D-cracking

Steel bar corrosion

Alkali–silica reaction

FIGURE 13.11 Some other distresses in concrete pavement.

13.4 Summary

This chapter discusses the common types of distresses in concrete pavements, their causes, and remedies. There are some other minor cracks not discussed here for being local or do not need any attention. The causes of the previously discussed cracks are very general. Exact causes of a crack may be difficult to tell unless detailed investigation is made. In addition, multiple reasons may lead a particular distress. The remedies mentioned in this chapter are also general. Before deciding the option, maintenance cost, budget, local materials, past success, etc. should be considered. In fact, every highway agency has its own maintenance plan for different types of distresses. The highway agencies develop it based on their long success history. In addition to general repair strategies, AASHTO (2015)–recommended repair and preventive treatments are summarized in Table 13.1.

Distress	Preventive treatments	Repair treatments
JPCP pumping	Reseal joints Restore joint load transfer Subsurface drainage Edge support (tied PCC should edge beam)	Subseal or mud-jack PCC slabs (effectiveness depends on materials and procedures)
JPCP joint faulting	Subseal joints Reseal joints Restore load transfer Subsurface drainage Edge support (tied PCC should edge beam)	Grind surface Structural overlay
JPCP slab cracking	Subseal (loss of support) Restore load transfer Structural overlay	Full- or partial-depth repair
JPCP joint or crack spalling	Reseal joints	Full- or partial-depth repair
CRCP punchouts	Polymer or epoxy grouting Subseal (loss of support)	Full-depth repair
PCC disintegration	None	Full-depth repair Thick overlay

Source: *From AASHTO (2015). Mechanistic-Empirical Pavement Design Guide: A Manual of Practice. Washington, DC: American Association of State Highway and Transportation Officials. Used with permission.*

TABLE 13.1 Candidate Repair and Preventive Treatments for Rigid Pavement

13.5 Fundamentals of Engineering (FE) Exam-Style Questions

FE13.1 Which of the following distresses is not considered as a design criterion in the rigid pavement design?

A. Transverse cracking
B. Transverse joint faulting
C. Punchouts
D. Polished aggregate

Solution D

Polished aggregate is not considered while designing rigid pavement. The other three distresses are considered as the design criteria.

FE13.2 The full form of D-cracking is:

A. Diagonal cracking
B. Distributed cracking
C. Divisional cracking
D. Durability cracking

Solution D

Durability cracking (D-cracking) is a series of closely spaced, crescent-shaped cracks near a joint, corner, or crack, caused by freeze-thaw expansion of the large aggregates.

FE13.3 Transverse joint faulting means:

A. The difference in elevation across a joint
B. The joint is distorted by crushing
C. Pumping in joints
D. Deformation at joints

Solution A

Transverse joint faulting means the difference in elevation across a joint. The other three distresses have no formal names.

13.6 Practice Problems

13.1 Name and define the load-related cracking in rigid pavement.

13.2 Name and define the non-load-related cracking in rigid pavement.

13.3 Differentiate between pumping and water bleeding.

13.4 Why is corner cracking the most common cracking in asphalt pavement?

13.5 What are the causes of transverse cracking?

CHAPTER 14

Distress Models in Rigid Pavement

14.1 Background

The performance of a pavement structure is dependent upon the interaction between pavement response and strength of the different layers. Wheel loads induce stresses and strains in each layer. This stress-strain causes deformation and cracking of pavement layer. Some equations are used to calculate the deformation and cracking of the pavement layers. The equations used for concrete pavement are discussed in this chapter. Jointed plain concrete pavement (JPCP), shown in Fig. 14.1, considers three types of distresses:

1. Transverse slab cracking (bottom-up and top-down)
2. Mean transverse joint faulting
3. Smoothness

Continuously reinforced concrete pavement (CRCP) considers two types of distresses:

1. Punchouts
2. Smoothness

Remember that these equations are mechanistic-empirical in nature. Local calibration is required for greater accuracy. In addition, the equations determine the distresses at mean level (50% reliability). These distress values are increased using the normality assumption for an increased reliability.

14.2 Jointed Plain Concrete Pavement

14.2.1 Transverse Slab Cracking (Bottom-Up and Top-Down)—JPCP

The transverse cracking of JPCP considers both the bottom-up and top-down cracking modes. The potential for either cracking mode is present in all slabs under normal service conditions. Any given slab can crack either from bottom-up or top-down, but not both directions. The predicted bottom-up and top-down cracking are therefore not particularly meaningful on their own, and combined cracking is stated to exclude the likelihood of both cracking methods occurring on the same slab. The percentage of slabs with transverse cracks (including all severities) in a given traffic lane is used as a measure of transverse cracking and is estimated for both bottom-up and top-down cracking using the following global equation:

$$\text{CRK} = \frac{1}{1 + C_4 (\text{DI}_F)^{C_5}} \tag{14.1}$$

where CRK = Predicted amount of bottom-up or top-down cracking (fraction)
DI_F = Fatigue damage calculated using the procedure described in this section
$C_{4,5}$ = Calibration factors; $C_4 = 1.0$, $C_5 = -1.98$

The general expression for fatigue damage accumulations considering all critical factors for JPCP transverse cracking is as depicted in Eq. (14.2) and referred to as Miner's hypothesis:

$$\text{DI}_F = \sum \frac{n_{i,j,k,l,m,n,o}}{N_{i,j,k,l,m,n,o}} \tag{14.2}$$

where DI_F = Total fatigue damage (top-down or bottom-up)
$n_{i,j,k,...}$ = Applied number of load applications at condition i, j, k, l, m, n
$N_{i,j,k,...}$ = Allowable number of load applications at condition i, j, k, l, m, n
i = Age (accounts for change in PCC modulus of rupture and elasticity, slab/base contact friction, deterioration of shoulder LTE)
j = Month (account for change in base elastic modulus and effective dynamic modulus of subgrade reaction)
k = Axle type (single, tandem, and tridem for bottom-up cracking; short, medium, and long wheelbase for top-down cracking)
l = Load level (incremental load for each axle type)
m = Equivalent temperature difference between top and bottom PCC surfaces
n = Traffic offset path
o = Hourly truck traffic fiction

The applied number of load applications ($n_{i,j,k,l,m,n}$) is the actual number of axle type k of load level l that passed through traffic path n under each condition (age, season, and temperature difference). The allowable number of load applications is the number of load cycles at which fatigue failure is expected (corresponding to 50% slab cracking) and is a function of the applied stress and PCC strength. Use Eq. (14.3) to determine the allowable number of load applications.

$$\log(N_{i,j,k,l,m,n}) = C_1 \left(\frac{M_{ri}}{\sigma_{i,j,k,l,m,n}} \right)^{C_2} \tag{14.3}$$

where $N_{i,j,k,...}$ = Allowable number of load applications at condition i, j, k, l, m, n
M_{ri} = PCC modulus of rupture at age i, psi
$\sigma_{i,j,k,...}$ = Applied stress at condition i, j, k, l, m, n
C_1 = Calibration constant, 2.0
C_2 = Calibrated constant, 1.22

The calculation of fatigue damage is a method of summing up damage from each increment of damage. Once the damage is measured from top-down and bottom-up, the resulting cracking is determined using Eq. (14.1), while the combined total cracking is estimated using the equation:

$$TCRACK = \left(CRK_{\text{Bottom-up}} + CRK_{\text{Top-down}} - CRK_{\text{Bottom-up}} \times CRK_{\text{Top-down}}\right) \times 100\% \quad (14.4)$$

where TCRACK = Total transverse cracking (%, all severities)
$CRK_{\text{Bottom-up}}$ = Predicted amount of bottom-up transverse cracking (fraction)
$CRK_{\text{Top-down}}$ = Predicted amount of top-down transverse cracking (fraction)

Note that Eq. (14.4) essentially assumes that a bottom-up or top-down slab may crack but not both. Calculating critical response using neural networks (for speed) involves integrating the slab and base path into an equivalent section based on equal stresses (load and temperature/moisture gradients) and contact friction between slab and base. As these factors change over time, this is performed weekly.

The above calculation is for the mean level (i.e., reliability of 50%). However, the designer needs a reliability of more than 50% for almost all projects as the design does not guarantee the performance requirements throughout the design. Also, in terms of the consequences of failure, the more critical the project is, the higher the reliability of the desired design. The AASHTOWare pavement ME design software calculates the test section's reliability according to the user-defined design criteria or threshold values. The mean distress or IRI value (reliability of 50%) is increased by the number of standard errors added to the specified level of reliability. For example, reliability of 90% uses a factor of 1.282 times the standard error, and a reliability of 95% uses a factor of 1.645. This factor is referred to as the standard normal deviate. Mathematically, the *R*% reliability distress can be written as:

$$\begin{aligned}\text{Distress at } R\% \text{ reliability} = {} & \text{Distress at 50\% reliability} + \text{Standard error} \\ & \times \text{Normal deviate at } R\% \text{ reliability}\end{aligned} \quad (14.5)$$

The standard error of predicted mean total transverse cracking of JPCP pavement ($S_{e(\text{CR})}$) is listed in the equation:

$$S_{e(\text{CR})} = (5.3116\ \text{TCRACK})^{0.3903} + 2.99 \quad (14.6)$$

where TCRACK = predicted transverse cracking based on mean inputs (corresponding to 50% reliability), percentage of slabs.

Example 14.1: Transverse Cracking in JPCP A JPCP concrete pavement section has the traffic and stress response data as listed in Table 14.1.

The yearly average modulus of rupture of concrete is 700 psi. Assume the produced stresses at the bottom of concrete and at the surface are equal.

a. Calculate the predicted amount of transverse cracking after 10 years at 50% reliability.

b. Calculate the predicted amount of transverse cracking after 10 years at 90% reliability. The 90% reliability has a standard normal deviate of 1.282.

Axle type	Annual average repetition	Produced stress (yearly average) (psi)
Single	50,000	250
Tandem	30,000	260
Tridem	13,000	280
Quad	1,750	300

TABLE 14.1 Load and Response Data for Example 14.1

Solution

Given:

Modulus of rupture of concrete, $M_{ri} = 700$ psi

Known coefficients:

$C_1 = 2.0$
$C_2 = 1.22$
$C_4 = 1.0$
$C_5 = -1.98$

Axle type	Annual average repetition	Produced stress (yearly average) (psi)	$N = 10^{2.0\left(\frac{M_{ri}}{\sigma_i}\right)^{1.22}}$	$DI = \frac{n}{N}$
Single	50,000	250	10559635	0.004735
Tandem	30,000	260	4960052	0.006048
Tridem	13,000	280	1308262	0.009937
Quad	1,750	300	419678.8	0.00417

TABLE 14.2 Load and Response Data Analysis for Example 14.1

$\sum DI = 0.02489$ per year from Table 14.2 (sum of the last column)

$\sum DI = 0.02489 \times 10 = 0.2489$ after 10 years

$$\text{CRK} = \frac{1}{1 + C_4(\text{DI}_F)^{C_5}} = \frac{1}{1 + (0.2489)^{-1.98}}$$
$$= 0.0599$$
$$= 0.0599$$

$$\text{TCRACK} = (\text{CRK}_{\text{Bottom-up}} + \text{CRK}_{\text{Top-down}} - \text{CRK}_{\text{Bottom-up}} \times \text{CRK}_{\text{Top-down}}) \times 100\%$$
$$= (0.0599 + 0.0599 - (0.0599)(0.0599)) \times 100\%$$
$$= 11.62\%$$

The predicted transverse cracking at 50% reliability = 11.62%.

The 90% reliability has a standard normal deviate of 1.282.
The standard error of predicted mean total transverse cracking:

$$\begin{aligned} S_{e(\text{CR})} &= (5.3116\ \text{TCRACK})^{0.3903} + 2.99 \\ &= (5.3116 \times 11.62)^{0.3903} + 2.99 \\ &= 7.99 \end{aligned}$$

Predicted transverse cracking at 90% reliability = 11.62 + 1.282 × 7.99 = 21.86%.

Answer *The predicted transverse cracking at 50% reliability is 11.62% and the predicted transverse cracking at 90% reliability is 21.86%.*

14.2.2 Mean Transverse Joint Faulting—JPCP

Using an incremental approach, mean transverse joint faulting (Fig. 14.1) is predicted month by month. Each month a faulting increment is determined, and the current faulting level affects the magnitude of increment. The fault is calculated monthly as a total of fault rises from all previous months in pavement life from the date of traffic opening using the following equations:

$$\text{Fault}_m = \sum_{i=1}^{m} \Delta\text{Fault}_i \tag{14.7}$$

$$\Delta\text{Fault}_i = C_{34}\left(\text{FAULTMAX}_{i-1} - \text{Fault}_{i-1}\right)^2 \times \text{DE}_i \tag{14.8}$$

FIGURE 14.1 JPCP transverse joint faulting.

$$\text{FAULTMAX}_i = \text{FAULTMAX}_0 + C_7 \times \sum_{j=1}^{m} \text{DE}_j \times Log(1 + C_5 \times 5.0^{\text{EROD}})^{C_6} \quad (14.9)$$

$$\text{FAULTMAX}_0 = C_{12} \times \delta_{\text{curling}} \times \left[\log(1 + C_5 \times 5.0^{\text{EROD}}) \times \log\left(\frac{P_{200} \times \text{Wet Days}}{P_s}\right)\right]^{C_6} \quad (14.10)$$

where Fault_m = Mean joint faulting at the end of month m, in.
ΔFault_m = Incremental change (monthly) in mean transverse joint faulting during month i, in.
FAULTMAX_i = Maximum mean transverse joint faulting for month i, in.
FAULTMAX_0 = Initial maximum mean transverse joint faulting, in.
EROD = Base/subbase erodibility factor
DE_i = Differential density of energy of subgrade deformation during month i
δ_{curling} = Maximum mean monthly slab corner upward deflection PCC due to temperature curling and moisture warping
P_s = Overburden on subgrade, lb
P_{200} = Percent subgrade material passing #200 sieve
Wet Days = Average annual number of wet days (greater than 0.1-in. rainfall)
$C_{1,2,3,4,5,6,7,12,34}$ = Global calibration constants
$C_1 = 1.0184$; $C_2 = 0.91656$; $C_3 = 0.0021848$; $C_4 = 0.0008837$; $C_5 = 250$; $C_6 = 0.4$; $C_7 = 1.83312$

C_{12} and C_{34} are defined as follows:

$$C_{12} = C_1 + C_2 \times \text{FR}^{0.25} \quad (14.11)$$

$$C_{34} = C_3 + C_4 \times \text{FR}^{0.25} \quad (14.12)$$

where FR = Base freezing index defined as percentage of time the top base temperature is below freezing (32°F) temperature.

Constants used for restored rigid pavement are: $C_1 = 0.6$; $C_2 = 1.2$; $C_3 = 0.002125$; $C_4 = 0.000884$; $C_5 = 400$; $C_6 = 0.4$; $C_7 = 1.83312$.

The standard error of predicted mean joint faulting of JPCP pavement ($S_{e(F)}$) is listed below:

$$S_{e(F)} = (0.0097 \times \text{Fault})^{0.5178} + 0.014 \quad (14.13)$$

where Fault = Predicted mean transverse joint faulting at any time in inches.

Example 14.2: Transverse Joint Faulting in JPCP A JPCP site has the following conditions:

Maximum mean monthly slab corner upward deflection due to temperature curling and moisture warping = 0.05 in.
Base freezing index = 1.2%
Base/subbase erodibility factor = 0.15

Subgrade materials passing the 0.075-mm (No. 200) sieve = 12.6%
Number of days rainfall greater than 0.1 in. in a year = 28
Overburden on subgrade = 250 lb

a. Determine the initial maximum mean transverse joint faulting at mean level (50% reliability).
b. Determine the initial maximum mean transverse joint faulting at 95% reliability. The reliability (R) of 95% has a standard normal deviate (Z_R) of 1.645.

Solution

$FAULTMAX_0 = ?$
$\delta_{curling}$ = Maximum mean monthly slab corner upward deflection of PCC due to temperature curling and moisture warping= 0.05 in.
FR = Base freezing index = 1.2%
EROD = Base/subbase erodibility factor = 0.15
P_{200} = Subgrade materials passing the 0.075-mm (No. 200) sieve = 12.6%
Wet Days = Number of days rainfall greater than 0.1 in. in a year = 28
P_s = Overburden on subgrade = 250 lb

$C_1 = 1.0184$
$C_2 = 0.91656$
$C_5 = 250$
$C_6 = 0.4$
$C_{12} = C_1 + C_2 \times FR^{0.25} = 1.0184 + 0.91656 \times 1.2^{0.25} = 1.978$

$$\begin{aligned}
FAULTMAX_0 &= C_{12} \times \delta_{curling} \times \left[\log(1 + C_5 \times 5.0^{EROD}) \times \log\left(\frac{P_{200} \times \text{Wet Days}}{P_s}\right)\right]^{C_6} \\
&= (1.978)(0.05)\left[\log(1 + 250 \times 5.0^{0.15}) \times \log\left(\frac{12.6 \times 28}{250}\right)\right]^{0.4} \\
&= 0.067
\end{aligned}$$

Therefore, the joint faulting at 50% reliability = 0.067 in.
The standard error of the predicted mean joint faulting:

$$\begin{aligned}
S_{e(F)} &= (0.0097 \times \text{Fault})^{0.5178} + 0.014 \\
&= (0.0097 \times 0.067)^{0.5178} + 0.014 \\
&= 0.036
\end{aligned}$$

Joint faulting at 95% reliability = 0.067 + 1.645 × 0.036 = 0.126 in.

Answer *The initial maximum mean transverse joint faulting at mean level is 0.067 in. and the initial maximum mean transverse joint faulting at 95% reliability is 0.126 in.*

Each passage of an axle may cause only one occurrence of critical loading (i.e., if DE has the maximum value) for faulting analysis. Since the optimum faulting development occurs at night when the slab is curled upwards and joints are opened, and the efficiencies of load transfer are lower, only axle-load repetitions are applied from 8:00 p.m. to 8 a.m. It is taken into account in the faulting analysis.

For faulting analysis, the relative linear nighttime temperature differential is calculated as the mean difference between the top and bottom PCC surfaces occurring from 8:00 p.m. to 8:00 a.m. for each calendar month. The corresponding temperature gradient for the month is then determined using Eq. (14.14) for each month of the year.

$$\Delta T_m = \Delta T_{t,m} - \Delta T_{b,m} + \Delta T_{sh,m} + \Delta T_{\text{PCW}} \tag{14.14}$$

where ΔT_m = Effective temperature difference for month m
$\Delta T_{t,m}$ = Mean PCC top-surface nighttime temperature (from 8:00 p.m. to 8:00 a.m.) for month m
$\Delta T_{b,m}$ = Mean PCC bottom-surface nighttime temperature (from 8:00 p.m. to 8:00 a.m.) for month m
$\Delta T_{sh,m}$ = Equivalent temperature difference due to reversible shrinkage for month m for old concrete (i.e., shrinkage is fully developed)
ΔT_{PCW} = Equivalent temperature difference due to permanent curl/warp

Using the available climate data, the temperature in the top PCC layer is computed at 11 evenly spaced points for each hour through the thickness of the PCC layer. Such temperature distributions are converted into the equivalent temperature difference between top and bottom PCC surfaces.

The corner deflections due to slab curling and shrinkage warping are determined for each month using the effective temperature differential for each calendar month and the corresponding effective k-value and base modulus for each month. Corner deflections are calculated using a methodology implemented in the AASHTOWare pavement ME design software using a finite element–based neural network rapid-response solution. Use the calculated corner deflections to determine the initial maximum fault.

Using past cumulative differential energy (i.e., differential energy generated from axle-load applications for all months before the current month), the maximum fault is adjusted for past traffic damage. For each increment, deflections at the loaded and unloaded corner of the slab are calculated using the neural networks for each axle type and axle load.

The load transfer efficiency (LTE) joint strongly influences the magnitude of loaded and unloaded slab corner deflections. The LTE from the aggregate interlock, dowels (if present), and base/subgrade are determined to evaluate the initial transverse joint LTE. Table 14.1 lists the LTE_{base} values that are included in the AASHTOWare pavement ME design software. The overall initial joint LTE is calculated using Eq. (14.15) after the contributions of the aggregate interlock, dowels, and base/subgrade are determined.

$$\text{LTE}_{\text{joint}} = 100\left[1 - (1 - \text{LTE}_{\text{dowel}}/100)(1 - \text{LTE}_{\text{agg}}/100)(1 - \text{LTE}_{\text{base}}/100)\right] \tag{14.15}$$

where $\text{LTE}_{\text{joint}}$ = Total transverse joint LTE, %
LTE_{dowd} = Joint LTE if dowels are the only mechanism of load transfer, %
LTE_{base} = Joint LTE if the base is the only mechanism of load transfer, %
LTE_{agg} = Joint LTE if aggregate interlock is the only mechanism of load transfer, %

For each calendar month the LTE is determined, and output can be monitored over time to see whether it retains a high level. If the mean mid-depth PCC nighttime temperature is below freezing point (32°F), then this month's joint LTE will be increased. This is achieved by assigning 90% of base LTE for the month. Each month, the LTE aggregate interlock and dowel portion are modified. Assumed effective base LTE listed below for different base types:

- Aggregate base: $LTE_{Base} = 20\%$
- ATB or CTB: $LTE_{Base} = 30\%$
- Lean concrete base: $LTE_{Base} = 40\%$

Using past cumulative differential energy (i.e., differential energy accrued from axle load applications for all months before the current month), the maximum fault is adjusted for past traffic damage. For each increment, deflections at the loaded and unloaded corner of the slab are determined using the neural networks for each axle type and axle load. Using these deflections, the differential energy of subgrade deformation, DE; shear stress at the slab corner, τ, and overall dowel bearing stress, σ_b (for doweled joints), are measured using the equations:

$$DE = \frac{k}{2}\left(\delta_L^2 - \delta_U^2\right) \tag{14.16}$$

$$\tau = \frac{AGG \times (\delta_L - \delta_U)}{h_{PCC}} \tag{14.17}$$

$$\sigma_b = \frac{\zeta_d(\delta_L - \delta_U)}{d \times dsp} \tag{14.18}$$

where DE = Differential energy, lb/in.
δ_L = Loaded corner deflection, in.
δ_U = Unloaded corner deflection, in.
AGG = Aggregate interlock stiffness factor
k = Coefficient of subgrade reaction, psi/in.
h_{PCC} = PCC slab thickness, in.
ζ_D = Dowel stiffness factor = $J_d \times k \times l \times dsp$
d = Dowel diameter, in.
dsp = Dowel spacing, in.
J_d = Non-dimensional dowel stiffness at the time of load application
l = Radius of relative stiffness, in.

The loss of shear capacity (Δs) due to repeated wheel load applications is defined in terms of the transverse joint width based on a method derived from the Portland Cement Association (PCA) load transfer test data analysis. The following shear loss occurs during the increase in time (month):

$$\Delta s = \begin{cases} 0 & \text{if } w < 0.001 h_{PCC} \\ \sum_j \dfrac{0.005}{1.0 + (jw/h_{PCC})^{-5.7}}\left(\dfrac{n_j}{10^6}\right)\left(\dfrac{\tau_j}{\tau_{ref}}\right) & \text{if } jw < 3.8 h_{PCC} \\ \sum_j \dfrac{0.068}{1.0 + 6.0 \times (jw/h_{PCC} - 3)^{-1.98}}\left(\dfrac{n_j}{10^6}\right)\left(\dfrac{\tau_j}{\tau_{ref}}\right) & \text{if } jw > 3.8 h_{PCC} \end{cases} \tag{14.19}$$

where n_j = Number of applied load applications for the current increment by load group (J)

w = Joint opening, mils (0.001 in.)

τ_j = Shear stress on the transverse crack from the response model for the load group(J), psi

$$\tau_j = \frac{AGG \times (\delta_L - \delta_U)}{h_{PCC}}$$

τ_{ref} = Reference shear stress derived from the PCA test results, psi

$$\tau_{ref} = 111.1 \times \exp\{-\exp[0.9988 \times \exp(-0.1089 \log J_{AGG})]\}$$

J_{AGG} = Joint stiffness on the transverse crack computed for the time increment

The dowel damage, DAM_{dow}, is determined using the equation:

$$DAM_{dow} = C_8 \sum_{j=1} \left(\frac{J_d \times (\delta_L - \delta_U) \times dsp}{df_c'} \right) \tag{14.20}$$

where DAM_{dow} = Damage at dowel–concrete interface

C_8 = Coefficient equal to 400

n = Number of load applications for the current increment by load group j (i.e., $j = 1$ to n)

J_d = Non-dimensional dowel stiffness at the time of load application

δ_L = Deflection at the corner of the loaded slab induced by the axle, in.

δ_U = Deflection at the corner of the unloaded slab induced by the axle, in.

dsp = Space between adjacent dowels in the wheel path, in.

d = Dowel diameter, in.

f'_c = PCC compressive strength, psi

The error increment developed during the current month is determined. The extent of the increase depends on the average faulting frequency, the level of faulting at the beginning of the month, and total differential energy, DE, collected from all axle loads moved from 8:00 p.m. to 8:00 a.m. for one month. The faulting will be determined at the end of the month. For the number of months in the pavement design life, these steps are repeated.

14.2.3 Smoothness—JPCP

AASHTOWare pavement ME design smoothness is predicted in the AASHTOWare pavement ME design as a feature of the pavement's initial as-constructed profile and any changes in the longitudinal profile over time and traffic due to distress and foundation movements. The IRI model was calibrated and validated using LTPP field data to confirm that it would yield valid results under a variety of climatic and field conditions. Eq. (14.21) is the calibrated model's final form:

$$IRI = IRI_I + C_1 \times CRK + C_2 \times SPALL + C_3 \times TFAULT + C_4 \times SF \tag{14.21}$$

where IRI = Predicted IRI, in./mile

IRI_I = Initial smoothness measured as IRI, in./mile

CRK = Percent slabs with transverse cracks (all severities)

SPALL = Percentage of joints with spalling (medium and high severities)
TFAULT = Total joint faulting cumulated per mile, in.
$C_1 = 0.8203$
$C_2 = 0.4417$
$C_3 = 1.4929$
$C_4 = 25.24$
SF = Site factor

$$\text{SF} = \text{AGE}(1 + 0.5556 \times \text{FI})(1 + P_{200}) \times 10^{-6} \tag{14.22}$$

where AGE = Pavement age, yr
FI = Freezing index, °F days
P_{200} = Percent subgrade material passing No. 200 sieve

Transverse cracking and faulting are obtained using the models previously described. The use of Eq. (14.23) determines the spalling of the transverse joint, calibrated using LTPP and other data.

$$\text{SPALL} = \left(\frac{\text{AGE}}{\text{AGE} + 0.01}\right)\left(\frac{100}{1 + 1.005^{(-12\times\text{AGE}+\text{SCF})}}\right) \tag{14.23}$$

where SPALL = Percentage joints spalled (medium and high severities)
AGE = Pavement age since construction, yr
SCF = Scaling factor (site-, design-, and climate-related)

$$\begin{aligned} SCF = &-1400 + 350 \times \text{AC}_{\text{PCC}} \times (0.5 + \text{PREFORM}) + 43.4\left(f_c'\right)^{0.4} \\ &-0.2(\text{FT}_{\text{cycles}} \times \text{AGE}) + 43H_{\text{PCC}} - 536\text{WC}_{\text{PCC}} \end{aligned} \tag{14.24}$$

where AC_{PCC} = PCC air content, %
AGE = Time since construction, yr
PREFORM = 1 if preformed sealant is present; 0 if not
f_c' = PCC compressive strength, psi
$\text{FT}_{\text{cycles}}$ = Average annual number of freeze-thaw cycles
H_{PCC} = PCC slab thickness, in.
WC_{PCC} = PCC water/cement ratio

The standard error of predicted mean IRI of JPCP pavement is listed in the equation:

$$S_{e(\text{JPCP\_IRI})} = 29.03 \ln(\text{IRI}) - 103.8 \tag{14.25}$$

Example 14.3: IRI in JPCP A JPCP site has the following condition:

Initial IRI = 63

After 20 years of service:

Freezing index = 1,220°F days
Subgrade materials passing the 0.075-mm (No. 200) sieve = 12.8%

Percent slabs with transverse cracks (all severities) = 7.5%
Total joint faulting cumulated per mile = 0.10 in.
Percentage of joints with spalling (medium and high severities) = 3.4%

Determine the IRI after 20 years of service.

Solution
IRI can be calculated using $IRI = IRI_I + C_1 \times CRK + C_2 \times SPALL + C_3 \times TFAULT + C_4 \times SF$

where IRI_I = 63 in./mile
FI = Freezing index = 1,220°F days
CRK = Percent slabs with transverse cracks (all severities) = 7.5%
SPALL = Percentage of joints with spalling (medium and high severities) = 3.4%
TFAULT = Total joint faulting cumulated per mile = 0.10 in.
P_{200} = Subgrade materials passing the 0.075-mm (No. 200) sieve = 12.8%
$C_1 = 0.8203$
$C_2 = 0.4417$
$C_3 = 1.4929$
$C_4 = 25.24$
SF = Site factor = ?
IRI = Predicted IRI, in./mi = ?

Let us find SF and SPALL first to determine IRI.

$$\begin{aligned} SF &= AGE(1 + 0.5556 \times FI)(1 + P_{200}) \times 10^{-6} \\ &= 20(1 + 0.5556 \times 1{,}220)(1 + 12.8) \times 10^{-6} \\ &= 0.184 \end{aligned}$$

$$\begin{aligned} \text{Therefore, } IRI &= IRI_I + C_1 \times CRK + C_2 \times SPALL + C_3 \times TFAULT + C_4 \times SF \\ &= 63 + 0.8203\,(7.5) + 0.4417\,(3.4) + 1.4929\,(0.1) + 25.24\,(0.184) \\ &= 75.44 \end{aligned}$$

Answer *The IRI after 20 years of service is 75.44 in./mile.*

Example 14.4: Spalling in JPCP A JPCP site has the following conditions:

PCC air content = 3.2%
PCC compressive strength = 4,000 psi
Average annual number of freeze-thaw cycles = 120
PCC slab thickness = 4.0 in.
PCC water/cement ratio = 0.40

Determine the percentage joints spalled (medium and high severities) after 10 years of service.

Solution
SPALL = Percentage joints spalled (medium and high severities) = ?

$$\text{SPALL} = \left(\frac{\text{AGE}}{\text{AGE} + 0.01}\right)\left(\frac{100}{1 + 1.005^{(-12\times\text{AGE}+\text{SCF})}}\right)$$

where $\text{SCF} = -1{,}400 + 350 \times \text{AC}_{\text{PCC}} \times (0.5 + \text{PREFORM}) + 43.4\left(f_c'\right)^{0.4}$
$-0.2(\text{FT}_{\text{cycles}} \times \text{AGE}) + 43H_{\text{PCC}} - 536\text{WC}_{\text{PCC}}$

AC_{PCC} = PCC air content = 3.2%
f_c' = PCC compressive strength = 4,000 psi
$\text{FT}_{\text{cycles}}$ = Average annual number of freeze-thaw cycles = 120
H_{PCC} = PCC slab thickness = 4.0 in.
WC_{PCC} = PCC water/cement ratio = 0.40
PREFORM = 0 (as nothing is told about the sealant)

$$\text{SCF} = -1{,}400 + 350 \times \text{AC}_{\text{PCC}} \times (0.5 + \text{PREFORM}) + 43.4\left(f_c'\right)^{0.4} - 0.2(\text{FT}_{\text{cycles}} \times \text{AGE}) + 43H_{\text{PCC}} - 536\text{WC}_{\text{PCC}}$$

$$\text{SCF} = -1{,}400 + 350(3.2)(0.5 + 0) + 43.4(4{,}000)^{0.4} - 0.2(120 \times 10) + 43(4) - 536(0.4) = 291$$

$$\begin{aligned}\text{SPALL} &= \left(\frac{\text{AGE}}{\text{AGE} + 0.01}\right)\left(\frac{100}{1 + 1.005^{(-12\times\text{AGE}+\text{SCF})}}\right)\\ &= \left(\frac{10}{10 + 0.01}\right)\left(\frac{100}{1 + 1.005^{(-12\times10+291)}}\right)\\ &= 29.85\end{aligned}$$

Answer *The percentage joints spalled after 10 years of service is 29.85%.*

14.3 Continuously Reinforced Concrete Pavement

14.3.1 CRCP Punchouts

The following globally calibrated model predicts CRCP punchouts (Fig. 14.2) as a function of accumulated fatigue damage due to top-down stresses in the transverse direction, as shown in the equation:

$$\text{PO} = \frac{A_{\text{PO}}}{1 + \alpha_{\text{PO}}\,\text{DI}_{\text{PO}}^{\beta_{\text{PO}}}} \tag{14.26}$$

where PO = Total predicted number of medium- and high-severity punchouts per mile
DI_{PO} = Accumulated fatigue damage (due to slab bending in the transverse direction)
$A_{\text{PO}}, \alpha_{\text{PO}}, \beta_{\text{PO}}$ = Calibration constants (216.8421, 33.15789, −0.58947, respectively)

CRCP design identifies more significant factors affecting the number of punchouts. These are the spacing of cracks and the width of cracks. Numerical models

Figure 14.2 CRCP punchouts.

determine such two variables. These are not mentioned here because of complexity. The AASHTOWare pavement ME design software, however, uses these to measure punchouts more accurately. Readers are encouraged to use the AASHTOWare pavement ME design manual of practice (2015).

For each time increase in the design life, the damage accumulated at the critical point on top of the slab is calculated. For each load level in each gear configuration or axle load spectra, the tensile stress at the top of the slab is used to calculate the number of allowable load repetition, $N_{i,j}$, as shown in Eq. (14.27) due to this load level at this time increase.

$$\log(N_{i,j}) = C_1\left(\frac{M_{r_i}}{\sigma_{i,j}}\right)^{C_2} - 1 \tag{14.27}$$

where M_{r_i} = PCC modulus of rupture at age i, psi
$\sigma_{i,j}$ = Applied stress at time increment i due to load magnitude j, psi
$C_{1,2}$ = Calibration constants; $C_1 = 2.0$; $C_2 = 1.22$

It has been found that the critical stress at the top of the slab, which is transverse and located near a transverse crack, is 40 to 60 in. from the edge. A value of 48 in. is very often used as the critical position. After observations that a very high percentage of punchouts was 2 ft or less, a crack spacing of 2 ft was used as the critical width. This stress is determined using neural net models based on slab thickness, edge offset traffic, PCC properties, base course properties and thickness, subgrade stiffness, temperature gradient equivalent, and other factors.

The standard error for the CRCP punchouts ($S_{e(PO)}$) can be calculated using the equation:

$$S_{e(\mathrm{PO})} = 2 + 2.2593\,\mathrm{PO}^{0.4882} \tag{14.28}$$

where PO = predicted mean medium- and high-severity punchouts, number per mile.

14.3.2 Smoothness—CRCP

CRCP's smoothness change is the result of a combination of the pavement's initial as-constructed profile and any change in longitudinal profile over time and traffic due to distress development and foundation movement. Main CRCP IRI distresses include punchouts. CRCP's global IRI model is given in the equation:

$$\text{IRI} = \text{IRI}_I + C_1\text{PO} + C_2\text{SF} \tag{14.29}$$

where IRI_I = Initial IRI, in./mile
PO = Number of medium- and high-severity punchouts per mile
$C_1 = 3.15$
$C_2 = 28.35$
SF = Site factor

$$\text{SF} = \text{AGE}(1 + 0.556\text{FI})(1 + P_{200}) \times 10^{-6} \tag{14.30}$$

where AGE = Pavement age, yr
FI = Freezing index, °F days
P_{200} = Percent subgrade material passing No. 200 sieve

The standard error of predicted mean IRI of CRCP pavement is listed in the equation:

$$S_{e(\text{CRCP\_IRI})} = 7.08\ln(\text{IRI}) - 11 \tag{14.31}$$

Example 14.5: Punchouts in CRCP A CRCP concrete pavement section has the traffic and stress response data as shown in Table 14.3.

Axle type	Annual average repetition	Produced stress (yearly average) (psi)
Single	50,000	250
Tandem	35,000	275
Tridem	19,000	295
Quad	9,000	310

Table 14.3 Load and Response Data for Example 14.5

The yearly average modulus of rupture of concrete is 900 psi. Assume that the produced stresses at the bottom of concrete and at the surface are equal.

a. Calculate the number of punchouts per mile after 20 years at 50% reliability.
b. Calculate the number of punchouts per mile after 20 years at 95% reliability. The reliability (R) of 90% has a standard normal deviate (Z_R) of 1.282.

Solution

PO = punchouts/mile =?
$A_{\text{PO}} = 216.8421$
$\alpha_{\text{PO}} = 33.15789$
$\beta_{\text{PO}} = -0.58947$

$$\log(N_{i,j}) = C_1\left(\frac{M_{r_i}}{\sigma_{i,j}}\right)^{C_2} - 1$$

where $\sigma_{i,j}$ = given
$C_1 = 2.0$
$C_2 = 1.22$
$M_{r_i} = 900$ psi

Axle type	Annual average repetition	Produced stress (yearly average) (psi)	$N = 10^{\left[2.0\left(\frac{M_{ri}}{\sigma_i}\right)^{1.22}-1\right]}$	$\mathrm{DI} = \frac{n}{N}$
Single	50,000	250	349,743,642	0.000143
Tandem	35,000	275	31,340,574	0.001117
Tridem	19,000	295	6,290,919	0.00302
Quad	9,000	310	2,191,898	0.004106
				Σ DI = 0.008386 per year
				Σ DI = 0.008386 × 20 = 0.1677 after 20 years

DI, damage index; *N*, fatigue life.

TABLE 14.4 Load and Response Data Analysis for Example 14.5

From Table 14.4:

$$\mathrm{PO} = \frac{A_{\mathrm{PO}}}{1 + \alpha_{\mathrm{PO}}\,\mathrm{DI}_{\mathrm{PO}}^{\beta_{\mathrm{PO}}}} = \frac{216.8421}{1 + (33.15789)(0.1677)^{-0.58947}} = 2.3 \text{ punchouts per mile}$$

a. Punchouts after 20 years at 50% reliability = 2.3 no./mi.

b. The reliability (*R*) of 90% has a standard normal deviate (Z_R) of 1.282.

Punchouts at 90% reliability = Punchouts at 90% reliability + $S_{e(\mathrm{PO})} \times 1.282$
$S_{e(\mathrm{PO})}$ can be calculated using the equation:

$$S_{e(\mathrm{PO})} = 2 + 2.2593\ \mathrm{PO} = 2^{0.4882} + 2.2593\ (2.3)^{0.4882} = 5.3929$$

$$\begin{aligned}\text{Punchouts at 90\% reliability} &= \text{Punchouts at 90\% reliability} + S_{e(\mathrm{PO})} \times 1.282\\ &= 2.3 + 5.3929 \times 1.282\\ &= 9.2137\end{aligned}$$

Answer *The number of punchouts per mile after 20 years at 50% reliability is 2.3 and at 90% reliability is 9.2.*

Example 14.6: IRI of CRCP A CRCP site has the following condition:

Initial IRI = 62
After 10 years of service:

Freezing index = 720°F days
Subgrade materials passing the 0.075-mm (No. 200) sieve = 13.6%
Number of punchouts in a mile = 20
Determine the IRI after 10 years of service.

Solution
IRI can be calculated using the equation $IRI = IRI_I + C_1 \times PO + C_2 \times SF$

where PO = 20 per miles
$C_1 = 3.15$
$C_2 = 28.35$
SF = ?

Let us find SF now.

$$\begin{aligned} SF &= AGE(1 + 0.556FI)(1 + P_{200}) \times 10^{-6} \\ &= 10(1 + 0.556 \times 720)(1 + 13.6) \times 10^{-6} = 0.0586 \end{aligned}$$

$$\begin{aligned} \text{Therefore, } IRI &= IRI_I + C_1 PO + C_2 SF \\ &= 62 + 3.15(20) + 28.35(0.0586) \\ &= 141.6 \end{aligned}$$

Answer *The IRI after 10 years is 142 in./mi.*

14.4 Recommended Design-Performance Criteria

Performance criteria are used to ensure that over its design life, a pavement design performs satisfactorily. Considering the type of pavement such as highways versus local

Performance criteria	Threshold value at end of design life
Mean joint faulting	Interstate: 0.15 in. Primary: 0.20 in. Secondary: 0.25 in.
Percent transverse slab cracking	Interstate: 10% Primary: 15% Secondary: 20%
IRI	Interstate: 160 in./mile Primary: 200 in./mile Secondary: 200 in./mile

Source: *From AASHTO (2015). Mechanistic-Empirical Pavement Design Guide: A Manual of Practice. Washington, DC: American Association of State Highway and Transportation Officials. Table 7-1. Used with permission.*

TABLE 14.5 Recommended Threshold Values for JPCP New, CPR, and Overlays roads, the designer chooses critical limits or threshold values. The software selects the default values unless any threshold values are set. Table 14.5 sets out the values for highway agencies' considerations, since these levels may differ between agencies based on their local conditions.

14.5 Summary

This chapter explains the equations that are used to calculate various major distresses and rigid pavement IRI throughout their service life using the AASHTOWare pavement ME design approach. These equations are calibrated for global conditions. Therefore, for better accuracy, local calibrations are recommended. Additionally, for higher reliability based on road types and location, the predicted distress is increased. For local roads in rural areas, the minimum reliability of 70% is assumed, and increases as the highway becomes more important (AASHTO, 2015). It is assumed that the measured distress and IRI are distributed roughly normally across distress and IRI ranges. As mentioned above, the standard deviation for each type of distress is determined by the calibration results used for each main distress from the model prediction error. This standard deviation is multiplied by the normal deviation and applied to the expected mean distress to achieve distress with certain reliability.

14.6 Fundamentals of Engineering (FE) Exam-Style Questions

FE14.1 The fatigue life of concrete increases with the increase in:

A. The PCC modulus of rupture
B. The applied stress
C. The amount of traffic
D. The ambient temperature

Solution A

Fatigue life of concrete is not affected by the amount of traffic or temperature. From the fatigue life equation, $N = 10^{2.0\left(\frac{M_{ri}}{\sigma}\right)^{1.22}}$, N increases with the increase in modulus of rupture (M_{ri}) and decreases with the decrease in applied stress (σ).

FE14.2 Continuously reinforced concrete pavement (CRCP) does not show:

A. Transverse slab cracking
B. Mean transverse joint faulting
C. Smoothness
D. Punchouts

Solutions A, B

CRCP shows only punchouts and smoothness. The other two distresses are eliminated by providing continuous reinforcement.

FE14.3 The models/equations used while analyzing rigid pavement using the AASHTOWare pavement ME design software are:

A. Empirical in nature
B. Mechanics based
C. The combination of mechanics and experience
D. Mechanistic-theoretical in nature

Solution C

The models/equations used while analyzing rigid or flexible pavement using the AASHTOWare pavement ME design software are mechanistic-empirical (meaning the combination of mechanics and experience).

14.7 Practice Problems

14.1 A JPCP concrete pavement section has the traffic and stress response data, as listed in Table P14.1.

Axle type	Annual average repetition	Produced stress (yearly average) (psi)
Single	120,000	225
Tandem	95,000	245
Tridem	37,000	265
Quad	13,000	285

TABLE P14.1 Load and Response Data for Prob. 14.1

The yearly average modulus of rupture of concrete is 700 psi. Assume that the produced stresses at the bottom of concrete and at the surface are equal.

a. Calculate the predicted amount of transverse cracking after 10 years.
b. Calculate the predicted amount of transverse cracking after 10 years at 90% reliability.

14.2 A JPCP site has the following conditions:

Maximum mean monthly slab corner upward deflection PCC due to temperature curling and moisture warping= 0.075 in.
Base freezing index = 0.97%
Base/subbase erodibility factor = 0.25
Subgrade materials passing the 0.075-mm (No. 200) sieve = 7.6%
Number of days rainfall greater than 0.1 in. in a year = 42
Overburden on subgrade = 250 lb

Determine the initial maximum mean transverse joint faulting.

14.3 A JPCP site has the following condition:

Initial IRI = 63

After 10 years of service:

Freezing index = 720°F days
Subgrade materials passing the 0.075-mm (No. 200) sieve = 3.8%
Percent slabs with transverse cracks (all severities) = 5.5%
Total joint faulting cumulated per mile = 0.10 in.

PCC air content = 4%
PCC compressive strength = 5,000 psi
Average annual number of freeze-thaw cycles = 50
PCC slab thickness = 8.0 in.
PCC water/cement ratio = 0.40

Determine the IRI after 10 years of service.

14.4 A CRCP concrete pavement section has the traffic and stress response data, as listed in Table P14.4.

Axle type	Annual average repetition	Produced stress (yearly average)
Single	190,000	195
Tandem	115,000	205
Tridem	62,000	215
Quad	1,200	235

TABLE P14.4 Load and Response Data for Problem 14.4

The yearly average modulus of rupture of concrete is 650 psi. Calculate the predicted amount of transverse cracking after 20 years. Assume that the produced stresses at the bottom of concrete and at the surface are equal.

14.5 A CRCP site has the following condition:

Initial IRI = 62

After 20 years of service:

Freezing index = 1,220°F days
Subgrade materials passing the 0.075-mm (No. 200) sieve = 19.3%
Number of punchouts in a mile = 5

Determine the IRI after 20 years of service.

14.6 A JPCP site has the following condition:

Initial IRI = 63

After 10 years of service:

Freezing index = 720°F days
Subgrade materials passing the 0.075-mm (No. 200) sieve = 3.8%
Percent slabs with transverse cracks (all severities) = 5.5%
Total joint faulting cumulated per mile = 0.10 in.
AC_{PCC} = PCC air content = 4%
f_c' = PCC compressive strength = 5000 psi
FT_{cycles} = Average annual number of freeze-thaw cycles = 50
H_{PCC} = PCC slab thickness = 8.0 in.
WC_{PCC} = PCC water/cement ratio = 0.40

Determine the IRI after 10 years of service.

CHAPTER 15

Rigid Pavement Design by AASHTOWare

15.1 Background

Rigid pavements compose of the reinforced or nonreinforced surface course of portland cement concrete (PCC). Such pavements may have steel reinforcing to minimize thermal cracking or remove transverse joints. Rigid pavements, as mentioned in Chap. 1, can be of different types. However, PCC, asphalt concrete (AC), and reinforcement may form various combinations, such as the following:

- Jointed plain concrete pavement (JPCP)
- Jointed reinforced concrete pavement (JRCP)
- Continuous reinforced concrete pavement (CRCP)
- Prestressed concrete pavement (PCP)
- Asphalt overlay on existing JPCP
- Asphalt overlay on existing CRCP
- Asphalt overlay on existing fractured JPCP
- Asphalt overlay on existing fractured CRCP
- Bonded PCC overlay on existing JPCP
- Bonded PCC overlay on existing CRCP
- JPCP overlay on existing asphalt pavement
- CRCP overlay on existing asphalt pavement
- Unbonded JPCP overlay on existing JPCP
- Unbonded JPCP overlay on existing CRCP
- Unbonded CRCP overlay on existing CRCP
- Unbonded CRCP overlay on existing JPCP
- JPCP restoration

The design technique is very similar to each other for the previously mentioned combinations. The fundamental difference is the material properties. Whatever the material of the layer (asphalt, CRCP, or JPCP) is, the inputs should be for this material. The first two types of rigid pavements (JPCP and JRCP) are commonly used and discussed in this chapter.

15.2 Pavement Structure

The base and subbase layers may not be used in the concrete pavement as the PCC slab is very rigid. Base course is defined as the layer directly below the slab of the PCC and the layers below the base layer. Usually, the concrete slab is placed over one or more sublayers. Nevertheless, it is also possible to place concrete slabs directly on a prepared subgrade for low-volume roads. A wide variety of materials and layers may be included in the base/subbase layer, including permeable drainage. For example, there may be two types of base layer:

1. Dense graded baseline, such as asphalt stabilized, cement stabilized, lean concrete, and unbound granular, may be considered. Many varieties of layer characteristics may be considered, but the designer must enter appropriate structural, thermal, and hydraulic parameters for these layers.
2. Permeable (drainage layer) base courses such as asphalt stabilized, cement stabilized, and unbound granular permeable layers may be considered.

Subbase layers may be conventional or asphalt stabilized, compacted reclaimed asphalt pavement, cement stabilized, lime stabilized, lime fly ash, lime cement fly ash, soil cement, and unbound granular materials. Bedrock may consist of any kind of bedrock, such as massive and continuous bedrock, or strongly fractured and weathered bedrock.

15.3 JPCP Design

Some key items required for JPCP design are discussed in this section.

Contact friction between JPCP and base course. An input to determine the cracking of the JPCP is the contact friction between the PCC slab and the underlying base/subgrade layer. The real friction varies with slippage from zero to complete. The global calibration, however, shows full contact friction for all base types throughout the life of the pavements.

Tied concrete shoulder. The load transfer efficiency (LTE) is defined as the deflection ratio of the unloaded side multiplied by 100 to the loaded side of the joint. The greater the LTE, the greater the decrease in the concrete slab in deflections and stresses. Recommended LTE long-term lanes/shoulders are as follows:

- Traffic lane and shoulder monolithically placed and tied with deformed bars: 50% to 70%.
- Traffic lane and shoulder separately placed and tied with deformed bars: 30% to 50%.
- Untied concrete shoulders or other shoulder types: zero

Joint LTE. At the transverse joints, JPCP can be designed with or without a dowel bar. The main inputs are the diameter and spacing of the dowel. The program's sensitivity analysis reveals that the use of ample-size dowels can virtually eliminate joint faulting as a problem:

- Dowel diameter of 1/8 of the slab thickness
- Dowel spacing of 12 in. recommended, although may vary from 10 to 14 in.

Joint spacing. This factor affects JPCP splitting, joint faulting, and IRI very significantly. The shorter the spacing, the less faulting and cracking occurrence.

Joint random spacing. If a JPCP has random spacing, to estimate the amount of transverse cracking, each range could be run separately. The most critical spacing is the longest.

Joint skew. When using dowels, joint skewing is not recommended. Nevertheless, if used, an additional 2 ft is applied to the joint spacing to account for the increase in effective joint spacing when joints are skewed.

Base erodibility. The potential for base or subbase erosion has a substantial impact on pavement distress initiation and propagation. The design input is the erodibility class, which is graded on the basis of long-term erodibility behavior of different base types as follows:

- Class 1: Extremely erosion-resistant materials
- Class 2: Very erosion-resistant materials
- Class 3: Erosion-resistant materials
- Class 4: Fairly erodible materials
- Class 5: Very erodible materials

Zero-stress temperature and ultimate shrinkage. These factors affect JPCP in terms of the joint opening which affects joint LTE and joint faulting.

Permanent curl/warp effective temperature difference. This input includes a built-in temperature gradient at a set time plus effective moisture warping gradient (dry on top and wet on bottom) plus any long-term slab creep and settling effect on the base as the value of −10°F is set to reduce cracking during the national calibration.

15.4 CRCP Design

The performance of CRCP is highly dependent upon several inputs which are as follows:

Tied concrete shoulder. The long-term transfer of load across the lane/shoulder joint is modeled so that the design can take into account the effect of a tied shoulder. The user selects the type of shoulder to be considered as follows:

- Monolithically placed lane and shoulder and tied with deformed reinforcing bars
- Separately placed lane and shoulder and tied with deformed reinforcing bars
- Untied concrete shoulders or other shoulder types

Bar diameter. Commonly used bar diameter varies from #4 to #9. High-volume highways may require #6 or #7 deformed bars. Bars are typically coated with epoxy, especially in areas of deicing salts.

Trial percentage of longitudinal reinforcement. This parameter can vary between 0.50% and 1.00%. Climate conditions influence the amount of longitudinal reinforcement required. Cold climates warrant higher amount of longitudinal reinforcement. Crack spacing and width decrease as the number of longitudinal reinforcement increases.

Reinforcement depth. The depth of reinforcing steel has a significant effect on holding the crack width tight at the top of the slab. A depth of at least 3.5 in. and a maximum depth at the slab mid-depth are recommended. Placement of the steel above mid-depth of the slab holds the cracks tighter and reduces punchouts.

Crack spacing. Crack spacing is input by the user if experience warrants. The recommended range of spacing is 3 to 6 ft.

Base/slab friction coefficient. This friction coefficient varies by base type. Recommended values and ranges are as listed in Table 15.1.

Zero-stress temperature. The zero-stress temperature when the slab becomes a solid is defined as the average concrete set temperature. It is either entered by the user or determined from the following inputs: average hourly ambient temperatures for the construction month and quality of cementitious materials. The temperature of zero stress is very important for the efficiency of CRCP. The lower the temperature, the tighter the transverse cracks will be over time and the lower the punchout occurrence.

Permanent curl and warp. Permanent curl/warp effective temperature difference (same recommendations as JPCP).

Ultimate shrinkage. Ultimate 40% relative humidity shrinkage is either input by the user or estimated from models. It depends on the type of curing (curing compound or water cure), type of cement (I, II, III), water content (through water/cement ratio), and compressive strength of 28 days. Use type II cement, cure with water, reduce water content, and generally increase concrete strength to minimize ultimate shrinkage.

Crack width. Crack width is estimated and is a very critical factor throughout the entire design life. This initially depends on the construction temperature. The user then chooses the intended construction month, which is then used to estimate the concrete's zero-stress temperature. The concrete's ultimate shrinkage also controls crack width over time. Therefore, CRCP is desirable for anything that decreases shrinkage.

Crack LTE. The crack LTE is initially 100% over the first 20 years or so, but could then deteriorate to an unacceptable level with time and loads when LTE reduces the chance of increasing punchouts when critical bending stress rises at the top of the CRCP.

Subbase/base type	Friction coefficient (low, mean, high)
Fine-grained soil	0.5, 1.1, 2
Sand	0.5, 0.8, 1
Aggregate	0.5, 2.5, 4.0
Lime-stabilized clay	3.0, 4.1, 5.3
Asphalt-treated base (ATB)	2.0, 8.5, 18.7
Cement-treated base (CTB)	2.9, 9.6, 20.9
Soil cement	6.0, 7.9, 23
Lean-concrete base (LCB)	6.0, 10.7, 21.5
Lean-concrete base not cured	<36

Table 15.1 Recommended Base/Slab Friction Coefficients for CRCP

Crack LTE relies heavily on the width of the crack over time and also on the number of heavy axles that cross the crack and inflict sheer vertical damage and potential damage. Holding LTE above 90% or 95% is, therefore, an important criterion, as this virtually guarantees that there is no or minimum punchout.

Erosion and loss of support along slab edge. This parameter depends on several inputs, particularly base type and quality.

- HMA base: volumetric asphalt content
- CTB/LCB: modulus of elasticity, E_c
- Unbound granular base: fines content (minus #200 sieve)
- Annual precipitation
- Type and quality of subbase/subgrade (strength, fines)

Erosion is calculated for 10 years but uniformly accumulated year by year with a maximum practical amount.

Initial surface smoothness. JPCP and CRCP's initial IRI falls within a 50 to 100 in./mile range with a typical 63 in./mile value. This value can be adjusted to the value normally obtained for these pavements by the local highway agency.

Narrow or widened slabs. This input is commonly called lane width, but it is actually slab width.

- JPCP slab width is assumed to be 12 ft and can be the maximum of 15 ft.
- CRCP slab width is assumed to be 12 ft, and there is no formal way to increase it.

Just to remind you that the AASHTOWare pavement ME design software is undergoing continuous improvement. Therefore, design-related values listed in the book may change with time or as soon as there is next revision of the AASHTO pavement design manual.

15.5 Usage of the Software

15.5.1 Jointed Plain Concrete Pavement

A JPCP is a plain portland cement concrete pavement:

- With no steel reinforcement for controlling random cracking;
- With or without transverse joint load transfer devices (e.g., dowels);
- With joints that are spaced relatively close (e.g., 10 to 20 ft) to minimize transverse cracking from temperature and drying gradient shrinkage stresses;
- With mostly PCC or asphalt concrete shoulders; and
- With or without tied lane to lane, or lane to PCC shoulder longitudinal joints.

Pavement design begins with users establishing a trial JPCP design by selecting appropriate inputs in the start-up window as shown in Fig. 15.1:

- Site conditions (i.e., traffic, climate, and subgrade);
- Design features (layer types and thicknesses, edge support, transverse joint load transfer mechanism, shoulder type, and so on);

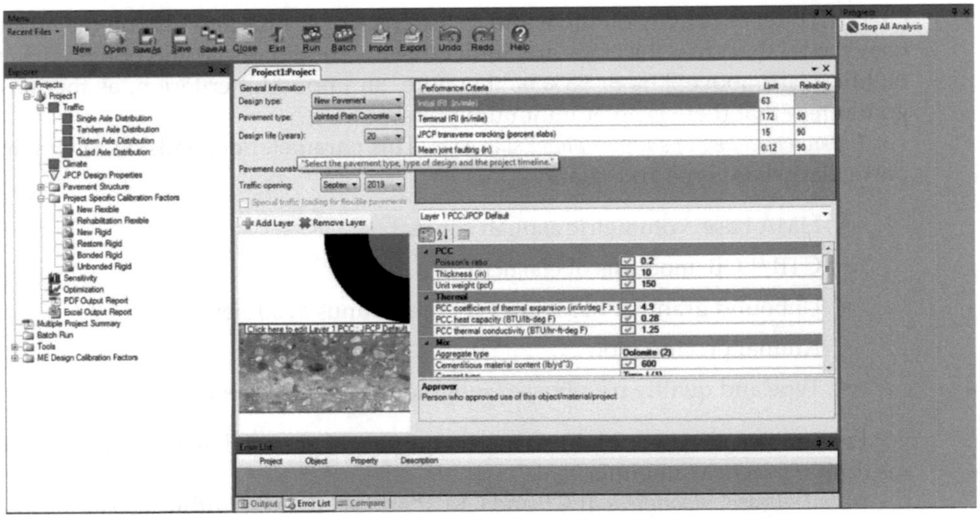

FIGURE 15.1 Assigning the JPCP pavement type in the software.

- Materials properties; and
- Performance criteria, among others.

Compared to flexible pavement, efficiency standards of JPCP are slightly different. The software asks about the JPCP transverse cracking and mean joint faulting in addition to the IRI, as shown in Fig. 15.1. These two will be addressed here:

Mean joint faulting (in.). The limit and reliability controls allow you to define the threshold value for transverse joint faulting at the end of the design life at a reliability level specified by the user or agency.

JPCP transverse cracking (percent slabs). The limit and reliability controls allow you to define the threshold value for transverse cracking at the end of the design life at a reliability level specified by the user.

Climate and traffic data are similar to those discussed in flexible pavement.

The pavement structure and materials of JPCP are quite different than the flexible pavement. The key PCC materials inputs required by the AASHTOWare pavement ME design software are as follows:

- Flexural strength (M_r)
- Elastic modulus (E_{PCC})
- Coefficient of thermal expansion (CTE)
- Ultimate shrinkage
- Concrete mix properties (e.g., cement type, cement content, aggregate type, etc.)

These inputs are needed to predict pavement responses to applied loads, long-term strength and elastic modulus, and the effect of climate (temperature, moisture, and humidity) on PCC expansion and contraction. Other PCC inputs include thickness, unit weight, Poisson's ratio, thermal factors, mixing factors, strength values, etc., as shown in Figs. 15.2 and 15.3.

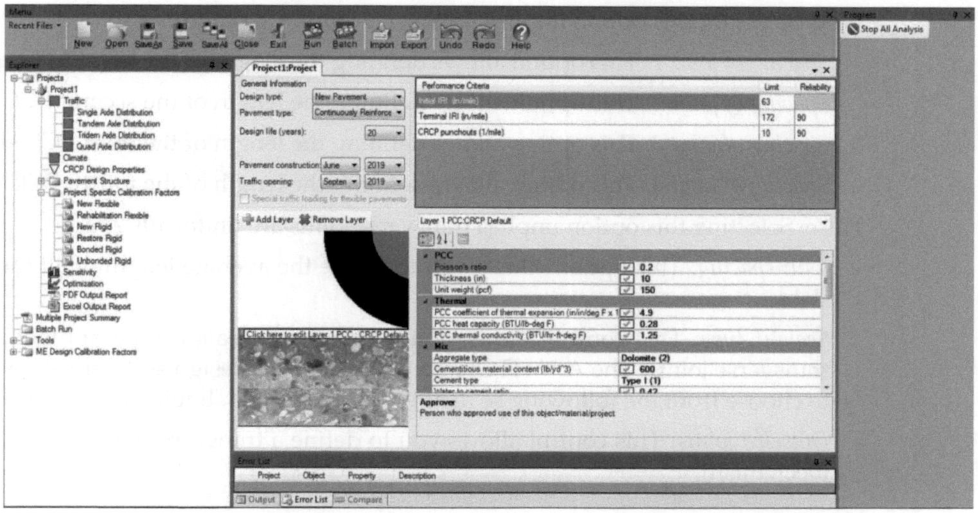

Figure 15.2 PCC properties.

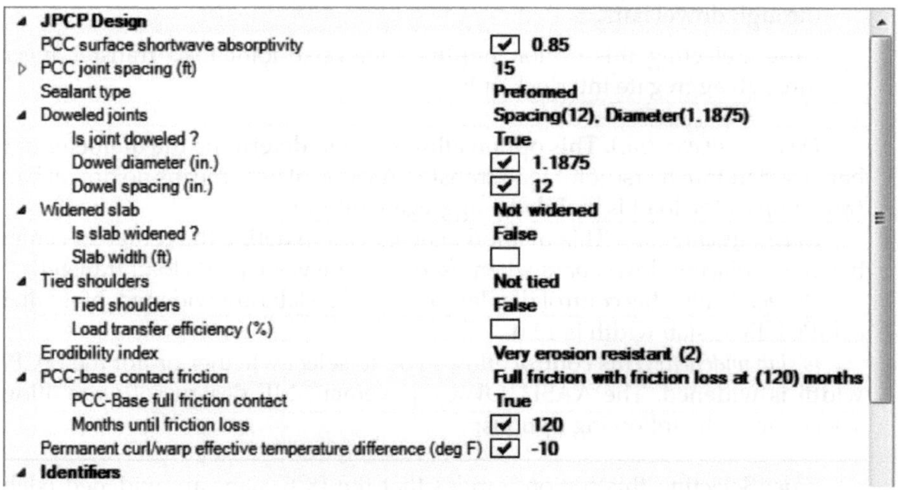

Figure 15.3 JPCP design properties.

PCC surface shortwave absorptivity. This control allows you to define the fraction of solar energy (sunshine) at the PCC surface that is absorbed by the PCC. The AASHTOWare pavement ME design software uses the default value of 0.85.

PCC joint spacing (ft). This control allows you to define whether transverse joints of the trial design are uniformly or randomly spaced.

Is joint spacing random? This control allows you to define whether transverse joints are uniformly spaced or randomly spaced. The AASHTOWare pavement ME design software allows for the following options:

True: Selecting this option implies transverse joints are randomly spaced.

Spacing of joint 1: This option allows defining the length of the first PCC slab.

Spacing of joint 2: This option allows defining the length of the second PCC slab.

Spacing of joint 3: This option allows defining the length of the third PCC slab.

Spacing of joint 4: This option allows defining the length of the fourth PCC slab.

False: Selecting this option implies transverse joints are uniformly spaced.

Joint spacing: This option allows you to define the average length of all the PCC slabs.

Sealant type: This control allows you to select the sealant type applied at the transverse joints. The AASHTOWare pavement ME design software allows you to choose from the following sealant types: preformed, liquid, and silicone.

Doweled joints: This control allows you to define a transverse joint load transfer mechanism.

Is joint doweled? This control allows defining whether transverse joint load transfer mechanism is through the dowel bars:

True: Selecting this option implies transverse joint load transfer mechanism is through dowel bars.

False: Selecting this option implies transverse joint load transfer mechanism is through aggregate interlock only.

Dowel diameter (in.). This option allows you to determine the diameter of the dowel bars used in transverse joint load transfer. A value of zero means no dowel bars, and the transition of the load is mainly by aggregate interlock.

Dowel spacing (in.). This method enables you to define the center-to-center distance between adjacent dowel bars when used for transverse joints load transfer.

Widened slab. This control displays if the PCC slabs are widened. Note that the typical JPCP PCC slab width is 12 ft.

Is slab widened? This control allows you to select whether or not the JPCP PCC slab width is widened. The AASHTOWare pavement ME design software allows you to select one of the following options:

True: Selecting this option implies that the PCC slabs are widened (slab width is greater than 12 ft, typically 13 or 14 ft)

False: Selecting this option implies that the PCC slabs are not widened (slab width is 12 ft)

Slab width (ft.). This option allows you to define the slab width. The AASHTOWare pavement ME design software assumes a standard 12-ft slab width.

Tied shoulders. This control displays if tied PCC shoulders are used.

Tied shoulders. This control allows you to select whether or not to use tied PCC shoulders. The AASHTOWare pavement ME design software allows you to select one of the following options:

True: Selecting this option implies the use of tied PCC shoulders.

False: Selecting this option implies the use of other shoulder types such as PCC (without tie bars), asphalt concrete, or gravel shoulders.

Load transfer efficiency (%). You can define the long-term or terminal deflection LTE at the lane (PCC outer-lane slab) to the longitudinal joint of the PCC shoulder with this option. The default value of the AASHTOWare pavement ME design software is 50%. Typical long-term LTE is 50% to 70% for monolithically constructed tied PCC shoulder and 30% to 50% for separately constructed tied PCC shoulder, as mentioned above.

Erodibility index. This control enables you to use an index on a scale of 1 to 5 to choose the base course resistance to erosion. Resistance to material erosion is determined by both strength and durability. You can select one of the following options from the AASHTOWare software:

- *Extremely erosion resistant.* For extremely erosion resistant materials such as concrete and lean materials, select an erodibility index of 1.
- *Very erosion resistant.* For very erosion resistant base materials such as asphalt-treated and cement-treated materials, select an erodibility index of 2.
- *Erosion resistant.* For erosion-resistant base materials, select an erodibility index of 3.
- *Fairly erodible.* For fairly erodible base materials such as weakly stabilized aggregate materials and subgrade soils, select an erodibility index of 4.
- *Very erodible.* For very erodible base materials such as unstabilized subgrade soils, select an erodibility index of 5.

PCC-base contact friction. The interface between the underlying base and PCC slab is modeled with or without full friction for JPCP design. The AASHTOWare pavement ME design software allows you to determine:

- Whether the PCC slab–base interface has full friction; and
- For how long full friction is available at the interface if present after construction. This control displays options available for modeling the PCC slab–base interface condition.

PCC-base full-friction contact. This option allows you to choose whether the PCC slab–base interface after construction has full friction or not. You can choose one of the following options using AASHTOWare software:

True: Selecting this option implies there is full friction at the PCC slab–base interface after friction.

False: Selecting this option implies there is relatively little to no friction at the PCC slab–base interface after friction.

Months until friction loss. This option allows you to define the number of months after the loss of full friction on the PCC slab–base interface. The AASHTO Practice Manual makes recommendations based on the base material type. The AASHTOWare pavement ME design software allows a range of 0 to 1,200 months.

Permanent curl/warp effective temperature difference (°F). This input describes the effect of the following:

- PCC built-in temperature gradient at time of set
- Effective gradient of moisture warping (dry on top and wet on bottom)
- Long-term creep of the PCC slab
- Settlement of the PCC into the base

The asphalt layer or aggregate base layer is provided below the PCC layer. If an asphalt layer is provided, the design input of the asphalt layer is similar to that of the new flexible pavement. However, the base, subbase, and subgrade layers require the same design inputs as for the new flexible pavement.

15.5.2 Continuously Reinforced Concrete Pavement

Continuously reinforced concrete pavement design begins with establishing a trial design by selecting appropriate inputs for:

- Site conditions (i.e., traffic, climate, and subgrade);
- Design features (layer types and thicknesses, reinforcing steel size and depth, etc.);
- Materials properties; and
- Performance criteria, among others.

The general window for CRCP pavement design is shown in Fig. 15.4.

For performance criteria (Fig. 15.5), the CRCP requires a new criterion, namely the CRCP punchouts (number per mile). Limit and reliability controls allow you to set the threshold value for the punchouts at the end of the design life at the level of reliability defined by the user or agency.

Climate and traffic data are similar to those discussed in flexible pavement.

A reinforced-concrete layer is given for the surface layer instead of asphalt or plain concrete. This layer requires similar inputs to the JPCP. The main inputs of the PCC materials required by the AASHTOWare pavement ME design software are as follows (Figs. 15.6 and 15.7):

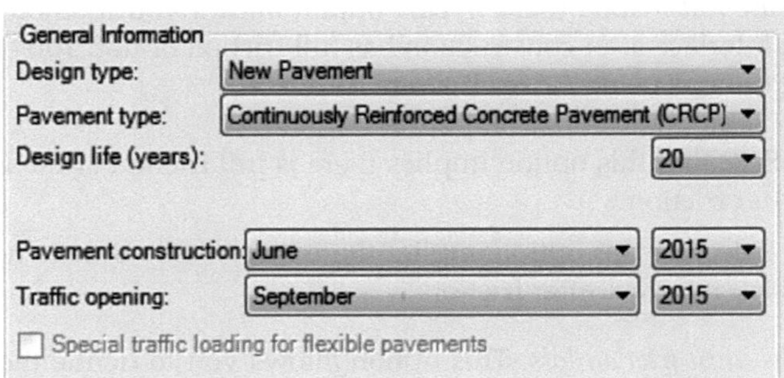

FIGURE 15.4 General information window.

Performance Criteria	Limit	Reliability
Initial IRI (in./mile)	63	
Terminal IRI (in./mile)	172	90
CRCP punchouts (1/mile)	10	90

Figure 15.5 Performance criteria for CRCP.

PCC		
Thickness (in.)	☑	9.5
Unit weight (pcf)	☑	150
Poisson's ratio	☑	0.2
Thermal		
PCC coefficient of thermal expansion (in./in./deg F x 10^-6)	☑	5.5
PCC thermal conductivity (BTU/hr-ft-deg F)	☑	1.25
PCC heat capacity (BTU/lb-deg F)	☑	0.28
Mix		
Cement type		Type I (1)
Cementitious material content (lb/yd^3)	☑	600
Water to cement ratio	☑	0.42
Aggregate type		Dolomite (2)
PCC zero-stress temperature (deg F)	☐	Calculated
Calculated internally?		True
User-specified PCC set temperature	☐	
Ultimate shrinkage (microstrain)	☐	632.3 (calculated)
Calculated internally?		True
User-specified PCC ultimate shrinkage	☐	
Reversible shrinkage (%)	☑	50
Time to develop 50% of ultimate shrinkage (days)	☑	35
Curing method		Curing Compound
Strength		

Figure 15.6 Portland cement concrete properties.

CRCP Design		
PCC surface shortwave absorptivity	☑	0.85
Shoulder type		Asphalt (2)
Permanent curl/warp effective temperature difference (deg F)	☑	-10
Steel (%)	☑	0.6
Bar diameter (in.)	☑	0.625
Steel depth (inch)	☑	4
Base/slab friction coefficient	☑	7.5
Crack spacing		Generate crack spacing using prediction model
Is crack spacing computed using prediction model?		True
Crack spacing (in.)	☐	
Identifiers		

Figure 15.7 CRCP design properties.

- Flexural strength (M_R)
- Elastic modulus (E_{PCC})
- Split tensile strength
- Coefficient of thermal expansion (CTE)

- Ultimate shrinkage
- Concrete mix properties (e.g., cement type, cement content, aggregate type, etc.)

These inputs are required to predict pavement reactions to applied loads, long-term strength and elastic modulus, and the effect of climate (temperature, moisture, and humidity) on PCC expansion and contraction.

Shoulder type. This control allows you to choose the shoulder of the CRCP. The AASHTOWare pavement ME design software offers a drop-down menu with the below choices for the type of shoulder, from which to choose one of the following:

Tied PCC–separate. This option defines a monolithically placed PCC shoulder tied with deformed bars at the traffic lane and shoulder longitudinal joint.

Tied PCC–monolithic. This option defines a PCC shoulder, separately placed and tied with deformed bars traffic lane and shoulder longitudinal joint.

Asphalt. This option defines an asphalt shoulder (note that untied PCC shoulders must be treated as an asphalt shoulder).

Gravel. This option defines a gravel shoulder.

Permanent cut/warp effective temperature difference (°F). This input describes the combined effect of the following:

- PCC built-in temperature gradient at time of set
- Effective gradient of moisture warping in the PCC (dry on top and wet on bottom)
- Long-term creep of the PCC slab
- Settlement of the PCC into the base

The AASHTOWare pavement ME design software uses the default value of –10°F.

Steel (%). This control allows the total area of all longitudinal steel reinforcement bars expressed as a percentage of the cross-sectional area of the PCC slab (i.e., slab width by thickness) to be defined.

Bar diameter (in.). This control allows you to define the diameter of the longitudinal steel reinforcement bars.

Steel depth (in.). This control allows you to define the depth of the longitudinal steel reinforcement bars within the PCC slab (measured from the top of the PCC slab).

Base/slab friction coefficient. This control allows you to define the level of friction at the PCC slab–base interface. The level of friction present is directly influenced by the following:

- Base material type
- Construction practices during PCC slab placement

This frictional resistance is influenced by the material type of the base layer and the friction coefficient's minimum, maximum, and mean values are discussed earlier in this chapter.

Crack spacing. This control displays the crack spacing, which is the average spacing between the transverse cracks formed in the CRCP mostly due to numerous factors such as PCC shrinkage, PCC slab/base friction, and PCC tensile strength.

Is crack spacing computed using a prediction model? This option allows you to select whether or not you want the AASHTOWare pavement ME design software to compute the crack spacing internally.

Crack spacing. This option allows you to define the crack spacing. The maximum recommended value is 72 in.

15.6 Interpretation and Analysis of the Trial Design

The predicted distress must be smaller than the threshold values set. If the predicted distress is greater than the predefined threshold values, the pavement of the trial will be revised. While the thickness of the layer is important, many other design factors often affect distress and IRI or smoothness. The designer will examine the performance prediction and decide which design feature to change to improve performance (e.g., layer thickness, material properties, combinations of layers, geometric features, and other inputs). Some of the parameters of the input are interrelated; changing one parameter may lead to a change to another. Some recommendations are listed in Tables 15.2 and 15.3 for the revision of the trial pavements.

Distress and IRI	Modifications to minimize or eliminate
Joint crack width	• Build JPCP to set at lower temperature (cool PCC, place cooler temperatures). • Reduce drying shrinkage of PCC (increase aggregate size, decrease water/cement ratio, decrease cement content). • Decrease joint spacing. • Reduce PCC coefficient of thermal expansion.
Joint LTE	• Use mechanical load transfer devices (dowels). • Increase diameter of dowels. • Reduce joint crack width (see joint crack width recommendations). • Increase aggregate size.
Joint faulting	• Increase slab thickness. • Reduce joint width over analysis period. • Increase erosion resistance of base. • Minimize permanent curl/warp through curing procedures that eliminate built-in temperature gradient. • PCC tied shoulder. • Widened slab (by 1 to 2 ft).
Slab cracking	• Increase slab thickness. • Increase PCC strength. • Minimize permanent curl/warp through curing procedures that eliminate built-in temperature gradient. • PCC tied shoulder (separate placement or monolithic placement). • Widened slab (1 to 2 ft). • Use PCC with lower coefficient of thermal expansion.
IRI JPCP	Require more stringent smoothness criteria and greater incentives.

Source: *From AASHTO (2015). Mechanistic-Empirical Pavement Design Guide: A Manual of Practice. Washington, DC: American Association of State Highway and Transportation Officials. Table 13-4. Used with permission.*

TABLE 15.2 Guidance on Modifying JCPC Trial Designs to Satisfy Performance Criteria

Distress and IRI	Modifications to minimize or eliminate
Crack width	• Build CRCP to set at lower temperature (cool PCC, place cooler temperatures). • Reduce drying shrinkage of PCC (increase aggregate size, decrease w/c ratio, decrease cement content). • Increase percent longitudinal reinforcement. • Reduce depth of reinforcement (minimum depth 3.5 in.). • Reduce PCC coefficient of thermal expansion.
Crack LTE	• Reduce crack width (see crack width recommendations). • Increase aggregate size. • Reduce depth of reinforcement.
Punchouts	• Increase slab thickness. • Increase percent longitudinal reinforcement. • Reduce crack width over analysis period. • Increase PCC strength. • Increase erosion resistance of base (specific recommendations for each type of base). • Minimize permanent curl/warp through curing procedures that eliminate built-in temperature gradient. • PCC tied shoulder or widened slab.
IRI CRCP	Require more stringent smoothness criteria and greater incentives.

Source: *From AASHTO (2015). Mechanistic-Empirical Pavement Design Guide: A Manual of Practice. Washington, DC: American Association of State Highway and Transportation Officials. Table 13-5. Used with permission.*

TABLE 15.3 Guidance on Modifying CRCP Trial Designs to Satisfy Performance Criteria

15.7 Summary

This chapter addresses background knowledge for rigid pavement design software usage, analysis of results, etc. When the software run is complete, the user can access Microsoft Excel and Adobe PDF formats with input and output summaries generated by the program. It also provides an output summary of both tabular and graphical formats of the distress and performance prediction. All charts are plotted in Microsoft Excel and can conveniently be included in electronic documents and reports. If all the predicted distresses in its service life are smaller than the predefined threshold values, then the pavement of the trial is considered sufficient. If any of the distresses in its service life are greater than the predefined threshold values, the trial pavement will be deemed insufficient and revised as discussed above.

15.8 Fundamentals of Engineering (FE) Exam-Style Questions

FE15.1 The layer directly below the PCC slab of a rigid pavement is called:

A. Surface layer
B. Base layer
C. Subbase layer
D. None of the above

Solution B

Base course is defined as the layer directly below the PCC slab and subbase layers are below the base layer.

FE15.2 Dowel bars at the transverse joints of a rigid pavement (select all that apply):

A. Reduces the pumping
B. Reduces the faulting
C. Reduces the transverse cracking
D. Reduces the linear cracking

Solutions A and B

Dowel bars at the transverse joints of a rigid pavement reduce faulting, which eventually reduces pumping.

FE15.3 While analyzing a rigid pavement, the AASHTOWare pavement ME design software allows the inclusion of (select all that apply):

A. Different subbase materials
B. Different base materials
C. Different dowel spacing
D. JPCP overlay on existing asphalt pavement

Solutions A, B, C, and D

The AASHTOWare pavement ME design software allows the inclusion of a variety of options for all layers of a pavement.

15.9 Practice Problems

15.1 A new JPCP geometry and materials are shown in Fig. P15.1. The project will start sometime next year.

The laboratory testing results are provided below.

PCC:

Unit weight = 145 pcf
Poisson's ratio = 0.25
Coefficient of thermal expansion = 6×10^{-6} per degree Fahrenheit
Modulus of rupture = 720 psi or Level 3
Young's modulus = 4,500,000 psi or Level 3

Climate: Choose

Traffic:

AADTT = 6,000
No. of lanes in each direction = 3
Design speed = 50 mph
Traffic wander = 8 in.
Lane width = 12 ft

Use the default values for the unmentioned data. After the AASHTOWare pavement ME design software analysis:

a. Compare the amount of JPCP transverse cracking (%) after 20 years if the PCC thicknesses are 4, 6, and 8 in. Use a bar chart or line graph.
b. Compare the mean joint faulting (in.) after 20 years if the PCC thicknesses are 4, 6, and 8 in. Use a bar chart or line graph.

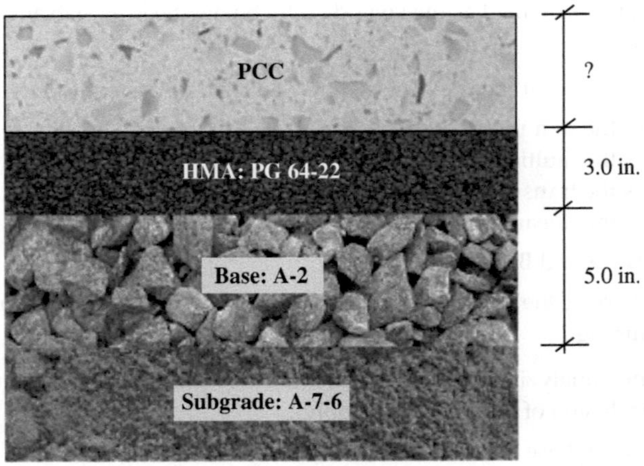

FIGURE P15.1 Geometry of the trial pavement for Prob. 15.1.

15.2 A new CRCP concrete geometry and materials are shown in Fig. P15.2. The project will start sometime next year.

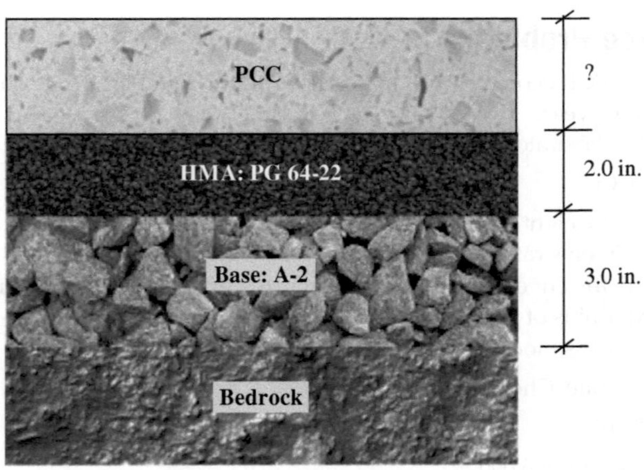

FIGURE P15.2 Geometry of the trial pavement for Prob. 15.2.

The laboratory testing results are provided below.
PCC:

Unit weight = 155 pcf
Poisson's ratio = 0.22
Coefficient of thermal expansion = 5×10^{-6} per degree Celsius
Modulus of rupture = 700 psi or Level 3
Young's modulus = 4,400,000 psi or Level 3

Climate: Choose

Traffic:

AADTT = 5,000
No. of lanes in each direction = 3
Design speed = 55 mph
Traffic wander = 10 in.
Lane width = 12 ft

Use the default values for the unmentioned data.
After the AASHTOWare pavement ME design software analysis:

a. Compare the amount of CRCP punchouts (per mile) after 20 years if the PCC thicknesses are 4, 6, and 8 in. Use a bar chart or line graph.
b. Compare the amount of CRCP punchouts (per mile) after 20 years if the PCC thickness is 6 in. and the HMA layer is not provided. Use a bar chart or line graph.

CHAPTER 16

Drainage Design in Pavement

16.1 Background

Pavement may get water broadly from surface and subsurface sources.

Surface water sources are rainfall, snow melting, etc. A major part of this water flows over the surface to the nearby channel and is discharged, as shown in Fig. 16.1. Therefore, the pavement must have longitudinal and transverse slopes in order to carry this water to the nearby channel under gravity. The discharge channel must be capable of efficiently discharging the surface water.

Subsurface water is caused by the infiltration of a small portion of surface water, frost melting (thawing of ice lenses), and groundwater seepage. The underlying pavement materials must be capable of discharging the subsurface water from the pavement laterally. It should be taken into account, however, that some subsurface water will flow vertically down by gravity, which need not be taken into account in the design. Another factor is that subsurface water sources may not occur simultaneously. For example, it is not expected that the frost-melting water and ground-seepage water would occur simultaneously as the frost-action soil has very low permeability when frozen (Garber and Hoel, 2015). Therefore, when designing surface drainage, an optimal relationship is considered.

Proper drainage of water from the pavement is an important design step in pavement design either from the surface water or from the subsurface water. During the traffic flow, surface water is splashed, which reduces safety and impedes smooth traffic flow, as shown in Fig. 16.2. Therefore, the longer the surface water remains on the surface, the greater the water that is infiltrated. The underlying base, subbase, and subgrade layers are also harmed by the infiltrated water.

16.2 Surface Drainage

The major tasks of surface drainage involve providing transverse slope, longitudinal slope, longitudinal channel, curbs, gutter, etc.

16.2.1 Transverse Slopes

Transverse slope (Fig. 16.3) works very well to quickly remove the surface water from the surface of the pavement. Cross-slopes on both sides of the pavement are crowned

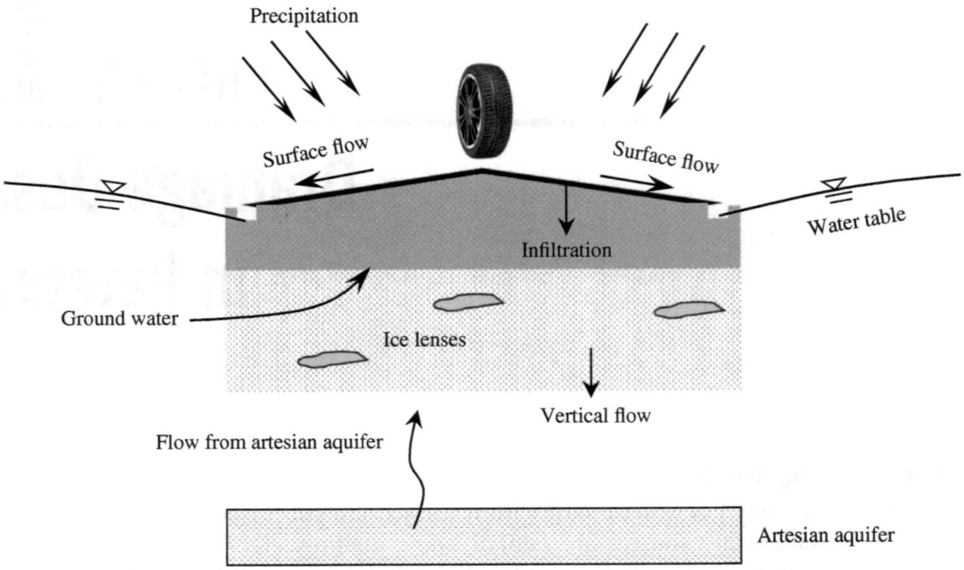

FIGURE 16.1 Drainage process in pavement.

FIGURE 16.2 Issues due to inadequate drainage in pavement. (*Courtesy of Trucking Info.*)

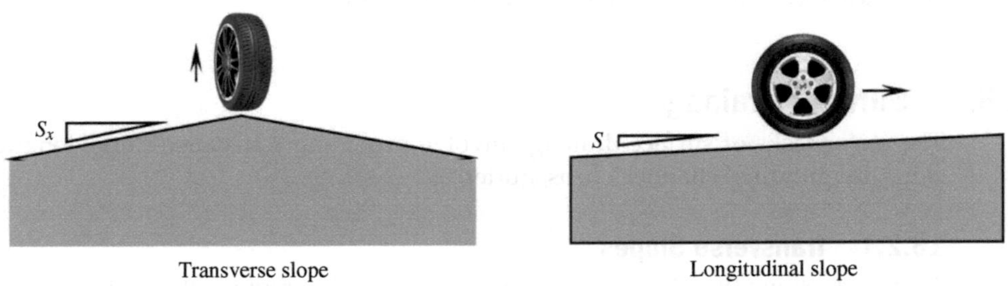

FIGURE 16.3 Transverse and longitudinal slopes in pavement.

Longitudinal slope

Longitudinal channel

Figure 16.4 Longitudinal slopes in pavement.

with the centerline of the pavement. A 2% cross-slope that is not so difficult to steer the vehicle is very often recommended.

16.2.2 Longitudinal Slopes and Channels

The longitudinal slope (Fig. 16.4) is another way to remove surface water from the pavement. The minimum slope of 0.2% for flat regions is recommended for uncurbed pavement. For curved pavement, however, the minimum slope of 0.5% (0.3% with crowned pavement) is recommended.

A longitudinal channel, also called ditch, is also very often provided to collect water from the surface, subsurface drains, etc. Figure 16.5 shows a roadside slope.

16.2.3 Curbs and Gutters

Curbs and gutters are primarily provided on urban roads to prevent traffic from encroaching with adjacent property and surface water passage. This is connected to the storm sewer very often. Figure 16.6 shows an example of a curb and a gutter.

16.2.4 Calculating the Runoffs by Rational Method

Two methods are very popular in calculating the surface runoff:

1. Rational method
2. U.S. Soil Conservation Services (SCS) method

The rational method is one of the most common methods used to calculate small-drainage peak flows of less than 200 acres. This approach is the most accurate for small-drainage runoff estimates with lots of impervious area. This is because the rational method needs the duration of the storm to be at least equal to the time of concentration. Time of concentration means the time required to flow the runoff from the farthest point of the catchment to the desired outlet. The rational formula to determine the peak discharge is given in the equation:

$$Q = CIA \tag{16.1}$$

where Q = Peak rate of discharge in cubic feet per second (cfs)

FIGURE 16.5 A drainage channel serving pavement and residential areas.

FIGURE 16.6 Curbs and gutter in a pavement.

I = Average rainfall intensity for the duration of the time of concentration (in./h). Rainfall intensity is related to rainfall duration and design storm recurrence interval. Rainfall intensity at a duration equal to the time of concentration (T_c) is used to calculate the peak flow in the rational method. The rainfall intensity can be selected from the appropriate intensity–duration–recurrence interval (I–D–R) curve. Figure 16.7 shows an IDR curve. The National Weather Service (NWS) has an automatic rainfall measuring device network that collects data across the United States on intensity and duration. Every country has its own guidelines for rainfall calculation. These data are used to evaluate the curves of rainfall intensity from which it is possible to obtain rainfall intensity for a given frequency. Duration is called the time in which the intensity is constant. The average number of years between the occurrence of a given intensity and duration is called frequency or occurrence interval. Estimating the intensity and frequency of rainfall is based on probability. For example, a "50-year" occurrence interval or frequency of a 1-h rainfall intensity of 2.0 in./h means that the rainfall intensity of 2.0 in./h for a duration of 1 h comes back on an average every 50 years.

A = Drainage area (acres). The field, measured in a horizontal plane, is defined as the drainage surface in acres. Typically the area is measured using a planimeter from plans or maps. The area includes all the surrounding land divided by the drainage. In the design of highway drainage, this area will often include upland property beyond the right-of-way highway.

C = Runoff coefficient representing the fraction of runoff to rainfall. This variable represents the ratio of runoff to rainfall. It is the most difficult input variable to estimate. It represents the interaction of many complex factors, including the storage of water in surface depressions, infiltration, antecedent moisture, ground cover, ground slopes, and soil types. In reality, the coefficient may vary with respect to prior wetting and seasonal conditions. The use of average values has been adopted to simplify the determination of this coefficient. Table 16.1 lists runoff coefficients for various combinations of ground cover and slope. Where a drainage area is composed of subareas with different runoff coefficients, a weighted coefficient for the total drainage area is computed by dividing the summation of the products of the subareas and their coefficients by the total area. Thus,

$$C_{\text{weighted average}} = \frac{C_1A_1 + C_2A_2 + C_3A_3 + \dots\dots + C_nA_n}{A_1 + A_2 + A_3 + \dots\dots A_n} \tag{16.2}$$

where $C_{1,2,3,\dots}$ = Runoff coefficients of first, second, third segment, etc., respectively
$A_{1,2,3,\dots}$ = Area of first, second, third segment, etc., respectively
n = Number of segment

The runoff coefficient is the most difficult input variable to estimate in the above equation. This represents the interaction of many complex factors such as surface depression, infiltration antecedent moisture, soil cover, soil slopes, permeability of the soil, prior wetting, etc. Table 16.1 lists some common C-values.

The coefficients in Table 16.1 are applicable for 10 years or less recurrence interval storms. Less frequent, higher-intensity storms need adjusted runoff coefficients because the impact of infiltration and other losses on runoff is proportionally smaller. Runoff coefficient adjustment factors are listed in Table 16.2 for storms at different intervals of recurrence.

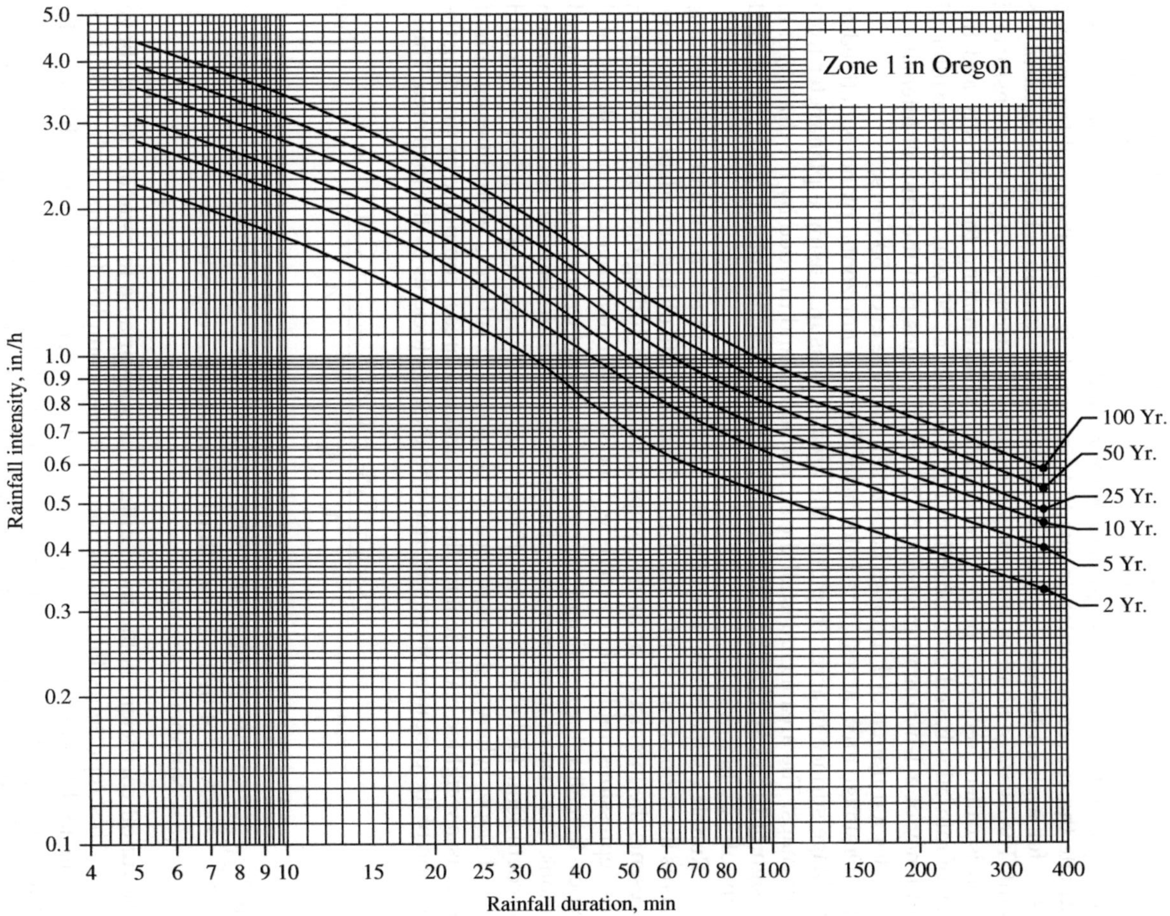

FIGURE 16.7 Example of intensity–duration–recurrence interval (I–D–R) curve (ODOT, 2014). This is an IDR curve from Zone I in Oregon. Every design agency has its own IDR curves for its regions.

Surface type	Flat	Rolling (slope 2–10%)	Hilly ground (slope >10%)
Pavement and roofs	0.90	0.90	0.90
Earth shoulder	0.50	0.50	0.50
Drives and walks	0.75	0.80	0.85
Gravel pavement	0.85	0.85	0.85
City business areas	0.80	0.85	0.85
Light residential: 1 to 3 units/acre	0.35	0.40	0.45
Normal residential: 3 to 6 units/acre	0.50	0.55	0.60
Dense residential: 6 to 15 units/acre	0.70	0.75	0.80
Lawns	0.17	0.22	0.35
Grass shoulders	0.25	0.25	0.25
Side slopes, earth	0.60	0.60	0.60
Side slopes, turf	0.30	0.30	0.30
Median areas, turf	0.25	0.30	0.30
Cultivated land, clay, and loam	0.50	0.55	0.60
Cultivated land, sand, and gravel	0.25	0.30	0.35
Industrial areas, light	0.50	0.70	0.80
Industrial areas, heavy	0.60	0.80	0.90
Parks and cemeteries	0.10	0.15	0.25
Playgrounds	0.20	0.25	0.30
Woodland and forests	0.10	0.15	0.20
Meadows and pasture land	0.25	0.30	0.35
Unimproved areas	0.10	0.20	0.30

TABLE 16.1 Some Common C-Values for 10 Years or Less Recurrence Interval Storms (ODOT, 2014)

Recurrence interval	Runoff coefficient adjustment factor
10 years or less	1.00
25 years	1.10
50 years	1.20
100 years	1.25

TABLE 16.2 Runoff Coefficient Adjustment Factors for Different Recurrence Intervals (ODOT, 2014)

Travel time (T) is the time it takes for water to travel from one location to another. Travel time between two points is determined using the following relationship:

$$T = \frac{L}{V} \tag{16.3}$$

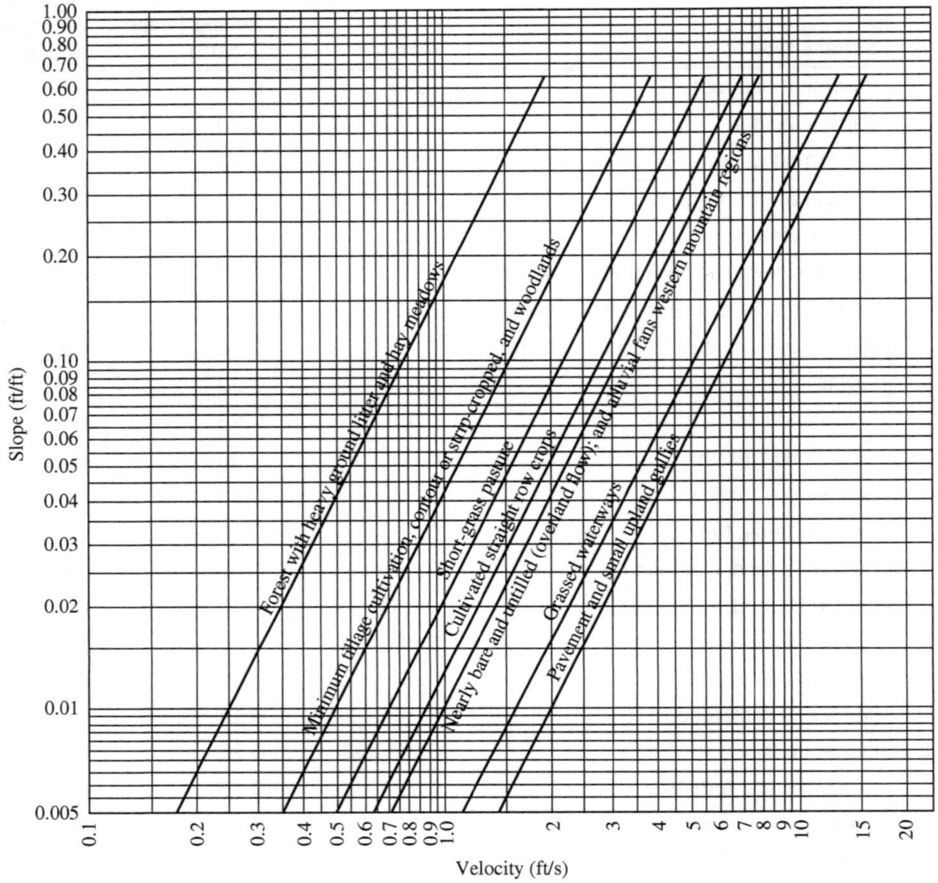

FIGURE 16.8 Average velocities for overland flow (NRCS, 2010).

where T = Travel time, s

L = Distance between the two points under consideration, ft

V = Average velocity of flow between the two points, ft/s (Fig. 16.8 can be used)

Time of concentration (T_c) is the time required for runoff to travel from the hydraulically most distant point in the watershed to the outlet. The hydraulically most distant point is the point with the longest travel time to the watershed outlet, and not necessarily the point with the longest flow distance to the outlet. Time of concentration is generally applied only to surface runoff and may be computed using many different methods. Depending on the slope and character of the river and the flow direction, the time of concentration will vary. It is determined by taking into account the sum of travel time within the watershed for various elements and can be presented as follows:

$$T_c = T_1 + T_2 + T_3 + \ldots\ldots + T_n \tag{16.4}$$

where $T_{1,2,3}$ = Travel time of first, second, third segment, etc., respectively
n = Number of segment

The steps of finding the peak discharge (Q) can be listed as follows:

Step 1. Divide the watershed into number of segments of equivalent hydraulic properties such as land type, slope, etc.

Step 2. Based on each segment's slope, find the overland flow velocity using Fig. 16.8.

Step 3. Calculate travel time for each segment using $T = \frac{L}{V}$.

Step 4. Calculate the time of concentration using $T_c = T_1 + T_2 + T_3 + \ldots\ldots + T_n$.

Step 5. From the IDR curve, determine the rainfall intensity using the time of concentration as the rainfall duration (say, 60 min) and desired time of occurrence (say, 50 years).

Step 6. Find out the weighted average C-value of the watershed.

Step 7. Calculate the peak discharge using the equation $Q = CIA$.

Example 16.1: Peak Flow by the Rational Method A 200-acre rural drainage area in Zone 1 of Oregon consists of four different watershed areas, as shown in Table 16.3.

Segment	Land type	Land slope	Segment area	Distance to the discharge point (ft)
1	Lawns (bare ground)	Rolling–5%	35%	4,000
2	Minimum tillage cultivated land	Flat–1.0%	15%	2,000
3	Forested area	Hilly–20%	40%	3,000
4	Paved area	Rolling–3%	10%	5,000

Table 16.3 Watershed Data for Example 16.1

Determine the peak flow for a 50-year time interval using the rational method.

Solution

Step 1. Divide the watershed into number of segments.

Already given in the problem.

Step 2. Based on each segment's slope, find the overland flow velocity as listed in Table 16.4 using Fig. 16.8.

Segment	Land type	Land slope	Flow velocity from Fig. 16.8 (ft/s)
1	Lawns (bare ground)	Rolling–5%	2.1
2	Minimum tillage cultivated land	Flat–1.0%	0.47
3	Forested area	Hilly–20%	1.15
4	Paved area	Rolling–3%	3.75

Table 16.4 Watershed Data Analysis for Example 16.1

Step 3. Calculate travel time for each segment using $T = \frac{L}{V}$, as listed in Table 16.5.

Segment	Land type	Flow velocity (ft/sec)	Distance to the discharge point (ft)	Travel time (s) $T = \frac{L}{V}$
1	Lawns (bare ground)	2.1	4,000	1,905
2	Minimum tillage cultivated land	0.47	2,000	4,255
3	Forested area	1.15	3,000	2,609
4	Paved area	3.75	5,000	1,333

TABLE 16.5 Travel Time Calculation for Example 16.1

Step 4. Calculate the time of concentration using $T_c = T_1 + T_2 + T_3 + \ldots\ldots + T_n$.

$$T_c = 1{,}905 + 4{,}255 + 2{,}609 + 1{,}333 = 10{,}102 \text{ s} \approx 168 \text{ min}$$

Step 5. From the IDR curve, determine the rainfall intensity.

For Zone 1 of Oregon, as shown in Fig. 16.7, for rainfall duration of 168 minute, $I \approx 0.7$ in./h.

Step 6. Find out the weighted average *C*-value of the watershed using Table 16.6. Table 16.1 is used to find out individual *C* for each segment.

Segment	Land type	Land slope	Segment area	Segment area (acre), *A*	*C* (from Table 16.1)
1	Lawns (bare ground)	Rolling–5%	35%	70	0.22
2	Minimum tillage cultivated land	Flat–1.0%	15%	30	0.50
3	Forested area	Hilly–20%	40%	80	0.20
4	Paved area	Rolling–3%	10%	20	0.90

TABLE 16.6 Watershed Data Summary for Example 16.1

$$C = \frac{C_1A_1 + C_2A_2 + C_3A_3 + C_4A_4}{A_1 + A_2 + A_3 + A_4}$$

$$= \frac{0.22(70) + 0.50(30) + 0.20(80) + 0.90(20)}{200}$$

$$= 0.322$$

For 50-year interval, runoff coefficient adjustment factor = 1.2

$$C = 1.2 \times 0.322 = 0.3864$$

Step 7. Calculate the peak discharge using the equation $Q = CIA$.

$$Q = CIA = 0.3864\ (0.7 \text{ in./h})(200 \text{ acres}) = 54.1 \text{ cfs}$$

Answer *The peak flow is 54.1 cfs.*

16.2.5 Calculating the Runoffs by U.S. Soil Conservation Service (SCS) Method

This method has been developed by the U.S. Soil Conservation Service (SCS), now known as the Natural Resources Conservation Service (NRCS), for relatively large drainage area. This method is also referred to as the TR-55 method of technical release (TR). The use of this method is not so popular; therefore, it is very concisely discussed in this chapter.

This method has two parts: the first part determines the runoff h in inches. The second part calculates the peak discharge using the value of h obtained in the first part. The depth h depends on the amount of rainfall (P) in in. and the initial abstraction (I_a). The initial abstraction (I_a) is the amount of rainfall absorbed by the soil and does not flow. Another term, the potential maximum retention (S), is the measure of the imperviousness of the watershed area. The following equation is used to calculate the runoff of a basin:

$$h = \frac{(P - I_a)^2}{P - I_a + S} = \frac{(P - 0.2S)^2}{P + 0.8S} \tag{16.5}$$

where h = Runoff (in.)
P = Precipitation (in.)
S = Potential maximum retention after runoff begins (in.)
I_a = Initial abstraction, 0.2S

The maximum basin retention (S) is related to the soil and cover conditions of the watershed through the CN. CN has a range of 0 to 100, and S is related to CN by:

$$S = \frac{1{,}000}{\text{CN}} - 10 \tag{16.6}$$

where CN = Curve number, varies from 0 (pervious) to 100 (impervious).

Once h is determined, the peak discharge (Q_p) is determined using the following equation:

$$Q_p = q_u A h \tag{16.7}$$

where Q_p = Peak discharge (ft^3/s)
A = Area of drainage basin (mile^2)
h = Runoff (in.)
q_u = Peak discharge for 24-h storm at time of concentration (ft^3/s/mile^2/in.)

The unit of q_u, ft^3/s/mile^2/in., is very often written as csm/in. or cfs per square mile of drainage area per inch of runoff. The CN procedure is less accurate when runoff is less than 0.5 in. and the weighted CN is less than 40 (NRCS, 1986).

The CN is related to the hydrologic soil group (HSG), the type of cover, the treatment, the hydrological condition, and the antecedent runoff condition (ARC). Another aspect considered is whether impervious areas outlet directly to the drainage system (connected) or whether the flow extends over pervious areas before reaching the drainage system (unconnected).

Hydrologic Soil Groups

To account for the ability of different soils to infiltrate, NRCS has divided soils into four HSGs. They are defined as follows:

- *HSG Group A (low runoff potential).* This group of soils has high infiltration rates even when thoroughly wetted. They mainly consist of thick, well-drained gravel and sand. Such soils have a high water transmission rate (the final rate of infiltration greater than 0.3 in./h).
- *HSG Group B.* This group of soils has moderate infiltration rates when thoroughly wetted. These consist chiefly of soils that are moderately deep to deep, moderately well drained to well drained with moderately fine to moderately coarse textures. Such soils have a moderate transmission rate of water (final infiltration rate from 0.15 to 0.30 in./h).
- *HSG Group C.* This group of soils has slow infiltration rates when thoroughly wetted. These consist chiefly of soils with a layer that impedes downward movement of water or soils with moderately fine to fine textures. Such soils have a slow transmission rate of water (end infiltration rate from 0.05 to 0.15 in./h).
- *HSG Group D (high runoff potential).* This group of soils has very slow infiltration rates when thoroughly wetted. They consist mainly of clay soils with a high swelling potential, soils with a permanent high water table, soils with clay or clay layer at or near the surface, and shallow soils over almost impervious materials. Such soils have a very slow transmission rate of water (final infiltration rate below 0.05 in./h).

Cover Type

Cover types include the effect of vegetation, bare soil, and impervious surfaces. There are several methods to determine the type of cover, such as field reconnaissance, aerial photographs, and land use maps.

Treatment

Treatment is a cover-type modifier to describe the management of cultivated agricultural lands. It includes mechanical practices, such as contouring and terracing, and management practices, such as crop rotations and reduced or no tillage.

Hydrologic Condition

Hydrologic condition indicates the effects of cover type and treatment on infiltration and runoff and is generally estimated from density of plant and residue cover on sample areas. Good hydrological condition means that for that specific HSG cover type, and treatment, the soil typically has low runoff potential. Some considerations to be taken into account when estimating the effect of the cover on infiltration and runoff are as follows:

1. Canopy or density of lawns, crops, or other vegetative areas
2. Amount of year-round cover
3. Amount of grass or close-seeded legumes in rotations
4. Percent of residue cover
5. Degree of surface roughness

Selection of a HSG should be done based on measured infiltration rates, soil survey, or judgment from a qualified soil science or geotechnical professional. Table 16.7 presents curve numbers for antecedent soil moisture condition II (average moisture condition). To alter the curve number based on moisture condition or other parameters, Table 16.8 can be used.

Table 16.9 provides seasonal rainfall limits to determine the antecedent moisture condition (AMC). When AMC I or III exists, CN is first calculated for AMC II and then revised accordingly using Table 16.8.

Peak Discharge Computation

For a selected rainfall frequency, the 24-h rainfall (P) is obtained from detailed local precipitation maps. CN and total runoff (h) for the watershed are computed. The CN is used to determine the initial abstraction (I_a). I_a/P is then computed.

Peak discharge per square mile per in. of runoff (q_u) is obtained from Figs. 16.10A, 16.10B, 16.10C, and 16.10D) using T_c, rainfall distribution type, and I_a/P ratio. If the

Cover description		Curve numbers for hydrologic soil group			
		A	B	C	D
Open space (lawns, parks, golf courses, cemeteries, etc.)	Poor condition (grass cover <50%)	68	79	86	89
	Fair condition (grass cover 50% to 75%)	49	69	79	84
	Good condition (grass cover >75%)	39	61	74	80
Impervious areas	Paved parking lots, roofs, driveways, etc. (excluding right-of-way)	98	98	98	98
Streets and roads	Paved; curbs and storm sewers (excluding right-of-way)	98	98	98	98
	Paved; open ditches (including right-of-way)	83	89	92	93
	Gravel (including right-of-way)	76	85	89	91
	Dirt (including right-of-way)	72	82	87	89
Western desert urban areas	Natural desert landscaping (pervious area only)	63	77	85	88
	Artificial desert landscaping (impervious weed barrier, desert shrub with 1- to 2-in. sand or gravel mulch and basin borders)	96	96	96	96
Urban districts	Commercial and business (85% impervious)	89	92	94	95
	Industrial (72% impervious)	81	88	91	93
Residential districts by average lot size	1/8 acre or less (townhouses) (65% impervious)	77	85	90	92
	¼ acre (38% impervious)	61	75	83	87
	1/3 acre (30% impervious)	57	72	81	86
	½ acre (25% impervious)	54	70	80	85
	1 acre (20% impervious)	51	68	79	84
	2 acres (12% impervious)	46	65	77	82

Table 16.7 Curve Number (CN) for Fully Developed Urban Areas (vegetation established) for Antecedent Soil Moisture Condition II (NRCS, 1986)

CN for condition II	Corresponding CN for condition I	III
0	0	0
5	2	17
10	4	26
15	7	33
20	9	39
25	12	45
30	15	50
35	19	55
40	23	60
45	27	65
50	31	70
55	35	75
60	40	79
65	45	83
70	51	87
75	57	91
80	63	94
85	70	97
90	78	98
95	87	99
100	100	100

TABLE 16.8 Adjustments to Select Curve Number (CN) for Soil Moisture Conditions

AMC	Dormant season	Growing season
I	<0.5	<1.4
II	0.5 to 1.1	1.4 to 2.1
III	>1.1	>2.1

TABLE 16.9 Determining AMC Using the Five-Day Antecedent Rainfall (in.)

computed I_a/P ratio is outside the range listed in Figs. 16.10A, 16.10B, 16.10C, and 16.10D for the rainfall distribution of interest, then the limiting value should be used.

Figures 16.10A, 16.10B, 16.10C, and 16.10D use type of rainfall distribution such as type I, IA, II, and III. NRCS (1968) developed four synthetic 24-h rainfall distributions (I, IA, II, and III) from available NWS duration-frequency or local storm data. Type IA is the least intense and type II the most intense short-duration rainfall. The four distributions depicted in Fig. 16.9 show their approximate geographic boundaries in the

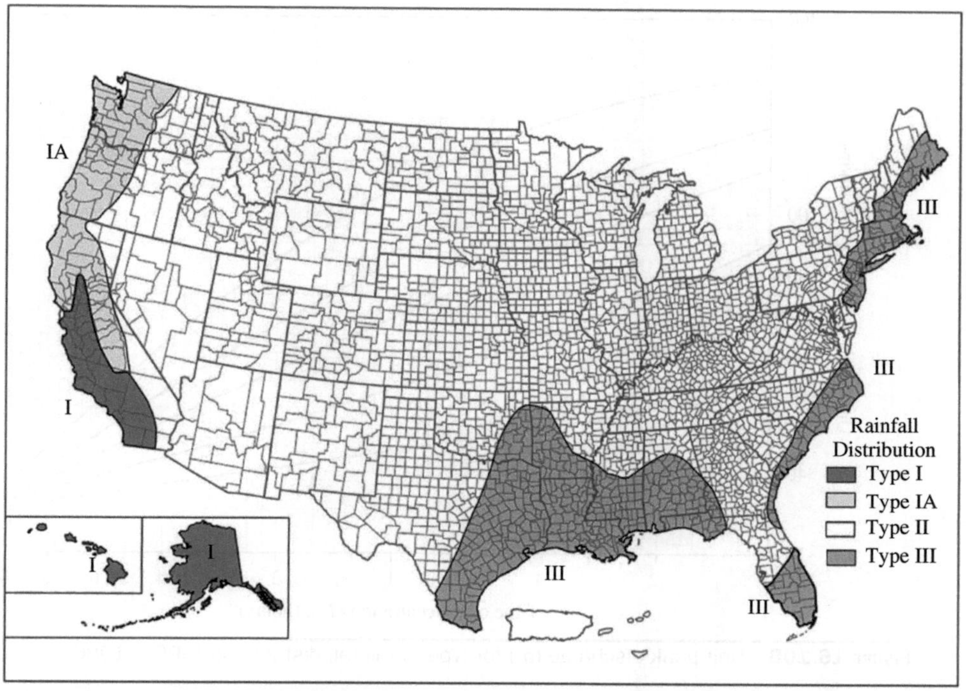

FIGURE 16.9 Geographic boundaries for rainfall distributions (NRCS, 1986).

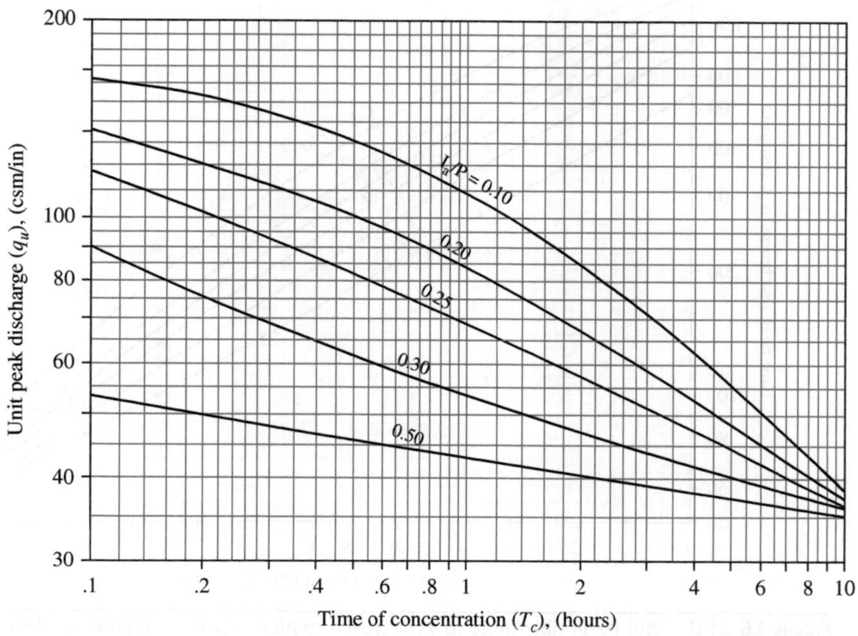

FIGURE 16.10A Unit peak discharge (q_u) for type I rainfall distribution (NRCS, 1986).

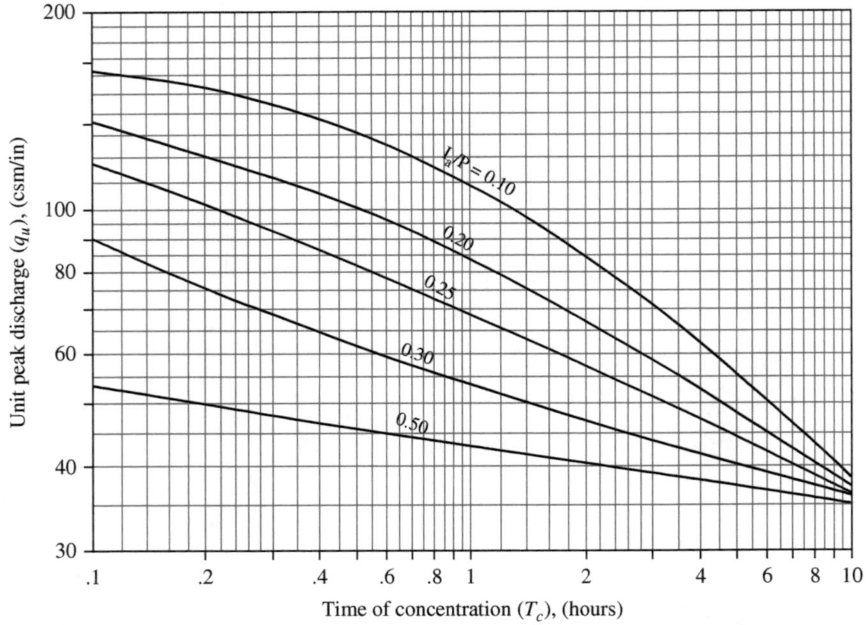

FIGURE 16.10B Unit peak discharge (q_u) for type IA rainfall distribution (NRCS, 1986).

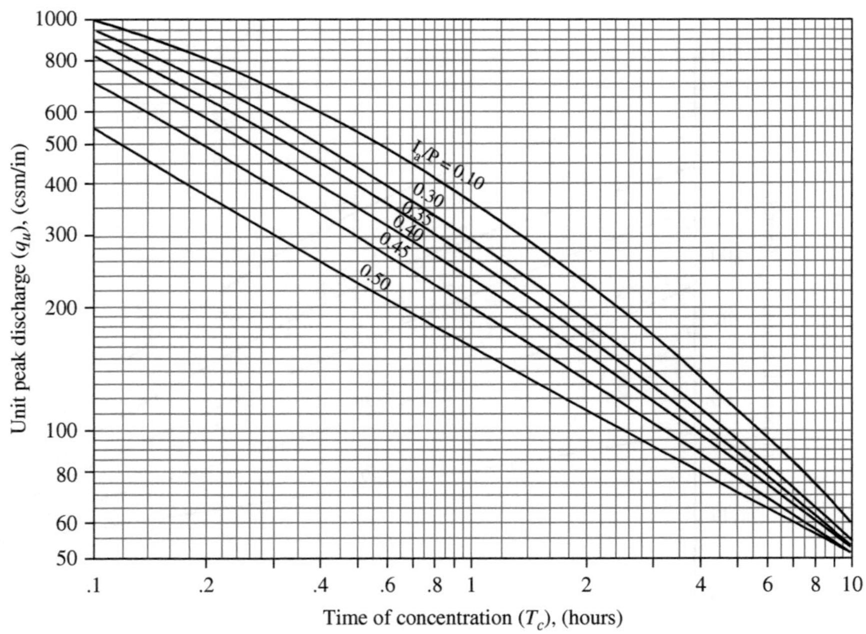

FIGURE 16.10C Unit peak discharge (q_u) for type II rainfall distribution (NRCS, 1986).

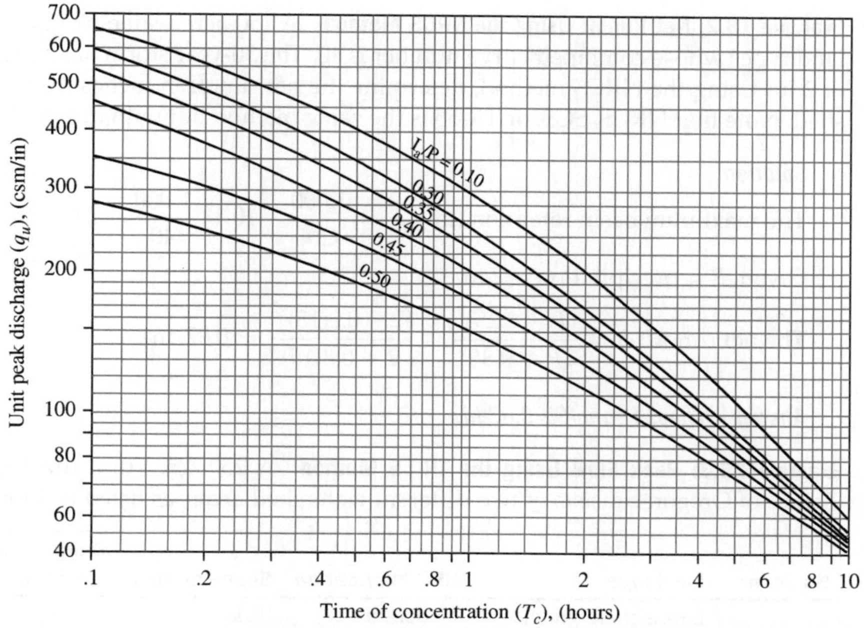

FIGURE 16.10D Unit peak discharge (q_u) for type III rainfall distribution (NRCS, 1986).

United States. Types I and IA represent the Pacific maritime climate with wet winters and dry summers. Type III represents Gulf of Mexico and Atlantic coastal areas where tropical storms bring large 24-h rainfall amounts. Type II represents the rest of the United States.

The steps to determine the runoffs/discharge volume can be summarized as follows:

Step 1. Determine the land use type (e.g., streets or residential areas) and type of cover (e.g., grass cover or dirt cover).

Step 2. Determine the HSG (A, B, C, or D) based on the description provided.

Step 3. Use Table 16.7 to determine the CN.

Step 4. Determine the AMC from Table 16.9.

Step 5. Revise the CN for the appropriate AMC from Table 16.8.

Step 6. Calculate T_c. You may need to use Fig. 16.8 to find out the flow velocity.

Step 7. Calculate I_a/P ratio.

Step 8. Find out the type of rainfall distribution from Fig. 16.9.

Step 9. Calculate peak discharge per square mile per in. of runoff (q_u) for 24-h storm using Figs. 16.10A, 16.10B, 16.10C, and 16.10D.

Step 10. Calculate the peak discharge (Q_p) using the equation $Q_p = q_u A h$.

Example 16.2: Net Runoff Using the NRCS Method A 150-acre watershed has 11 distinct land types whose combined curve number is 40. The 30-year storm has a gross rainfall of 2 in. Using the NRCS method, determine the net runoff in inches from this watershed. Note that NRCS, SCS, or TR-55 is the same method but with different names.

Solution

The maximum basin retention (in.), $S = \dfrac{1{,}000}{\text{CN}} - 10 = \dfrac{1{,}000}{40} - 10 = 15$

Given, P = precipitation = 2 in.

The net runoff, $h = \dfrac{(P - 0.2S)^2}{P + 0.8S} = \dfrac{(2 - 0.2(15))^2}{2 + 0.8(15)} = 0.07$ in.

Answer *The net runoff is 0.07 in.*

Example 16.3: Peak Flow Using the TR-55 Method A 250-acre rural drainage area in Zone 1 of Oregon consists of two different watershed areas, as listed in Table 16.10.

Segment	Land type	HSG information	Segment area	Distance to the discharge point (ft)
1	Lawns (bare ground with full grass cover)	Moderate infiltration	35%	800
2	Paved area with open ditches	High infiltration	65%	1,200

Table 16.10 Watershed Data for Example 16.3

The design rainfall for 50-year storm event is 3.0 in. Five-day antecedent rainfall (in.) in growing season is 1.0 in. The average land slope is about 2%. One acre equals 0.0015625 mi^2.

Determine the peak flow for a 50-year time interval using the TR-55 method.

Solution

Step 1. Determine the land use type (e.g., streets or residential areas) and type of cover (e.g., grass cover or dirt cover).

Already given: lawns (bare ground with grass cover) and paved areas.

Step 2. Determine the HSG (A, B, C, or D) based on the description provided.

Information is given about the HSG. High infiltration means HSG A and moderate infiltration means HSG B.

Step 3. Use Table 16.7 to determine the CN.

Using AMC type II, CN for lawn areas with full grass cover = 61
CN for paved areas with open ditches = 89

$$\text{Weighted CN} = \frac{\text{CN}_1 A_1 + \text{CN}_2 A_2}{A_1 + A_2} = \frac{(61)(35) + (89)(65)}{(35) + (65)} = 79.2$$

Step 4. Determine the AMC from Table 16.9.

Five-day antecedent rainfall (in.) in growing season is 1.0 in. Thus, AMC is type I.

Step 5. Revise the CN for the appropriate AMC from Table 16.8.

Using interpolation, CN for AMC type I $= 63 - \frac{(63-57)}{80-75}(80-79.2) \approx 62$

(A rough interpolation should be okay.)

Step 6. Calculate h.

$$S = \frac{1{,}000}{\text{CN}} - 10 = \frac{1{,}000}{62} - 10 = 6.13$$

$$h = \frac{(P - I_a)^2}{P - I_a + S} = \frac{(P - 0.2S)^2}{P + 0.8S} = \frac{(3.0 - 0.2(6.13))^2}{3.0 + 0.8(6.13)} = 0.398 \text{ in.}$$

Step 7. Calculate T_c as listed in Table 16.11.

Segment	Land type	Flow velocity (ft/sec) from Fig. 16.8	Distance to the discharge point (ft)	Travel time (sec) $T = \frac{L}{V}$
1	Lawns	2.25	800	355.6
2	Paved area	2.75	1,200	436.4

Table 16.11 Travel Time Calculation for Example 16.3

From Fig. 16.8, for 2% slope, flow velocity in lawn areas = 2.25 ft/s and velocity in paved area 2.75 ft/s.

$$T_c = 355.6 + 436.4 = 792 \text{ s} \approx 0.22 \text{ h}$$

Step 8. Calculate I_a/P ratio.

$$I_a = 0.2S = 0.2\,(6.13) = 1.226$$

$$I_a/P = 1.226/3 = 0.4$$

Step 9. Find out the type of rainfall distribution from Fig. 16.9.

For Oregon (northwest-west of the United States), rainfall distribution is type IA.

Step 10. Calculate peak discharge per square mile per inch of runoff (q_u) for 24-h storm using Figs. 16.10A, 16.10B, 16.10C, and 16.10D.

From Fig. 16.10B, for rainfall distribution type IA, $q_u \approx 63$ csm/in.

Step 11. Calculate the peak discharge (Q_p) using the equation $Q_p = q_u A h$.

$$Q_p = q_u A h = (63 \text{ csm/in.})\,(250 \times 0.0015625 \text{ mi}^2)\,(0.398 \text{ in.}) = 9.79 \text{ cfs}$$

Answer *The peak flow is 9.79 cfs.*

16.2.6 Designing of Open Channel

With a given depth of flow in a uniform channel, the mean velocity v may be computed by the Manning's equation:

$$v = \text{Velocity of flow} = Q/A = \frac{1.486}{n} R_H^{2/3} S^{1/2} \qquad (16.8)$$

where Q = Total discharge
A = Cross-sectional area of flow
P = Wetted perimeter of the flow
v = Velocity of flow
R_H = Hydraulic radius of flow = A/P
S = Longitudinal slope of the channel
n = Manning's coefficient (Table 16.12)

The Manning's equation can also be written as:

$$Q = Av = \frac{1.486}{n} AR_H^{2/3} S^{1/2}$$

For circular conduit flowing full, $R_H = D/4$, where D is the diameter of the conduit.

The Manning's equation gives a reliable estimate of velocity only if the discharge, channel cross section, roughness, and slope are constant for uniform flow. The Manning's equation can be solved for the average velocity v in a trial channel. The channel size and shape can be adjusted to have the desired flow velocity.

Hazen–Williams' equation can also be used instead of Manning's equation for designing the channel. This method is not so popular. Hazen–Williams' equation is presented below:

$$v = 0.849 C R_H^{0.63} S^{0.54}$$

where C is the Hazen–Williams' coefficient. All other symbols are similar to the Manning's equation. Do not confuse the Hazen-Williams' coefficient (C) used in this equation with the Runoff coefficient (C) used in the Rational method in Section 16.2.4. These are completely different parameters.

Surface description	Manning's coefficient, *n*
Rooftops	0.011
Concrete	0.013
Asphalt	0.015
Bare soil	0.018
Sparse vegetation*	0.100
Grass: Short grass prairie, lawn	0.150
Grass: Dense grasses†, meadow (good condition)	0.240
Range (natural)	0.130
Woods‡: Light underbrush	0.400
Woods‡: Dense underbrush	0.800

*Areas where vegetation is spotty and consists of less than 50% vegetative cover.
†Species such as weeping love grass, bluegrass, buffalo grass, blue grama grass, and native grass mixtures.
‡Consider cover to a height of 0.1 ft. This is part of the plant cover that will obstruct sheet flow.

Table 16.12 Some Common Values of Manning's Coefficient Related to Pavement

Example 16.4: Design of Open Channel

A trapezoidal brick sewer channel with 1:1 side slopes shown in Fig. 16.11 carries a flow of 100 m^3/s under gravity. Calculate the required longitudinal slope of the channel. Hazen–Williams' coefficient is 100 for brick sewer.

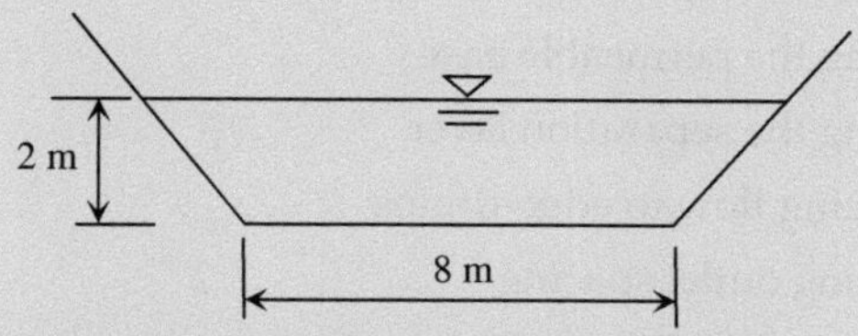

FIGURE 16.11 A trapezoidal brick sewer channel for Example 16.4.

Solution

Area of flow, A = Average width × Depth = (8 m + 2 m) (2 m) = 20 m^2

Hydraulic radius of flow,

$$R_H = \frac{\text{Wetted area}}{\text{Wetted perimeter}} = \frac{A}{P} = \frac{(8\,\text{m} + 2\,\text{m})(2\,\text{m})}{8\,\text{m} + 2\sqrt{(2\,\text{m})^2 + (2\,\text{m})^2}} = \frac{20\,\text{m}^2}{13.66\,\text{m}} = 1.46\,\text{m}$$

Hazen–Williams' coefficient, C = 100 for brick sewer

$$\text{Flow velocity, } v = \frac{Q}{A} = \frac{100\frac{\text{m}^3}{\text{s}}}{20\ \text{m}^2} = 5\ \text{m/s}$$

$$\text{Therefore, } v = 0.849CR_H^{0.63}S^{0.54}$$

$$5\,\text{m/s} = 0.849(100)(1.46)^{0.63}S^{0.54}$$

$$S = 0.00339 = 0.34\%$$

Answer *The longitudinal slope of the channel is 0.34%.*

Example 16.5 Design of Open Channel

Determine the design flow rate in a 24-in.-diameter storm sewer with the longitudinal slope of 0.003 and Manning's coefficient of 0.01, when it is flowing full under the gravity.

Solution

For flowing full, hydraulic radius of flow, $R_H = D/4$ = 24 in./4 = 6 in. = 0.5 ft.

$$\text{Cross-sectional area, } A = \frac{\pi D^2}{4} = \frac{\pi(2\ \text{ft})^2}{4}$$

Design flow rate,

$$Q = \frac{1.486}{n}AR_H^{2/3}S^{1/2} = \frac{1.486}{0.01}\left(\frac{\pi(2\ \text{ft})^2}{4}\right)(0.5\ \text{ft})^{2/3}(0.003^{1/2}) = 16\ \text{ft}^3/\text{s}$$

Answer *The flow rate is 16 cfs.*

16.3 Subsurface Drainage

The major steps of subbase drainage design are given below:

1. Quantifying water inflow
2. Designing the permeable base
3. Designing the separation layer
4. Quantifying flow to edge-drains
5. Computing outlet spacing
6. Checking outlet flow

16.3.1 Drainage Geometry and Permeability

Drainage Geometry

Geometric design decisions such as maximum and minimum slopes, pavement and shoulder interface, cross sections, location of filter fabrics, overlap of fabrics, joints, separation layer location, trench dimensions, and so on are critical to pavement performance. Filter fabrics should be selected only when the subgrade provides adequate support for compaction of the drainage layer. Requirements for pipe cover are a result of loading and depth of frost. In freeze-thaw areas, trench walls should not be steeper than 10 vertical on 1 horizontal to minimize the depth of frost penetration. In nonfrost areas, trench sloping is not required when high-speed traffic is subject to the pavement over the trench. Trench backfill is a critical component and it is necessary to select backfill materials and construction procedures properly.

The permeable foundation is usually placed directly below the portland cement concrete (PCC) for rigid pavements. To avoid fines from spreading into the permeable base, a separator layer is put between the permeable base and the subgrade. The permeable base can be directly below the layers of asphalt concrete (AC) for flexible pavements. In order to protect the relatively weak permeable base layer, adequate surface thickness is necessary. Placing the open-graded material under a thin AC sheet, however, could lead to early failures. The design pavement thickness and location of the permeable base should be in line with the AASHTO 1993 or the mechanical-empirical approach.

In designing the drainage of a permeable base, it is important to use the true slope and width of the permeable layer. When the longitudinal slope (S) is combined with the pavement cross slope (S_x), shown in Fig. 16.3, the true or resultant slope (S_R) of the flow path is determined by the equation:

$$S_R = \sqrt{(S^2 + S_X^2)} \tag{16.9}$$

The resultant length (L_R) of the flow path is:

$$L_R = W\sqrt{\left(1 + \left(\frac{S}{S_X}\right)^2\right)} \tag{16.10}$$

where S_R = Resultant slope
S = Longitudinal slope
S_X = Transverse slope
W = Width of permeable base, ft

Permeability

The coefficient of permeability depends primarily on the characteristics of the permeable base materials. The most significant properties affecting permeability are effective grain size, D_{10}; porosity, n; and percent passing the No. 200 sieve, P_{200}. These parameters accounted for over 91% of the variation in the hydraulic conductivity measured. However, proper gradation and density are vital to the stability of granular materials. In order to obtain the desired permeability, it is evident that the fine portion of the aggregate must be removed; this may adversely affect the stability of the drainage layer. This adverse effect can be compensated by stabilizing with a small amount of asphalt or portland cement, particularly for the more open-graded materials. The addition of the stabilizer only slightly decreases the permeability.

Due to the complexity of quantifying the coefficient of permeability, it was estimated using different empirical methods. There are a variety of close relationships between permeability and grain size. The most common relationship is given below (for filter sands):

$$k = C_k D_{10}^2 \tag{16.11}$$

where k = Permeability, mm/s
C_k = Experimental coefficient
D_{10} = Effective grain size (size corresponding to passing 10%)

16.3.2 Computation of Subsurface Water

Surface Infiltration

In the evaluation of surface infiltration, two techniques were used extensively: the method of infiltration ratio and the method of crack infiltration. The method of infiltration ratio is highly empirical based on both the ratio of infiltration and the rate of rainfall. The method of crack infiltration is based on infiltration test results. The infiltration was found to be directly related to cracking. A significant difference was observed in infiltration predicted from these two methods. Since the method of crack infiltration is more reasonable and is focused on field measures, it is more often used to evaluate surface infiltration, but the method of infiltration ratio can be used as a check. The larger of the two results can be used in the design inflow, if necessary.

Infiltration Ratio Method The method of infiltration ratio is a simplified way of estimating the inflow of water. The method assumes that a fixed portion of rain falling on a pavement will enter the pavement. Thus, all that are needed are an infiltration ratio and rainfall rate. For application of the 1-h duration, 2-year frequency rainfall is recommended. The equation for the inflow is:

$$q_i = CRF \tag{16.12}$$

where q_i = Pavement infiltration, m^3/day/m^2 (ft^3/day/ft^2)
C = Infiltration ratio
R = Rainfall rate, mm/h (in./h)
F = Conversion factor, 2.0 for in./h, 0.24 for mm/h

Cedegren (1974) has recommended infiltration ratios varying from 0.50 to 0.67 for PCC pavements and 0.33 to 0.50 for AC pavements.

Crack Infiltration Method Ridgeway (1976) recommended an inflow rate estimated by the water-carrying capacity of a pavement crack or joint and by an estimated joint or crack length. Ridgeway's research indicated that the condition of the crack or joint (i.e., sealed or unsealed and debris filled, wide or narrow cracks or joints) and the type of base layer underlying the pavement surface (i.e., open-graded or dense-graded) both play a role in defining the infiltration capacity of the joint/crack. An equation to compute the infiltration rate for "normal" conditions of uncracked pavement is:

$$q_i = I_C\left[\frac{N_C}{W} + \frac{W_C}{WC_S}\right] + k_P \tag{16.13}$$

where q_i = Rate of pavement infiltration, m³/day/m² (ft³/day/ft²)
I_C = Crack infiltration rate, can be assumed 0.223 m³/day/m (2.4 ft³/day/ft)
W_C = Length of contributing transverse joints or cracks, m (ft)
W = Width of permeable base, m (ft)
C_S = Spacing of contributing transverse cracks, m (ft)
k_p = Pavement permeability, m/day (ft/day)
N_C = Number of contributing longitudinal joints (commonly number of lanes plus 1.0)

Example 16.6: Amount of Infiltration Using the Infiltration Ratio An asphalt pavement is having a rainfall of 1.2 in./h. Determine the amount of infiltration using the infiltration ratio method.

Solution

Infiltration ratio, $C = 0.4$ (consider the range of 0.33 to 0.50 for asphalt pavement)
Rainfall rate, $R = 1.2$ in./h
Conversion factor, $F = 2.0$ for in./h
Rate of pavement infiltration, $q_i = ?$
Rate of pavement infiltration, $q_i = CRF = 0.4$ (1.2 in./h) (2.0) = 0.96 ft³/day/ft²

Answer *The amount of infiltration is 0.96 ft³/day/ft².*

Example 16.7: Infiltration Rate Using the Crack Infiltration Method A new two-lane asphalt pavement has the following characteristics:

Lane width = 12 ft
Shoulder width = 6 ft
Number of contributing longitudinal cracks = 3
Length of contributing transverse crack = 20 ft
Rate of infiltration = 0.03 ft/day
Spacing of transverse crack = 40 ft

Determine the surface infiltration rate for this pavement using the crack infiltration method.

Solution

Crack infiltration rate, $I_C = 2.4$ ft³/day/ft (default value)
Length of contributing transverse crack, $W_C = 20$ ft
Width of permeable base, $W = 2$ lanes and 2 shoulders $= 2(12\text{ ft}) + 2(6\text{ ft}) = 36$ ft
Spacing of contributing transverse cracks, $C_S = 40$ ft
Pavement permeability, $k_p = 0.03$ ft/day
Number of contributing longitudinal joints, $N_C = 2$ lanes $+ 1 = 3$
Rate of pavement infiltration, $q_i = ?$

$$q_i = I_C\left[\frac{N_C}{W} + \frac{W_C}{WC_S}\right] + k_p = (2.4)\left[\frac{3}{36} + \frac{20}{36(40)}\right] + 0.03 = 0.263\ \frac{\text{ft}}{\text{day}}$$

Answer *The surface infiltration rate is 0.263 ft³/day/ft².*

Frost Meltwater

Flow from frost melt should also be included in the inflow rate in areas of frost. Chart for estimating the inflow of meltwater from ice lenses, as shown in Fig. 16.12. The rate

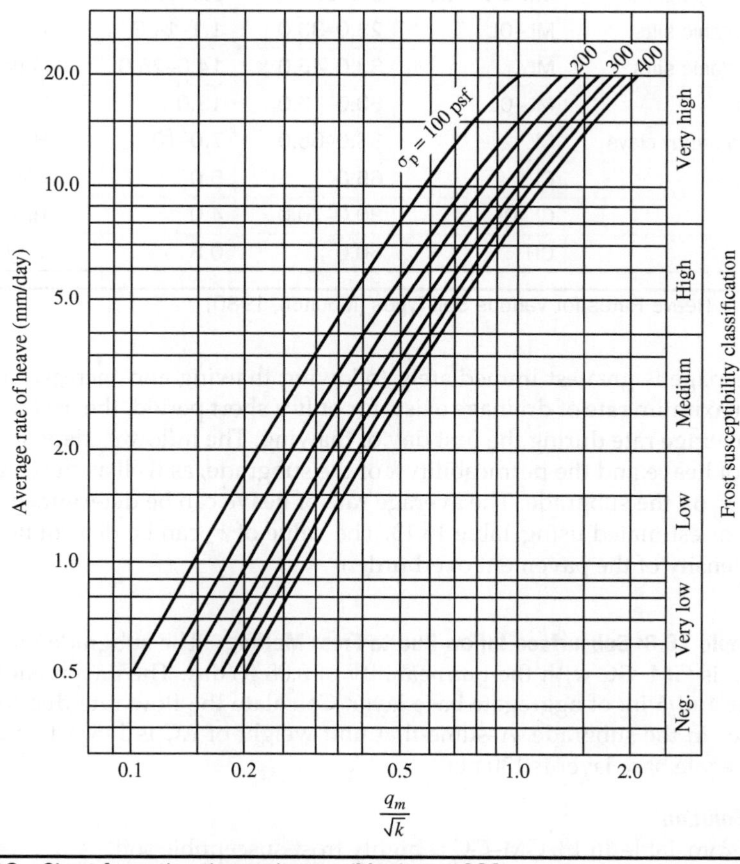

FIGURE 16.12 Chart for estimating meltwater (Moulton, 1980).

	Symbol	Percent <0.02 mm	Heave rate (mm/day)	Frost susceptibility
Gravels and sand gravels	GP	0.4	3.0	Medium
Gravels and sand gravels	GW	0.7–1.0	0.3–1.0	Low
Gravels and sand gravels	GW	1.0–1.5	1.0–3.5	Low to medium
Gravels and sand gravels	GW	1.5–4.0	3.5–2.0	Medium
Silty and sandy gravels	GP–GM	2.0–3.0	1.0–3.0	Low to medium
Silty and sandy gravels	GW–GM & GM	3.0–7.0	3.0–4.5	Medium to high
Clayey and silty gravels	GW–GC	4.2	2.5	Medium
Clayey and silty gravels	GM–GC	15.0	5.0	High
Clayey and silty gravels	GC	15.0–30.0	2.5–5.0	Medium to high
Sands and gravelly sands	SP	1.0–2.0	0.8	Very low
Silty and gravelly sands	SW	2.0	3.0	Medium
Silty and gravelly sands	SP–SM	1.5–2.0	0.2–1.5	Low
Silty and gravelly sands	SW–SM	2.0–5.0	1.5–6.0	Low to high
Silty and gravelly sands	SM	5.0–9.0	6.0–9.0	High
Clayey and silty sands	SM–SC & SC	9.5–35.0	5.0–7.0	High
Silts and organic silts	ML–OL	23.0–33.0	1.1–14.0	Low to high
Silts and organic silts	ML	33.0–45.0	14.0–25.0	Very high
Clayey silts	ML–CL	60.0–75.0	13.0	Very high
Gravelly and sandy clays	CL	38.0–65.0	7.0–10.0	High
Lean clays	CL	65.0	5.0	High
Lean clays	CL–OL	30.0–70.0	4.0	High
Flat clays	CH	60.0	0.8	Very low

Table 16.13 Heave Rates for Various Soil Types (Moulton, 1980)

of seepage is greatest immediately following thawing and increases rapidly. Because the maximum rate of drainage exists for only a short period, the design inflow rate q_m is the average rate during the first day of thawing. The inflow q_m depends on the average rate of heave and the permeability k of the subgrade, as well as the consolidation pressure σ_p on the subgrade. The average rate of heave can be determined from laboratory tests or estimated using Table 16.13. The value of σ_p can be determined by calculating the density of the pavement overburden.

Example 16.8: Subsurface Inflow Due to Frost Melting The subgrade material of a pavement is GM–GC with the permeability of 0.08 ft/day. The pavement has a 10-in. AC layer and 6 in. of aggregate base layer. Calculate the flow rate due to the melted ice lenses in the subgrade. Assume that unit weight of AC is 145 pcf and unit weight of aggregate base layer is 130 pcf.

Solution

From Table 16.13, GM–GC is highly frost-susceptible soil.

$$\sigma_P = \left(\frac{10}{12}\text{ ft}\right)\left(145\frac{\text{lb}}{\text{ft}^2}\right) + \left(\frac{6}{12}\text{ ft}\right)\left(130\frac{\text{lb}}{\text{ft}^2}\right) = 186\,\frac{\text{lb}}{\text{ft}^2}$$

Using Fig. 16.12, for the stress level of 186 psf at the middle of high susceptible soil:

$$\frac{q_m}{\sqrt{k}} \approx 0.55$$

$$q_m = 0.55\sqrt{k} = 0.55\sqrt{0.08} = 0.16\,\frac{\text{ft}}{\text{day}}$$

Answer *The flow rate due to the melted ice lenses is about 0.16 ft³/day/ft².*

Groundwater

Two types of groundwater may reach the pavement structure:

1. Groundwater under gravity
2. Groundwater from artesian flow

Groundwater under gravity is only possible if the groundwater table is located above the pavement drainage channel. The following steps can be followed to determine the groundwater inflow under gravity (q_g).

Step 1. Determine the parameter, L_i, radius of influence: $L_i = 3.8(H - H_0)$.

Step 2. Estimate q_1, inflow from above the bottom of the drainage layer:

$$q_1 = \frac{k(H - H_0)^2}{2L_i}$$

where k is the permeability of the soil in the cut slope.

Step 3. Determine the parameter $\dfrac{(L_i + 0.5W)}{H_0}$.

Step 4. Determine the parameter W/H_0.

Step 5. Use the chart in Fig. 6.13 to estimate the parameter $\dfrac{k(H - H_0)}{2q_2}$.

Step 6. Knowing k, permeability of the soil in the subgrade, H, and H_0, determine q_2, the seepage inflow from the bottom of the drainage layer.

Step 7. Determine the total lateral inflow into the drainage pipe: $q_d = q_1 + q_2$, for collector pipes on both sides, and $q_d = 2(q_1 + q_2)$, for a collector pipe on one side of the pavement only.

Step 8. Determine the groundwater inflow (q_g) into the drainage layer per unit area:

$$q_g = \frac{q_2}{0.5W} \text{ for collector pipes on both sides}$$

$$q_g = \frac{q_1 + 2q_2}{W} \text{ for collector pipe on one side}$$

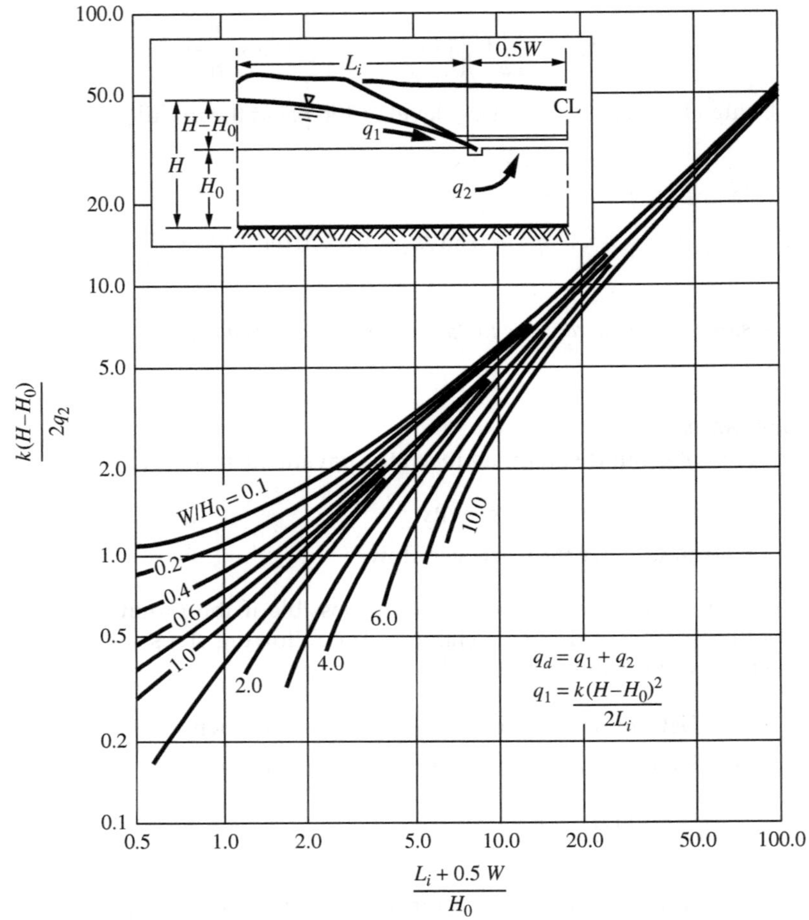

FIGURE 16.13 Chart for determining flow rate in horizontal drainage blanket (Moulton, 1980).

An artesian aquifer is a confined aquifer containing groundwater under positive pressure. Water from the artesian well reaches the ground surface under the natural pressure of the aquifer. For seepage of water from artesian aquifers, the following equation is used to estimate the inflow from artesian source (q_a):

$$q_a = k\frac{\Delta H}{H_0} \tag{16.14}$$

where q_a = Inflow from artesian source, ft³/day/ft² of drainage area
k = Coefficient of permeability, ft/day
ΔH = Excess hydraulic head, ft
H_0 = Thickness of the subgrade soil between the drainage layer and the artesian aquifer, ft

Example 16.9: Subsurface Inflow Due to Groundwater under Gravity A pavement is built cutting the natural ground and the groundwater table is above the perforated drainage pipe, as shown in Fig. 16.14. Due to this construction, the water–table drawdown is 10 ft and the bedrock is located 40 ft below the drainage pipe. The permeability of the pavement materials used is around 0.0075 ft/day. There are collector pipes on both sides of the pavement.

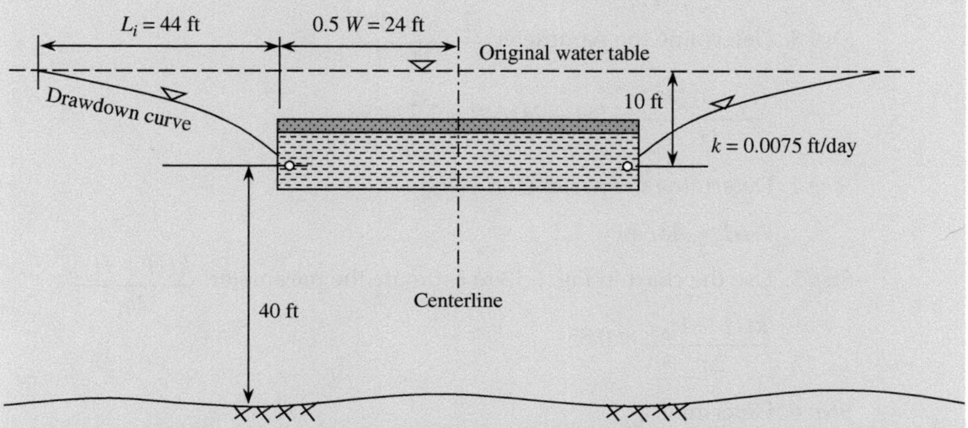

FIGURE 16.14 Cross section of a pavement for Example 16.9.

Calculate the following:

a. Inflow from above the bottom of the drainage layer.
b. The seepage inflow from the bottom of the drainage layer.
c. The total lateral inflow into the drainage pipe.
d. The groundwater inflow into the drainage layer per unit area.

Solution

a. Inflow from above the bottom of the drainage layer, $q_1 = ?$
b. The seepage inflow from the bottom of the drainage layer, $q_2 = ?$
c. The total lateral inflow into the drainage pipe: $q_d = q_1 + q_2 = ?$
d. The groundwater inflow into the drainage layer per unit area, $q_g = ?$

Given,

Radius of influence, $L_i = 44$ ft

Coefficient of permeability, $k = 0.0075$ ft/day

$H - H_0 = 10$ ft

$0.5W = 24$ ft; $W = 24 \text{ ft}/0.5 = 48$ ft

Thickness of the subgrade, $H_0 = 40$ ft

Step 1. Determine the parameter, L_i.

Given: $L_i = 44$ ft

Step 2. Estimate q_1.

$$q_1 = \frac{k(H - H_0)^2}{2L_i} = (0.0075)\,(10)^2/(2 \times 44) = 0.00852 \text{ ft}^3/\text{day per linear foot}$$

Step 3. Determine the parameter $\frac{(L_i + 0.5W)}{H_0}$.

$$\frac{(L_i + 0.5W)}{H_0} = (44 + 24)/40 = 1.7$$

Step 4. Determine the parameter: W/H_0.

$$W/H_0 = 48/40 = 1.2$$

Step 5. Use the chart in Fig. 6.13 to estimate the parameter: $\frac{k(H - H_0)}{2q_2}$.

$$\frac{k(H - H_0)}{2q_2} \approx 0.85$$

Step 6. Determine q_2.

$$\frac{k(H - H_0)}{2q_2} \approx 0.85$$

$q_2 = 0.044$ ft³/day per linear foot

Step 7. Determine the total lateral inflow q_d.

$q_d = q_1 + q_2$, for collector pipes on both sides

$q_d = q_1 + q_2 = 0.00852 + 0.044 = 0.0525$ ft³/day per linear foot

Step 8. Determine the groundwater inflow (q_g).

$$q_g = \frac{q_2}{0.5W} \quad \text{for collector pipes on both sides}$$

$$q_g = \frac{q_2}{0.5W} = 0.044/24 = 0.00183 \text{ ft}^3/\text{day}/\text{ft}^2$$

Answer *The inflow from above the bottom of the drainage layer is 0.00852 ft³/day/ft. The seepage inflow from the bottom of the drainage layer is 0.044 ft³/day/ft. The total lateral inflow into the drainage pipe is 0.0525 ft³/day/ft. The groundwater inflow into the drainage layer per unit area is 0.00183 ft³/day/ft².*

Example 16.10: Subsurface Inflow from an Artesian Aquifer An artesian aquifer is located 30 ft below the drainage outlet of a pavement. The piezometric level in the artesian aquifer is 4 ft above the drainage outlet, as shown in Fig. 16.15.

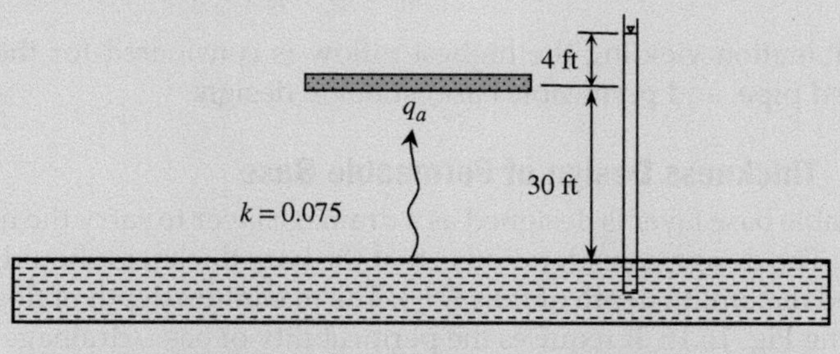

FIGURE 16.15 Cross section of a pavement for Example 16.10.

The coefficient of permeability of the soil below the pavement is 0.075 ft/day. Determine the inflow water to the pavement structure due to the presence of this artesian aquifer.

Solution

Given:

Excess hydraulic head, $\Delta H = 4.0$ ft

Thickness of the subgrade, $H_0 = 30$ ft

Coefficient of permeability, $k = 0.075$ ft/day

Inflow from artesian source, $q_a = k\dfrac{\Delta H}{H_0} = \left(0.075\dfrac{\text{ft}}{\text{day}}\right)\dfrac{4.0\text{ft}}{30.0\text{ ft}} = 0.01\dfrac{\text{ft}}{\text{day}}$

Answer *The inflow water is 0.01 ft³/day/ft².*

Vertical Downward Outflow

Downward vertical outflow (q_v) happens when there is no groundwater inflow. This flow should be deducted from the pavement site's total possible inflow. There is, however, no well-established formulation for calculating the vertical down outflow and therefore very frequently discarded.

Calculation of Net Inflow

The different types of subsurface inflow discussed previously may not be applicable to all pavements. For example, if groundwater table is located way below the pavement then the groundwater seepage (q_1, q_2, or q_g) is not applicable. Similarly, if the pavement site is not located in freezing zones, then there should not be any melting ice consideration.

Therefore, the net inflow is calculated, taking into account all the possible inflows in those regions. Nonetheless, there is no simultaneous occurrence of different types of subsurface inflow described above. Garber and Hoel (2015) have established a set of relationships to estimate the net inflow rate (q_n), which are given below:

$$q_n = q_i \tag{16.15}$$

$$q_n = q_i + q_g \tag{16.16}$$

$$q_n = q_i + q_a \tag{16.17}$$

$$q_n = q_i + q_m \tag{16.18}$$

$$q_n = q_i - q_v \tag{16.19}$$

The combination yielding the highest inflow is considered for the drainage channel, perforated pipe, and permeable base/subbase design.

16.3.3 Thickness Design of Permeable Base

A permeable base layer is designed as a drainage layer to carry the net inflow to the outlet point. The design consideration is that the base thickness should equal or exceed the computed maximum depth of flow (H_m). The maximum depth of flow (H_m) can be calculated using Fig. 16.16. It requires the permeability of base/drainage layer (k_d), slope (S),

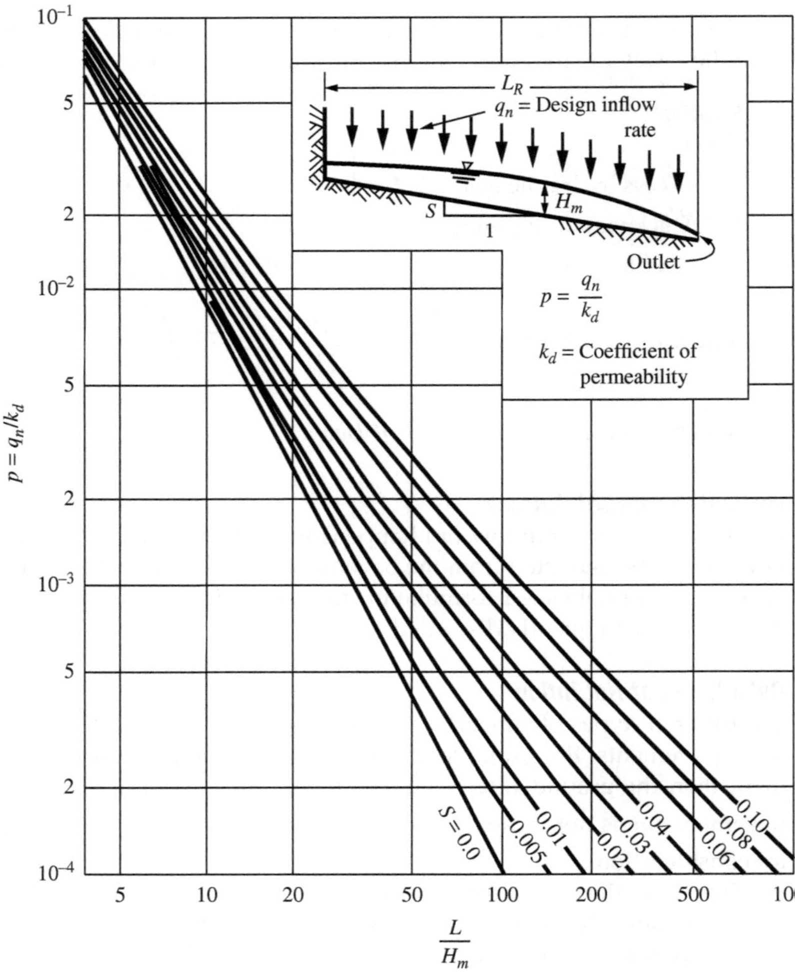

FIGURE 16.16 Chart for estimating maximum depth of flow caused by steady inflow (Moulton, 1980).

length of drainage (L_R), and rate of uniform net inflow (q_n). In other words, it is possible to determine the required coefficient of permeability k_d of the base/drainage layer if the maximum depth of flow, H_m, and the other parameters are known.

Example 16.11: Thickness Design of Permeable Base A pavement system uses a permeable base with permeability of 1,800 ft/day and the slope of 4.0%. The length of the flow path is 60 ft. The subsurface has the following inflows:

$q_a = 0.018$ ft³/day/ft²

$q_i = 1.263$ ft³/day/ft²

$q_g = 0.96$ ft³/day/ft²

Design the thickness of drainage base layer.

Solution

Given:

$q_a = 0.018$ ft³/day/ft²

$q_i = 1.263$ ft³/day/ft²

$q_g = 0.96$ ft³/day/ft²

Permeability, $k_d = 1{,}800$ ft/day

Length of drainage, $L_R = 60$ ft

Slope, $S = 4\% = 0.04$

The net inflow rate (q_n) is the largest of the following combination:

$q_n = q_i = 1.263$ ft³/day/ft²

$q_n = q_i + q_g = 1.263 + 0.96 = 2.223$ ft³/day/ft²

$q_n = q_i + q_a = 1.263 + 0.018 = 1.281$ ft³/day/ft²

$q_n = q_i + q_m = 1.263 + 0 = 1.263$ ft³/day/ft²

Therefore, $q_n = 2.223$ ft³/day/ft² (the largest one)

Parameter, $p = q_n/k_d = 2.223/1{,}800 = 0.0012 = 1.2 \times 10^{-3}$

From Fig. 16.16, for $p = 1.2 \times 10^{-3}$ and $S = 0.04$, $L_R/H_m \approx 70$

Therefore, $H_m = L_R/70 = 60$ ft/70 = 0.857 ft = 10.28 in. (say, 10.5 in. to round off)

Answer *The thickness of base layer is 10.50 in.*

16.3.4 Materials Requirements for Permeable Base

The most important influence on permeability is the gradation of aggregates comprising the permeable base. The recommended minimum coefficient of permeability is 1,000 ft/day for use in high-type highways. The gradations in Table 16.14 are considered to provide permeabilities on the order of 1,000 to 5,000 ft/day in the unstabilized form. However, the AASHTO No. 57 and AASHTO No. 67 gradations presented in the table are typically modified with either asphalt or cement. Other typical

State	Percent passing sieve size											
	2 in.	1.5 in.	1 in.	¾ in.	½ in.	⅜ in.	No. 4	No. 8	No. 16	No. 40	No. 50	No. 200
AASHTO #57		100	95–100		25–60		0–10	0–5				
AASHTO #67			100	90–100		20–55	0–10	0–5				
Iowa			100					10–35			0–15	0–6
Minnesota			100	65–100		35–70	20–45			2–10		0–3
New Jersey		100	95–100		60–80		40–55	5–25	0–12		0–5	
Pennsylvania	100			52–100		33–65	8–40		0–12			0–5

TABLE 16.14 Typical Unstabilized Permeable Base Gradations (FHWA, 1999)

State	Sieve size					
	1 in.	½ in.	⅜ in.	No. 4	No. 8	No. 200
California	100	90–100	20–45	0–10	–	0–2
Florida	100	90–100	20–45	0–10	0–5	0–2
Illinois	90–100	84–100	40–60	0–12	–	–
Kansas	100	90–100	20–45	0–10	0–5	0–2
Ohio	95–100	–	–	25–60	0–10	–
Texas	100	95–100	20–45	0–15	0–5	2–4
Wisconsin	95–100	80–95	25–50	35–60	20–45	3–10
Wyoming	90–100	20–50	–	20–50	10–30	0–4

TABLE 16.15 Typical Asphalt-Stabilized Permeable Base Gradation (Percent Passing) (FHWA, 1999)

State	Sieve size						
	1½ in.	1 in.	¾ in.	½ in.	⅜ in.	No. 4	No. 8
California	100	88–100	$X \pm 15$		$X \pm 15$	0–16	0–6
Virgina		100		25–60		0–10	0–5
Wisconsin		100	90–100		20–55	0–10	0–5

TABLE 16.16 Typical Cement-Stabilized Permeable Base Gradation (Percent Passing) (FHWA, 1999)

gradations used in asphalt-stabilized applications are presented in Table 16.15, and those used in cement-stabilized applications are presented in Table 16.16.

16.3.5 Design of Separator Layer

The separation layer means a small layer used between the permeable base and the subgrade so that fines from the permeable base are not clogged. If clogging occurs, the permeability of the permeable base is dramatically reduced and the drainage layer is less effective. This clogging can be avoided if the coarse material includes other filter materials, granular materials, or a suitable geotextile. The separator layer is required to perform the following:

1. Prevent fines from pumping up from the subgrade into the permeable base
2. Provide a stable platform to facilitate the construction of the permeable base and other overlying layers
3. Provide a shield to deflect infiltrated water over to its edge-drain, thereby providing protection for the subgrade, and
4. Distribute live loads to the subgrade without excessive deflection. Only an aggregate separator layer can satisfactorily accomplish (2) and (4)

The granular separator layer is preferred to the fabric since the granular layer provides the construction platform and distribution of loads to the subgrade. When geotextiles are used as separator layers, they are most often used about stabilized subgrades, which provide the construction platform and load distribution. Both granular and geotextile materials can prevent pumping of fines if they satisfy the filter requirements. The thickness of the granular separator ranges from 4 to 12 in.

Aggregate Separator Layer

Separator layer is designed based on the filter requirements, not the permeability requirements. In general rule, the maximum percent of fines passing the No. 200 sieve should not exceed 12%, and the coefficient of uniformity (C_u) should be between 20 and 40. The aggregate separator layer should consist of durable, crushed, angular aggregate. Typical gradation of dense-graded separator layer is presented in Table 16.17.

Separator layer must follow the following filter requirements to ensure that the separator layer prevents subgrade fines migrating up, and to ensure that fines in the separator layer do not move into the permeable base:

At the separator and subgrade interface:

$$D_{15}\text{ (separator layer)} \leq 5\,D_{85}\text{ (subgrade)} \tag{16.20}$$

$$D_{50}\text{ (separator layer)} \leq 25\,D_{50}\text{ (subgrade)} \tag{16.21}$$

At the separator and permeable base:

$$D_{15}\text{ (base)} \leq 5\,D_{85}\text{ (separator layer)} \tag{16.22}$$

$$D_{50}\text{ (base)} \leq 25\,D_{50}\text{ (separator layer)} \tag{16.23}$$

where D_i is the i-th percent size on the grain-size distribution curve of the soil.

Sieve size	Percent passing
1.5 in.	100
¾ in.	95–100
No. 4	50–80
No. 40	20–35
No. 200	5–12

TABLE 16.17 Typical Gradation Requirement for Separator Layer

Geotextile Separator Layer

For geotextiles as separator layers, the following criteria is recommended where steady-state flow is anticipated:

$$\text{AOS or } O_{95}\text{ (geotextile)} \leq B\, D_{85}\text{ (soil)} \tag{16.24}$$

The coefficient B ranges from 0.5 to 2.0 and is a function of the soil type to be filtered, soil density, uniformity coefficient C_u if soil is granular, and geotextile type (woven or nonwoven). For sands, gravelly sands, silty sands, and clayey sands (less than 50% passing 0.075 mm), B is a function of C_u. For:

- $C_u \leq 2$ or ≤ 8, $B = 1$
- $2 \leq C_u \leq 4$, $B = 0.5\,C_u$
- $4 < C_u < 8$, $B = 8/C_u$

where AOS = Apparent opening size, mm
O_{95} = Opening size in the geotextile for which 95% are smaller, mm AOS $\approx O_{95}$
B = Dimensionless coefficient
D_{85} = Soil particle size for which 85% are smaller, mm
$C_u = D_{60}/D_{10}$

Sandy soils, which are not uniform, tend to bridge across the openings; thus, the larger pores may actually be up to twice as large as the larger soil particles because two particles cannot pass through the same hole at the same time. Therefore, use of $B = 1$ would be conservative for retention. For silts and clays, B is a function of the type of geotextile:

- For woven geotextiles $B = 1$; $O_{95} \leq D_{85}$
- For nonwoven geotextiles $B = 1.8$; $O_{95} \leq 1.8\,D_{85}$
- For both AOS or $O_{95} \leq 0.3$ mm

To ensure that the geotextile will survive the construction process, certain minimum geotextile strength and endurance properties are required. These minimum requirements are presented in Table 16.18. Note that these values are not based on any systematic

Property	ASTM test method	Units	Geotextile class 2[1] Elongation[2] <50%	Geotextile class 2[1] Elongation[2] >50%
Grab strength	D 4632	N	1,100	700
Sewn seam strength[3]	D 4632	N	990	630
Tear strength	D 4533	N	400	250
Puncture strength	D 4833	N	400	250
Burst strength	D 3786	kPa	2,700	1,300

[1] The geotextile class is based on the AASHTO M 288 specification.
[2] As measured in accordance with ASTM D 4632.
[3] When seams are required.

Table 16.18 Geotextile Strength Property Requirements for Drainage Geotextiles (Holtz et al., 1998).

research but were derived from empirical observations of satisfactorily performing geotextiles used in drainage applications. They are not intended to replace site-specific evaluation, testing, and design.

16.3.6 Design of Longitudinal Collector Pipes

Longitudinal collector pipes or edge drains are commonly circular pipes and are usually constructed of porous concrete, perforated corrugated metal, or vitrified clay. The pipes are laid in trenches located at required depths that will allow the drainage of the subsurface water from the pavement structure. The trenches are then backfilled with granular material to facilitate free flow of the subsurface water into the drains. Longitudinal collector design involves determining the size of the pipe, the location, and the identification of appropriate backfill material.

Design

Edge-drains can be designed for the following:

- Pavement infiltration flow rate
- Peak flow from the permeable base
- Average flow rate during the time to drain the permeable base

For design based on the pavement infiltration flow rate, the design flow capacity of the edge-drain is:

$$Q = Q_p L_o = (q_i W) L_o \tag{16.25}$$

For design based on peak flow, the design flow capacity is:

$$Q = Q_p L_o = (kS_x H) L_o \tag{16.26}$$

For design based on average flow to drain the permeable base, the design flow capacity is:

$$Q = Q_p L_o = [WHn_e U(24/t)] L_o \tag{16.27}$$

where Q = Pipe flow capacity, m³/day (ft³/day)
Q_p = Design pavement discharge rate, m³/day/m (ft³/day/ft)
L_o = Outlet spacing, m (ft)
q_i = Pavement infiltration, m³/day/m² (ft³/day/ft²)
W = Width of the granular layer, m (ft)
k = Permeability of granular layer, m³/day (ft³/day)
S_x = Transverse slope, m/m (ft/ft)
H = Thickness of granular layer, m (ft)
n_e = Effective porosity of granular material
U = Percent drainage (as 1% = 0.01)
t = Time for drainage, U, to be reached, days

In designing the edge-drain, it is usually assumed that all the flow is to be handled by the pipe. For systems having very permeable backfill in the edge-drain trench, this results in conservative designs that can provide a factor of safety against settlement of material in the pipes. The capacity of a pipe edge-drain is computed using the Manning's equation:

$$Q = \frac{K}{n} S^{1/2} R^{2/3} A \tag{16.28}$$

where Q = Pipe flow capacity
$K = 1$
n = Pipe coefficient of roughness
S = Slope of pipe
R = Hydraulic radius = A/P
A = Pipe cross-sectional area
P = Wetted perimeter

Filter Requirements for Edge-Drains

The filter material must be sufficiently fine to prevent the adjacent soil from piping or flowing into the edge-drains but sufficiently coarse to enable water to pass without significant resistance. It may be appropriate to use several different aggregates, one adjacent to the other, to satisfy the filter requirements. It is difficult to construct this procedure without contamination and can be replaced with geotextiles.

Geotextiles can be used as a trench drain cover, a pipe drain wrapping, or as a permeable bases filter to satisfy the filter requirements; a geotextile can be chosen to replace the aggregate filter. Because of the relative ease of installation (in contrast with the difficulty of installing a filter aggregate and a coarse aggregate in separate layers without contamination), the use of geotextiles can be more cost-effective.

The filter requirements for aggregates (clogging/permeability criterion) were originally developed based on the ratio of the filter material's grain size to the soil size. The obstruction criterion ensures that the filter material is sufficiently fine to prevent the adjacent finer material from piping or flowing into the filter material, and the permeability criterion ensures that the material is sufficiently coarse to allow water to flow.

Prefabricated Edge-Drains

The requirements for prefabricated edge-drain filters are listed below:

- The core must be able to sustain a certain amount of stress.
- The core must have a required flow rate.
- The geotextile must be able to pass through the flow.
- The geotextile must be able to retain the adjacent soil.
- The geotextile must sustain the normal stress between core protrusion locations.

Design of Outlet Drain

The outlet drain design consists mainly of measuring the outlet pipe's capacity to ensure that this pipe's size is at least as large as the edge-drain. The slope is dictated by the geometry of the roadway, but should not be less than 3% if possible. The efficiency of the outlet pipes is regulated using the formula of Manning.

Retrofit Edge-Drains

There are differences of opinion about the effectiveness and design of various remedial pavement drainage techniques. Such variations may be due to the lack of sufficient evidence to substantiate the different views and the many variables that influence the drainage of the pavement. The design and effectiveness criteria of edge-drains can be described as follows:

- The edge-drain should be directly adjacent to the pavement/shoulder joint under the shoulder.
- By eliminating the filter fabric at the subbase–edge-drain interface, eroded fines cannot clog the filter fabric.
- Trench backfill should be permeable enough to transmit water to the longitudinal edge-drain pipe.
- The most commonly used trench width was 12 in.; locating the top of the pipe at the bottom of the layer to be drained was recommended.
- The spacing of the outlet should not exceed 500 ft; additional outlets at the bottom of the sag vertical curves should be given.
- Because of the tendency of flexible corrugated plastic pipe to curl, it was suggested to use rigid PVC pipe for lateral outlets; rigid PVC pipe helps maintain the appropriate grade of outlet pipe and protects against crushing.
- Headwalls prevent damage to the outlet pipe, prevent slope erosion, and encourage the location of the outlet pipe.

Measures showed that 25% to 45% of the rainfall infiltrated the pavement and passed through the edge-drains of the retrofit. It was found that after installation of the system, water flow from the pavement joints and cracks stopped. Surface seepage and frost heave issues did not occur after installation of the subdrainage.

16.3.7 DRIP Software

The original version of the microcomputer software "Drainage Requirements in Pavements (DRIP) version 1.0" was developed by Applied Research Associates, Inc. (ARA) under a contract from the FHWA. DRIP was finalized in 1997. The system was updated and enhanced in 2000 by ARA's ERES Department. Figure 16.17 displays the sample of DRIP program.

Each of the following features can be executed independently from within the program shown in Fig. 16.17:

- Roadway geometry calculations
- Sieve analysis calculations
- Inflow calculations
- Permeable base design
- Separator layer design
- Edge-drain design

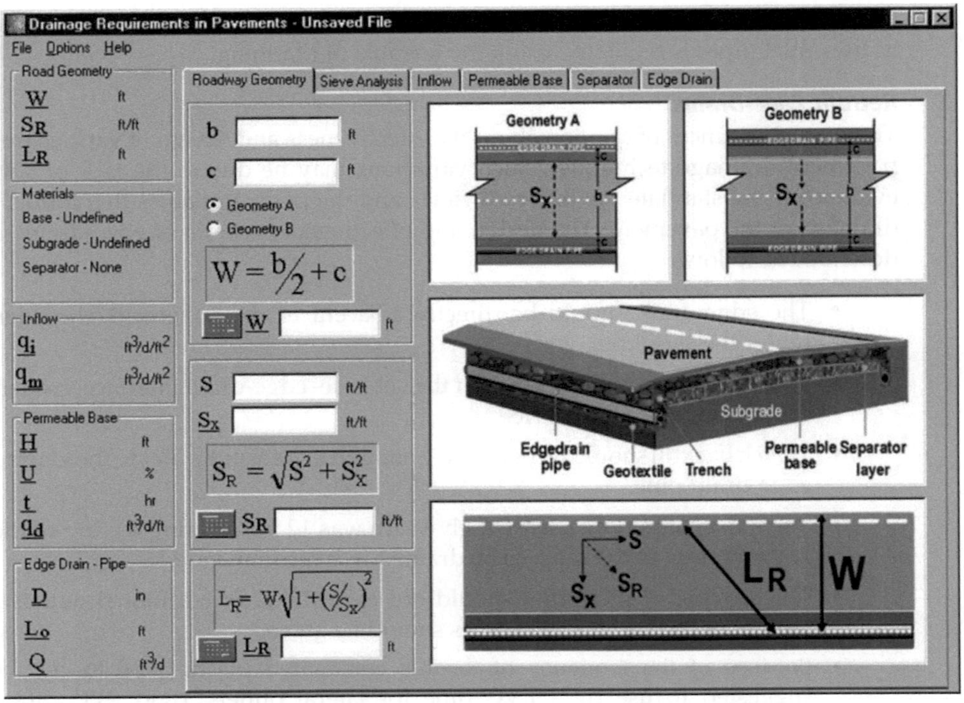

FIGURE 16.17 The DRIP client window.

16.4 Summary

This chapter discusses the drainage properties, requirement, and drainage design of pavement. Pavement gets water broadly in two ways: surface water and subsurface water. The surface water comes from rainfall, snow melting, etc. and can be drained by providing transverse slope, longitudinal slope, longitudinal channel, curbs, gutter, etc. Two methods are very popular in calculating the surface runoff: rational method and U.S. SCS method. Rational method is one of the most commonly used procedures for calculating peak flows from small drainages less than 200 acres. This method is the most accurate for runoff estimates from small drainages with lots of impervious area. U.S. SCS, now known as the Natural Resources Conservation Service (NCRS), developed this method for relatively large drainage areas.

The subsurface water comes from the infiltration of a minor part of the surface water, frost melting (thawing of ice lenses), and seepage of groundwater. The underlying pavement materials must be designed to discharge the subsurface water laterally out of pavement. The major steps of subbase drainage design are quantifying water inflow, designing the permeable base, designing the separation layer, quantifying flow to edge-drains, computing outlet spacing, and checking outlet flow. The FHWA-developed DRIP software can also be used to perform the subsurface drainage design.

16.5 Fundamentals of Engineering (FE) Exam-Style Questions

FE16.1 Which of the following cannot be used to drain surface water from the pavement?

A. Transverse slope
B. Geotextile layer
C. Providing longitudinal channel
D. Curbs and gutter

Solution B

Geotextile layer is used in subsurface drainage system.

FE16.2 Rational method is applicable for calculating peak flows from drainage area:

A. Less than 200 acres
B. More than 200 acres
C. Of any size of watershed
D. Of negligible slopes

Solution A

Rational method is one of the most commonly used procedures for calculating peak flows from small drainages less than 200 acres.

FE16.3 Which of the following is a source of subsurface water in pavement? (Check all that apply)

A. Surface infiltration
B. Frost meltwater
C. Groundwater under gravity
D. Groundwater from artesian flow

Solutions A, B, C, and D

16.6 Practice Problems

16.1 A 150-acre watershed has three distinct land types with runoff coefficient as shown in Table P16.1. The 25-year rainfall intensity is 2 in./h. Using the rational method, calculate the net runoff from this watershed.

Type	Area (acre)	Runoff coefficients
Area 1	40	0.3
Area 2	80	0.1
Area 3	30	0.05

TABLE P16.1 Runoff Coefficients Data for Prob. 16.1

16.2 A 100-acre rural drainage area in Zone 1 of Oregon consists of four different watershed areas, as shown in Table P16.2.

Determine the peak flow for a 100-year time interval using the rational method.

Segment	Land type	Land slope	Segment area	Distance to the discharge point (ft)
1	Pavement	Flat—1.0%	35%	4,000
2	Earth shoulder	Rolling—5%	15%	2,000
3	Drives and walks	Hilly—20%	40%	3,000
4	Gravel pavement	Rolling—3%	10%	5,000

Table P16.2 Watershed Data for Prob. 16.2

Segment	Land type	HSG information	Segment area	Distance to the discharge point (ft)
1	Urban districts with 85% impervious	Slow infiltration	45%	1,000
2	Paved area with open ditches	High infiltration	55%	1,800

Table P16.4 Watershed Data for Prob. 16.4

16.3 Using the NRCS method, determine the depth of runoff for a 24-h, 100-year precipitation event of 9 in. (228.6 mm) if the soil can be classified as group B and the watershed is paved roadway including right-of-way and an antecedent soil condition III.

16.4 A 250-acre rural drainage area in Zone 1 of Oregon consists of four different watershed areas, as shown in Table P16.4.

The design rainfall for a 100-year storm event is 8.0 in. Five-day antecedent rainfall in dormant season is 1.5 in. The average land slope is about 2%. One acre equals 0.0015625 mile^2.

Determine the peak flow for a 100-year time interval using the TR-55 method.

16.5 A concrete pavement is having a rainfall of 15 mm/h. Determine the amount of infiltration using the infiltration ratio method.

16.6 Determine the groundwater infiltration rate for a new two-lane pavement using the crack infiltration method with the following characteristics:

Lane width = 12 ft
Shoulder width = 10 ft
Length of contributing transverse cracks (W_C) = 20 ft
Rate of infiltration (K_p) = 0.015 ft^3/day/ft^2
Spacing of transverse cracks = 30 ft

16.7 The subgrade material of a pavement is silty and gravelly sands with the permeability of 0.07 ft/day. The pavement has a 10-in. AC layer and 12 in. of aggregate base layer. Calculate the flow rate due to the melted ice lenses in the subgrade. Assume that the unit weight of AC is 145 pcf and unit weight of aggregate base layer is 130 pcf.

16.8 A pavement is built cutting the natural ground and the groundwater table is above the perforated drainage pipe, as shown in Fig. P16.8. Due to this construction, the water-table drawdown is 8 ft and the bedrock is located 40 ft below the original water table. The permeability of the pavement materials used is about 0.5 ft/day. There is only one collector pipe in this pavement.

Calculate the following:

a. Inflow from above the bottom of the drainage layer
b. The seepage inflow from the bottom of the drainage layer
c. The total lateral inflow into the drainage pipe
d. The groundwater inflow into the drainage layer per unit area

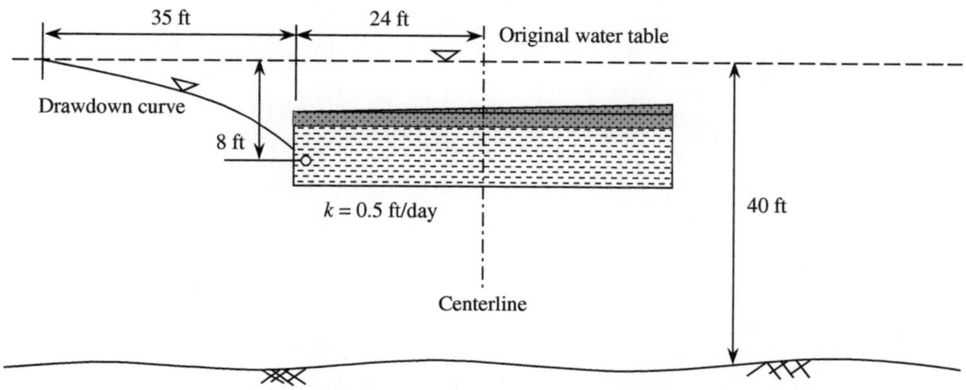

FIGURE P16.8 Profile of a pavement section for Prob. 16.8.

16.9 An artesian aquifer is located 20 ft below the drainage outlet of a pavement. The piezometric level in the artesian aquifer is 3 ft above the drainage outlet, as shown in Fig. P16.9.

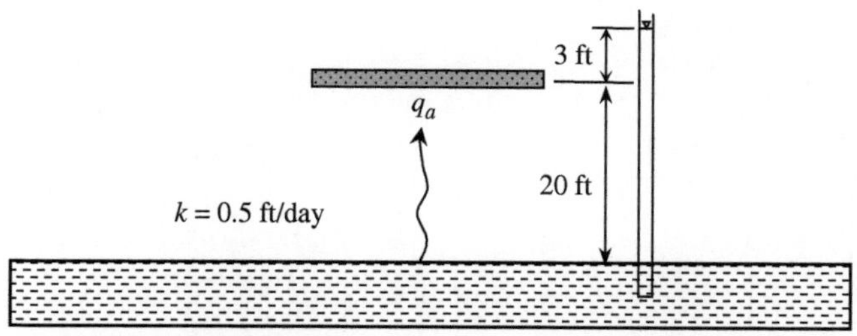

FIGURE P16.9 Artesian aquifer for Prob. 16.9.

The coefficient of permeability of the soil below the pavement is 0.5 ft/day. Determine the inflow water to the pavement structure due to the presence of this artesian aquifer.

16.10 A pavement system uses an 8-in.-thick permeable base with the slope of 2.0%. The length of the flow path is 40 ft. The subsurface has the following inflows:

$q_i = 1.263$ ft³/day/ft²
$q_m = 0.96$ ft³/day/ft²

Design the permeability requirement of the base layer.

CHAPTER 17

Sustainable Pavement Design

17.1 Concept of Sustainability

Sustainability is the concept of satisfying the needs of the current demand without compromising potential consumers' ability to meet their own needs, taking into account the three main components: economic, environmental, and social impacts. The significance of each of these factors depends on the goals, demands, characteristics location, materials, and constraints of a given project.

17.2 Role of Pavement in Sustainability

Sustainability in pavement refers to a pavement's ability to (Van Dam et al., 2015):

- Attain the engineering requirements in the pavement structures and materials.
- Preserve and improve surrounding ecosystems.
- Economical use of resources.
- Provide satisfaction to users.

The United States has over 4 million miles of public roads (FHWA, 2013a). Nearly 3 trillion vehicle miles traveled (VMT) over these roadways were logged in 2010, consuming over 169 billion gallons of fuel in the process (FHWA, 2010a). Total highway investment in the United States in 2008 was $182.1 billion (FHWA, 2010a). A variety of vehicles (cars, trucks, buses, bicycles) and users (commuters, commercial motor carriers, suppliers and service providers, local users, leisure travelers) benefit from sustainable pavements. Below are some examples of how pavements can affect sustainability:

- Environmental benefits by saving energy consumption, greenhouse gas emissions, noise, air quality, water treatment, and so on
- Social benefits by improving safety (fatalities, injury, damage to property), smoothness, vehicle operating costs, access mobility, aesthetics, etc.
- Economic benefits from reducing costs of construction, repair and rehabilitation, operating costs of vehicles, accident costs, etc.

According to AASHTO (2009), the aim of the transport system is no longer merely to move people and goods but to achieve sustainability in economic social and environmental sustainability. Transportation helps and enhances the quality of our lives.

17.3 Pavement Life Cycle

To better understand the effects of pavements on sustainability, it is useful to divide a pavement's life cycle into several significant phases as shown in Fig. 17.1.

Material production. This applies to all processes involved in the procurement of pavement products (e.g., mining, crude oil extraction) and processing (e.g., refining, manufacturing, mixing). Materials production affects such sustainability factors as air and water quality, health of the environment, human health and safety, depletion of non-renewable resources, and the life-cycle costs.

Pavement design. This refers to the process of defining a pavement's structural and functional requirements for specific site conditions (subgrade, climate, existing pavement structure, traffic loadings) and then determining the structural composition of the pavement and accompanying materials. Structural design affects sustainability factors such as performance life, longevity, cost of the life cycle, construction (e.g., constructability, sequencing schedule), and material use.

Construction. Pavement construction refers to initial construction and subsequent maintenance and rehabilitation efforts. Construction activities affect sustainability factors such as air and water quality, human health and safety, durability and traffic disruptions in the working area, as well as project costs and time.

Use. Pavement use refers to many key pavement factors (e.g., roughness, viscoelastic energy dissipation, deflection, macrotexture) that can have a significant impact on most sustainability metrics including fuel economy, vehicle operating costs, and related greenhouse gas emissions and energy consumption. Environmental interactions (e.g., stormwater disposal, heat/conductivity, and reflectivity) may also have an impact on

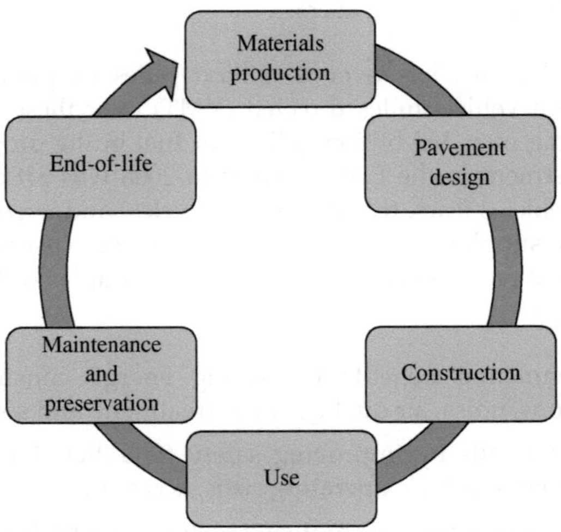

Figure 17.1 Pavement life cycle.

other sustainability factors such as human health and safety, the effect of urban heat islands, and global scale radiative forcing.

Maintenance and preservation. Pavement maintenance and preservation refers to actions that help slow a pavement's rate of deterioration by identifying and addressing specific deficiencies in the pavement that contribute to overall deterioration. Maintenance and preservation have an impact on sustainability factors such as performance life, reliability, life-cycle costs, construction (e.g., constructability, sequencing scheduling), and material usage. Appendix B addresses different maintenance and preservation techniques.

End-of-life. Pavement end-of-life refers to recycling any part of a pavement system that has reached the end of its useful life. Considerations of end-of-life affect sustainability factors such as waste generation and disposal quality of air and water, and use of materials. Appendix C addresses different recycling techniques.

17.4 Materials Considerations for Sustainability

17.4.1 Aggregates

Aggregates are relatively cheaper and have a low environmental impact compared with other pavement materials. Aggregates, however, constitute the more significant part of the structure of the pavement. Because aggregates are consumed in such large quantities, they can have a significant impact on the sustainability of the pavement. For example, in 2012, the United States produced about 1,324 million tons of crushed stone worth about $12 billion. Of the total crushed stone, 82% were used as construction material, mostly for road construction and maintenance. Approximately 927 million tons of sand and gravel worth about $6.4 billion have been produced for sand and gravel, of which 26% for road base, road covering, and road stabilization and 12% for asphalt concrete aggregates and other asphalt aggregate products (USGS, 2013a). Reducing the use of raw materials and using locally available materials and recycled materials would improve overall sustainability. Some strategies for the adoption of sustainability in pavements for aggregate use are shown in Fig. 17.2.

17.4.2 Asphaltic Materials

Asphalt binder is used to manufacture asphalt mix between 3% and 6% of the total mix. The United States used about 130 million barrels (23 million tons) of asphalt binder and road oil in 2011, worth $7.7 billion (EIA, 2011). The amount of U.S.-produced asphalt paving mixtures was estimated at $11.5 billion in 2007 (United States Census Bureau, 2007a). This data shows the importance of asphalt binder in the sustainability of pavements. In recent years, asphalt technology has greatly improved with the use of reclaimed asphalt pavement (RAP) and recycled asphalt shingles (RAS). Also, polymer and rubber are used to improve binder properties. Some other sustainable practices, such as the use of warm-mix and cold-mix asphalt mixtures, are used to reduce energy consumption and greenhouse gas emissions, reduce the impacts of transporting materials, etc. Some strategies for improving asphalt materials production are listed below.

- Reduce the amount of virgin asphalt binder and virgin aggregate in asphalt concrete by plant recycling
- Reduce energy needed and emissions from mixing asphalt concrete

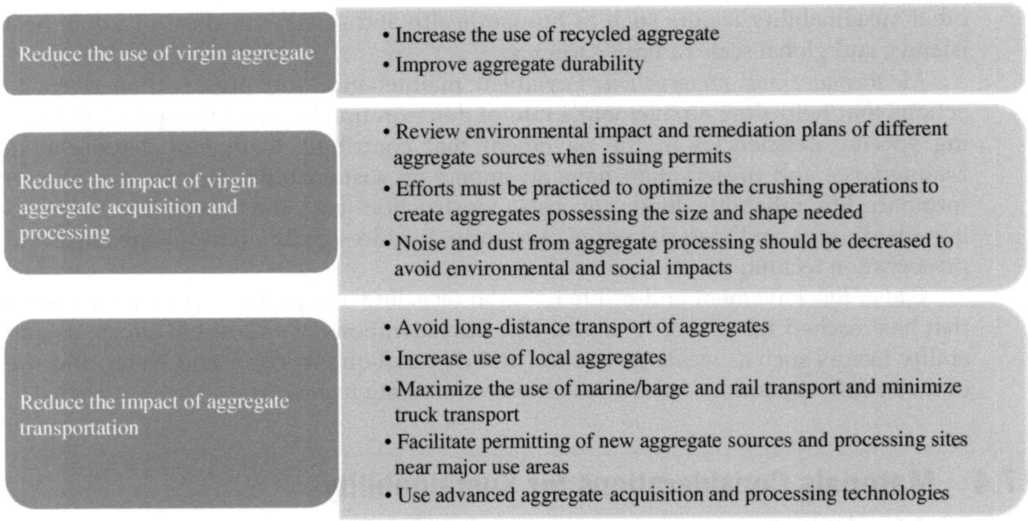

FIGURE 17.2 Strategies to adopt sustainability in using aggregates in pavements.

- Extend the life of asphalt concrete materials
- Reduce the need for virgin materials and transportation through in-place recycling
- Develop alternatives to petroleum-based binders

17.4.3 Concrete Materials

According to the U.S. Geological Survey, the United States used about 111 million tons of hydraulic cement in 2005, worth about $9.1 billion. Approximately 5% of cement is used for road paving purposes in the United States in 2011 (USGS, 2013b). The United States has 5,500 ready-mixed concrete plants in 2011 (United States Census Bureau, 2013). The value of ready-mixed concrete manufactured in the United States in 2007 was estimated at $34.7 billion, indicating that the value of concrete used for road paving was approximately $1.7 billion based on 5% of cement used (United States Census Bureau, 2007b). However, portland cement production uses huge amounts of energy and produces greenhouse gasses. Reductions can be made in these energy consumption and emission levels by reducing the use of portland cement in paving mixtures. Several techniques for concrete paving are listed in Table 17.1.

17.4.4 Other Materials

Steel reinforcing fibers, geosynthetics, etc. are some other materials used in pavement steel production consumes a significant amount of energy and emits greenhouse gasses. It is even worse to make steel from raw materials. Steel production using electric arc furnaces has less impact on the environment as it produces steel from recycled steel. To increase the strength of concrete, fibers are used as reinforcement. This reduces the concrete slab thicknesses and extends the spacing of the joint. Some

Strategy	Sustainability improving approach
Reduce non-renewable energy consumption and greenhouse gas emissions in cement manufacturing	Improve cement plant efficiency through better energy harvesting and improved grinding
	Utilize renewable energy including wind and solar
	Utilize more efficient fossil fuels
	Utilize waste fuels
	Utilize biofuels
	Minimize clinker content in portland cement using limestone and inorganic processing
	Increase production of blended cement or supplementary cementitious materials (SCMs)
Reduce energy consumption and emission in concrete production	Increase concrete mixing plant efficiency and reduce emissions
	Utilize renewable energy
	Use electrical energy from the grid
	Use less cement in concrete mixtures without compromising performance
	Use more blended cements without compromising performance
	Increase addition rate of SCMs at concrete plant without compromising performance
Reduce water use in concrete production	Recycle washout water
	Recycle water used to process aggregates
Increase the use of recycled aggregates in concrete	Change specifications to allow more significant amounts of recycled, co-product or waste materials to be used in concrete without compromising performance
	Use RCWMs and marginal aggregates in lower-lift of two-lift pavement
Improve the durability of concrete	Lower water use through admixture use
	Utilize an effective quality assurance program throughout material production

TABLE 17.1 Strategies for Improving Pavement Sustainability for Concretes

fibers can reduce plastic cracking and cracking severity. Geosynthetics can reinforce soil and unstabilized subbase and base materials to minimize the thicknesses necessary for other layers of pavement. It can also reduce or monitor reflection cracking development.

17.5 Rehabilitation Design for Sustainability

There is a greater opportunity to consider alternate materials, pavement structures, and construction procedures with the introduction of the AASHTOWare pavement ME design approach. As discussed earlier in this book, AASHTOWare pavement ME design allows to evaluate changing input parameters in a matter of minutes to assess their effects on the final design. After trying various alternative approaches, it is possible to select the most sustainable design. The implementation of the AASHTOWare pavement ME design procedure thus contributes to the overall sustainability of the

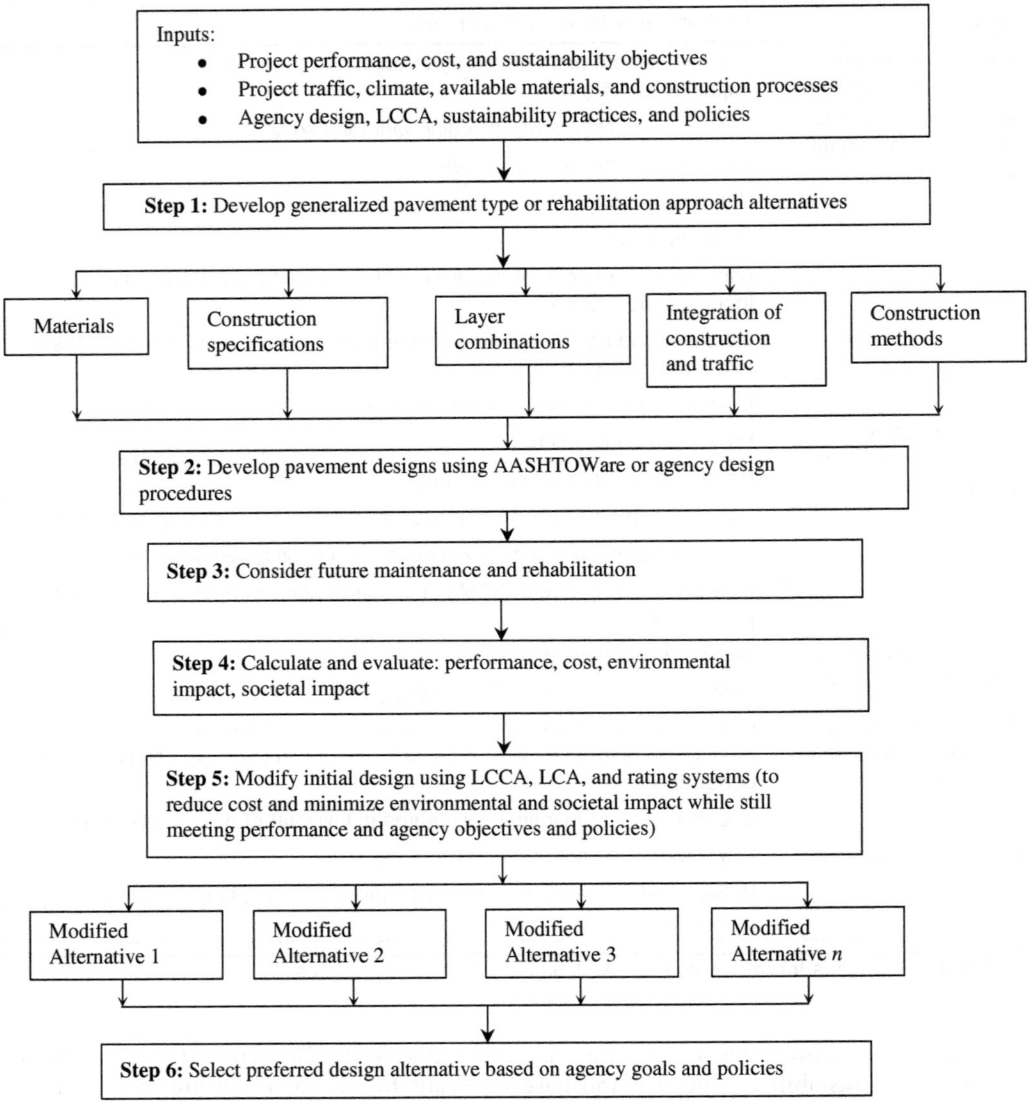

FIGURE 17.3 Process for considering sustainability in pavement design. (*From Van Dam, T. J., Harvey, J. T., Muench, S. T., Smith, K. D., Snyder, M. B., Al-Qadi, I. L., Ozer, H., et al. (2015). Towards Sustainable Pavement Systems: A Reference Document. Report FHWA-HIF-15-002. Washington, DC: Federal Highway Administration (FHWA).*)

resulting design. Some other strategies for adopting sustainability in pavement design include optimizing the use of materials and cross sections, and evaluating pavement designs using life-cycle assessment (LCA), life-cycle cost analysis (LCCA), and rating systems to evaluate their environmental and societal impacts to improve them. The complete cycle can be summarized, as shown in Fig. 17.3.

Objectives	Sustainability improving approach
Reduce fuel consumption and emission	Minimize haul distances
	Select appropriate equipment type and size for the job
	Reduce idle time
	Use alternative fuels
	Retrofit construction equipment, use hybrid equipment, or both
Reduce noise	Introduce construction time restrictions
	Ensure equipment maintenance or modification
Accelerate construction	Implement effective traffic control and lane closure strategies
	Establish performance goals and measures for work zones
	Use computer software for construction sequencing and managing traffic delays
	Implement intelligent transportation warning systems
Control erosion, water runoff, and sedimentation	Use perimeter control barriers (fences, straw bales, etc.)
	Minimize the extent of disturbed areas
	Apply erosion control matting or blankets
	Store/stockpile away from the watercourse

Table 17.2 Strategies to Improve Sustainability of Pavement Construction Operations

17.6 Construction Considerations for Sustainability

Pavement construction affects the sustainability of a pavement construction in different ways:

- Fuel consumption during material transport
- Exhaust emissions such as carbon monoxide, carbon dioxide, sulfur oxide, nitrogen oxide, etc.
- Traffic delays and noise emissions during construction
- Constructed characteristics of the pavement surface such as surface friction, noise, and possibly fuel efficiency during the service time
- Pavement performance and overall life as a result of construction quality

Strategies to improve sustainability of general pavement construction operations are listed in Table 17.2.

17.7 Maintenance for Sustainability

Pavement maintenance or preservation is all about reducing the agency's life-cycle cost of the pavement. The agency considers only the cost of materials and pavement construction in its life cycle. The use-phase costs (primarily vehicle operating costs) are not taken into account by the agency and are mostly borne by users of pavement. Pavement maintenance or preservation uses gravel, water, cement, and asphalt combinations as construction materials and engines for combustion (e.g., transportation, removal, and application) to place.

Critical factors for selecting appropriate maintenance or preservation treatment include treatment performance history, overall performance requirements, construction constraints, LCCA, and LCA.

The sustainability value of any maintenance or preservation treatment is difficult to judge, as there are multiple factors at work. In general, treatments using the least quantity of material to maintain smoothness over the longest period have the greatest positive effect. Understanding the full impact of the life cycle is a crucial element in establishing the advantages and disadvantages of any maintenance or preservation treatment. Some of the commonly used maintenance or preservation treatments are listed below.

- Crack filling/sealing
- Asphalt patching
- Fog seals/rejuvenators
- Chip seals
- Slurry seals
- Microsurfacing
- Ultra-thin and thin asphalt concrete overlays
- Hot in-place recycling (HIR)
- Cold in-place recycling (CIR)
- Ultra-thin bonded wearing course
- Bonded concrete overlays

General approaches for enhancing sustainability include the use of thinner cross sections, the use of local materials, the maintenance of smoothness, and the increased quality of construction, all of which minimize the environmental burden and lead to more efficient treatment prospects for improving sustainability in the following areas:

- Improved maintenance materials that require the use of less content or last longer
- Improved approaches for optimizing treatment selection and timing
- Improved construction

17.8 End-of-Life Considerations for Sustainability

17.8.1 Asphalt Pavement

Asphalt pavement recycling can be done through central plant or in-place recycling techniques as follows:

- Central plant recycling (hot and cold)
- Full-depth reclamation
- Landfilling

Central plant recycling is the method of producing hot or cold asphalt mixtures in a central plant by combining virgin aggregates, new asphalt binders, recyclers, and a certain amount of RAP. RAP is milled off from old pavements, transported to plants, screened, and then used. In hot central plant recycling, heat is used to mix RAP, virgin aggregates, and new asphalt binders. Cold central plant recycling combines RAP with an emulsified asphalt/recycling agent and new aggregates, if necessary, without the use of oil. Full-depth reclamation (FDR) is a method in which the full thickness of the current asphalt pavement and the predetermined portion of the underlying materials (base, subbase, and subgrade) are uniformly pulverized and blended to produce a single material. All of these approaches are discussed in detail in Appendix C.

Asphalt and concrete pavements are usually recycled and reused as construction materials (EPA, 2009). According to industry data, less than 1% of RAP was transported to landfills in 2012, with 68.3 million tons of RAP being used in new asphalt concrete mixtures. This represents a 22% increase in the use of RAP in 2012 relative to 2009 (Hansen and Copeland, 2013). The total amount of recycled concrete used in the United States was estimated to be 140 million tons in 2014, including recycled materials from both pavements and other sources (CDRA, 2014). Such recycled materials can be used either in new asphalt or concrete mixtures, or as aggregates in base layers, or in a variety of other applications, such as filling, riprap, and ballast.

Several approaches to enhance sustainability with respect to end-of-life pavement recycling, along with related environmental benefits, are summarized as follows:

- Improve plant technology such as improved dust control systems
- Increase initial quality of pavement materials and construction, which will increase the level of performance and the overall pavement life and reduce the total cost of pavement
- Use rejuvenators or softening agents to reduce the brittleness of these materials, and reduce the cracking
- Maintain and manage rap stockpiles fractionated and moisture free to meet the design volumetrics and extended performance
- Select proper type and amount of additives or stabilizers for the expected improvement in performance and service life
- Implement structural overlays to protect the recycled layers from direct exposure to weathering and slow down the deterioration rate
- Improve construction quality to improve the long-term performance of pavements

17.8.2 Concrete Pavement

Concrete recycling includes demolishing, removing, and crushing hardened concrete from the old concrete pavement for the production of recycled concrete aggregate (RCA). The U.S. Geological Survey (USGS) reported that the cost of RCA in 2005 ranged from \$3.41/ton in New Jersey to more than \$8.09/ton in California, Louisiana, and Hawaii at an average cost of \$6.93/ton. Virgin aggregates were estimated to cost an average of \$6.52/ton, ranging from \$3.54/ton in Michigan to more than \$10.01/ton in

Mississippi and Hawaii (Van Dam et al., 2015). There are three main end-of-life choices for concrete pavement surfaces:

1. Recycling
2. Reuse
3. Landfilling

The use of recycled concrete aggregates instead of new aggregates is inherently sustainable when all other considerations are equal. The following subsections discuss strategies for enhancing the sustainability of concrete recycling by optimizing the production and use of materials, which are also summarized as follows:

- Optimizing use of recycled materials through testing and characterization helps better understanding of the RCA properties
- Adjustment of RCA production operations to reduce fuel and wastage
- Customizing preparation and breaking of source concrete for higher production, reduce waste, reduce fuel, preserve natural sources, etc.
- Customizing crushing and sizing operations to increase production, reduce waste, preserve natural sources, etc.
- Sequestration of carbon dioxide to reduce impact on climate change
- On-site processing to reduce fuel, labor, and traffic congestion

17.9 Measuring Pavement Sustainability

Sustainability measurement is the first step toward establishing benchmarks and assessing progress. The reasons for measuring sustainability can be categorized into three broad categories:

1. *Accounting*. To quantify the parameters such as cost saving and energy saving
2. *Decision support*. To take organizational or project decisions
3. *Process improvement*. To provide feedback in support of refining and updating the overall methodology

Currently, four general measurement tools, or methods, can be used to quantify sustainability:

1. Performance assessment
2. Life-cycle cost analysis (LCCA)
3. Life-cycle assessment (LCA)
4. Sustainability rating systems

17.9.1 Performance Assessment

Performance evaluation measures and assesses the performance of the pavement for its intended function and physical characteristics. Various performance parameters such as conventional distress (roughness, rutting, fatigue cracking, top-down cracking, and thermal cracking), composite condition ratings, and pavement structural capacity can be assessed.

Performance is most often assessed on the basis of current standard practice. For example, if the current standard asphalt pavement surfacing is supposed to last 10 years, the value of alternative surfacing (say, open-grade, stone matrix, or rubber asphalt) is calculated on the basis of how their expected service life corresponds to the standard 10 years (Van Dam et al., 2015). Alternatives are expected to perform in a manner that is equal to or better than the current standard practice or to produce some other economic, social, or environmental benefits.

17.9.2 Life-Cycle Cost Analysis

Life-cycle cost analysis uses an economic analysis to determine the total cost of a project over its life cycle. This offers a cost comparison between two or more similar design solutions with equivalent benefits. The initial cost of the project and all additional projected costs (such as maintenance costs, annual fees, and salvage value) are converted into current costs and added up to produce a net present value (NPV) or a net present cost (NPC). When several alternatives with similar benefits are considered over identical analysis periods, the NPVs or costs may be compared to determine the alternative that is most cost-effective (FHWA, 2013b).

The value of salvage is also included in the LCCA. It is the value of the materials required to be extracted from the pavement structure once it has expired (no remaining service life; the pavement is to be removed and replaced).

The discount rate parameter is used while using LCCA, which represents the combined effect of interest and inflation rates. The choice of discount rate is very important for the calculation of the NPV. Higher discount rates minimize the present value of future costs by more than lower discount rates. A zero discount rate means that future costs are the same as current costs. Negative discount rates increase the present value of future costs over current costs. Positive discount rates lower the present value of future costs below current costs. For example, for a positive discount rate, if you need to pay $1 after 10 years, the current cost of $1 is more than $1. Nevertheless, if you're going to get a $1 salvage value after 20 years, the present value (gain) is less than $1. It is the opposite of a negative worth that is not common.

Let's take an example for better understanding. Say there are some initial costs (materials, labor, paving, etc.) for the construction of a pavement. There will be some annual maintenance costs including crack sealing, patching, etc. once it is in operation. The recycled millings will return some interest to us after their end of life. Then you can calculate the present value as follows:

$$\text{Present value (\$)} = \text{Initial cost (\$)} + \text{Annual cost} \times (F_1)\ (\$) - \text{Salvage value} \times (F_2)\ (\$)$$

where F_1 and F_2 are two factors to convert annual cost and salvage value, respectively, into present values.

- For positive discount rate, $F_1 > 1.0$ and $F_2 < 1.0$.
- For negative discount rate, $F_1 < 1.0$ and $F_2 > 1.0$.
- For zero discount rate, $F_1 = 1.0$ and $F_2 = 1.0$.

The most popular LCCA tool for pavement applications is the FHWA's RealCost Software, originally developed in 1997 and undergone numerous improvements

(FHWA, 2010b). It is available at the FHWA website at www.fhwa.dot.gov/infrastructure asstmgmt/lccasoft.cfm.

17.9.3 Life-Cycle Assessment

Life-cycle assessment quantifies the pavement system's life-cycle environmental impacts. From a number of key environmental factors, such as energy use and greenhouse gas emissions, results are expressed. It is possible to use some other units, but they are not common. The International Organization for Standardization (ISO, 2006) states that LCA is a mechanism that discusses the environmental aspects and potential environmental impacts (e.g., resource usage and environmental consequences of releases) throughout the life cycle of a product from the procurement of raw materials through manufacturing, use, end-of-life treatment recycling, and final disposal (i.e., cradle to disposal). Nevertheless, multiple software programs (e.g., Athena, Gabi, SimaPro, PaLATE) can be used to build LCA models.

More recently, the project emissions estimator (PE-2) method was built to provide a measure of greenhouse gas emissions for pavement construction maintenance and use (Mukherjee et al., 2013). In addition, AASHTO has published a GreenDOT method calculating carbon dioxide emissions from highway agency activities, construction, and maintenance (Gallivan, Ang-Olson, and Papson, 2010).

17.9.4 Rating System

A rating system for sustainability is basically a list of best practices for sustainability with a specific associated metric commonly expressed in ratings. In a common unit, it quantifies any best practice. In this way, a specific unit (points) can be used to compare the diverse measuring units of sustainability best practices (e.g., stormwater runoff pollutant, pavement design life, amount of recycled materials, energy consumed/saved, pedestrian accessibility, environment connectivity, and art value). A rating system should count all best practices equally in its simplest form (e.g., all worth one point), in which case the rating system is a list of the number of best practices used. Rating systems assess best practices in more complex forms (usually in relation to their impact on sustainability or priority), which can help to choose the most impactful best practices to be used within a limited scope or budget. In the transportation sector, a number of national and international rating systems are currently available.

A number of rating systems relevant to pavements are INVEST, Greenroads®, Envision™, GreenLITES, etc. The FHWA created INVEST as a self-evaluation tool; version 1.0 is available at www.sustainablehighways.org. It is point-based with a focus on state Departments of Transportation (DOTs) and Metropolitan Planning Organizations (MPOs).

17.10 Summary

Sustainability is a continuous constant cycle of improvement. This chapter provides the pavement group with information on how to continue this process by considering sustainability throughout the life cycle of the pavement. In design, material selection, construction, use phase, maintenance, and recycling processes, the principle of pavement sustainability is discussed. Sustainability scopes are available in each pavement life-cycle phase. Several innovative technologies are being used to enhance the sustainability of pavement such as recycled materials, warm-mix, cold-mix, thickness optimization, and mechanical-empirical pavement design software. There are some emerging trends in the sustainability of the pavement, such as a growing understanding of the use phase.

There are several ways in which pavement sustainability can be assessed and measured, such as performance assessment LCCA, LCA, and sustainability rating systems. Most of the methods of assessment are used in pavement. However, whichever approach is used, the trend toward improving sustainability should be decided to examine and optimize the use of various sustainability parameters.

Finally, sustainability is highly context-sensitive and depends on the agency's regions, projects, resources, technologies available, and economic, environmental, and societal goals. It needs collaborations among key stakeholders, effective education and outreach, development of focused research techniques, development and implementation of useful LCA tools, etc. to introduce sustainability in pavement engineering. Van Dam et al. (2015)'s research is an excellent resource for pavement system sustainability.

17.11 Fundamentals of Engineering (FE) Exam–Style Questions

FE17.1 Sustainability consideration means:

A. Satisfying the needs of the present demand
B. Ensuring the ability of future users to satisfy their own needs
C. Both of them
D. None of them

Solution C

FE17.2 Which of the following requirements does not need to be satisfied to ensure sustainability in pavement design?

A. Attain the engineering requirements in the pavement structures and materials
B. Preserve and improve surrounding ecosystems
C. Economical use of resources
D. Provide the highest level of satisfaction to users

Solution D

All three options are true except for the last option. Providing the highest level of satisfaction to users is rarely possible considering the economic, environmental, and social constraints.

FE17.3 Some strategies for improving pavement sustainability concerning asphalt materials production are: (Check all that apply)

A. Reduce the amount of virgin asphalt and virgin aggregate by plant recycling
B. Reduce energy needed and emissions from mixing asphalt concrete
C. Extend the life of asphalt concrete materials
D. Reduce the need for virgin materials and transportation through in-place recycling
E. Use concrete pavement instead of asphalt pavement

Solutions A, B, C, and D

Usage of concrete pavement instead of asphalt pavement may not be a sustainable option as cement production is much costly and risky to environments.

17.12 Practice Problems

17.1 Explain the pavement life cycle.

17.2 List some strategies to adopt sustainability in pavements in terms of aggregate usages.

17.3 List some strategies for improving pavement sustainability with regard to asphalt materials.

17.4 List some strategies for improving pavement sustainability with regard to concrete materials.

17.5 List strategies to improve the sustainability of general pavement construction operations.

17.6 List strategies to improve sustainability with regard to pavement maintenance or preservation.

17.7 What are the options for the end-of-life considerations of asphalt pavement? Explain them.

17.8 What are the options for the end-of-life consideration of concrete pavement? Explain them.

17.9 Why is the measurement of sustainability important? Discuss the methods of measuring sustainability.

CHAPTER 18

Pavement Rehabilitation Design

18.1 Background

After design and construction of pavement, the pavement is referred to another department (commonly known as *Pavement Management System* or PMS) of highway agency that monitors the performance of the pavement, performs maintenance works, and performs subsequent operation. The pavement system undergoes deterioration and damage due to application of repeated loads and adverse climate. It thus requires continuous supervision. Based on the pavement's damage level, it may require any one of the following items:

1. Routine maintenance
2. Preventive maintenance
3. Corrective maintenance
4. Major/minor rehabilitation

Maintenance activities that are planned and performed on a routine basis to preserve the condition of highway system are defined as routine maintenance. Routine maintenance is also conducted to respond to specific conditions and events that restore the highway system to a desired level of riding comfort. Routine maintenance includes mowing, cleaning roadsides, cleaning ditches, sealing cracks in the pavement, painting pavement markings, and pruning trees.

Preventive maintenance is conducted commonly on good pavements to preserve the system without (or negligible) improving the quality but retards future deterioration, as shown in Fig. 18.1. Applications as seal coats, crack-joint filling, micro-surfacing, etc. are examples of preventive maintenance.

Corrective (or reactive) maintenance is performed in response to a deficiency or deficiencies that negatively impact the operations of the pavements, as shown in Fig. 18.1. Patching potholes may be considered a corrective maintenance as pothole is a deficiency of a pavement.

Major rehabilitation is the extension of service life and/or increasing the load-carrying capacity of an existing pavement. Minor rehabilitation is the functional improvement (riding comfort and safety) of an existing pavement with increasing the structural capacity and life, as shown in Fig. 18.1. Full-depth reclamation, thick overlay,

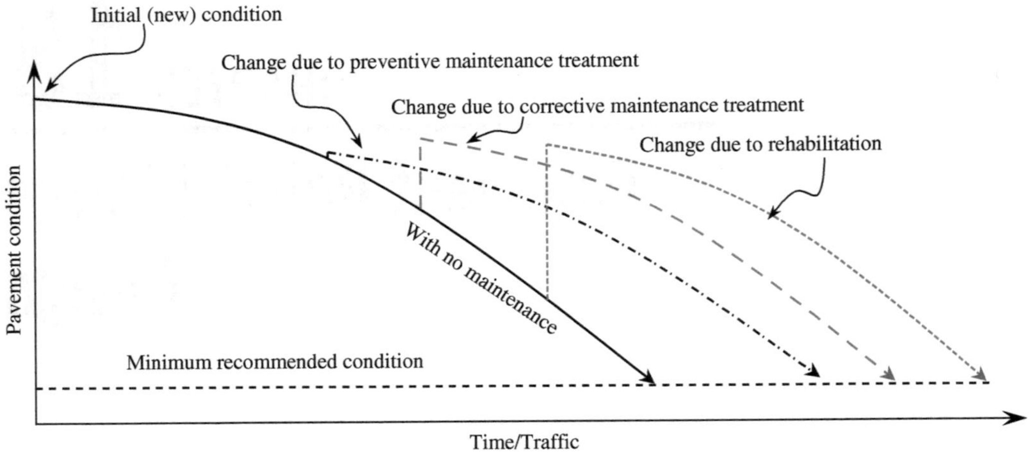

FIGURE 18.1 Pavement condition deterioration with and without maintenance.

or even reconstruction are examples of major rehabilitation, whereas applying thin overlay or surface recycling is minor rehabilitation.

Maintenance is not a part of pavement or highway design. It is performed by PMS. However, one component of PMS is rehabilitation design which is contributed by the pavement design group. Therefore, in this text, only the rehabilitation design section is discussed. For more information about PMS, readers are referred to *Pavement Management Guide* published by the AASHTO, Washington DC (latest edition is the second edition published in 2012). Appendix B of this textbook also discusses the PMS processes and their activities.

Rehabilitation design requires an evaluation of the existing pavement to provide key information. The AASHTOWare provides detailed and specific guidelines for conducting a pavement evaluation program and taking the results from that program to establish inputs to the AASHTOWare software.

It is important to note that the AASHTOWare inputs of existing pavement layers for overlay design are similar to those required for new or reconstructed pavements except that the values may be different due to depreciation or load and climate-caused deterioration of the existing layers and materials with the passage of time. The most important task in pavement evaluation is to determine the extent of damage and material properties of the in-place layers. This section provides a brief summary of the overall pavement evaluation process, followed by guidelines to obtain inputs to the AASHTOWare for use in rehabilitation design.

Steps of rehabilitation of design include:

1. Overall condition assessment
2. Fully defining condition assessment
3. Analysis of pavement evaluation data
4. AASHTOWare software analysis

18.2 Overall Condition Assessment

The first step in the pavement rehabilitation design process is to determine the overall condition of the existing pavement. Overall pavement condition could be determined by evaluating the following eight major categories of the existing pavement:

1. Structural adequacy (load related)
2. Functional adequacy (user related)
3. Subsurface drainage adequacy
4. Material durability
5. Shoulder condition
6. Extent of maintenance activities performed in the past
7. Variation of pavement condition or performance within a project
8. Miscellaneous constraints (e.g., bridge and lateral clearance and traffic control restrictions)

Three types of cases may arise:

1. If the pavement has a severe level of distresses, exceeding the threshold value, extensive field, and laboratory testing to characterize the pavement layers becomes less important.
2. If the pavement has no structural distress, field and laboratory testing become important to determine the condition of the existing pavement layers. For this case, results from the field [ground penetrating radar (GPR), falling weight deflectometer (FWD), and dynamic cone penetration (DCP) tests] and laboratory tests could be used to determine the condition of the existing layers.
3. If the pavement has a marginal level of distress, the results from the visual distress survey may be used to determine the location and frequency of the field tests and cores.

18.3 Fully Defining Condition Assessment

The steps for detailed assessment on the condition of the existing pavement for selecting a proper rehabilitation strategy are shown in Fig. 18.2. All steps to complete a detailed assessment of the pavement and individual layers are not always needed.

The fully defining condition assessment starts with the historical data, such as data collected in the overall condition assessment. Then, the field evaluation plan is made. It consists of a detailed condition survey, nondestructive testing, destructive sampling, and testing. Pavement visual surveys are performed to identify the types, magnitudes, and severities of different distresses. The visual survey needs to be performed on the pavement, adjacent shoulders, and any drainage feature along the project site. Automated distress surveys are adequate for rehabilitation design purposes, for most cases. GPR has been used successfully to determine the condition of the existing pavement structure, identify areas with subsurface voids, locate areas with severe stripping in hot-mix asphalt (HMA), and locate interfaces with weak bonds between two HMA layers.

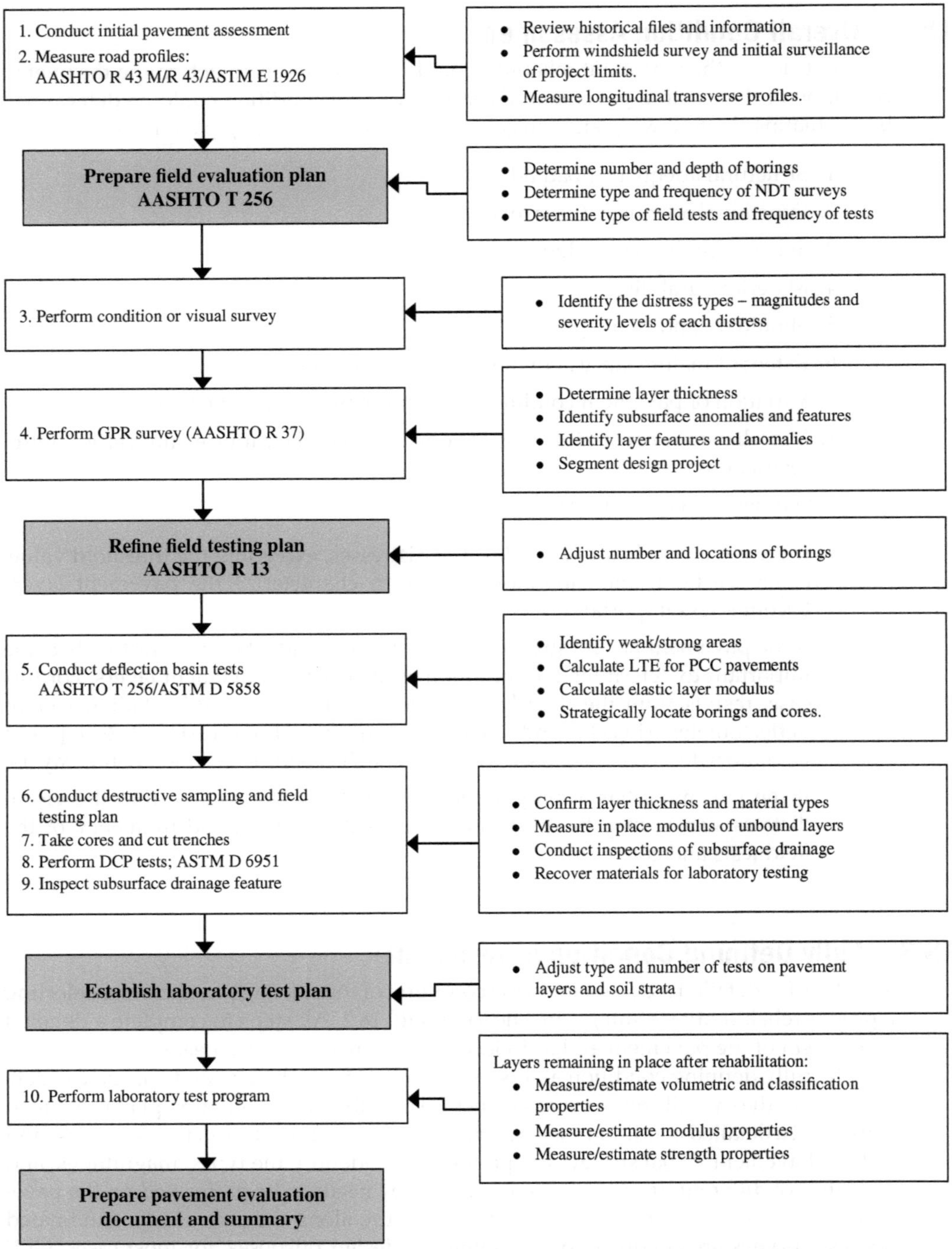

FIGURE 18.2 Steps for assessing condition of existing pavements for rehabilitation design. (*From AASHTO (2015). Mechanistic-Empirical Pavement Design Guide: A Manual of Practice. Washington, DC: American Association of State Highway and Transportation Officials. Figure 9-1. Used with permission.*)

Nondestructive testing could be performed prior to any destructive tests, such as cores and materials excavation, to better select the locations of such tests. The most widely used deflection testing device is the FWD. Destructive tests require the physical removal or damage of the pavement layer to observe the condition of the material. The cores could be used to confirm the layer thicknesses and material types, examine the pavement materials for material durability problems, and collect samples for laboratory tests. Some cores could be drilled through any cracks observed at the surface of the pavement. These cores could be used to determine the depth of cracking and whether the cracks initiated at the surface (top-down) or bottom of the asphalt layer (bottom-up). Knowing the depth and origin of cracking, a proper rehabilitation strategy for the project could be selected.

The DCP or the California bearing ratio (CBR) can be used to measure the strength of unbound layers and materials in pavement evaluations. It may also be used for estimating soil layer thickness by identifying sudden changes in strength within the pavement structure and foundation. If the existing pavement has subsurface drains that may remain in place, the outlets need to be found and inspected. Mini-camera can also be used to ensure that the edge drains and lateral lines flow freely and do not restrict the removal of water from the pavement structure. The engineer must develop an adequate laboratory test program to estimate the material properties of each layer needed as inputs into the AASHTOWare pavement ME design software.

18.4 Analysis of Pavement Evaluation Data

The pavement structural evaluation for determining the condition of the existing pavement layers is based on an analysis of the visual distress surveys, deflection basin, other field tests, and laboratory tests. It is recommended that the highest input level available be used to design high-volume roadways for rehabilitation. Once all data is at hand, the structural adequacy is evaluated as listed in Tables 18.1 and 18.2 for rigid pavement and flexible pavement, respectively.

18.5 General Overview of Rehabilitation Design Using AASHTOWare

The data obtained in the previous steps are used as inputs to design any rehabilitation option in the AASHTOWare software. The AASHTOWare pavement ME design software can evaluate a wide range of rehabilitation designs for flexible, rigid, and composite pavements. The AASHTOWare pavement ME design software starts with a trial rehabilitation strategy similar to developing the initial trial design for new pavements. A considerable amount of analysis and engineering judgment is required when determining specific treatments required to design a feasible rehabilitation strategy for a given pavement condition. The AASHTOWare pavement ME design software considers five major strategies, as listed below, which may be applied singly or in combination to obtain the most effective rehabilitation option.

1. Reconstruction without lane additions—considered under new pavement design strategies
2. Reconstruction with lane additions—considered under new pavement design strategies

Load-related distress		Inadequate (Poor)	Marginal (Fair)	Adequate (Good)
JPCP Deteriorated Cracked Slabs (medium- and high-severity transverse and longitudinal cracks and corner breaks), % slabs	Interstate, Freeway	>10	5–10	<5
	Primary	>15	8–15	<8
	Secondary	>20	10–20	<10
JRCP Deteriorated Cracked Slabs (medium- and high-severity transverse cracks and corner breaks), #/lane-mi	Interstate, Freeway	>40	15–40	<15
	Primary	>50	20–50	<20
	Secondary	>60	25–60	<25
JPCP Mean Transverse Joint/ Crack Faulting, in.	Interstate, Freeway	>0.15	0.10–0.15	<0.1
	Primary	>0.20	0.12–0.20	<0.125
	Secondary	>0.30	0.15–0.30	<0.15
CRCP Punchouts (medium and high severity), #/lane-mi.	Interstate, Freeway	>10	5–10	<5
	Primary	>15	8–15	<8
	Secondary	>20	10–20	<10

Source: *From AASHTO (2015). Mechanistic-Empirical Pavement Design Guide: A Manual of Practice. Washington, DC: American Association of State Highway and Transportation Officials. Table 9-9. Used with permission.*

Table 18.1 Distress Types and Severity Levels Recommended for Assessing Rigid Pavement Structural Adequacy

3. Structural overlay may include removal and replacement of pavement layers
4. Non-structural overlay
5. Restoration without overlays

The AASHTOWare pavement ME design software provides detailed guidelines on the use and design of rehabilitation strategies depending on the type and condition of the existing pavement and offers specific details on the use of material-specific overlays for existing flexible and rigid pavements. This section provides an overview of strategies for the rehabilitation of existing flexible, rigid, and composite pavements. Figure 18.3 shows the steps that are suggested for use in determining a preferred rehabilitation strategy.

18.6 Rehabilitation Design with HMA Overlays

The AASHTOWare includes specific details for selecting and designing HMA overlays to improve the surface condition or to increase the structural capacity of the following pavements:

Load-related distress		Inadequate (Poor)	Marginal (Fair)	Adequate (Good)
Fatigue Cracking, percent of total lane area	Interstate, Freeway	>20	5–20	<5
	Primary	>45	10–45	<10
	Secondary	>45	10–45	<10
Longitudinal Cracking in Wheel Path, ft/mi	Interstate, Freeway	>1,060	265–1,060	<265
	Primary	>2,650	530–2,650	<530
	Secondary	>2,650	530–2,650	<530
Reflection Cracking, % of total lane area	Interstate, Freeway	>20	5–20	<5
	Primary	>45	10–45	<10
	Secondary	>45	10–45	<10
Transverse Cracking Length, ft/mi	Interstate, Freeway	>800	500–800	<500
	Primary	>1,000	800–1,000	<800
	Secondary	>1,000	800–1,000	<800
Rutting, mean depth, maximum between both wheel paths, in.	Interstate, Freeway	>0.45	0.25–0.45	<0.25
	Primary	>0.60	0.35–0.60	<0.35
	Secondary	>0.80	0.40–0.80	<0.40
Shoving, percent of wheel path area	Interstate, Freeway	>10	1–10	None
	Primary	>20	10–20	<10
	Secondary	>50	20–45	<20

Source: *From AASHTO (2015). Mechanistic-Empirical Pavement Design Guide: A Manual of Practice. Washington, DC: American Association of State Highway and Transportation Officials. Table 9-10. Used with permission.*

Table 18.2 Distress Types and Levels Recommended for Assessing Current Flexible Pavement Structural Adequacy

- HMA overlays of existing HMA-surfaced pavements, both flexible and semi-rigid
- HMA overlays of existing portland cement concrete (PCC) pavements that have received fractured slab treatments: crack and seat, break and seat, and rubblization
- HMA overlays of existing intact PCC pavements [jointed plain concrete pavement (JPCP) and continuously reinforced concrete pavement (CRCP)], including composite pavements or second overlays of original PCC pavements

Figure 18.4 presents a generalized flow chart for pavement rehabilitation with HMA overlays of HMA-surfaced flexible, semi-rigid, or composite pavements, fractured PCC pavements, and intact PCC pavements.

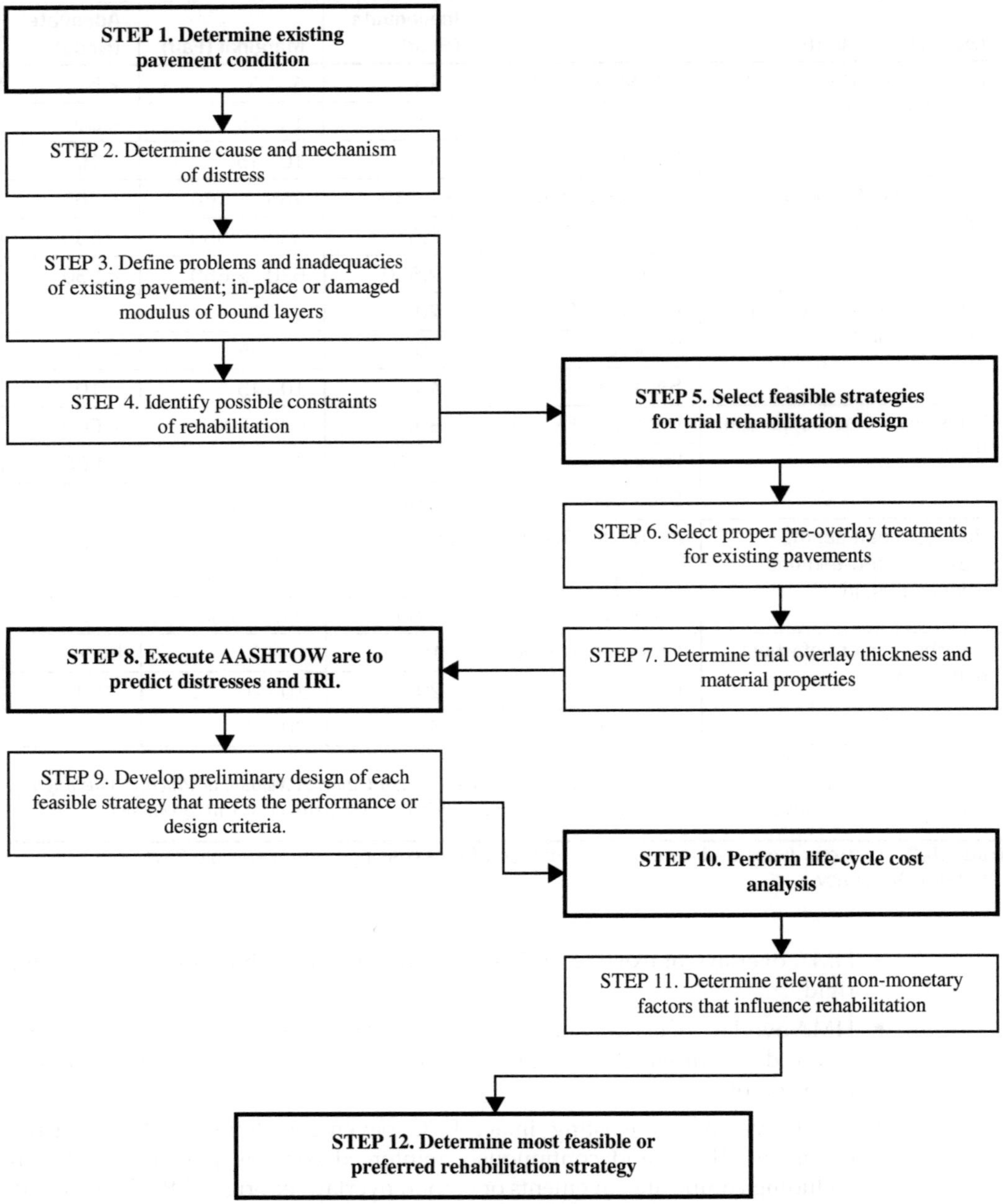

FIGURE 18.3 Steps for determining a preferred rehabilitation strategy. (*From AASHTO (2015). Mechanistic-Empirical Pavement Design Guide: A Manual of Practice. Washington, DC: American Association of State Highway and Transportation Officials. Figure 12-1. Used with permission.*)

For existing flexible or semi-rigid pavements, the designer needs to first decide on what, if any, pre-overlay treatment is needed for minimizing the effect of existing pavement distresses on the HMA overlay. Pre-overlay treatments may include the following:

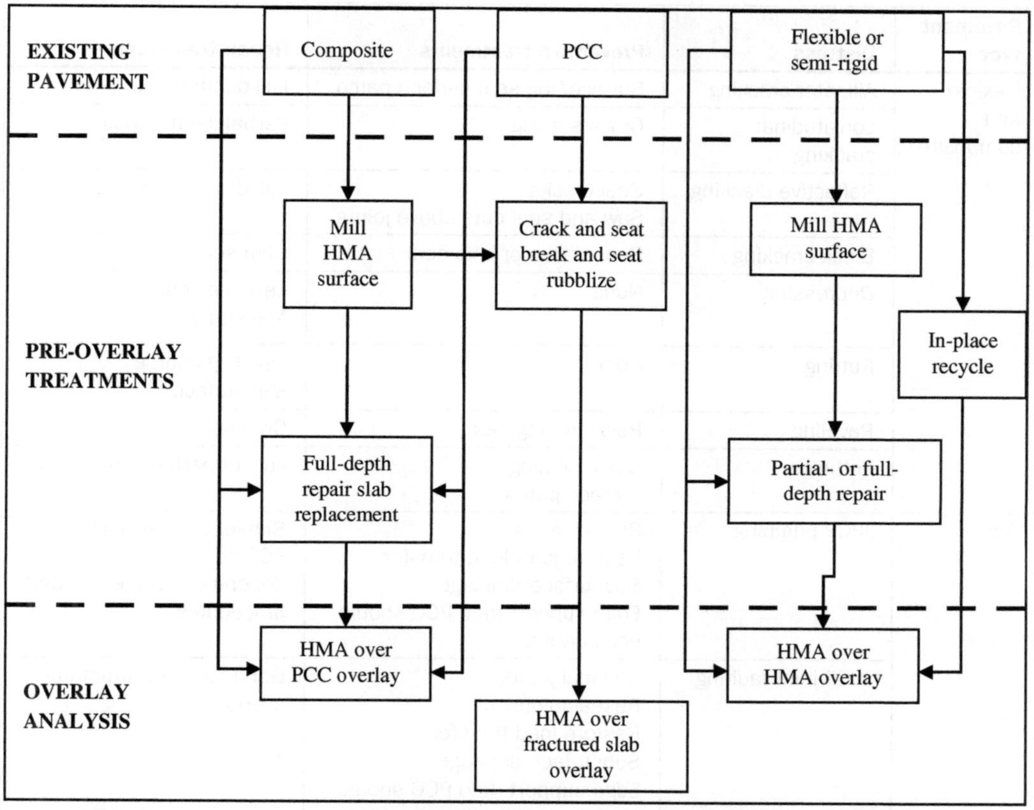

FIGURE 18.4 Flow chart of rehabilitation design options using HMA overlays. (*From AASHTO (2015). Mechanistic-Empirical Pavement Design Guide: A Manual of Practice. Washington, DC: American Association of State Highway and Transportation Officials. Figure 12-2. Used with permission.*)

- Do nothing
- A combination of milling full- or partial-depth repairs, or in-place recycling

In either case, the resulting analysis is an HMA overlay of an existing HMA-surfaced pavement. Similarly, the analysis for existing PCC pavements may be either an HMA over PCC analysis or an HMA over fractured slab analysis depending on whether or not crack and seat, break and seat, or rubblization techniques are applied to the existing PCC pavement. Existing composite pavements may result in either an HMA over PCC analysis or an HMA over fractured slab analysis depending on whether or not the existing HMA surface is removed and the underlying PCC pavement is fractured.

Determining how much of the distress or damage could be repaired before the HMA overlay is placed requires a careful mix of experience and engineering judgment. Table 18.3 lists some of the candidate repair or pre-overlay treatments for all types of pavements, while Table 18.4 lists the major rehabilitation treatments of existing HMA and HMA over PCC pavements. Deciding on the pre-overlay treatment to be used could be based more on experience and historical data, rather than on the distresses and IRI predicted with the AASHTOWare pavement ME design software.

Pavement type	Distress	Preventive treatments	Repair treatments
Flexible and composite	Alligator cracking	Surface/fog seal surface patch	Full-depth repair
	Longitudinal cracking	Crack sealing	Partial-depth repair
	Reflective cracking	Seal cracks Saw and seal cuts above joints	Full-depth repair
	Block cracking	Seal cracks or chip seal	Chip seal
	Depression	None	Leveling course Mill surface
	Rutting	None	Leveling course Mill surface
	Raveling	Rejuvenating seal	Chip seal/surface seal
	Potholes	Crack sealing Surface patch	Full- or partial-depth repair
Rigid	JPCP pumping	Reseal joints Restore joint load transfer Subsurface drainage Edge support (tied PCC should edge beam)	Subseal or mud-jack PCC slabs (effectiveness depends on materials and procedures)
	JPCP joint faulting	Subseal joints Reseal joints Restore load transfer Subsurface drainage Edge support (tied PCC should edge beam)	Grind surface; structural overlay
	JPCP slab cracking	Subseal (loss of support) Restore load transfer Structural overlay	Full- or partial-depth repair
	JPCP joint or crack spalling	Reseal joints	Full- or partial-depth repair
	CRCP punchouts	Polymer or epoxy grouting Subseal (loss of support)	Full-depth repair
	PCC disintegration	None	Full-depth repair thick overlay

Source: *From AASJTO (2015). Mechanistic-Empirical Pavement Design Guide: A Manual of Practice. Washington, DC: American Association of State Highway and Transportation Officials. Table 12-2. Used with permission.*

Table 18.3 Candidate Repair and Preventive Treatments for Different Pavements

The maximum number of overlay layers that may be specified is four. This includes up to three HMA layers, and one unbound or chemically stabilized layer. The total number of layers of the existing pavement and the overlay is limited to 14. For the initial design, however, it is suggested that the total number of layers be limited to no more than eight to reduce the number of required inputs and run time. Depending on the general pavement condition rating, suggested rehabilitation options for level 3 input level are listed in Table 18.5.

Pavement condition	Distress types	Candidate treatments for developing rehabilitation design strategy											
		Full-depth HMA repair	Partial-depth HMA repair	Cold milling	Hot or cold in-place recycling	Cracking sealing	Chip seal	HMA overlay	HMA overlay of fractured PCC slab	Bonded PCC overlay	Unbounded PCC overlay	Subsurface drainage Improvement	Reconstruction (HMA or PCC)
Structural	Alligator cracking	✓			✓		✓	✓	✓	✓	✓		✓
	Longitudinal cracking (low severity)		✓	✓	✓	✓		✓		✓	✓		✓
	Thermal cracking	✓		✓	✓	✓		✓		✓	✓		✓
	Reflection cracking	✓	✓	✓				✓	✓	✓	✓		✓
	Rutting—subsurface			✓	✓			✓		✓	✓		✓
	Shoving—subsurface	✓						✓					✓
Functional	Excessive patching							✓			✓		✓
	Smoothness			✓				✓					
Drainage, Moisture Damage	Raveling		✓	✓				✓					
	Stripping	✓	✓					✓		✓	✓	✓	✓
	Flushing/bleeding		✓				✓	✓					
Durability	Raveling		✓	✓	✓		✓	✓					
	Flushing/bleeding		✓	✓	✓		✓	✓					
	Shoving—HMA		✓	✓	✓			✓					
	Rutting—HMA			✓	✓			✓					
	Block cracking			✓	✓	✓	✓	✓					
Shoulders	Same as traveled lanes	Same treatments as recommended for the traveled lanes.											

Source: *From AASHTO (2015). Mechanistic-Empirical Pavement Design Guide: A Manual of Practice. Washington, DC: American Association of State Highway and Transportation Officials. Table 12-3. Used with permission.*

Table 18.4 Summary of Major Rehabilitation Strategies and Treatments Prior to Overlay Placement for Existing HMA and HMA/PCC Pavements

Condition	General pavement condition rating		Rehabilitation options to consider
Adequate (has remaining life)	Excellent	No cracking, minor rutting, and/or minor mixture-related distresses (e.g., raveling); little to no surface distortions or roughness	• Surface repairs without overlays (not analyzed with the AASHTOWare) • Pavement preservation strategy (not analyzed with the AASHTOWare) • Non-structural overlay • Overlay designed for future truck traffic levels
	Good	Limited load- and/or non-load-related cracking, minor to moderate rutting, and/or moderate mixture-related distresses; some surface distortions and roughness	• Pavement preservation strategy (not analyzed with the AASHTOWare) • Overlays designed for future truck traffic levels, with or without milling and surface repairs
Marginal (may or may not have remaining life)	Fair	Moderate load and/or non-load-related cracking, moderate rutting, moderate amounts of mixture-related distresses, and/or some roughness (IRI > 120 in./mi)	• Pre-overlay treatments recommended • Structural overlay, with or without milling and surface repairs • Replace surface layer prior to overlay • In-place recycling prior to overlay
Inadequate (no remaining life)	Poor	Extensive non-load-related cracking, moderate load-related cracking, high rutting, extensive mixture–related distresses, and/or elevated levels of roughness (IRI > 170 in./mi).	• Pre-overlay treatment recommended if not reconstructed • Structural overlay, with milling or leveling course, and surface repairs. • Replace existing layers prior to overlay • In-place recycling prior to overlay • Reconstruction
	Very good	Extensive load-related cracking and/or very rough surfaces (IRI > 220 in./mi).	• Pre-overlay treatment recommended if not reconstructed • Structural overlay with milling and surface repairs • Replace existing layers prior to overlay • In-place recycling prior to overlay • Reconstruction

Source: *From Mechanistic-Empirical Pavement Design Guide: A Manual of Practice. Washington, DC: American Association of State Highway and Transportation Officials. Table 12-1. Used with permission.*

TABLE 18.5 Definitions of Surface Condition for Input Level 3 Pavement Condition Ratings and Suggested Rehabilitation Options

18.7 Rehabilitation Design with PCC Overlays

PCC rehabilitation design process requires nine steps listed below.

Steps 1–4. Evaluation of the existing pavement.

1. Determine existing pavement condition.
2. Determine causes and mechanism of distress.
3. Define problems and inadequacies of existing pavement.
4. Identify possible constraints.

Step 5. Rehabilitation strategy selection.
Step 6. Rehabilitation design.
Step 7. Perform life-cycle cost analysis (as desired).
Step 8. Determine non-monetary factors that influence rehabilitation (as desired).
Step 9. Determine preferred rehabilitation strategy (as desired).

Figure 18.5 presents the design process for major PCC rehabilitation strategies included in the AASHTOWare pavement ME design software.

As with the new pavement design, the first step in rehabilitation design is to select a trial design with defined layers, material types and properties, and relevant design features based on the future level of traffic anticipated. This is followed by the selection of the design performance criteria (used for evaluating the adequacy of the trial design) and the desired level of reliability. Next, the AASHTOWare pavement ME design software is used to process the input data. Data processing includes estimating

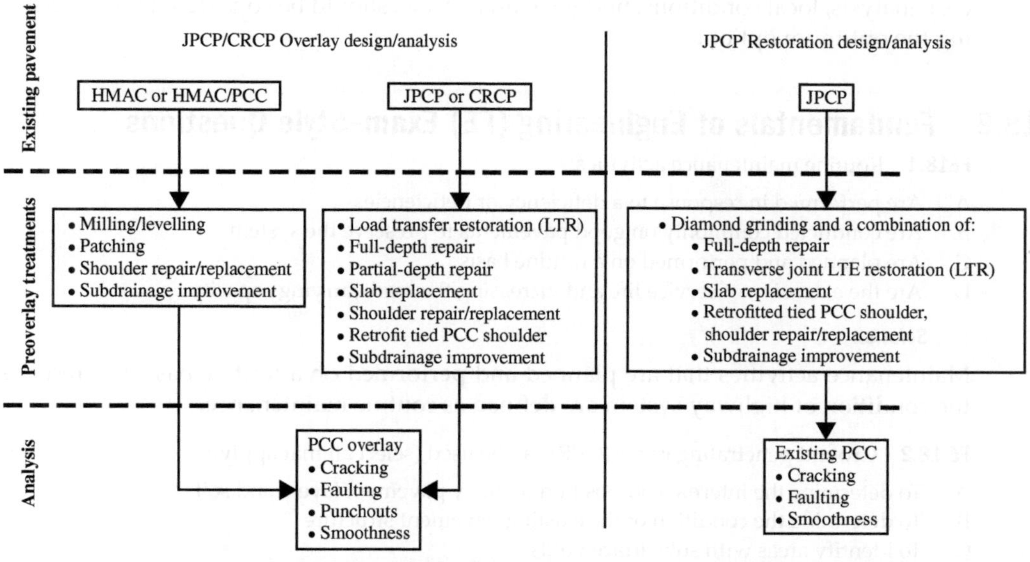

Figure 18.5 Overall design process for major PCC rehabilitation strategies. (*From AASHTO (2015). Mechanistic-Empirical Pavement Design Guide: A Manual of Practice. Washington, DC: American Association of State Highway and Transportation Officials. Figure 12-7. Used with permission.*)

climate-related aspects such as pavement temperature profile for each analysis period using the ICM and computing long-term PCC flexural strength.

Next, the processed data is used to perform a design analysis by computing pavement structural responses (stress, deflections) required for each distress type incrementally. Computed structural responses are used in transfer functions to estimate distress and smoothness.

The trial rehabilitation design is then evaluated for adequacy using prescribed performance criteria at the given reliability level. Trial designs deemed inadequate are modified and reevaluated until a suitable design is achieved. Design modifications could range from making simple changes to JPCP overlay thickness, varying joint spacing, varying PCC strength, or adopting a new rehabilitation strategy altogether. The design process for rehabilitation design with JPCP overlays or CPR of existing JPCP is very similar to new or reconstructed JPCP design.

18.8 Summary

This chapter discusses how to design rehabilitation using the AASHTOWare pavement ME design software. The rehabilitation design is very time-consuming as it requires so much field investigation, data collection, and analysis. It also requires an expert engineer as there are so many areas to make engineering judgments besides statistical analysis. Sometimes panel discussion is required to make a decision. For example, it is very difficult to judge the coefficient of friction between layers (1.0 for full bonding and 0 for no bonding). In old pavement, the friction coefficient is neither 0 nor 1. It is somewhat in between these two values (i.e., partial bonding). Another thing is the options for rehabilitation, as there are many different ways to rehabilitate the pavement. Life-cycle cost analysis, local conditions, budget available, etc. should be considered while selecting the optimum option.

18.9 Fundamentals of Engineering (FE) Exam–Style Questions

FE18.1 Routine maintenance activities:

A. Are performed in response to a deficiency or deficiencies
B. Are conducted commonly on good pavements to preserve the system
C. Are planned and performed on a routine basis
D. Are the extension of service life and increasing the load-carrying capacity

Solution C

Maintenance activities that are planned and performed on a routine basis to preserve the condition of highway system are defined as routine maintenance.

FE 18.2 Ground penetrating radar (GPR) can be used (select all that apply):

A. To determine the internal composition of many pavement layers and soils
B. To determine the condition of the existing pavement structure
C. To identify areas with subsurface voids
D. To locate areas with severe stripping in HMA
E. To locate interfaces with weak bonds between two HMA layers

Solutions A to E

FE 18.3 Which of the following operations is not the example of rehabilitation?

A. Full-depth reclamation
B. Thick/thin overlay
C. Surface recycling
D. Patching potholes

Solution D

Patching potholes is considered a corrective maintenance as a pothole is a deficiency of a pavement.

18.10 Practice Problems

18.1 Discuss briefly what happens if no maintenance of pavement is performed over its service life.

18.2 What types of data are collected while assessing the overall condition of the existing pavement?

18.3 What types of data are collected while assessing the fully defining condition of the existing pavement?

18.4 Discuss briefly the general overview of rehabilitation design using the AASHTOWare pavement ME design software.

18.5 What are the types of HMA overlays that can be analyzed using the AASHTOWare pavement ME design software?

18.6 What are the nine steps required for PCC rehabilitation design using the AASHTOWare pavement ME design software?

CHAPTER 19

Geometric Design of Highways

19.1 Background

The geometric design of roads is the branch of highway engineering dealing with the physical elements of the roadway to optimize efficiency and safety while minimizing cost and environmental damage (AASHTO, 2018). Geometric design is also carried out to support broader community goals such as providing access to employment, schools, businesses, and residences; accommodating a range of travel modes such as walking, bicycling, transit, and automobiles, and minimizing fuel use, emissions, and environmental damage. Geometric roadway design can be divided into three parts:

1. Cross section
2. Alignment
3. Profile

The cross section shows the position and number of vehicle and bicycle lanes and sidewalks, along with their cross-slope or banking. Cross sections also show drainage features, pavement structure, and other items outside the category of geometric design. The alignment is the route of the road such as horizontal curves. The profile is the vertical aspect of the road such as crest and sag curves, and the straight grade lines connecting them.

19.2 Cross Section of Highways

The cross section of highways can be of many shapes depending on the topography, available land space, drainage requirement, sidewalk requirement, etc. However, the mandatory considerations are transverse slopes, shoulders, longitudinal drainage system, sidewalks, etc.

Transverse slopes are provided in the center of the roadway at 2% to 6% forming crown. Each way can be crowned separately on divided roads as on two-lane highways. It may also have a unidirectional cross-slope across the entire width of the traveled way. A cross section with each roadway crowned separately has an advantage in rapidly draining the pavement during rainstorms. However, the difference in the cross section between high and low points is minimal. Disadvantages are that more inlets and underground drainage lines are needed, and treatment of intersections is more difficult because of the number of high and low points on the cross section. The use of such sections should be limited to regions with high rainfall or where snow and ice are important factors. Sections having no curbs and a wide depressed median are particularly well-suited for these conditions (AASHTO, 2018).

Drainage away from the median may provide a savings in drainage structures, minimize drainage across the inner, higher-speed lanes, and simplify treatment of intersecting streets. Drainage to the median is beneficial in that the outer lanes used by most of the traffic are of less surface water. This surface runoff, however, should then be collected into a single conduit under the median. Where curbed medians are present, drainage is concentrated next to or on higher speed lanes. Where the median is narrow, this concentration may result in splashing on the windshields of opposing traffic.

The shoulder slope should be designed in conjunction with the drainage system to prevent ponding upon the roadway. The slopes of the shoulder range from 2% for a paved or impervious surface to 8% for a turf surface. Frequent outlets are very often needed for drainage.

Superelevation is particularly provided at the horizontal curves to balance the centrifugal force developed at the car due to the motion in curve. At the curved segment, transverse slope is provided by the superelevation.

The drainage channel should have enough cross section to provide sufficient drainage capacity and roadbed stability. A depth of 0.3 to 1.2 m (1 to 4 ft) below the shoulder break is recommended.

Sidewalks should be separated as far as possible from the roadway. In areas fully developed with retail stores and offices, it may not be practical to offset the sidewalk from the roadway because of the right-of-way considerations. In such cases, curbs are used to separate the sidewalk from the edge of the roadway. In high-traffic speed areas, it should be located behind the guardrail.

19.3 Lane Widths

The lane width of a roadway affects the comfort of driving and decreases the number of crashes. Lane widths of 2.7 to 3.6 m (9 to 12 ft) are generally used, with a 3.6-m (12-ft) lane predominant on most high-speed, high-volume highways. The extra cost of providing a 3.6-m (12-ft) lane width over the cost of providing a 3.0-m (10-ft) lane width is offset to some extent by a reduction in cost of shoulder maintenance and a reduction in surface maintenance due to decreased wheel concentrations at the lane edges. The wider 3.6-m (12-ft) lane provides desirable clearances between large commercial vehicles traveling in opposite directions on two-lane, two-way rural highways (AASHTO, 2018).

Lane widths also affect highway level of service (LOS). Narrow lanes force drivers to operate their vehicles closer to each other laterally than they would normally desire. Restricted clearances have a similar effect. In fact, the effective width of the traveled path is limited by adjacent obstacles such as retaining walls, bridge trusses, or headwalls, and parked cars that limit lateral clearance.

In urban areas where pedestrian crossings, right-of-way, or existing development makes it difficult to increase lane widths, the use of 3.3-m (11-ft) lanes may be appropriate. Lanes 3.0 m (10 ft) wide are acceptable on low-speed facilities; and lanes 2.7 m (9 ft) wide may be appropriate on low-volume roads in rural and residential areas.

Example 19.1: Lane Width Recommend the lane width for the following conditions:

a. Interstate highway

b. School zones

c. Residential areas

Solution

a. The interstate is always busy and high volumes of traffic use it. Therefore, a 12-ft lane width is recommended.

b. School zones are low-speed zones. Therefore, an 11-ft lane width is recommended.

c. Residential areas have low volumes of traffic. Therefore, a 10-ft lane width is recommended.

19.4 Shoulders

It is desirable that a vehicle stopped on the shoulder should clear the edge of the travel lane by at least 1 ft (0.3 m), and better to be 2 ft (0.6 m). This is why the adoption of 10 ft (3.0 m) as the shoulder width is preferred in high-speed, high-volume roadways. In difficult areas and on low-volume highways, a minimum shoulder width of 2 ft (0.6 m) should be considered and a 6- to 8-ft (1.8- to 2.4-m) shoulder width is preferable. High-traffic and high-speed highways and highways carrying large numbers of trucks should have usable shoulders at least 10 ft (3.0 m) wide and preferably 12 ft (3.6 m) wide. However, widths greater than 10 ft (3.0 m) may encourage unauthorized use of the shoulder as a travel lane. Where bicyclists and pedestrians are to be accommodated on the shoulders, a minimum usable shoulder width (i.e., clear of rumble strips) of 4 ft (1.2 m) should be considered (AASHTO, 2018).

19.5 Rumble Strips

Rumble strips are raised, or grooved, patterns constructed on or in travel lane and shoulder pavements as shown in Fig. 19.1. Rumble strips may also be placed as part of the edge line or centerline. Vehicle tires passing over rumble strips produce a sudden audible sound, cause the vehicle to vibrate, and indicate that the driver needs to take corrective action. There are several basic rumble strip designs or types:

- Milled-in
- Rolled-in
- Formed
- Raised

There are three common uses of rumble strips:

- The most common use is the continuous shoulder rumble strip located on the roadway shoulder to alert drivers to possible roadway departures.

Figure 19.1 Rumble strips.

- Centerline rumble strips may be used on some two-lane rural highways to reduce the possibility for head-on collisions.
- Transverse rumble strips may be installed on approaches to intersections, toll plazas, horizontal curves, and work zones.

Rumble strips have been found to be effective in reducing crashes but may include complaints from nearby residents about noise levels, bicyclists' and motorcyclists' concerns about potential loss of control, and roadway maintenance issues (AASHTO, 2018).

19.6 Curbs

Curb means the raised or vertical element in a roadway that causes an increase in a driver's ability to control a vehicle that strikes or overrides the other vehicle. Curbs also serve the following purposes:

- Drainage control
- Roadway edge delineation
- Right-of-way reduction
- Aesthetics
- Delineation of pedestrian walkways
- Reduction of maintenance operations
- Assistance in orderly roadside development

Sloping curbs with heights up to 100 mm (4 in.) may be considered for use on high-speed roadway when necessary due to drainage considerations and restricted right-of-way. Sloping curbs with 150-mm (6-in.) heights may be considered for use on high-speed urban/suburban facilities with frequent access points and intersecting streets (AASHTO, 2018).

19.7 Drainage Channels

Drainage channels (Fig. 19.2) collect and convey surface water from the highway surface and carry it to discharge point. It must have sufficient capacity for the design runoff,

FIGURE 19.2 Drainage channel in roadway.

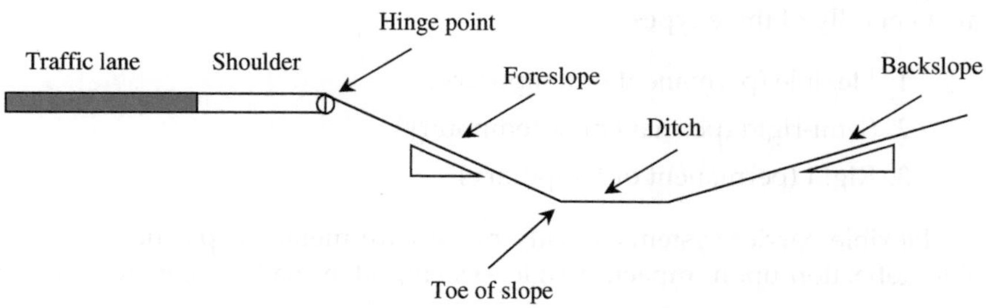

FIGURE 19.3 Designation of roadside regions.

provide unusual storm water with minimal damage to the highway, and be positioned and shaped to ensure smooth water outflow. Channels should be protected against erosion with the least expensive protective lining, and kept clean and free of debris that can lead to a reduction in the channel capacity.

19.8 Sideslopes

Sideslopes should be designed to enhance roadway stability and to provide a reasonable opportunity for recovery for an out-of-control vehicle. Three regions of the roadside are used to reduce the potential for loss of control for vehicles that run off the road (Fig. 19.3):

1. The top of the slope (hinge point)

2. The foreslope
3. The toe of the slope (intersection of the foreslope with level ground or with a backslope, forming a ditch)

The hinge point adds to the loss of steering control because while crossing this point, vehicles begin to get airborne. Rounding the hinge point can increase the general safety of the roadside. The foreslope region is important in the design of high slopes where a driver could try for a recovery maneuver or reduce speed before impacting the ditch area. The toe of the slope is often within the clear zone on the roadside and the probability of an out-of-control vehicle entering the ditch is therefore high. In this case, a smooth transition between fore- and backslopes should be provided (AASHTO, 2018).

19.9 Traffic Barriers

According to AASHTO (2018), traffic barriers are used to prevent vehicles leaving the traveled way from colliding with objects or barrier itself. Barriers themselves are a source of potential crashes. Therefore, their use should be carefully considered. Traffic barriers include both longitudinal barriers and crash cushions. The primary function of longitudinal barriers is to redirect errant vehicles and of crash cushions is to decelerate errant vehicles to a stop.

In addition to traffic barriers, sometimes noise barriers, fall-protection barriers, etc. are also used on roadways as shown in Fig. 19.4.

19.9.1 Longitudinal Barriers

Longitudinal barriers are located along the roadside and in medians (Fig. 19.5). They are generally of three types:

1. Flexible (permanent or temporary)
2. Semi-rigid (permanent or temporary)
3. Rigid (permanent or temporary)

Flexible barrier systems, mostly made with metals or plastics, undergo considerable deflection upon impact. A rigid system, often made of concrete, does not deflect

(a) Noise barrier (b) Fall-protection barrier

FIGURE 19.4 Noise barrier and fall-protection barrier.

(a) Roadside barriers (flexible and permanent)

(b) Roadside barriers (rigid and temporary)

(c) Median barriers

FIGURE 19.5 Longitudinal barriers.

FIGURE 19.6 Bridge railings.

substantially upon impact. During collisions with a rigid system, energy is dissipated by the increase and decrease of the vehicle and the deformation of the vehicle body. Installation of a rigid system is most appropriate along narrow medians or shoulders, etc. Curved areas are also very often used as longitudinal barriers.

19.9.2 Bridge Railings

Bridge railings (Fig. 19.6) are designed to redirect an impacting vehicle and minimize the potential for the vehicle to penetrate the railing. Bridge railings also reduce the potential for vehicles, pedestrians, or cyclists to fall from the structure (AASHTO, 2018). Bridge railings are longitudinal barriers to traffic that vary mainly in their foundations from other traffic barriers. These railings are a structural extension of a bridge while other traffic barriers are usually set in or on soil. Very often bridge railings also include a fall-protection wall especially if the bridge has a sidewalk.

19.9.3 Crash Cushions

Crash cushions (Fig. 19.7) are protective systems that prevent errant vehicles from impacting roadside obstacles by decelerating the vehicle to a safe stop when hit

Permanent cushion | Water-filled semi-permanent cushion

Figure 19.7 Crash cushions.

head-on or redirecting vehicles away from the obstacle. A common application of a crash cushion is at the end of a bridge rail. Where conditions on-site warrant, a crash cushion should also be regarded as an alternative to a roadside barrier to protect rigid objects such as bridge piers, overhead sign supports, abutments, and retaining-wall ends. Crash cushions may also be used to shield roadside and median barrier terminals (AASHTO, 2018). Crash cushions can be permanent or semi-permanent, such as water-filled or sand-filled cushions.

19.10 Medians

A median (Fig. 19.8) is the depressed, raised, or flush portion of a highway separating opposing directions of the traveled way. Medians are highly desirable on two-way roadways carrying four or more lanes. The principal functions of a median are as follows:

- Separates opposing traffic
- Provides a recovery area for out-of-control vehicles
- Provides a stopping area in case of emergencies
- Allows space for speed changes and for storage of left-turning and u-turning vehicles
- Diminishes headlight glare, and provides width for future lanes

Figure 19.8 A median in a local road.

- Provides a refuge area for pedestrians crossing the street
- Controls the location of intersection traffic conflicts

Most of the median widths range from 4 to 80 ft (1.2 to 24 m) and in some instances even wider media are used (AASHTO, 2018).

19.11 Pedestrian Facilities

19.11.1 Sidewalks

Sidewalks (Fig. 19.9) are rarely provided in rural areas, although the potential for collisions with pedestrians is higher in rural areas due to the higher speeds and absence of lighting. Sidewalks in rural areas are effective in reducing pedestrian collisions. Sidewalks in rural and suburban areas are more often justified at points of pedestrian concentrations, such as residential areas, schools, businesses, and industrial plants (AASHTO, 2018).

19.11.2 Grade-Separated Pedestrian Crossings

A grade-separated pedestrian facility (Fig. 19.10) allows pedestrians and motor vehicles to cross at different levels, either over or under a roadway. It provides a safe path for pedestrians to cross the road without disrupting traffic. Pedestrian separations should be provided where pedestrian volume, traffic volume, intersection capacity, and other conditions favor their use such as at central business districts, factories, schools, and athletic fields (AASHTO, 2018).

19.11.3 Curb Ramps

Curb ramps (Fig. 19.11) are necessary to provide access at pedestrian crossings between the sidewalk and the street. Several federal laws require that pedestrian facilities be readily accessible and accessible to individuals with disabilities. When designing a project that includes curbs and adjacent sidewalks, proper attention should be given to the needs of persons with disabilities, such as those with mobility or visual impairment. Detectable warnings are needed where the curb has been removed to alert visually impaired pedestrians that they have arrived at the street/sidewalk interface (AASHTO, 2018).

FIGURE 19.9 Two sidewalks.

FIGURE 19.10 A grade-separated pedestrian facility.

FIGURE 19.11 A sidewalk curb ramps.

19.12 Bicycle Facilities

Bicycle travel facilities (Fig. 19.12) are provided by the street and highway system as it presently exists. To provide travel facilities effectively for bicycle traffic, the designer should be familiar with the size of the bicycle, operating characteristics, speed, and

Bicycle facility Bicycle-cum-pedestrian facility

FIGURE 19.12 Bicycle facilities on roads.

needs. These factors determine acceptable turning radii, grades, and sight distance (AASHTO, 2018).

19.13 On-Street Parking

A roadway network should be designed and developed to ensure that vehicles operating on the system travel safely and efficiently. Although the movement of vehicles is the primary function of a roadway network, segments of the network may also provide on-street parking for better land use (Fig. 19.13). On-street parking generally decreases

FIGURE 19.13 On-street parking.

through traffic capacity, impedes traffic flow, and increases crash potential. Within urban areas and in rural communities located on arterial highway routes, on-street parking should be considered to accommodate existing and developing land uses. When a proposed roadway improvement is to include on-street parking, parallel parking should be considered. Under certain circumstances, angle parking is an allowable form of street parking. Angle parking presents specific problems due to the varying lengths of vehicles and the problems of sight distance associated with vans and recreational vehicles (AASHTO, 2018).

Example 19.2: Highway Parameters List the common dimensions of the following:

a. Width of medians
b. Heights of sloping curbs
c. Minimum usable shoulder width when bicyclists and pedestrians are to be accommodated
d. Minimum shoulder width for low-volume highways

Solution

a. Most median widths are in the range of 4 to 80 ft (1.2 to 24 m).
b. Heights of sloping curbs range from 4.0 to 6.0 in.
c. Minimum usable shoulder width when bicyclists and pedestrians are to be accommodated is 4.0 ft.
d. A minimum shoulder width of 2 ft (0.6 m) should be considered for low-volume highways.

19.14 Horizontal Curves

Horizontal curves (Fig. 19.14) are required at locations where the roadway needs to change its horizontal alignment to the left or right to accommodate some geographic conditions such as mountain or physical obstruction such as buildings. In other words, using a horizontal curve, two roadways converge, resulting in a gradual transition

FIGURE 19.14 Horizontal curves.

between these two that allows a vehicle to navigate a turn at a gradual rate rather than a sharp turn. The design of a horizontal curve depends on the desired speed limit and friction of the roadway. Sometimes drainage provision is also considered. Horizontal curves are designed considering a part of circles as to provide the driver with a constant turning rate with radii determined by the laws of physics surrounding centripetal force.

19.14.1 Types of Horizontal Curves

Horizontal curves are of five types:

1. Simple curve
2. Compound curve
3. Broken-back curve
4. Reverse curve
5. Spiral curve

A simple curve is a circular arc that connects two tangents. The most widely used curve is a simple curve. A compound curve is composed of two or more circular arcs of different radii tangent to each other, with their centers on the same side of the alignment. The combination of a short length of tangent (less than 100 ft) connecting two circular arcs that have centers on the same side is called a broken-back curve. A reverse curve consists of two circular arcs tangent to one another, with centers on the opposite sides of the alignment.

Compound, broken-back, and reverse curves are unsuitable for modern high-speed highway, rapid transit, and railroad traffic and should be avoided if possible. In mountainous terrain, however, they are sometimes necessary to avoid excessive grades or very deep cuts and fills. Compound curves are often used on exit and entrance ramps of interstate highways and expressways, although easement curves are generally a better choice for these situations. Easement curves are desirable, especially for railroads and rapid transit systems, to lessen the sudden change in curvature at the junction of a tangent and a circular curve.

A spiral makes an excellent easement curve because its radius decreases from infinity in the tangent to the curve it meets uniformly. Spirals are used to connect a tangent with a circular curve, a tangent with a tangent (double spiral), and a circular curve with a circular curve.

19.14.2 Simple Curve

The rate of curvature of circular curves can be designated either by their radius (e.g., a 1,500-m curve or a 1,000-ft curve) or by their degree of curve. There are two different designations for degree of curve, the arc definition and the chord definition, both of which are defined using the English system of units. By the arc definition, degree of curve is the central angle subtended by a circular arc of 100 ft (see Fig. 19.15). This definition is preferred for highway work. By the chord definition, degree of curve is the angle at the center of a circular arc subtended by a chord of 100 ft. This definition is convenient for very gentle curves and hence is preferred for railroads.

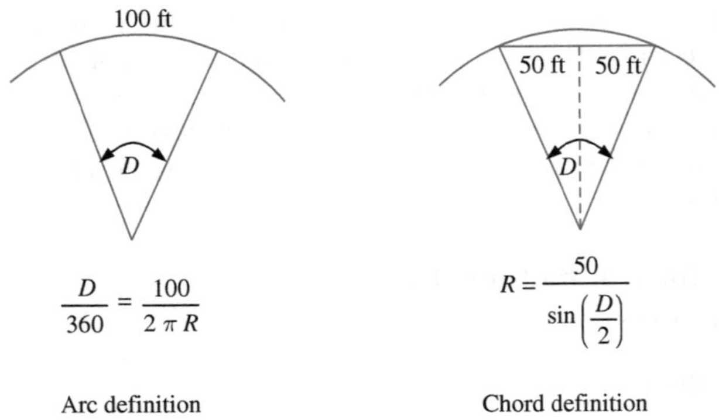

Figure 19.15 Degree of circular curve.

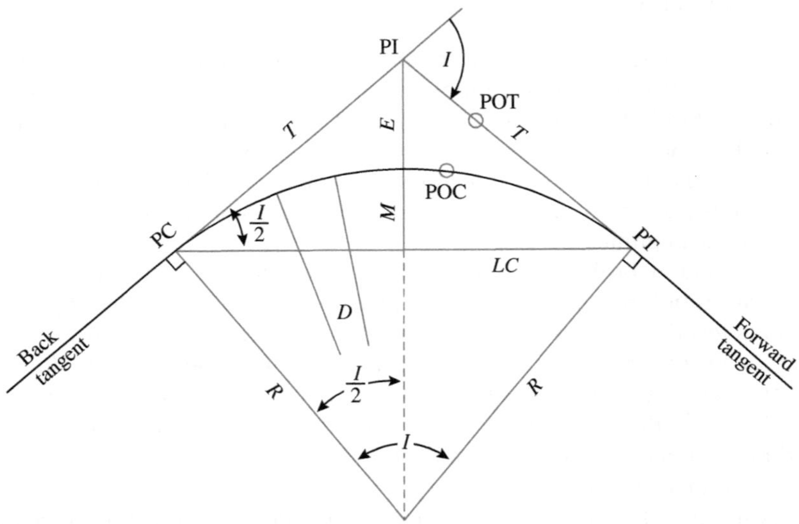

Figure 19.16 Circular curve elements.

19.14.3 Simple Curve Formulas

Circular curve elements are shown in Fig. 19.16. The point of intersection, PI, of the two tangents is also called the vertex, V. In stationing, the back tangent precedes the PI, the forward tangent follows it. The beginning of the curve, or point of curvature, PC, and the end of the curve, or point of tangency, PT, are also sometimes called BC and EC, respectively. Other expressions for these points are tangent to curve, TC, and curve to tangent, CT. The curve radius is R. Note that the radii at the PC and PT are perpendicular to the back tangent and forward tangent, respectively.

The distance from PC to PI and from PI to PT is called the tangent distance, T. The line connecting the PC and PT is the long chord, LC. The length of the curve, L, is the distance from PC to PT, measured along the curve for the arc definition, or by 100-ft chords for the chord definition.

The external distance, E, is the length from the PI to the curve midpoint on a radial line. The middle ordinate M is the (radial) distance from the midpoint of the long chord to the curve's midpoint. Any point on curve is POC, and any point on tangent is POT. The degree of any curve is D_a (arc definition) or D_c (chord definition). The change in direction of two tangents is the intersection angle, I, which is also equal to the central angle subtended by the curve.

By definition, and from inspection of Fig. 19.16, the following relationships can be derived:

$$\text{Radius of curve by arc definition, } \frac{D}{360} = \frac{100}{2\pi R} \text{ or } R = \frac{5{,}729.58}{D} \text{ or } R = \frac{LC}{2\sin\left(\frac{I}{2}\right)} \quad (19.1)$$

$$\text{Radius of curve by chord definition, } R = \frac{50}{\sin\left(\frac{D}{2}\right)} \quad (19.2)$$

$$\text{Tangent length, } T = R\tan\left(\frac{I}{2}\right) = \frac{LC}{2\cos\left(\frac{I}{2}\right)} \quad (19.3)$$

$$\text{Length of curve, from PC to PT, } L = RI\left(\frac{\pi}{180}\right) = \frac{I}{D}(100) \quad (19.4)$$

$$PC = PI - T \quad (19.5)$$

$$PT = PI + L \quad (19.6)$$

$$\text{Length of middle ordinate, } M = R\left[1 - \cos\left(\frac{I}{2}\right)\right] \quad (19.7)$$

$$\text{Length of sub-chord, } c = 2R\sin\left(\frac{d}{2}\right) \quad (19.8)$$

$$\text{Curve length for sub-chord, } l = Rd\left(\frac{\pi}{180}\right) \quad (19.9)$$

$$\text{External distance, } E = R\left[\frac{1}{\cos\left(\frac{I}{2}\right)} - 1\right] \quad (19.10)$$

Remember that in highway construction the unit of length is most often used as station for being very convenient and 1 station = 100 ft.

Example 19.3: Parameters of Horizontal Curve A horizontal curve has an external distance of 620 ft and a radius of 3,200 ft. Determine the middle ordinate and the tangent length of the horizontal curve.

Solution

External distance, $E = 620$ ft

Radius, $R = 3{,}200$ ft

Middle Ordinate

$$M = R\left[1 - \cos\left(\frac{I}{2}\right)\right] = R\left[1 - \frac{R}{E+R}\right] = 3{,}200\,\text{ft}\left[1 - \frac{3{,}200\,\text{ft}}{620\,\text{ft} + 3{,}200\,\text{ft}}\right] = 519\,\text{ft} \approx 520\,\text{ft}$$

Tangent Length

$$T = R\tan\left(\frac{I}{2}\right)$$

$R = 3{,}200$ ft; however, the term $I/2$ is unknown. Let us find out $I/2$.

$$\text{Known, } M = R\left[1 - \cos\left(\frac{I}{2}\right)\right]$$

$$\text{Or, } \cos\left(\frac{I}{2}\right) = 1 - \frac{M}{R}$$

Therefore, $I/2 = 33$

$$\text{Now, } T = R\tan\left(\frac{I}{2}\right) = 3{,}200 \text{ ft} \tan(33) = 2{,}078 \text{ ft}$$

Answer *The middle ordinate and the tangent length of the horizontal curve are 520 ft and 2,078 ft, respectively.*

Example 19.4: Parameters of Horizontal Curve A horizontal curve has a radius of 1,300 ft and an intersection angle of 22°. The point of intersection (PI) is located at 15 + 33 station. Determine the positions (stations) of the PC and PT.

Solution

Radius, $R = 1{,}300$ ft

Intersection angle, $I = 22°$

Location of PI = 15 + 33 stations

PC

$$\text{Tangent distance, } T = R\tan\left(\frac{I}{2}\right) = (1{,}300 \text{ ft})\tan\left(\frac{22}{2}\right) = 253 \text{ ft}$$

15 + 33 stations = 1,533 ft

$PC = PI - T = 1{,}533 \text{ ft} - 253 \text{ ft} = 1{,}280 \text{ ft} = 12 + 80$ stations

PT

$$L = RI\left(\frac{\pi}{180}\right) = (1{,}300\text{ ft})(22°)\left(\frac{\pi}{180}\right) = 499\text{ ft}$$

$PT = PC + L = 1{,}280 + 499 = 1{,}779\text{ ft} = 17 + 79$ stations

Answer *The position of the* PC *is 12 + 80 stations and the position of the* PT *is 17 + 79 stations.*

Example 19.5: Parameters of Horizontal Curve A horizontal curve is designed with a 1,500 ft radius and a tangent length of 300 ft. The PI is located at 15 + 80 stations. Determine the positions (stations) of the PC and PT.

Solution

Radius, $R = 1{,}500$ ft

Tangent length, $T = 300$ ft

Location of PI = 15 + 80 stations

PC

$PC = PI - T = 1{,}580\text{ ft} - 300\text{ ft} = 1{,}280\text{ ft} = 12 + 80$ stations

PT

$PT = PC + L$

PC is known; however, L is unknown.

Now, $L = RI\left(\frac{\pi}{180}\right)$

I is still unknown. Let us first find out I.

We know, $T = R\tan\left(\frac{I}{2}\right)$

$$\tan\left(\frac{I}{2}\right) = \frac{T}{R} = \frac{300}{1{,}500} = 0.2$$

Therefore, $I = 22.62°$

Now, $L = RI\left(\frac{\pi}{180}\right) = 1{,}500(22.62°)\left(\frac{\pi}{180}\right) = 592\text{ ft}$

Finally, $PT = PC + L = PC + 592\text{ ft} = 1{,}280 + 592 = 1{,}872\text{ ft} = 18 + 72$ stations

Answer *The position of the* PC *is 12 + 80 stations and the position of the* PT *is 18 + 72 stations.*

Example 19.6: Parameters of Horizontal Curve A horizontal curve is to be designed with a 1,500-ft radius and angle at intersection of 30°. Determine the external distance and degree of curve by the arc definition.

Solution

Radius, $R = 1{,}500$ ft

Intersection angle, $I = 30°$

$$\text{External distance, } E = R\left[\frac{1}{\cos\left(\frac{I}{2}\right)} - 1\right] = (1{,}500 \text{ ft})\left[\frac{1}{\cos 15} - 1\right] = 52.9 \text{ ft}$$

$$\text{Degree of curve by arc definition, } D = \frac{5{,}729.58}{R} = \frac{5{,}729.58}{1{,}500} = 3.82°$$

Answer *The external distance of the curve is 52.9 ft and the degree of the curve is 3.82°.*

Example 19.7: Parameters of Horizontal Curve A horizontal curve is to be designed with a tangent length of 250 ft and an intersection angle of 40°. Determine the length of the long chord and the curve length.

Solution

Tangent length, $T = 250$ ft

Intersection angle, $I = 40°$

$$\text{Long chord, } LC = 2T\cos\left(\frac{I}{2}\right) = 2(250 \text{ ft})\cos 20 = 470 \text{ ft}$$

$$\text{Curve length, } L = RI\left(\frac{\pi}{180}\right)$$

However, R is unknown. Let us find R first.

$$\text{We know, } T = R\tan\left(\frac{I}{2}\right)$$

$$\text{Therefore, } R = \frac{T}{\tan\left(\frac{I}{2}\right)} = \frac{250 \text{ ft}}{\tan(20)} = 687 \text{ ft}$$

$$\text{Finally, } L = RI\left(\frac{\pi}{180}\right) = (687 \text{ ft})(40°)\left(\frac{\pi}{180}\right) = 480 \text{ ft}$$

Answer *The length of long chord is 470 ft and the length of the curve is 480 ft.*

19.14.4 Design of Simple Curve

When a vehicle moves in a circular path, it undergoes a centripetal acceleration that acts toward the curvature center. Due to this centripetal acceleration, an imaginary force that motorists feel push them outward which is called centrifugal acceleration. To balance the centrifugal acceleration, the outside of the roadway is raised with respect to the inner side of the road so that the component of the weight of the motor counterbalances it. This raising is called the superelevation. Considering this phenomenon, the basic equations governing the determination of curve radius, R, are as follows:

$$0.01e + f = \frac{V^2}{15R} \text{ for U.S. customary units} \tag{19.11}$$

$$0.01e + f = \frac{V^2}{127R} \text{ for metric units} \tag{19.12}$$

where e = Rate of roadway superelevation (%)
f = Side friction factor (in decimal)
V = Vehicle speed (mph for U.S. customary units or kmph for metric units)
R = Radius of curvature measured to vehicle's center of gravity

This equation is referred to as the simplified curve formula and yields slightly larger (and, thus, more conservative) estimates of friction demand than would be obtained using the analytical curve formula.

19.14.5 Design Parameters

Side Friction Factor, f

The side friction factor at impending skid depends on a number of other factors, among which the most important are the speed of the vehicle, the type and condition of the roadway surface, and the type and condition of the vehicle tires. The maximum side friction factors developed between new tires and wet concrete pavements range from about 0.5 at 30 kmph (20 mph) to approximately 0.35 at 100 kmph (60 mph). For normal wet concrete pavements and smooth tires, the maximum side friction factor at impending skid is about 0.35 at 70 kmph (45 mph). In all cases, the studies show a decrease in friction values as speeds increase, as shown in Fig. 19.17.

One series of tests found coefficients of friction for ice ranging from 0.05 to 0.20, depending on the condition of the ice (i.e., wet, dry, clean, smooth, or rough). Tests on loose or packed snow show coefficients of friction ranging from 0.20 to 0.40.

Superelevation, e

The maximum rates of superelevation used on highways are controlled by four factors:

1. Climate conditions (i.e., frequency and amount of snow and ice)
2. Terrain conditions (i.e., flat, rolling, or mountainous)
3. Type of area (i.e., rural or urban)
4. Frequency of very slow-moving vehicles

The highest superelevation rate for highways in common use is 10%, although 12% is used in some cases. Superelevation rates above 8% are only used in areas without snow and ice. Although higher superelevation rates offer an advantage to those drivers traveling at high speeds, current practice considers that rates in excess of 12% are beyond practical limits.

Minimum Radius

The minimum radius is a limiting value of curvature for a given design speed and is determined from the maximum rate of superelevation and the maximum side friction factor selected for design. The minimum radius of curvature is based on a threshold of driver comfort that is sufficient to provide a margin of safety against skidding and vehicle rollover. Figure 19.18 shows the consequence of inadequate curve radius.

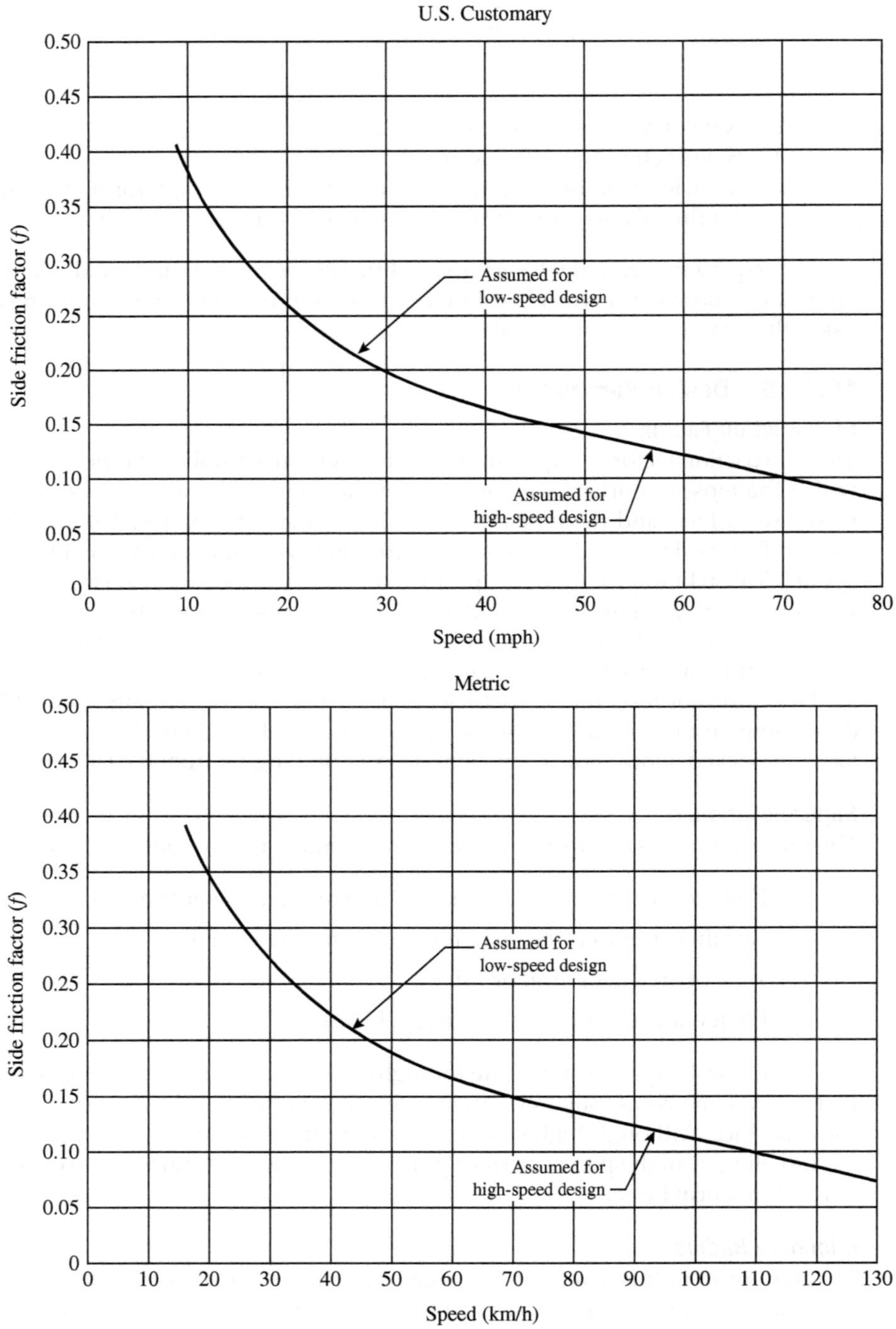

FIGURE 19.17 Variation of side friction factor with vehicle speed. (*From AASHTO (2018). A Policy on Geometric Design of Highways and Streets. Washington, DC: American Association of State Highway and Transportation Officials. Figure 3-6. Used with permission.*)

Figure 19.18 Inadequate sharp curve. (*Courtesy of azfamily.*)

19.14.6 Sight Distance on Horizontal Curves

Where there are sight obstructions (such as walls, cut slopes, buildings, and longitudinal barriers), a design may need adjustment in the normal highway cross section or the alignment. Because of the many variables in alignment, in cross section, and in the number, type, and location of potential obstructions, specific study is usually needed for each individual curve. Two types of stopping sight distances are considered while designing horizontal curves:

1. Horizontal sight line offset (HSO)
2. Passing sight distance

Horizontal Sight Line Offset (HSO)

A stopping sight distance sightline obstruction is any roadside object (Fig. 19.19) within the horizontal sightline offset (HSO) distance (such as median barriers, guardrails, bridges, walls, cut slopes, buildings, or wooded areas), 2.0 feet or greater above the roadway surface at the centerline of the lane on the inside of the curve.

For example, with an 80-kmph (50-mph) design speed and a curve with a 350-m (1,150-ft) radius, a clear sight area with a horizontal sight line offset of approximately 6.0 m (20 ft) is needed for stopping sight distance. As another example, for a sight obstruction at a distance HSO equal to 6.0 m (20 ft) from the centerline of the inside lane on a curve with a 175-m (575-ft) radius, the sight distance needed is approximately at the upper end of the range for a speed of approximately 60 kmph (40 mph).

The HSO can be calculated using the equation:

$$\text{HSO} = R\left[1 - \cos\left(\frac{28.65S}{R}\right)\right] \tag{19.13}$$

where HSO = Horizontal sightline offset, ft or m
S = Stopping sight distance, ft or m
R = Radius of curve, ft or m

Where sufficient stopping sight distance is not available because a railing or a longitudinal barrier constitutes a sight obstruction, alternative designs should be considered. The alternatives are as follows:

- Increase the offset to the obstruction

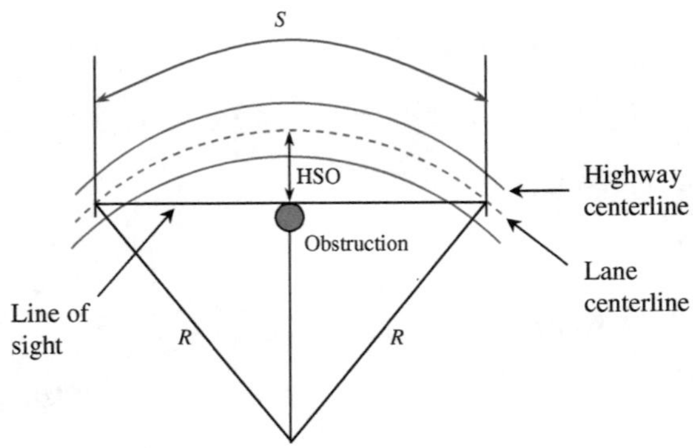

Figure 19.19 Horizontal sightline offset (HSO).

- Increase the radius
- Reduce the design speed

The above requires stopping sight distance (SSD) or simply, S. It is the sum of two distances:

1. The distance traversed by the vehicle from the instant the driver sights an object necessitating a stop to the instant the brakes are applied
2. The distance needed to stop the vehicle from the instant brake application begins. These are referred to as brake reaction distance and braking distance, respectively.

Brake reaction time is the interval from the instant that the driver identifies the existence of an obstacle on the roadway ahead which requires braking until the moment that the driver actually applies the brakes. The recommended design criterion of 2.5 s for brake reaction time exceeds the 90th percentile of reaction time for all drivers.

S or SSD is calculated using the following equations:

$$\text{SSD} = 1.47Vt + \frac{V^2}{30\left(\frac{a}{32.2} \pm G\right)} \quad \text{for U.S. customary units} \tag{19.14}$$

$$\text{SSD} = 0.278Vt + \frac{V^2}{254\left(\frac{a}{9.81} \pm G\right)} \quad \text{for metric units} \tag{19.15}$$

where SSD = S = Stopping sight distance, ft or m
t = Driver reaction time (commonly, 2.5 s)
V = Vehicle approach speed (mph or kmph)
$\pm G$ = Percent grade divided by 100 (uphill grade "+", downhill grade "−")
a = Driver's deceleration (commonly, 11.2 ft/s^2 or 3.4 m/s^2)

Passing Sight Distance

The minimum passing sight distance for a two-lane road or street is about twice the minimum stopping sight distance at the same design speed. To conform to those greater sight distances, clear sight areas on the inside of curves should have widths in excess of those discussed.

Example 19.8: Horizontal Sightline Offset A horizontal curve has a radius of 1,000 ft and connects the tangents of a two-lane highway at a location whose speed limit is 55 mph. If the highway curve is not super-elevated, determine the horizontal sightline offset that a large signboard can be placed from the centerline of the inside lane of the curve.

Solution

Radius of curvature, $R = 1{,}000$ ft
Speed, $V = 55$ mph
Driver reaction time, $t = 2.5$ s (commonly used)
Driver's deceleration, $a = 11.2\ \text{ft/s}^2$ (commonly used)
Grade, $G = 0$ (as no grade is specified)
Horizontal sightline offset, HSO = ?

To calculate HSO, we need SSD or S. Let us calculate SSD or S.

$$\text{SSD} = 1.47Vt + \frac{V^2}{30\left(\left(\frac{a}{32.2}\right) + G\right)} = 1.47(55\ \text{mph})(2.5\ \text{s}) + \frac{(55\ \text{mph})^2}{30\left(\frac{11.2}{32.2} + 0\right)} = 492\ \text{ft}$$

$$\text{HSO} = R\left[1 - \cos\left(\frac{28.65\ S}{R}\right)\right] = 1{,}000\ \text{ft}\left[1 - \cos\left(\frac{28.65(492\ \text{ft})}{1{,}000\ \text{ft}}\right)\right] = 30.1\ \text{ft}$$

Answer *The horizontal sightline offset is 30.1 ft.*

19.14.7 Setting Simple Curve

There are several methods of setting a horizontal curve in a field. Computing deflection angles and chords is one of them, and may be easiest to the authors. Figure 19.20 shows the basics of computing deflection angles and chords. Let A is a point on the curve with PC being the point of curve. Also, let the curve length (arc) from PC to A be l, chord length be c, and the subtended angle be d. Then, by geometry of the circle it can be shown that the deflection angle (angle between tangent and chord) is $d/2$.

The chord length (c) and the arc length (l) can be calculated as follows:

$$\text{Chord length, } c = 2R\sin\left(\frac{d}{2}\right) \tag{19.16}$$

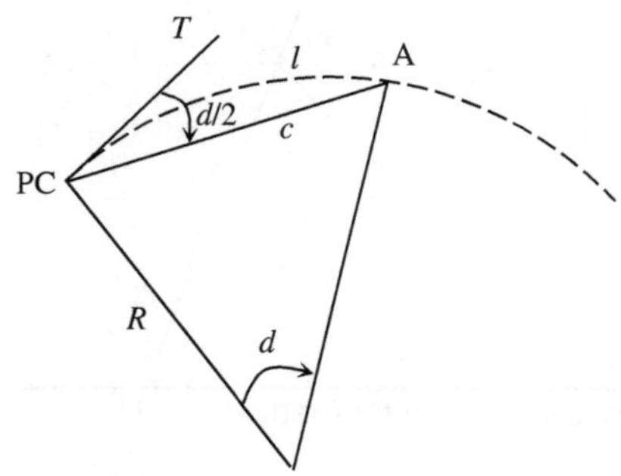

FIGURE 19.20 Calculating deflection angle and chord length.

$$\text{Arc length, } l = Rd\left(\frac{\pi}{180}\right) \tag{19.17}$$

$$\text{or, } d = \frac{180l}{\pi R} \tag{19.18}$$

Step 1. Calculate R based on given data.

Step 2. Calculate the length of curve, L.

Step 3. Divide L into stations; the first and the last division may be a fraction of a station (say, 45.5 ft). Call this l.

Step 4. For each l, calculate d using the equation, $d = \frac{180l}{\pi R}$. Then, calculate deflection angle $d/2$ and chord length (c) using $c = 2R\sin\left(\frac{d}{2}\right)$.

That means after the PC, *the first coordinate is at c ft with angle of d/2 with respect to the tangent.*

Step 5. For next l (should be 1 station = 100 ft), calculate $d = D$. The cumulative deflection angle = $d/2 + D/2$.
That means after the PC, *the first coordinate is at c ft with angle of d/2 with respect to the tangent. The second coordinate is at (c + 100) ft with angle of (d/2 + D/2) with respect to the tangent.*

Step 6. Similarly continue up to the last station (last round division).

Step 7. If the last division is fraction, calculate the deflection angle for that fraction curve length. Then, calculate the coordinate of PT. This step is optional, but will verify the calculation.

Example 19.9: Computation of Horizontal Curve A proposed horizontal curve in a highway has two tangents with the bearings of N 45°24′16″ E and N 5°14′22″ W, as shown in Fig. 19.21. The highway design engineer wants to obtain the best fit for the simple circular curve to join these tangents, deciding that the external ordinate is to be 35 ft. The PI is located at the station of 217 + 55.22.

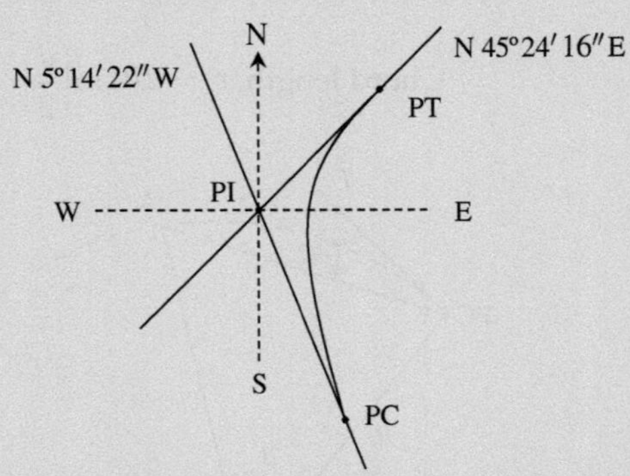

FIGURE 19.21 The horizontal curve for Example 19.9.

Determine the following:

a. Central angle of the curve
b. Radius of the curve
c. Length of the tangent of the curve
d. Station of the PC
e. Length of the curve
f. Station of the PT
g. Deflection angle and chord from the PC to the first full-station on the curve

Solution

a. Central angle of the curve (from Fig. 19.22)
Intersection angle, $I = 5°14'22'' + 45°24'16'' = 50°38'38'' = 50.411°$

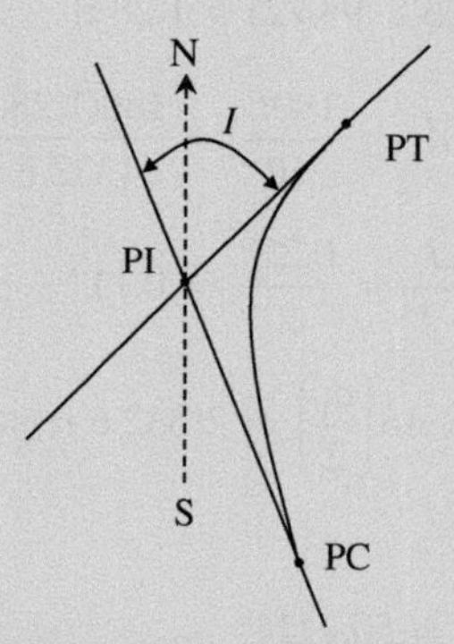

FIGURE 19.22 Analysis of the horizontal curve for Example 19.9.

b. Radius of the curve
External ordinate, $E = 35$ ft

$$E = R\left[\frac{1}{\cos\left(\frac{I}{2}\right)} - 1\right]$$

$$R = \frac{E}{\left[\frac{1}{\cos\left(\frac{I}{2}\right)} - 1\right]} = \frac{35 \text{ ft}}{\left[\frac{1}{\cos\left(\frac{50.411}{2}\right)} - 1\right]} = 332.6 \text{ ft}$$

c. Length of the tangent of the curve

$$T = R\tan(I/2) = (332.6 \text{ ft})\tan(50.411/2) = 156.5 \text{ ft}$$

d. Station of the PC

$$PC = PI - T = (217 + 55.22) - (1 + 56.5) = 215 + 98.72$$

e. Length of the curve

$$L = RI\left(\frac{\pi}{180}\right) = (332.6\ \text{ft})(50.411)\left(\frac{\pi}{180}\right) = 292.6\ \text{ft}$$

f. Station of the PT

$$PT = PC + L = (215 + 98.72) + (2 + 92.6) = 218 + 91.32$$

g. Deflection angle and chord from the PC to the first full-station on the curve

Reference point = PC = 215 + 98.72

First full-station is located at 216 + 00. The distance to the first full-station = (216 + 00) − (215 + 98.72) = 1.28 ft

$$d_1 = \frac{180 l_1}{\pi R} = \frac{180(1.28\ \text{ft})}{\pi(332.6\ \text{ft})} = 0.22°$$

$$\frac{d_1}{2} = \frac{0.22°}{2} = 0.11°$$

$$c_1 = 2R\sin\left(\frac{d_1}{2}\right) = 2(332.6\ \text{ft})\sin\left(\frac{0.22}{2}\right) = 1.277\ \text{ft}$$

Answers

a. *Intersection angle, I = 50.411°*

b. *Radius of the curve, R = 332.6 ft*

c. *Tangent length of the curve, T = 156.5 ft*

d. *Station of the PC = 215 + 98.72*

e. *Length of the curve, L = 292.6 ft*

f. *Station of the PT = 218 + 91.32*

g. *Deflection angle and chord from the PC to the first full-station are 0.11° and 1.277 ft, respectively*

Note that the curve parameters are rounded to practical values in real design.

Example 19.10: Computation of Horizontal Curve A simple-circular horizontal curve has the following features (Fig. 19.23):

- Degree of curve = 15°
- Bearing on incoming (back) tangent is N 85°24′16″ E
- Bearing on outgoing (forward) tangent is S 25°14′22″ E
- Station of the PI = 178 + 33.87

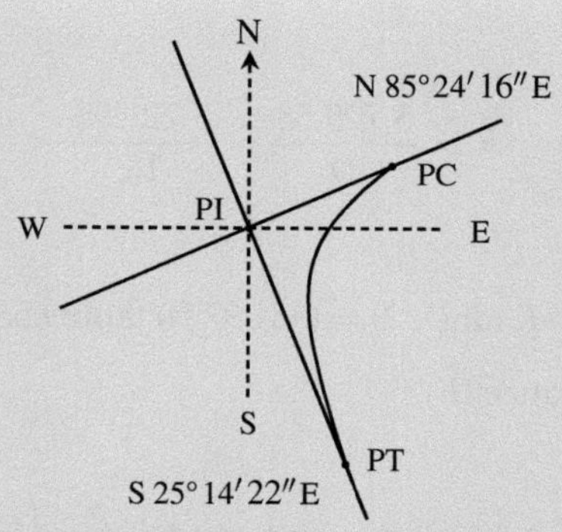

Figure 19.23 The horizontal curve for Example 19.10.

Determine the following:

a. Intersection angle (I)
b. Radius (R)
c. Tangent (T)
d. External distance (E)
e. Middle ordinate (M)
f. Long chord (LC)
g. Length of the curve (L)
h. Station of the PC
i. Station of the PT

Solution

a. Intersection angle (I) from Fig. 19.24

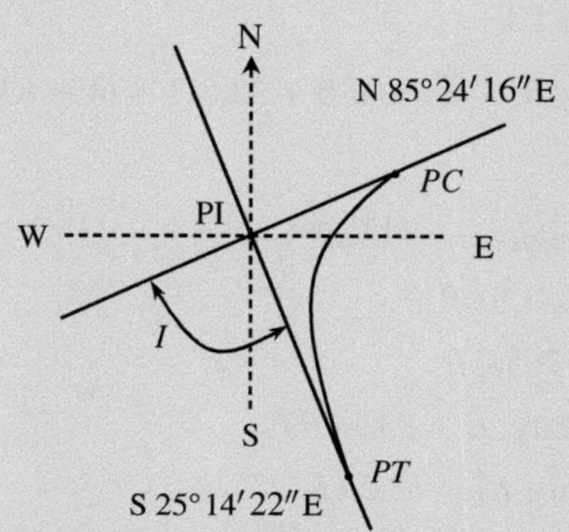

Figure 19.24 Analysis of the horizontal curve for Example 19.10.

$I = 85°24'16'' + 25°14'22'' = 110°38'38'' = 110.411°$
$I = 180° - 110.411° = 69.589°$

b. Radius (R)

$$R = \frac{5{,}729.58}{D} = \frac{5{,}729.58}{15} = 381.97 \text{ ft}$$

c. Tangent (T)

$$T = R\tan(I/2) = (381.97 \text{ ft})\tan(69.589/2) = 265.4 \text{ ft}$$

d. External distance (E)

$$E = R\left[\frac{1}{\cos\left(\frac{I}{2}\right)} - 1\right] = (381.97 \text{ ft})\left[\frac{1}{\cos\left(\frac{69.589}{2}\right)} - 1\right] = 83.16 \text{ ft}$$

e. Middle ordinate (M)

$$M = R\left[1 - \cos\left(\frac{I}{2}\right)\right] = (381.97 \text{ ft})\left[1 - \cos\left(\frac{69.589}{2}\right)\right] = 68.29 \text{ ft}$$

f. Long chord (LC)

$$LC = 2R\sin\left(\frac{I}{2}\right) = 2(381.97 \text{ ft})\sin\left(\frac{69.589}{2}\right) = 435.93 \text{ ft}$$

g. Length of the curve (L)

$$L = RI\left(\frac{\pi}{180}\right) = (381.97 \text{ ft})(69.589)\left(\frac{\pi}{180}\right) = 463.93 \text{ ft}$$

h. Station of the PC

$$PC = PI - T = (178 + 33.87) - (2 + 65.4) = 175 + 68.47$$

i. Station of the PT

$$PT = PC + L = (178 + 33.87) + (4 + 63.93) = 182 + 97.8$$

Answers

a. Intersection angle, $I = 69.589°$
b. Radius, $R = 381.97$ ft
c. Tangent, $T = 265.4$ ft
d. External distance, $E = 83.16$ ft
e. Middle ordinate, $M = 68.29$ ft
f. Long chord, $LC = 435.93$ ft
g. Length of the curve, $L = 463.93$ ft
h. Station of the PC = 175 + 68.47
i. Station of the PT = 182 + 97.8

Example 19.11: Computation of Horizontal Curve The intersection angle of a 6° curve is 60°50′, and the PC is located at 140 + 22.75. Determine the deflection angles and chord lengths for setting out the curve.

Solution

Step 1. Calculate R based on given data.

$$R = \frac{5729.58}{D} = \frac{5729.58}{6} = 954.93 \text{ ft}$$

Note: You should determine R-required based on superelevation, and friction factor. But those data are not given. This is why the above equation has been used.

Step 2. Calculate the length of the curve, L.

$$I = 60°50° = \left(60 + \frac{50}{60}\right)^{\circ} = 60.833°$$

$$L = RI\left(\frac{\pi}{180}\right) = (954.93 \text{ ft})(60.833)\left(\frac{\pi}{180}\right) = 1{,}013.89 \text{ ft}$$

Step 3. Divide L into stations; the first and the last division may be fraction of a station (say, 45.5 ft). Call this l.

The PC is located at 140 + 22.75

The PT is located at (140 + 22.75) + (10 + 13.89) = 150 + 36.64

The distance to first station from the PC = 141 − (140 + 22.75) = 77.25 ft

Step 4. For each l, calculate d. Then, calculate deflection angle $d/2$ and chord length (c) using $c = 2R\sin\left(\frac{d}{2}\right)$ as listed in Table 19.1.

Reference point = PC = 140 + 22.75

First station:

$$d_1 = \frac{180 l_1}{\pi R} = \frac{180(77.25 \text{ ft})}{\pi(954.93 \text{ ft})} = 4.64°$$

$$c_1 = 2R\sin\left(\frac{d_1}{2}\right) = 2(954.93 \text{ ft})\sin\left(\frac{4.64}{2}\right) = 77.31 \text{ ft}$$

$$\frac{d_1}{2} = \frac{4.64°}{2} = 2.32° = 2°19'12''$$

Second station:

$l_2 = 100$ ft

$$d_2 = \frac{180 l_2}{\pi R} = \frac{180(100 \text{ ft})}{\pi(954.93 \text{ ft})} = 6.0° \text{ (or simply } D\text{, degree by 100-ft arc)}$$

$$c_2 = 2R\sin\left(\frac{d_2}{2}\right) = 2(954.93 \text{ ft})\sin\left(\frac{6.0}{2}\right) = 99.95 \text{ ft}$$

Cumulative deflection angle, $d/2 + D/2 = 4.64/2 + 6.0/2 = 5.32° = 5°19'12''$

Last station:

$$l_{PT} = 36.64 \text{ ft}$$

$$d_{PT} = \frac{180 l_{PT}}{\pi R} = \frac{180(36.64 \text{ ft})}{\pi(954.93 \text{ ft})} = 2.198° = 2°11'53''$$

$$\frac{d_{PT}}{2} = \frac{2°11'53''}{2} = 1°5'57''$$

$$c_{PT} = 2R\sin\left(\frac{d_{PT}}{2}\right) = 2(954.93 \text{ ft})\sin\left(\frac{2.198}{2}\right) = 36.63 \text{ ft}$$

Station	Deflection angle	Chord length (ft)
140 + 22.75	0	0
141 + 00	0 + 2°19′12″ = 2°19′12″	77.31
142 + 00	2°19′12″ + 3° =5°19′12″	99.95
143 + 00	5°19′12″ + 3° = 8°19′12″	99.95
144 + 00	8°19′12″ + 3° = 11°19′12″	99.95
145 + 00	11°19′12″ + 3° = 14°19′12″	99.95
146 + 00	14°19′12″ + 3° = 17°19′12″	99.95
147 + 00	17°19′12″ + 3° = 20°19′12″	99.95
148 + 00	20°19′12″ + 3° = 23°19′12″	99.95
149 + 00	23°19′12″ + 3° = 26°19′12″	99.95
150 + 00	26°19′12″ + 3° = 29°19′12″	99.95
150 + 36.64	29°19′12″ + 1°5′57″ = 30°25′9″	36.64

Table 19.1 Chord Length Calculation for Example 19.11

Note: Some rounding off error may occur in the calculation. This will not affect the design and construction.

Check: The deflection angle to PT is half the intersection angle (I). This relationship can be a checking tool for whether the calculations are correct.

$$\frac{I}{2} = \frac{60°50'}{2} = 30°25' \approx 30°25'9''(\textit{calculated})$$

Therefore, the calculation is correct.

19.14.8 Spiral Curves

A spiral curve (Fig. 19.25) provides a gradual transition from moving in a straight line to moving in a curve around a point (or vice versa). The use of a spiral is about making the road or track follow the same form that the vehicle naturally takes. In a car, you don't go directly from going straight to fully turning. There is a transition area where you slowly turn the steering wheel. Lateral acceleration is slowly increased as the spiral is entered, or it is slowly decreased as the spiral is exited.

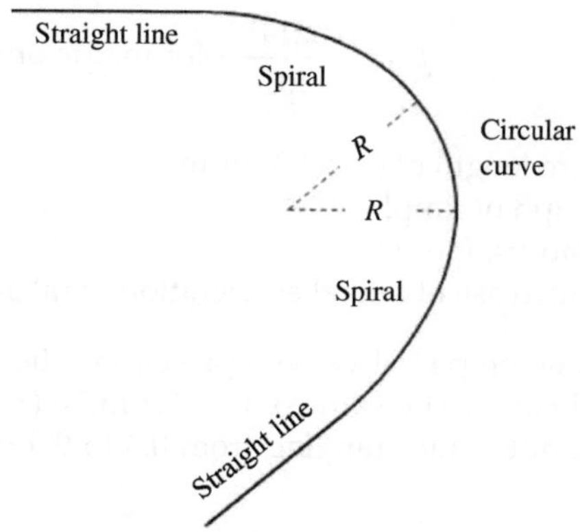

FIGURE 19.25 Spiral curve.

Spirals are widely used in the railroads. Cars on a road have the freedom to move from one side of the lane to the other, but a train does not have that luxury. A train is guided rigidly by the rails and has no room to deviate. On railroad tracks, all but very low-speed curves have spirals on both sides of the curve. This provides a gentle transition in lateral acceleration. This greatly improves passenger comfort. Spirals are used on freight lines because they also reduce the forces on the track components themselves. One disadvantage of the use of spirals is that it increases the amount of space required for each curve, i.e., lengthen the curve.

The principal advantages of transition curves in horizontal alignment are the following:

- A properly designed transition curve provides a natural, easy-to-follow path for drivers.
- Transition curves minimize encroachment on adjoining traffic lanes and tend to promote uniformity in speed.
- The transition curve length provides a suitable location for the superelevation runoff. The transition from the normal pavement cross-slope on the tangent to the fully superelevated section on the curve can be accomplished along the length of the transition curve in a manner that closely fits the speed–radius relationship for vehicles traversing the transition.
- A spiral transition curve also facilitates the transition in width where the traveled way is widened on a circular curve.
- The appearance of the highway or street is enhanced by applying spiral transition curves.

Standard Length of Spiral

The following equations [Eqs. (19.19) and (19.20)] for gradual attainment of lateral acceleration on railroad track curves are the basic expressions used by some highway agencies for computing minimum length of a spiral transition curve:

$$L = \frac{3.15V^3}{RC} \text{ for U.S. customary units} \tag{19.19}$$

$$L = \frac{0.0214V^3}{RC} \text{ for metric units} \tag{19.20}$$

where L = Minimum length of spiral, ft or m
V = Speed, mph or kmph
R = Curve radius, ft or m
C = Rate of increase of lateral acceleration, ft/s³ or m/s³

The factor C is an empirical value representing the comfort and safety levels provided by the spiral curve. The value of $C = 0.3$ m/s³ (1 ft/s³) is generally accepted for railroad operation, but values ranging from 0.3 to 0.9 m/s³ (1 to 3 ft/s³) are used for highways.

Minimum Length of Spiral

The minimum length of the spiral is calculated using two considerations: driver comfort and shifts in the lateral position of vehicles. Criteria based on driver comfort are intended to provide a spiral length that allows for a comfortable increase in lateral acceleration as a vehicle enters a curve. The criteria based on lateral shift are intended to provide a spiral curve that is sufficiently long to result in a shift in a vehicle's lateral position within its lane that is consistent with that produced by the vehicle's natural spiral path. For U.S. customary units, $L_{s,min}$ is the larger of the following equations:

$$L_{s,min} = \sqrt{24(p_{min})R} \tag{19.21}$$

$$L_{s,min} = \frac{3.15V^3}{RC} \tag{19.22}$$

For metric units, $L_{s,min}$ is the larger of the equations:

$$L_{s,min} = \sqrt{24(p_{min})R} \tag{19.23}$$

$$L_{s,min} = \frac{0.0214V^3}{RC} \tag{19.24}$$

where $L_{s,min}$ = Minimum length of spiral, ft or m
p_{min} = Minimum lateral offset between the tangent and circular curve, 0.66 ft or 0.20 m
V = Speed, mph or kmph
R = Curve radius, ft or m
C = Rate of increase of lateral acceleration, 4 ft/s³ or 1.2 m/s³

Maximum Length of Spiral

Spirals should not be so long (relative to the length of the circular curve) that drivers are misled about the sharpness of the approaching curve. A conservative maximum length of spiral can be computed using the equation:

$$L_{s,max} = \sqrt{24(p_{max})R} \tag{19.24}$$

where $L_{s,max}$ = Minimum length of spiral, ft or m
p_{max} = Minimum lateral offset between the tangent and circular curve, 3.3 ft or 1.0 m
V = Speed, mph or kmph
R = Curve radius, ft or m

19.14.9 General Controls for Horizontal Alignment

Besides the theoretical derivations described previously, there are some practical considerations in practice for efficient and smooth operations. Excessive curvature or poor traffic capacity causes economic losses due to excessive travel time. Some controls may be followed where appropriate to avoid these poor design practices (AASHTO, 2018):

- Alignment should be as practical as possible and should be consistent with the topography to preserve developed properties and community values.
- The designer should attempt to use generally flat curves, saving the minimum radius for the most critical conditions.
- Consistent alignment should always be sought. Sharp curves should not be introduced at the ends of long tangents. Sudden changes from areas of flat curvature to areas of sharp curvature should be avoided.
- For small deflection angles, curves should be sufficiently long to avoid the appearance of a kink.
- Sharp curvature should be avoided on long, high fills. In the absence of cut slopes, shrubs, and trees that extend above the level of the roadway, it is difficult for drivers to perceive the extent of curvature and adjust their operation accordingly.
- Abrupt reversals in alignment should be avoided.
- To avoid the appearance of inconsistent distortion, the horizontal alignment should be coordinated carefully with the profile design.
- Changing median widths on tangent alignments should be avoided, where practical, so as not to introduce a distorted appearance.

19.15 Vertical Curves

19.15.1 Background

Vertical curves are needed to provide smooth transitions between straight segments (tangents) of grade lines for highways and railroads in vertical planes (Fig. 19.26).

Two basic types of vertical curves exist, *crest* and *sag*. Crest type undergoes a negative change in grade; that is, the curve turns downward. Sag type undergoes a positive change in grade; that is, the curve turns upward. There are several factors that must be taken into account when designing a grade line of tangents and curves on any highway or railroad project. They include:

- Providing a good fit with the existing ground profile, thereby minimizing the depths of cuts and fills,
- Balancing the volume of cut material against fill,

FIGURE 19.26 A vertical curve.

- Maintaining adequate drainage,
- Not exceeding maximum specified grades, and
- Meeting fixed elevations such as intersections with other roads.

In addition, the curves must be designed to:

- Fit the grade lines they connect,
- Have lengths sufficient to meet specifications covering a maximum rate of change of grade (which affects the comfort of vehicle occupants), and
- Provide sufficient sight distance for safe vehicle operation.

19.15.2 Equation of an Equal Tangent Vertical Curve

Figure 19.27 shows a parabola that joins two intersecting tangents of a grade line. The parabola is essentially identical to that in Fig. 19.27, except that the terms used are those commonly employed by surveyors and engineers.

In the figure, PVC denotes the point of vertical curve, sometimes called the BVC (beginning of vertical curve), or VPC (vertical point of curvature); VPI is the vertex, often called vertical point of intersection; and PVT is the point of vertical tangency, interchangeably called the VPT (vertical point of tangency), or EVC (end of vertical curve). The percent grade of the back tangent (straight segment preceding PVI) is g_1, and that of the forward tangent (straight segment following PVI) is g_2. The curve length L is the horizontal distance (in stations) from the PVC to the PVT. The curve of Fig. 19.27 is called equal tangent because the horizontal distances from the PVC to PVI and from PVI to the PVT are equal, each being $L/2$. On the xy axis system, x values are horizontal

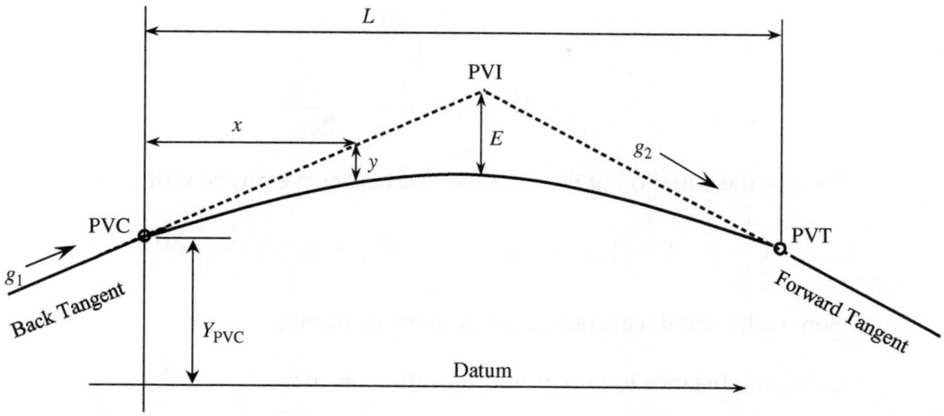

FIGURE 19.27 Terms for a vertical curve.

distances measured from the PVC, and y values are elevations measured from the vertical datum of reference. The following formulas can be derived from the above figure using trigonometry:

The general equation of parabola is $Y = c + bx + ax^2$ (19.25)

Now, let us find out how this equation can be applied to a vertical curve. First of all, when $x = 0$, $Y = c$. That means $c = Y_{PVC}$ (i.e., the elevation of the PVC) as $x = 0$ at the PVC.

Now let us find out b and a. On differentiation of the above equation,

$$\frac{d(Y)}{dx} = \frac{d}{dx}(c + bx + ax^2)$$

$$\frac{dY}{dx} = 0 + b + 2ax = b + 2ax \tag{19.26}$$

Now recall that $\frac{dY}{dx}$ means the slope of the curve at x. When $x = 0$, $\left.\frac{dY}{dx}\right|_{x=0} = b + 2a(0) = b$

That means, slope at $x = 0$ (at the PVC) is b, i.e., $b = g_1$.

Differentiating the equation one more time, $\frac{d}{dx}\frac{d(Y)}{dx} = \frac{d}{dx}(b + 2ax)$

$$\text{Or, } \frac{d}{dx}\frac{dY}{dx} = 2a$$

The above second-differentiation means the rate of change of slope = $2a$

$$\text{i.e., } \frac{\text{Change in slope}}{L} = 2a$$

$$\text{or, } \frac{g_2 - g_1}{L} = 2a$$

$$\text{Therefore, } a = \frac{g_2 - g_1}{2L} \tag{19.27}$$

Finally, the equation of the parabolic vertical curve can be written as:

$$Y = Y_{\text{PVC}} + g_1 x + ax^2 = Y_{\text{PVC}} + g_1 x + \left[\frac{g_2 - g_1}{2L}\right] x^2 \tag{19.28}$$

Some additional features can be derived as follows:

$$\text{Horizontal distance to min or max elevation on curve, } x_m = -\frac{g_1}{2a} = \frac{g_1 L}{g_1 - g_2} \tag{19.29}$$

$$\text{Tangent elevation} = Y_{\text{PVC}} + g_1 x \text{ and } = Y_{\text{PVI}} + g_2\left(x - \frac{L}{2}\right) \tag{19.30}$$

$$\text{Coefficient, } a = \frac{g_2 - g_1}{2L} \tag{19.31}$$

$$\text{Tangent offset at PV, } E = a\left(\frac{L}{2}\right)^2 \tag{19.32}$$

$$\text{Rate of change of grade, } r = \frac{g_2 - g_1}{L} \tag{19.33}$$

$$\text{Parameter, } K = \frac{L}{A} \tag{19.34}$$

$$\text{Station of PVC} = \text{Station of PVI} - L/2 \tag{19.35}$$

$$\text{Station of PVT} = \text{Station of PVI} + L/2 \tag{19.36}$$

$$\text{Elevation of PVC} = \text{Elevation of PVI} - g_1\,(L/2) \tag{19.37}$$

$$\text{Elevation of PVT} = \text{Elevation of PVI} + g_2\,(L/2) \tag{19.38}$$

where L = Length of curve
PVC = Point of vertical curvature
PVI = Point of vertical intersection
PVT = Point of vertical tangency
g_1 = Grade of back tangent (decimal)
g_2 = Grade of forward tangent (decimal)
x = Horizontal distance from PVC to point on curve
a = Parabola constant
y = Tangent offset
A = Absolute value of algebraic difference in grades (%)
E = Tangent offset at PVI = $\frac{AL}{800}$
r = Rate of change of grade
K = Rate of vertical curvature

Example 19.12: Tangent Slope of Sag Vertical Curve A sag vertical curve connects −2.3% and +3% tangents, and has a curve length of 300 ft. The station of the PVC is 12 + 00. Determine the tangent slope at station 14 + 00.

Solution

Length of curve, $L = 300$ ft
Grade of back tangent, $g_1 = -2.3\% = -0.023$
Grade of forward tangent, $g_2 = 3.0\% = 0.03$
Distance to the point of concern, x = station (14 + 00) − station (12 + 00) = 200 ft

$$Y = Y_{PVC} + g_1x + \left[\frac{g_2 - g_1}{2L}\right]x^2$$

$$\frac{d}{dx}(Y) = \frac{d}{dx}\left(Y_{PVC} + g_1x + \left[\frac{g_2 - g_1}{2L}\right]x^2\right)$$

$$\frac{dY}{dx} = 0 + g_1 + 2\left(\frac{g_2 - g_1}{2L}\right)x$$

$$\frac{dY}{dx} = g_1 + \left(\frac{g_2 - g_1}{L}\right)x$$

$$\left.\frac{dY}{dx}\right|_{x=200\text{ ft}} = -0.023 + \left(\frac{0.03 - (-0.023)}{300}\right)(200\text{ ft}) = 0.0123 = 1.23\%$$

Answer *The tangent slope at station 14 + 00 is 1.23%.*

Example 19.13: Parameters of Vertical Curve In a vertical curve, a back tangent with a −3% grade meets at station 1200 + 50 with a forward tangent of 4%, as shown in Fig. 19.28.

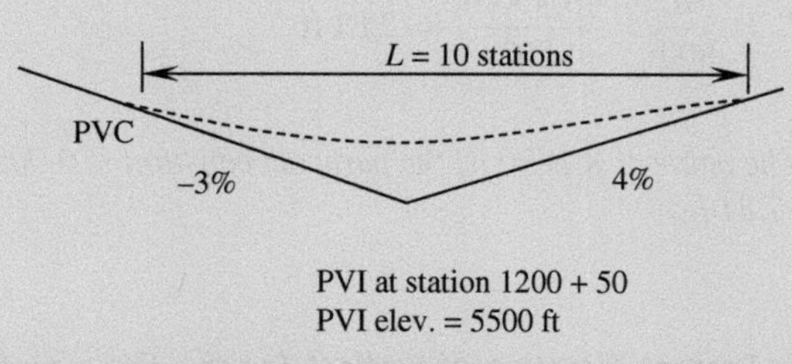

FIGURE 19.28 A sag vertical curve for Example 19.13.

Calculate the horizontal distance to the lowest point of the sag curve, and the tangent offset at the PVI.

Solution

Length of curve, $L = 10$ stations
Grade of back tangent, $g_1 = -3\% = -0.03$

Grade of forward tangent, $g_2 = 4.0\% = 0.04$

$$x_m = \frac{g_1 L}{g_1 - g_2} = \frac{-0.03(10 \text{ stations})}{-0.03 - 0.04} = 4.29 \text{ stations} \approx 4.3 \text{stations}$$

Absolute value of algebraic difference in grades (%), $A = |g_2 - g_1| = |4 - (-3)| = 7$

$$E = \frac{AL}{800} = \frac{7(10 \text{ stations})}{800} = 0.088 \text{ station} = 0.088 \text{ station} \left(100 \frac{\text{ft}}{\text{station}}\right) = 8.8 \text{ ft}$$

Answers *The horizontal distance to the lowest point of the sag curve is 4.3 stations and the tangent offset at the PVI is 8.8 ft.*

Example 19.14: Parameters of Vertical Curve A vertical 450-ft curve has the initial and the final grades of 2% and −3% respectively. Determine the value of K, the parabola constant, a, and the tangent offset at PVI.

Solution

Absolute value of algebraic difference in grades (%), $A = |g_2 - g_1| = |-3 - 2| = 5$
Grade of back tangent, $g_1 = 2\% = 0.02$
Grade of forward tangent, $g_2 = -3.0\% = -0.03$

$$K = \frac{L}{A} = \frac{450 \text{ ft}}{5} = 90 \text{ ft}$$

$$a = \frac{g_2 - g_1}{2L} = \frac{-0.03 - 0.02}{2(450 \text{ ft})} = -0.000055 \text{ per ft} = -0.0055\% \text{ per ft}$$

$$E = \frac{AL}{800} = \frac{(5)450 \text{ ft}}{800} = 2.81 \text{ ft}$$

Answers *The value of K is 90 ft, the parabola constant is 0.0055% per ft, and the tangent offset at* PVI *is 2.81 ft.*

Example 19.15: Tangent Elevation of Vertical Curve The point of the vertical curve is located at 3,200-ft elevation. The uphill grade is 2%. Determine the tangent elevation at 500 ft from the PVC.

Solution

Grade, $g_1 = 2\% = 0.02$
Point of vertical curve $Y_{PVC} = 3{,}200$ ft
Tangent elevation $= Y_{PVC} + g_1 x = 3{,}200 \text{ ft} + (0.02)(500 \text{ ft}) = 3{,}210 \text{ ft}$

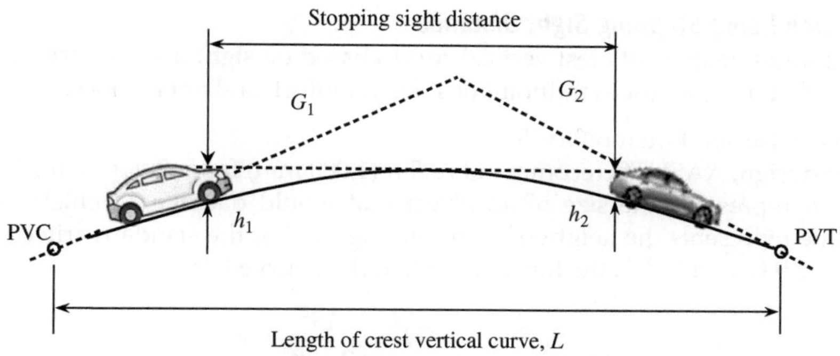

FIGURE 19.29 Parameters considered in determining the length of a crest vertical curve.

19.15.3 Sight Distances Related to Crest Vertical Curve

The vertical alignments of highways should provide ample sight distance for safe vehicular operation. Two types of sight distances are involved:

Stopping Sight Distance—the distance required for a given design speed to safely stop a vehicle, thus avoiding a collision with an unexpected stationary object in the roadway ahead (Fig. 19.29)

Passing Sight Distance—the distance required for a given design speed on two-lane two-way highways to safely overtake a slower moving vehicle, pass it, and return to the proper lane of travel leaving suitable clearance for an oncoming vehicle in the opposing lane (Fig. 19.29)

The basic equations for length of a crest vertical curve in terms of algebraic difference in grade and sight distance (stopping or passing) follow:

Case I: General Equation for $S \leq L$
The formula for length of curve L necessary to provide sight distance S on a crest vertical curve, where S is equal or less than L, is:

$$L = \frac{AS^2}{100\left(\sqrt{2h_1} + \sqrt{2h_2}\right)^2} \quad \text{(general equation for } S \leq L\text{)} \qquad (19.39)$$

where the units of S and L are stations if the English system is used, and one-tenth stations in the metric system. Also, the units of h_1 (the height of the driver's eye) and h_2 (the height of an object sighted on the roadway ahead) are in feet for the English system and meters for the metric system.

Case II: General Equation for $S > L$
The formula for length of curve L necessary to provide sight distance S on a crest vertical curve, where S is greater than L, is:

$$L = 2S - \frac{200\left(\sqrt{h_1} + \sqrt{h_2}\right)^2}{A} \quad \text{(general equation for } S > L\text{)} \qquad (19.40)$$

Design Using Stopping Sight Distance

Minimum lengths of crest vertical curves based on sight distance criteria generally are satisfactory from the standpoint of safety, comfort, and appearance.

Case I: Design Equation for $S \leq L$

For design, AASHTO recommends 3.5 ft (1,080 mm) for h_1, 2.0 ft for h_2. The lower value for h_2 represents the size of an object that would damage a vehicle and the higher value represents the height of an oncoming car. For the standard criteria of h_1 =3.50 ft, h_2 = 2.0 ft, and $S \leq L$, the formula for length of curve L is:

$$L = \frac{AS^2}{2{,}158} \tag{19.41}$$

Case II: Design Equation for $S > L$

For the standard criteria of h_1 =3.50 ft, h_2 =2.0 ft, and $S > L$, the formula for length of curve L is:

$$L = 2S - \frac{2{,}158}{A} \tag{19.42}$$

While calculating the length of curve required, any one case is assumed first, say $S \leq L$. After the computation, if it is found that $S > L$, then the calculation is redone using the other case.

Table 19.2 shows the computed K values for lengths of crest vertical curves corresponding to the stopping sight distances discussed previously for each design speed. For direct use in design, values of K are rounded as shown in the right column. These rounded values of K are higher than computed values, but the differences are not significant. The values of K listed in Table 19.2 uses the assumption "S is less than L," it can also be used without significant error where S is greater than L.

Design Using Passing Sight Distance

Design values of crest vertical curves for passing sight distance differ from those for stopping sight distance because of the different sight distance and object height criteria. Using the 1.08-m (3.50-ft) height of object results in the following specific formulas with the same terms as shown above:

Case I: Design Equation for $S \leq L$

$$L = \frac{AS^2}{2{,}800} \text{ for U.S. customary units} \tag{19.43}$$

$$L = \frac{AS^2}{864} \text{ for metric units} \tag{19.44}$$

Case II: Design Equation for $S > L$

$$L = 2S - \frac{2{,}800}{A} \text{ for U.S. customary units} \tag{19.45}$$

$$L = 2S - \frac{864}{A} \text{ for metric units} \tag{19.46}$$

Design speed (mph)	Stopping sight distance (ft)	Rate of vertical curvature, K^* design	Design speed (kmph)	Stopping sight distance (m)	Rate of vertical curvature, K^* design
15	80	3	20	20	1
20	115	7	30	35	2
25	155	12	40	50	4
30	200	19	50	65	7
35	250	29	60	85	11
40	305	44	70	105	17
45	360	61	80	130	26
50	425	84	90	160	39
55	495	114	100	185	52
60	570	151	110	220	74
65	645	193	120	250	95
70	730	247	130	285	124
75	820	312	–	–	–
80	910	384	–	–	–

*Rate of vertical curvature, K, is the length of curve per percent algebraic difference in intersecting grades (A), $K = L/A$.

Source: *From AASHTO (2018). A Policy on Geometric Design of Highways and Streets. Washington, DC: American Association of State Highway and Transportation Officials. Table 3-34. Used with permission.*

TABLE 19.2 Design K for Crest Vertical Curves Based on Stopping Sight Distance

For the minimum passing sight distances listed in Table 19.3, the minimum lengths of crest vertical curves are substantially longer than those for stopping sight distances. The extent of difference is evident by the values of K, or length of vertical curve per percent change in A, for passing sight distances listed in Table 19.3.

It is impractical to design a crest vertical curve with scope for passing sight distance because of high cost, where crest cuts are involved and there is the difficulty of fitting the resulting long vertical curves to the terrain, particularly for high-speed roads. Passing sight distance on crest vertical curves may be practical on roads with unusual combinations of low design speeds and gentle grades or higher design speeds with very small algebraic differences in grades. Ordinarily, passing sight distance is provided only at locations where combinations of alignment and profile do not need significant grading.

Example 19.16: Computation of a Crest Vertical Curve A crest vertical curve connects a +3.33% grade and a −5.55% grade. The PVI is at station 121 + 22 at an elevation of 423.89 ft. The design speed is 60 mph.

Determine the following:

a. Length of the vertical curve using the AASHTO method ("K" factors)

b. Station of the PVC

c. Elevation of the PVC

Design speed (mph)	Passing sight distance (ft)	Rate of vertical curvature, K* design	Design speed (kmph)	Passing sight distance (m)	Rate of vertical curvature, K* design
20	400	57	30	120	17
25	450	72	40	140	23
30	500	89	50	160	30
35	550	108	60	180	38
40	600	129	70	210	51
45	700	175	80	245	69
50	800	229	90	280	91
55	900	289	100	320	119
60	1,000	357	110	355	146
65	1,100	432	120	395	181
70	1,200	514	130	440	224
75	1,300	604	–	–	–
80	1,400	700	–	–	–

*Rate of vertical curvature, K, is the length of curve per percent algebraic difference in intersecting grades (A), $K = L/A$.
Source: *From AASHTO (2048). A Policy on Geometric Design of Highways and Streets. Washington, DC: American Association of State Highway and Transportation Officials. Table 3-35. Used with permission.*

Table 19.3 Design K for Crest Vertical Curves Based on Passing Sight Distance

d. Station of the PVT
e. Elevation of the PVT
f. Station of the high point
g. Elevation of the high point
h. Elevation of station 123 + 50

Solution

Grade of back tangent, $g_1 = 3.33\% = 0.033$
Grade of forward tangent, $g_2 = -5.55\% = -0.055$
Design speed = 60 mph
Location of PVI = 121 + 22
Elevation of PVI, $Y_{PVI} = 423.89$ ft

a. Length of the vertical curve using the AASHTO method ("K" factors)
$K = 151$ from the Table 19.2 for stopping sight distance criterion
$K = 357$ from the Table 19.3 for passing sight distance criterion
Use $K = 151$ for economic design (passing sight distance criterion is rarely used)
Absolute value of algebraic difference in grades (%), $A = |g_2 - g_1| = |-5.55-3.33| = 8.88$
$L = KA = 151(8.88) = 1{,}340.88$ ft

b. Station of the PVC
PVC = PVI − $L/2$
$L/2$ = 1,340.88 ft /2 = 670.44 ft
PVC = PVI − $L/2$ = (121 + 22) − (6 + 70.44) = 114 + 51.56

c. Elevation of the PVC
PVC = PVI − g_1 ($L/2$)
PVC = 423.89 ft − (0.033) (670.44 ft) = 401.77 ft

d. Station of the PVT
PVT = PVI + $L/2$
PVT = PVI + $L/2$ = (121 + 22) + (6 + 70.44) = 127 + 92.44

e. Elevation of the PVT
PVT = PVI + g_2 ($L/2$)
PVT = 423.89 ft + (−0.055) (670.44 ft) = 387.0 ft

f. Station of the high point
Distance from the PVC to the maximum point of curve:

$$x_m = \frac{g_1 L}{g_1 - g_2} = \frac{0.033(1{,}340.88\text{ ft})}{0.033 - (-0.055)} = 502.83\text{ ft}$$

Station of x_m = PVC + x_m = (114 + 51.56) + (5 + 2.83) = 119 + 54.39

g. Elevation of the high point

$$Y_{\text{high}} = Y_{\text{PVC}} + g_1 x_m + \left[\frac{g_2 - g_1}{2L}\right] x_m^2 = 401.77\text{ ft} + 0.033(502.83\text{ ft})$$
$$+ \left(\frac{-0.055 - 0.033}{2(1{,}340.88\text{ ft})}\right)(502.83\text{ ft})^2$$
$$= 410.1\text{ ft}$$

h. Elevation of station 123 + 50
Distance from the PVC = (123 + 50) − (114 + 51.56) = 898.44 ft

$$Y = Y_{\text{PVC}} + g_1 x + \left[\frac{g_2 - g_1}{2L}\right] x^2 = 401.77\text{ ft} + 0.033(898.44\text{ ft})$$
$$+ \left(\frac{-0.055 - 0.033}{2(1{,}340.88\text{ ft})}\right)(898.44\text{ ft})^2$$
$$= 404.93\text{ ft}$$

Answers

a. *Length of the vertical curve = 1,340.88 ft*
b. *Station of the PVC = 114 + 51.56*
c. *Elevation of the PVC = 401.77 ft*
d. *Station of the PVT = 127 + 92.44*
e. *Elevation of the PVT = 387.0 ft*
f. *Station of the high point = 119 + 54.39*
g. *Elevation of the high point = 410.1 ft*
h. *Elevation of station 123 + 50 = 404.93 ft*

19.15.4 Sight Distances Related to Sag Vertical Curve

At least four different criteria for establishing lengths of sag vertical curves are recognized to some extent, which are as follows:

1. Headlight sight distance
2. Passenger comfort
3. Drainage control
4. General appearance

Headlight Sight Distance

Headlight sight distance has been used directly by some agencies and for the most part is the basis for determining the length of sag vertical curves recommended here. When a vehicle traverses a sag vertical curve at night, the portion of highway lighted ahead is dependent on the position of the headlights and the direction of the light beam. A headlight height of 0.60 m (2 ft) and a 1-degree upward divergence of the light beam from the longitudinal axis of the vehicle are commonly assumed. The upward spread of the light beam above the 1-degree divergence angle provides some additional visible length of roadway, but is not generally considered in design. The following equations show the relationships between S, L, and A, using S as the distance between the vehicle and point where the 1-degree upward angle of the light beam intersects the surface of the roadway:

Case I: Design Equation for $S \leq L$

$$L = \frac{AS^2}{400 + 3.5S} \text{ for U.S. customary units} \tag{19.47}$$

$$L = \frac{AS^2}{120 + 3.5S} \text{ for metric units} \tag{19.48}$$

Case II: Design Equation for $S > L$

$$L = 2S - \frac{400 + 3.5S}{A} \text{ for U.S. customary units} \tag{19.49}$$

$$L = 2S - \frac{120 + 3.5S}{A} \text{ for metric units} \tag{19.50}$$

where L = Length of sag vertical curve, ft or m
A = Algebraic difference in grades, %
S = Light beam distance, ft or m

Passenger Comfort

Comfort due to change in vertical direction is not easily measured because it is affected appreciably by vehicle body suspension, vehicle body weight, tire flexibility, and other factors. Limited attempts at such measurements have led to the broad conclusion that riding is comfortable on sag vertical curves when the centripetal acceleration does not exceed 0.3 m/s^2 (1 ft/s^2). The general expression for such a criterion is:

$$L = \frac{AV^2}{46.5} \text{ for U.S. customary units} \tag{19.51}$$

$$L = \frac{AV^2}{395} \text{ for metric units} \tag{19.52}$$

where L = Length of sag vertical curve, ft or m
A = Algebraic difference in grades, %
V = Design speed, mph or kmph

Drainage Control

Drainage affects design of vertical curves where curbed sections are used. An approximate criterion for sag vertical curves is the same as that expressed for the crest conditions [i.e., a minimum grade of 0.30% should be provided within 15 m (50 ft) of the level point]. This criterion corresponds to K of 51 m (167 ft) per percent change in grade. The drainage criterion differs from other criteria in that the length of sag vertical curve determined for it is a maximum, whereas the length for any other criterion is a minimum. The maximum length of the drainage criterion is greater than the minimum length for other criteria up to 100 kmph (65 mph).

General Appearance

For improved appearance of sag vertical curves, previous guidance used a rule of thumb for minimum curve length of 30A in m (100A in ft), or K = 30 m (K = 100 ft) per percent change in grade. This approximation is a generalized control for small or intermediate values of A. Compared with headlight sight distance, it corresponds to a design speed of approximately 80 kmph (50 mph).

From the preceding discussion, it is evident that design controls for sag vertical curves differ from those for crests, and separate design values are needed. The headlight sight distance appears to be the most logical criterion for general use, and the values determined for stopping sight distances are within the limits recognized in current practice. The use of this criterion to establish design values for a range of lengths of sag vertical curves is recommended. As in the case of crest vertical curves, it is convenient to express the design control in terms of the K rate for all values of A. This entails some deviation from the computed values of K for small values of A, but the differences are not significant. Table 19.4 shows the range of computed values and the rounded values of K selected as design controls.

Example 19.17: Length of a Sag Vertical Curve Determine the minimum length of a sag vertical curve between a −2.7% grade and a +3.5% grade for a road with a 60-mph design speed. The vertical curve must provide 700 ft of headlight sight distance, the AASHTO passenger comfort, the AASHTO drainage condition, and the AASHTO appearance standard. Round up to the next greatest 10-ft interval.

Design speed (mph)	Stopping sight distance (ft)	Rate of vertical curvature, K* design	Design speed (kmph)	Stopping sight distance (m)	Rate of vertical curvature, K* design
15	80	10	20	20	3
20	115	17	30	35	6
25	155	26	40	50	9
30	200	37	50	65	13
35	250	49	60	85	18
40	305	64	70	105	23
45	360	79	80	130	30
50	425	96	90	160	38
55	495	115	100	185	45
60	570	136	110	220	55
65	645	157	120	250	63
70	730	181	130	285	73
75	820	206	–	–	–
80	910	231	–	–	–

*Rate of vertical curvature, K, is the length of curve per percent algebraic difference in intersecting grades (A), $K = L/A$.

Source: *From AASHTO (2018). A Policy on Geometric Design of Highways and Streets. Washington, DC: American Association of State Highway and Transportation Officials. Table 3-36. Used with permission.*

TABLE 19.4 Design Controls for Sag Vertical Curves

Solution

Headlight Sight Distance Criterion

Grade of back tangent, $g_1 = -2.7\%$

Grade of forward tangent, $g_2 = 3.5\%$

Design speed = 60 mph

Required stopping sight distance, $S = 700$ ft

Absolute value of algebraic difference in grades (%), $A = |g_2 - g_1| = |3.5 - (-2.7)| = 6.2$

Assume S is less than L, then $L = \dfrac{AS^2}{400 + 3.5S} = \dfrac{(6.2)(700\text{ ft})^2}{400 + 3.5(700\text{ ft})} = 1{,}066$ ft

S is not less than L; assumption not okay.

$$\text{Then, } L = 2S - \frac{400 + 3.5S}{A} = 2(700\text{ ft}) - \frac{400 + 3.5(700\text{ ft})}{6.2} = 940\text{ ft}$$

Passenger Comfort Criterion

$$L = \frac{AV^2}{46.5} = \frac{(6.2)(60\text{ ft})^2}{46.5} = 480\text{ ft}$$

Drainage Criterion

$K = 167$ ft per % change in grade

$L = KA = 167\ (6.2) = 1{,}035.4$ ft

General Appearance Criterion

$K = 100$ ft

$L = KA = 100\ (6.2) = 620$ ft

Combining all four criteria, drainage criterion controls, L should be 1,035.4 ft ≈ 1,040 ft.

Answer *The minimum length of the sag vertical curve is 1,040 ft.*

19.15.5 Sight Distances Related to Sag Vertical Curve at Undercrossing

Sight distance on the highway through a grade separation should be at least as long as the minimum stopping sight distance and preferably longer. Design of the vertical alignment is the same as at any other point on the highway except in some cases of sag vertical curves underpassing a structure, as illustrated in Fig. 19.30. While not a frequent concern, the structure fascia may cut the line of sight and limit the sight distance to less than otherwise is attainable. In general, it is practical to provide the minimum length of sag vertical curve at the grade separation structures, and even where the recommended grades are exceeded, the sight distance should not need to be reduced below the minimum recommended values for stopping sight distance.

The general equations for sag vertical curve length at undercrossing are as follows:

Case 1—Sight distance greater than length of vertical curve ($S > L$):

$$L = 2S - \frac{800\left[C - \left(\frac{h_1 + h_2}{2}\right)\right]}{A} \tag{19.53}$$

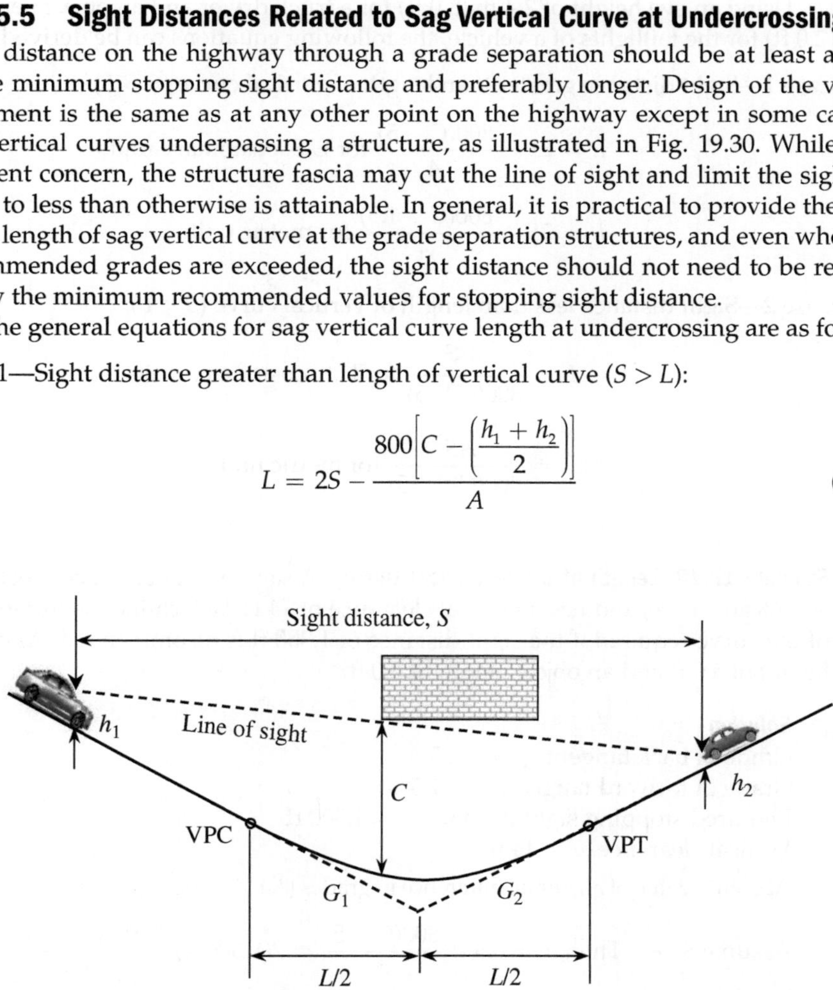

Figure 19.30 Parameters considered in determining the length of a sag vertical curve under an overhead structure.

Case 2—Sight distance less than length of vertical curve ($S < L$):

$$L = \frac{AS^2}{800\left[C - \left(\frac{h_1 + h_2}{2}\right)\right]} \tag{19.54}$$

where L = Length of sag vertical curve, ft or m
A = Algebraic difference in grades, %
S = Sight distance, ft or m
C = Vertical clearance, ft or m
h_1 = Height of eye, ft or m
h_2 = Height of object, ft or m

Using an eye height of 2.4 m (8.0 ft) for a truck driver and an object height of 0.6 m (2.0 ft) for the taillights of a vehicle, the following equations can be derived:

Case 1—Sight distance greater than length of vertical curve ($S > L$):

$$L = 2S - \frac{800(C - 5)}{A} \text{ for U.S. customary units} \tag{19.55}$$

$$L = 2S - \frac{800(C - 1.5)}{A} \text{ for metric units} \tag{19.56}$$

Case 2—Sight distance less than length of vertical curve ($S < L$):

$$L = \frac{AS^2}{800(C - 5)} \text{ for U.S. customary units} \tag{19.57}$$

$$L = \frac{AS^2}{800(C - 1.5)} \text{ for metric units} \tag{19.58}$$

Example 19.18: Length of a Sag Vertical Curve A sag vertical curve connects grades of −2.5% and 1.5%, and has a vertical clearance of 14 ft. Determine the minimum length of the curve required if the sight distance of 1,300 ft is recommended. Assume an eye height of 8.0 ft and an object height of 2.0 ft.

Solution

Grade of back tangent, $g_1 = -2.5\%$
Grade of forward tangent, $g_2 = 1.5\%$
Required stopping sight distance, $S = 1{,}300$ ft
Vertical clearance, $C = 14$ ft
Absolute value of algebraic difference in grades (%), $A = |g_2 - g_1| = |1.5 - (-2.5)| = 4.0$

Assume $S > L$. Then, $L = 2S - \frac{800(C - 5)}{A} = 2(1{,}300 \text{ ft}) - \frac{800(14 - 5)}{4.0} = 800$ ft

As $S > L$, the calculation is valid.

Answer *The minimum length of the curve is 800 ft.*

Example 19.19: Length of the Crest Vertical Curve A highway is to be designed with the required stopping sight distance of 700 ft. The equal-tangent crest vertical curve must be designed to connect grades of 1.5% and −2.5%. Determine the minimum length of the curve required. Assume $S \leq L$ and standard criteria of h_1 and h_2.

Solution

Grade of back tangent, $g_1 = 1.5\%$

Grade of forward tangent, $g_2 = -2.5\%$

Required stopping sight distance, $S = 700$ ft

Absolute value of algebraic difference in grades (%), $A = |g_2 - g_1| = |-2.5 - 1.5| = 4.0$

Length of the curve, $L = \dfrac{AS^2}{2{,}158} = \dfrac{4(700 \text{ ft})^2}{2{,}158} = 908 \text{ ft}$

Answer *The minimum length of the curve is 908 ft.*

Example 19.20: Length of the Crest Vertical Curve A highway is to be designed with the required stopping sight distance of 700 ft. The equal-tangent crest vertical curve must be designed to connect grades of 1.5% and −1.5% with $S > L$. Determine the minimum length of the curve required assuming standard criteria of h_1 and h_2.

Solution

Grade of back tangent, $g_1 = 1.5\%$

Grade of forward tangent, $g_2 = -1.5\%$

Required stopping sight distance, $S = 700$ ft

Absolute value of algebraic difference in grades (%), $A = |g_2 - g_1| = |-1.5 - 1.5| = 3.0$

Length of the curve,

$$L = 2S - \frac{2{,}158}{A} = 2(700 \text{ ft}) - \frac{2{,}158}{3.0} = 681 \text{ ft}$$

Answer *The minimum length of the curve is 681 ft.*

Example 19.21: Length of a Sag Vertical Curve A highway is to be designed with the minimum stopping sight distance of 700 ft. The sag vertical curve is to be designed based on the standard headlight criteria connecting −1.5% and 1.5% curves. Assuming the length of the curve is less than the sight distance, determine the required curve length.

Solution

Grade of back tangent, $g_1 = -1.5\%$

Grade of forward tangent, $g_2 = 1.5\%$

Required stopping sight distance, $S = 700$ ft

Absolute value of algebraic difference in grades (%), $A = |g_2 - g_1| = |1.5 - (-1.5)| = 3.0$

$$L = 2S - \frac{400 + 3.5S}{A} = 2(700 \text{ ft}) - \frac{400 + 3.5(700 \text{ ft})}{3.0} = 450 \text{ ft}$$

Answer *The minimum length of the curve is 450 ft.*

19.15.6 Setting Vertical Curve

Equation (19.28) shows the elevation of any point on the curve can be determined using the following equation:

$$Y = Y_{PVC} + g_1 x + ax^2 = Y_{PVC} + g_1 x + \left[\frac{g_2 - g_1}{2L}\right] x^2$$

This equation shows that while setting a vertical curve, the elevation at the beginning of curve (PVC), proposed grades (g_1 and g_2), and the length of the curve (L) are required. The following steps can be followed for setting vertical curve:

Step 1. Determine the minimum length of the curve, L. You may need to satisfy sight distance required based on selected design criteria (comfort, appearance, overhead obstruction, etc.). Design tables can also be used for simplicity.

Step 2. Compute the elevation of the beginning of the curve, PVC. If the elevation of PVI is known, then compute the elevation of PVC.

Step 3. Locate the distance (x) from the PVC along the length of curve where the elevations are to be computed. Commonly, x is chosen as a station (1 station = 100 ft). The first and the last distances (x) may be a fraction of a station (say, 42 ft, 77 ft).

Step 4. Use Eq. (19.28) and compute the elevation (Y) for different values of x.

Example 19.22: Computation of a Crest Vertical Curve An equal-tangent crest vertical curve joining 3% and −2% grade is to be designed. If the tangents intersect at station (248 + 00) at an elevation of 2,250 ft, calculate the elevations of intermediate points on the curve at the whole stations. Assume rate of vertical curvature, $K = 300$.

Solution

Grade of back tangent, $g_1 = 3\%$
Grade of forward tangent, $g_2 = -2\%$
Location of PVI = (248 + 00) station
Elevation of $Y_{PVI} = 2{,}250$ ft
Rate of vertical curvature, $K = 300$

Step 1. Determine the minimum length of the curve, L.

$$K = \frac{L}{A}$$

Absolute value of algebraic difference in grades (%), $A = |g_2 - g_1| = |-2.0 - 3.0| = 5.0$

$L = KA = 300(5.0) = 1{,}500$ ft

Step 2. Compute the elevation of the beginning of the curve, PVC.

Elevation of PVC = 2,250 − 3% of 750 ft = 2,227.5 ft

Station of PVC = (248 + 00) − (15 + 00)/2 = 240 + 50
Station of PVT = (240 + 50) + (15 + 00) = 255 + 50

Step 3. Locate the points along the length of curve where the elevations are to be computed.

Therefore, if the PVC is the reference point, then the first point should be at 50 ft (station 241). After that, there will be 14 full stations up to station 255, and then another 50 ft to reach the PVT.

Step 4. Use the equation of vertical curve and compute the elevation (Y) for different values of x as listed in Table 19.5.

$$Y = Y_{\text{PVC}} + g_1x + \left[\frac{g_2 - g_1}{2L}\right]x^2$$

$$Y_0 = 2{,}227.5 + (0.03)(0) + \left[\frac{-0.02 - 0.03}{2(1{,}500)}\right](0)^2 = 2{,}227.5$$

$$Y_{50} = 2{,}227.5 + (0.03)(50) + \left[\frac{-0.02 - 0.03}{2(1{,}500)}\right](50)^2 = 2{,}228.96$$

Station	Distance from PVC	Elevation, $Y = Y_{\text{PVC}} + g_1x + \left[\frac{g_2 - g_1}{2L}\right]x^2$
240 + 50	0	2,227.50
241 + 00	50	2,228.96
242 + 00	150	2,231.63
243 + 00	250	2,233.96
244 + 00	350	2,235.96
245 + 00	450	2,237.63
246 + 00	550	2,238.96
247 + 00	650	2,239.96
248 + 00	750	2,240.63
249 + 00	850	2,240.96
250 + 00	950	2,240.96
251 + 00	1,050	2,240.63
252 + 00	1,150	2,239.96
253 + 00	1,250	2,238.96
254 + 00	1,350	2,237.63
255 + 00	1,450	2,235.96
255 + 50	1,500	2,235.00

Table 19.5 Elevation Calculation for Example 19.22

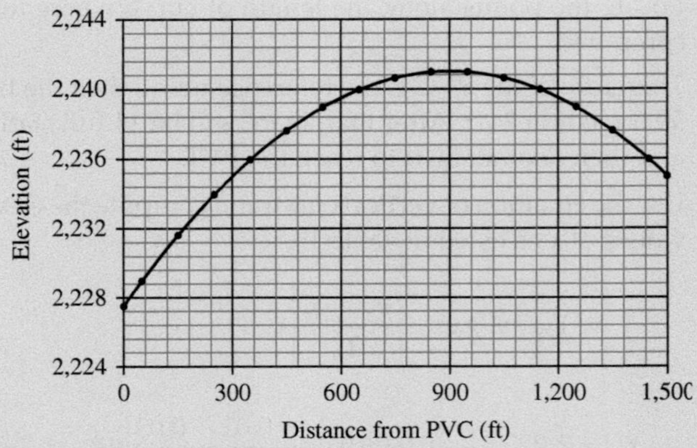

FIGURE 19.31 Plot of elevation for Example 19.22.

Check:
Distance from the PVC to the maximum point of curve:

$$x_m = \frac{g_1 L}{g_1 - g_2} = \frac{0.03(1{,}500 \text{ ft})}{0.03 - (0.02)} = 900 \text{ ft}$$

Figure 19.31 shows the maximum point on the curve is at 900 ft from the PVC.

One thing to learn from this problem is that the main aspect of setting a vertical curve is to find out the length of curve. The rest of the part is following the calculation.

Example 19.23: Computation of a Sag Vertical Curve A sag vertical curve joining −3% and 4% grade is to be designed for design speed of 70 mph. If the elevation of the PVC is 123.72 ft located at station 122 + 77, calculate the elevations of intermediate points on the curve at the whole stations.

Solution

Grade of back tangent, $g_1 = -3\%$
Grade of forward tangent, $g_2 = 4\%$
Location of PVC = (122 + 77) station
Elevation of $Y_{PVC} = 123.72$ ft
Design speed = 70 mph

Step 1. Determine the minimum length of the curve, L.

$$K = \frac{L}{A}$$

$K = 181$ for speed of 70 mph from Table 19.4.

Absolute value of algebraic difference in grades (%), $A = |g_2 - g_1| = |4.0 - (-3.0)| = 7.0$

$$L = KA = 181(7.0) = 1{,}267 \text{ ft.}$$

Step 2. Compute the elevation of the beginning of the curve, PVC.

Elevation of PVC = 123.72 ft

Step 3. Locate the points along the length of the curve where the elevations are to be computed.

Station of PVC = 122 + 77
Station of PVT = (122 + 77) + (12 + 67) = 135 + 44

Therefore, if the PVC is the reference point, then the first point should be at 23 ft (station 123). After that, there will be 12 full stations up to station 135, and then another 44 ft to reach the PVT.

Step 4. Use Eq. (19.28) and compute the elevation (Y) for different values of x as listed in Table 19.6.

$$Y = Y_{\mathrm{PVC}} + g_1 x + \left[\frac{g_2 - g_1}{2L}\right]x^2$$

$$Y_0 = 123.72 + (-0.03)(0) + \left[\frac{0.04 - (-0.03)}{2(1{,}267)}\right](0)^2 = 123.72 \text{ ft}$$

$$Y_{23} = 123.72 + (-0.03)(23) + \left[\frac{0.04 - (-0.03)}{2(1{,}267)}\right](23)^2 = 123.04 \text{ ft}$$

Station	Distance from PVC	Elevation, $Y = Y_{\mathrm{PVC}} + g_1 x + \left[\frac{g_2 - g_1}{2L}\right]x^2$
122 + 77	0	123.72
123 + 00	23	123.04
124 + 00	123	120.45
125 + 00	223	118.40
126 + 00	323	116.91
127 + 00	423	115.97
128 + 00	523	115.59
129 + 00	623	115.75
130 + 00	723	116.47
131 + 00	823	117.74
132 + 00	923	119.56
133 + 00	1,023	121.94
134 + 00	1,123	124.87
135 + 00	1,223	128.35
135 + 44	1,267	130.06

TABLE 19.6 Elevation Calculation for Example 19.23

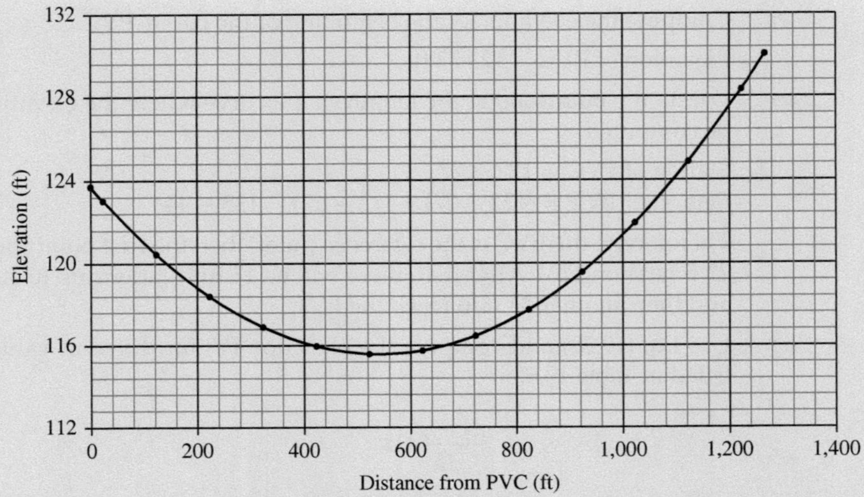

FIGURE 19.32 Plot of elevation for Example 19.23.

Check:
Distance from the PVC to the maximum point of the curve:

$$x_m = \frac{g_1 L}{g_1 - g_2} = \frac{-0.03(1{,}267\,\text{ft})}{-0.03 - (0.04)} = 543\ \text{ft}$$

Figure 19.32 shows the maximum point on the curve is at 543 ft from the PVC.

19.15.7 General Controls for Vertical Alignment

In addition to the specific controls for vertical alignment discussed previously, there are several general controls that should be considered in design.

- A smooth grade line with gradual changes, as consistent with the type of highway, road, or street and the character of terrain, should be sought for in preference to a line with numerous breaks and short lengths of grades.
- The roller-coaster or the hidden-dip type of profile should be avoided. Such profiles generally occur on relatively straight, horizontal alignment where the roadway profile closely follows a rolling natural ground line.
- A "broken-back" gradeline (two vertical curves in the same direction separated by a short section of tangent grade) generally should be avoided.
- On long grades, it may be preferable to place the steepest grades at the bottom and flatten the grades near the top of the ascent.
- Where at-grade intersections occur on roadway sections with moderate to steep grades, it is desirable to reduce the grade through the intersection. Such profile changes are beneficial for vehicles making turns and reduce crashes.
- Sag vertical curves should be avoided in cuts unless adequate drainage can be provided.

19.16 Other Features Affecting Geometric Design

In addition to the design elements discussed previously, several other features affect or are affected by the geometric design of a roadway. Some are discussed below:

19.16.1 Erosion Control and Landscape Development

Erosion prevention is one of the main factors in road design, construction, and maintenance of highways. It should be considered early in the location and design stages. Some degree of erosion control can be incorporated into the geometric design, particularly in the cross-sectional elements. The most direct application of erosion control is, of course, in the drainage design and the writing of landscaping and slope planting specifications.

19.16.2 Rest Areas, Information Centers, and Scenic Overlooks

Rest areas, information centers, and scenic overlooks are functional and desirable elements of the complete highway facility and are provided to reduce driver fatigue and for the convenience of highway users. A safety rest area is a roadside area with parking facilities separated from the roadway, so that travelers can stop and rest for a short period of time. The area may provide drinking water, restrooms, tables and benches, telephones, information displays, and other facilities for travelers. A rest area is not intended for social or civic gatherings or for active recreational activities such as boating, swimming, or organized games. An information center is a staffed or unstaffed facility at a rest area for the purpose of furnishing travel and other information or services to travelers. A scenic overlook is a roadside area provided for motorists to park their vehicles, beyond the shoulder, primarily for viewing the scenery or for taking photographs in a location removed from through traffic. Scenic overlooks do not provide comfort and convenience facilities.

19.16.3 Lighting

Lighting may reduce nighttime crashes on a highway or street and improve the ease and comfort of operation thereon. Where pedestrian concentrations and intersectional interferences on the roadside exist, fixed source lighting tends to reduce accidents. Lighting of rural highways may be desirable, but the need for it is much less than on streets and highways in urban areas.

19.16.4 Utilities

Utilities have little effect on the geometric design of the highway or street. However, full consideration, reflecting sound engineering principles and economic factors, should be given to measures needed to preserve and protect the integrity and visual quality of roads, its efficiency, and safety. Depending on the location of a project, the utilities involved could include the following:

- Sanitary sewers
- Water supply lines
- Oil, gas, and petroleum product pipelines
- Overhead and underground power and communications lines
- Cable television

- Wireless communication towers
- Drainage and irrigation lines
- Heating mains
- Special tunnels for building connections

19.17 Summary

The geometric design of the highway focuses primarily on the driving comfort, safety, and efficiency of the passenger. It does not deal with materials, structural strength, and durability. Due to vast changes in landscape in the United States, it is always not possible to provide all these design features discussed in this chapter. The design engineer balances the theoretical aspects, the conditions of the land, and the community's cost-effectiveness. However, strict application of these guidelines will not guarantee obtaining a good design. The following key ingredients are also required:

- *Consistency*. Geometric design should provide positive guidance to the drivers to achieve safety and efficiency and should avoid abrupt changes in guidelines. Highways must be designed to conform to driver expectations.
- *Esthetics*. Visual quality can be achieved by careful attention to coordinating horizontal and vertical alignments and to landscape developments. The process can be greatly aided by using computer perspectives and physical models.
- *Engineering judgment*. Experience and skills of the designer are important in producing a good design. Considerable creativity is required in developing a design that addresses environmental and economic concerns.

19.18 Fundamentals of Engineering (FE) Exam-Style Questions

FE19.1 A horizontal curve has the following data:

$I = 36°50'$
$R = 750$ ft
Station of PI $= 8 + 00$

The station of the PT is most nearly:

A. $9 + 52$
B. $9 + 77$
C. $10 + 05$
D. $10 + 32$

Solution D

$I = 36°50' = 36.83°$

$$T = R\tan\left(\frac{I}{2}\right) = 750\tan\left(\frac{36.83}{2}\right) = 250 \text{ ft}$$

$$L = \text{RI}\left(\frac{\pi}{180}\right) = 750(36.83)\left(\frac{\pi}{180}\right) = 482 \text{ ft}$$

$\text{PC} = \text{PI} - T = 8 + 00 - 250 = 550 \text{ ft} = 5 + 50$
$\text{PT} = \text{PC} + L = 5 + 50 + 482 = 10 + 32$

FE19.2 The advantage of using a parabola in the vertical design of highways and railroads is:

A. Easy construction
B. Easy calculation
C. Looks beautiful
D. Constant rate of change of grade

Solution D

Because parabolas provide a constant rate of change of grade, they provide the best comfort to the drivers.

FE19.3 A two-lane rural road has a 10-degree horizontal curve extending for 750 ft along its centerline. The road is 20 ft wide with 2-ft-wide shoulders. The design speed for this road is 50 mph. The required horizontal sight-line offset for this curve so that tress can be removed from the edge of the road is:

A. 28.7 ft
B. 38.7 ft
C. 48.7 ft
D. 18.7 ft

Solution A

$$R = \frac{5{,}729.58}{10} = 572.958 \text{ ft}$$

$$\text{SSD} = 1.47Vt + \frac{V^2}{30\left(\left(\frac{a}{32.2}\right) + G\right)} = 1.47(50)(2.5\text{s}) + \frac{(50)^2}{30\left(\frac{11.2}{32.2} + 0\right)} = 423.3\,\text{ft}$$

$$\text{HSO} = R\left[1 - \cos\left(\frac{28.65S}{R}\right)\right] = 572.9\left[1 - \cos\left(\frac{28.65(423.3)}{572.9}\right)\right] = 38.7 \text{ ft}$$

The question is asking from the edge of the road. So, 38.7 − 10 ft = 28.7 ft.

FE19.4 A 3° simple curve has the PI at station 177 + 50 with an intersection angle of 13°25′. The station of the point of curvature of the horizontal curve is most nearly:

A. 175 + 25
B. 174 + 50
C. 176 + 25
D. 172 + 75t

Solution A

$13°25' = 13 + 25/60 = 13.42°$

$\text{PC} = \text{PI} - T$

$$T = R\tan\left(\frac{I}{2}\right)$$

Therefore, we need to find out R first, then T and then PC.

$$l = Rd\left(\frac{\pi}{180}\right)$$

$$R = \frac{180l}{\pi d} = \frac{180(100 \text{ ft})}{\pi(3)} = 1{,}910 \text{ ft}$$

$$T = R\tan\left(\frac{I}{2}\right) = 1{,}910\tan\left(\frac{13.42}{2}\right) = 225 \text{ ft}$$

$$\text{PC} = \text{PI} - T = 17{,}750 - 225 = 17{,}525 = 175 + 25$$

FE19.5 For a sag vertical curve shown in Fig. FE19.5, the tangent slope (%) at station 9 + 00 is most nearly:

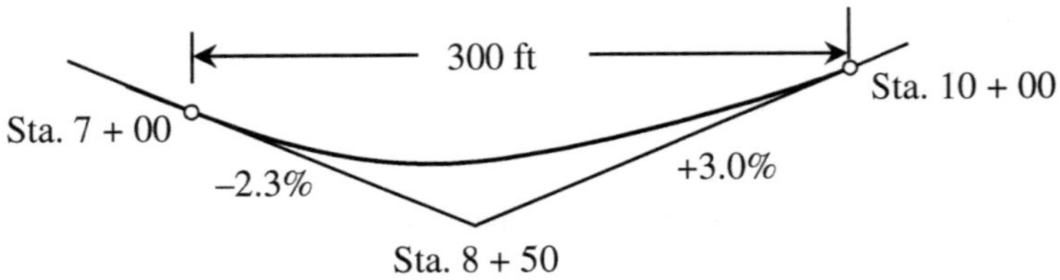

Figure FE19.5 The sag vertical curve for Question FE19.5.

A. 0.99
B. 1.23
C. 1.39
D. 1.52

Solution B

$$Y = Y_{PVC} + g_1 x + \left(\frac{g_2 - g_1}{2L}\right)x^2$$

$$\frac{dY}{dx} = \frac{d}{dx}(Y_{PVC}) + \frac{d}{dx}(g_1 x) + \left(\frac{g_2 - g_1}{2L}\right)\frac{d}{dx}(x^2)$$

$$Y' = g_1 + \left(\frac{g_2 - g_1}{L}\right)x$$

$x = (9 + 00) - (7 + 00)$ station $= 2 + 00$ station $= 200$ ft

Therefore,

$$Y' = g_1 + \left(\frac{g_2 - g_1}{L}\right)x = -0.023 + \left(\frac{0.03 - (-0.023)}{300}\right)(200) = 0.0123 = 1.23\%$$

19.19 Practice Problems

19.1 A horizontal curve has a radius of 800 ft and an intersection angle of 30°. Determine the tangent distance of the curve.

19.2 A horizontal curve has an external distance of 800 ft and a radius of 2,200 ft. Determine the middle ordinate of the curve.

19.3 A horizontal circular curve has an intersection angle of 38°, as shown in Figure P19.3. The station of point of curve (PC) is 12 + 00. The external distance (E) is 42 ft. Determine the radius (ft) of the curve.

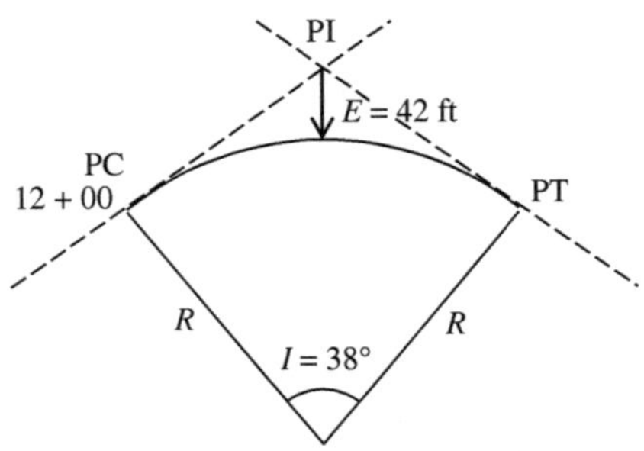

Figure P19.3 The horizontal circular curve for Prob. 19.3.

19.4 A 7-degree horizontal curve covers an angle of 63°15′34″. Determine its radius, the length of the curve, and the length of middle ordinate.

19.5 A horizontal curve is to be designed for a two-lane road in Colorado Springs. The following data are known:

Intersection angle: 40°
Tangent length = 450 ft
Station of PI: 3,300 + 12.65
$e = 10\%$
$f = 0.15$

Determine the following:

a. Maximum design speed feasible
b. Station of the PC
c. Station of the PT
d. Deflection angle and chord length to the first full-station from the PC

19.6 Given a circular horizontal curve connecting two tangents that intersect at an angle of 48°. The PI is at station (948 + 20) and the design speed of the highway is 70 mph. Determine the point of the tangent and the deflection angles from the PC to full stations for laying out the curve. Assume the superelevation of 8%.

19.7 Given a circular curve connecting two tangents that intersect at an angle of 64°. The PI is at station (948 + 20) and the design speed of the highway is 40 mph. Determine the point of the tangent and the deflection angles from the PC to full stations for laying out the curve. Assume the superelevation of 10%.

19.8 In a vertical curve, a back tangent with a −3% grade meets at station 1,200 + 50 station with a forward tangent of 4%, as shown in Figure P19.8.

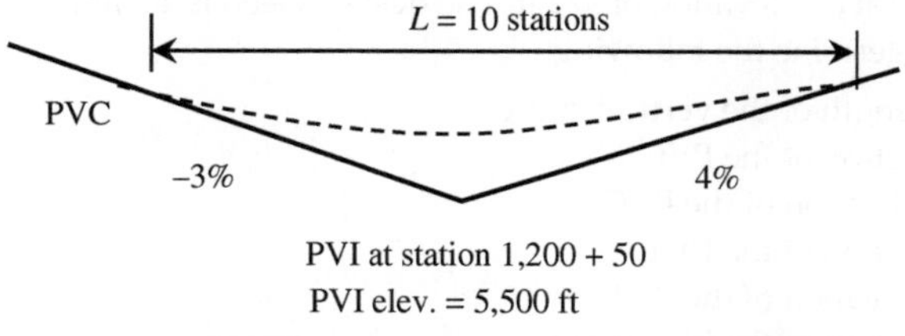

FIGURE P19.8 The sag vertical curve for Prob. 19.8.

Determine the rate of vertical curvature of the sag curve.

19.9 In a vertical curve, a back tangent with a −3% grade meets at station 1,200 + 50 station with a forward tangent of 4%, as shown in Figure P19.9.

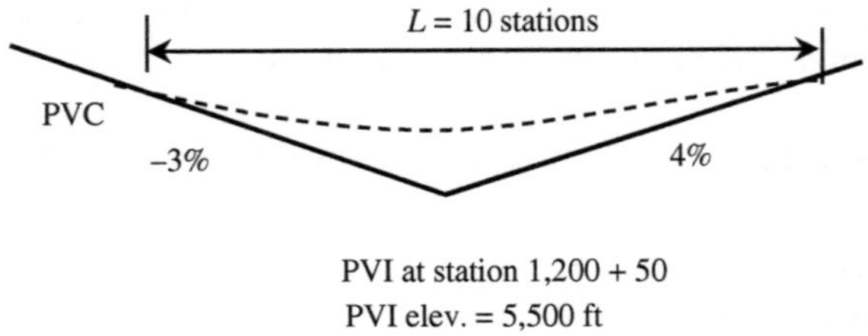

FIGURE P19.9 The sag vertical curve for Prob. 19.9.

Determine the parabola constant, *a*.

19.10 A vehicle is traveling at 17 mph on a horizontal curve. The radius of the curve is 25 ft and the rate of super elevation is 6%. Determine the road's side friction factor.

19.11 A sag vertical curve joining −2.3% and 3.33% grade is to be designed for design speed of 55 mph. If the elevation of the PVC is 4,123.75 ft located at station 3,045 + 22, calculate the elevations of intermediate points on the curve at the whole stations.

19.12 A sag vertical curve joining −3.33% and 2.3% grade is to be designed for design speed of 55 mph. If the elevation of the PVC is 4,123.75 ft located at station 3,045 + 22, calculate the elevations of intermediate points on the curve at the whole stations.

19.13 A crest vertical curve joining 3.9% and −1.9% grade is to be designed for design speed of 65 mph. If the elevation of the PVC is 1,722.45 ft located at station 0 + 00, calculate the elevations of intermediate points on the curve at the whole stations. Consider the criterion of stopping sight distance.

19.14 A crest vertical curve joining 1.6% and −4.1% grade is to be designed for design speed of 45 mph. If the elevation of the PVC is 645.22 ft located at station 10 + 55, calculate the elevations of intermediate points on the curve at the whole stations. Consider the criterion of passing sight distance.

19.15 A crest vertical curve connects a +4.20% grade and a −3.80% grade. The PVI is at station 10 + 00 at an elevation of 972 ft. The design speed is 40 mph.

Determine the following:

a. Length of the vertical curve
b. Station of the PVC
c. Elevation of the PVC
d. Station of the PVT
e. Elevation of the PVT
f. Station of the high point
g. Elevation of the high point
h. Elevation of station 15 + 09.22

19.16 A simple circular curve exists with a degree of curve $D = 10°$ and $e = 0.10$. A structure is proposed on land on the inside of the curve. Assume the road is on level grade.

Determine the following:

a. Radius of the curve
b. Current maximum safe speed of the curve
c. Minimum distance allowable between the proposed structure and the centerline of the curve such that the current maximum safe speed of the curve would not need to be reduced

19.17 A −2.5% grade is connected to a +1.0% grade by means of a 180-m vertical curve. The PI station is 100 + 00 and the PI elevation is 100.0 m above sea level. Determine the station and elevation of the lowest point on the vertical curve.

19.18 Determine the minimum length of a crest vertical curve between a +0.5% grade and a −1.0% grade for a road with a 60-mph design speed. The vertical curve must provide 550-ft stopping sight distance with AASHTO 2011 height criteria. Round up to the next greatest 10 ft interval. Show detailed calculations.

19.19 Determine the minimum length of a sag vertical curve between a −3.5% grade and a +2.7% grade for a road with a 50-mph design speed. The vertical curve must provide 1,300 ft headlight sight distance, the AASHTO passenger comfort, the AASHTO drainage condition, and the AASHTO appearance standard. Round up to the next greatest 10 ft interval.

19.20 A sag vertical curve connects grades of 1.5% and −2.5%, and has a vertical clearance of 14 ft. What is most nearly the minimum length of the curve required if the sight distance of 1,850 ft is recommended? Assume an eye height of 8.0 ft and an object height of 2.0 ft. Round up to the next greatest 10 ft interval.

APPENDIX A
Global Contexts of Pavement Design

A.1 Background

The functions of pavements are the same irrespective of the political boundaries. Thus, the design philosophy of pavement design remains the same. That is, pavement must support the repeated load along with comfort, safety, economy, and sustainability. However, the computational tools used by different regions of the world are slightly different. This appendix discusses a few international methods of pavement design.

A.2 U.K. Flexible Pavement Design

The pavement design procedure for both flexible and rigid pavements used in the United Kingdom was originally developed in 1970, revised in 1987, and the current form took shape in 2006 (Nunn, 2004). The design method is mostly empirical in nature.

A.2.1 Materials

Several layers are used in flexible pavement with different names, as shown in Fig. A.1. The bottom layer used over the subgrade is called the foundation layer. The foundation layer is similar to the subbase layer and can have four types of materials:

1. Unbound granular materials (modulus of equal or less than 50 MPa)
2. Hydraulically bound materials (HBMs) (modulus of equal or less than 100 MPa)
3. Fast-setting cement-treated materials (modulus of equal or less than 200 MPa)
4. Low-setting cement-treated stabilized materials (modulus of equal or less than 400 MPa)

The specification and properties of these foundation materials can be obtained from *Design Manual for Roads and Bridges (DMRB)* (DMRB, 2009).

Sometimes a capping layer is used before placing the foundation layer, especially when the subgrade strength is extremely poor (CBR less than 2.5%). The capping layer uses granular materials, unbound mixtures of aggregates, or treated soil. This capping layer improves the subgrade's bearing strength and provides a smooth bed for the foundation layer.

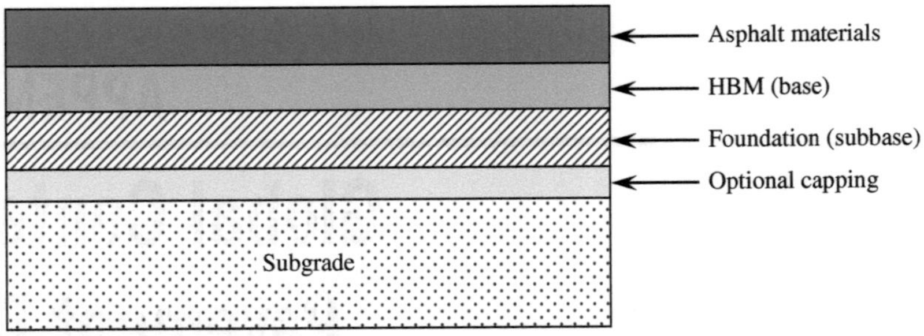

FIGURE A.1 Typical layers used in U.K. flexible pavement design method.

The thickness of the foundation depends on the strength of the subgrade, which is calculated mainly using CBR tests. The surface layer has two parts: the upper layer as the asphalt layer and the lower layer as the base layer. The base layer [also called hydraulically bound material (HBM)] is applied on top of the subbase. Hydraulically bound means the natural or artificial stones are stabilized using binding materials such as cement, lime, fly ash, and slags. The Base layer can be asphalt-treated granular material as well.

The thickness of capping and subbase layers is solely dependent on the modulus (E) of the subgrade, and it can be determined using the equation:

$$E = 17.6 \times \text{CBR}^{0.64} \tag{A.1}$$

where E is in MPa, and CBR is in percentage. This equation is most appropriate for fine soil. For coarser soil, plate bearing test may be required.

Base material is selected by the designer, and the final asphalt layer is found out using the traffic volume and the base material type.

A.2.2 Traffic

Volume of traffic is calculated similar to the ESAL calculation as follows:

$$T = \sum T_i \tag{A.2}$$

$$T_i = 365 \times F \times Y \times G \times W \times P \times 10^{-6} \tag{A.3}$$

where T = Total traffic volume
T_i = Different commercial traffic volume
F = Average annual daily traffic for each class
Y = Design years
G = Growth factor
W = Wear factor for each traffic class
P = Percentage of commercial vehicle in the heaviest loaded lane

The factors W and P can be obtained from DMRB (2006).

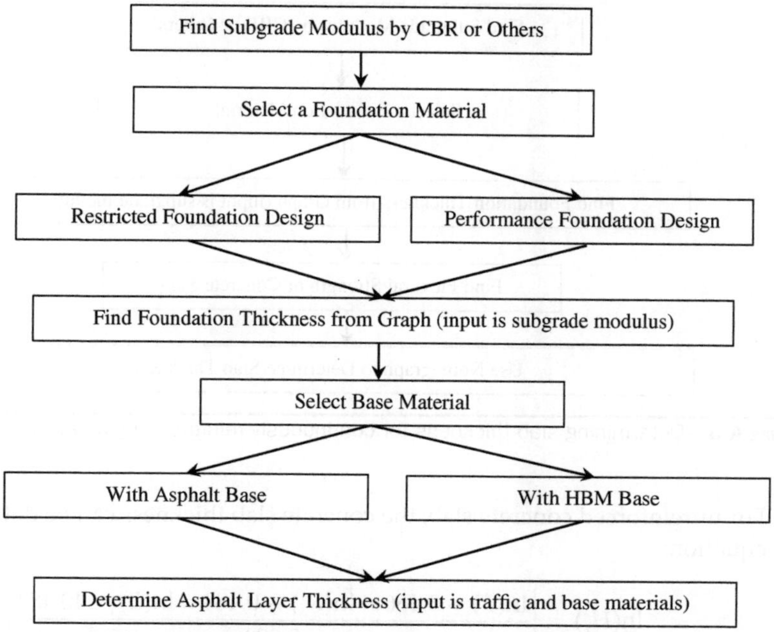

FIGURE A.2 Flowchart for U.K. flexible pavement design method.

A.2.3 Thickness Design

The thickness design process is shown in Fig. A.2. At the beginning, the modulus of subgrade is determined from the CBR test data. Then, the need of capping layer is evaluated. Capping is required if the CBR is less than 2.5%. Then, a foundation (subbase) material is selected. The thickness of the foundation is determined in two ways: restricted foundation design and performance foundation design. Restricted design is performed for areas where compliance testing is not feasible and thus not so popular. Performance foundation design method is less conservative and offers more flexibility to the designer by selecting a wide range of options for materials and yield economic foundation design.

Thickness of foundation (or subbase) can be determined from empirical graphs if the subgrade modulus is known. Upon completion of the foundation (or subbase) design, the type of base materials to be used (asphalt treated or HBM) is selected. Then, the thickness of the surface asphalt layer is determined from a nomograph using the traffic volume and base type.

A.3 U.K. Rigid Pavement Design

Concrete pavement design in the United Kingdom is empirical and the overall steps are presented in Fig. A.3. Slab thickness is determined using a nomograph for continuous reinforced rigid pavement. For unreinforced or jointed-reinforced pavement, closed-formed equations are used.

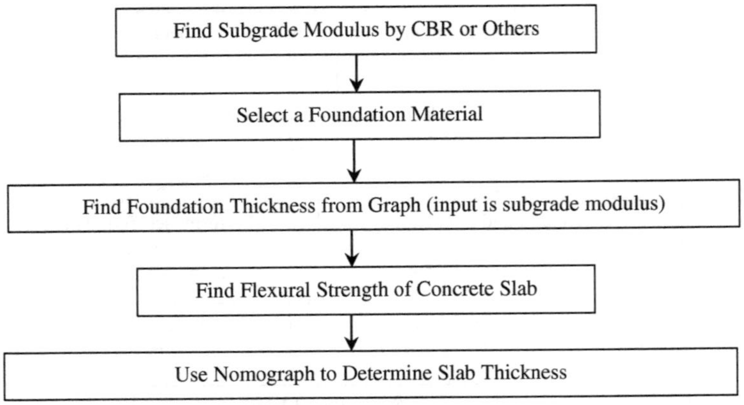

FIGURE A.3 Determining slab thickness for continuously reinforced rigid pavement.

For unreinforced concrete slab, the concrete slab thickness can be determined using the equation:

$$\ln(H_1) = \left[\frac{\ln(T) - 3.466 \times \ln(R_c) - 0.484 \times \ln(E) + 40.483}{5.094}\right] \quad \text{(A.4)}$$

where H_1 = Thickness of the concrete slab without a tied or 1-m (39.36-in.) edge strip
T = Design standard axles (millions)
R_c = Mean compressive cube strength (MPa)
E = Foundation stiffness (MPa)

For jointed-reinforced pavement, the concrete slab thickness can be determined using the equation:

$$\ln(H_1) = \left[\frac{\ln(T) - R - 3.171 \times \ln(R_c) - 0.326 \times \ln(E) + 45.15}{4.786}\right] \quad \text{(A.5)}$$

where R = 8.812 for 500 mm^2/m, 9.071 for mm^2/m, 9.289 for mm^2/m, and 9.479 mm^2/m of reinforcement. The thickness H_1 can be replaced by H_2 *if* there is a tied shoulder or 1-m edge strip, as follows:

$$H_2 = 0.934\,H_1 - 12.5 \quad \text{(A.6)}$$

A.4 Australian Flexible Pavement Design

The Australian pavement design methodology (called Austroads) was first implemented in 1979, improved in 1987, 1992, 2004, 2008, and 2012, and the current form was implemented in 2017 (Austroads, 2017). The current form is mechanistic-empirical in nature and the overall steps are presented in Fig. A.4.

For flexible pavement, a trial pavement is assumed, and the expected load groups are assigned with the project reliability. Then, the trial pavement is analyzed using linear elastic analysis by a program called CIRCLY. The pavement surface materials are

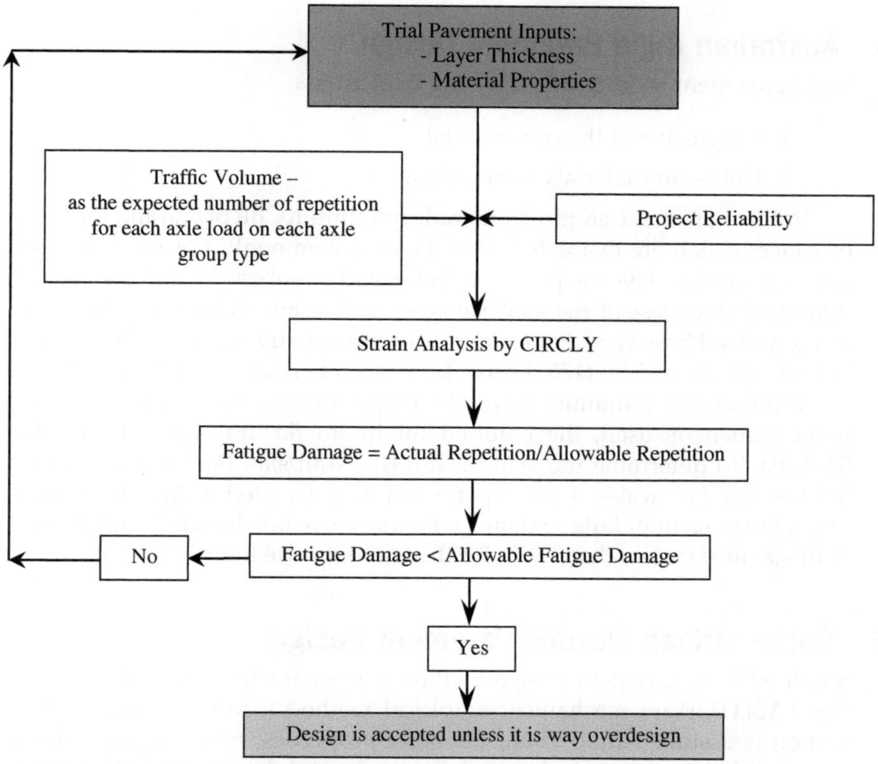

FIGURE A.4 Australian flexible pavement design flowchart.

assumed homogeneous, elastic, and isotropic. The unbound materials such as base, subbase, and subgrades are assumed anisotropic.

The fatigue life of asphalt mixture can be calculated using the equation:

$$N = \frac{\text{SF}}{\text{RF}}\left[\frac{6{,}918(0.856V_b + 1.08)}{E^{0.36}\varepsilon}\right]^5 \tag{A.7}$$

where N = Fatigue life, number of load repetition for certain axle
SF = Site or local factor (default value is 6.0)
RF = Reliability factor (1.0 for 50%, 3.9 for 90%, 9.0 for 97.5%, etc.)
E = Modulus of asphalt layer (MPa)
ε = Horizontal tensile strain at the bottom of asphalt layer (microstrain)
V_b = Binder content in the asphalt mix

The fatigue damage is calculated, as shown in the equation:

$$\text{Fatigue damage} = \frac{\text{Actual repetitions}}{\text{Allowable repetitions}} \times 100 \tag{A.8}$$

A.5 Australian Rigid Pavement Design

Rigid pavement is designed following two criteria:

1. Fatigue life of the concrete slab
2. Unbound materials erosion factor

In addition, the strength of subgrade (minimum CBR of 5%) and the flexural strength of concrete slab are evaluated. Base layer is commonly not used in rigid pavement, rather a subbase layer is placed in between the subgrade and the concrete slab. The minimum thickness of the subbase layer is determined from a table using the design traffic and subbase type. For example, for a design traffic of 5 million, 6 in. (150 mm) of bound subbase or 5 in. (125 mm) of lean concrete is selected (Austroads, 2012).

The required minimum flexural strength of concrete slab is 4.5 MPa, and if steel reinforcement is used, the required minimum flexural strength of concrete slab is 5.5 MPa. To determine the slab thickness, a trail slab thickness is selected. Then the fatigue life (allowable load repetitions) is calculated using the fatigue equation. Then, the percent of fatigue damage for each axle is calculated, and the sum of fatigue damage must be less than 100% for the entire service life.

A.6 South African Flexible Pavement Design

South African pavement design method is mechanistic-empirical in nature, similar to the AASHTOWare mechanistic-empirical method (SAPDM, 2013). A trial pavement section is assumed first. Then, the layer properties, traffic inputs, climate data, etc. are provided as shown in Fig. A.5. Then, structural responses are determined by the design software, which then calculates the damage in materials. Finally, the possible distresses are predicted using the empirical functions. The field distresses considered are as follows:

- Bottom-up fatigue crack in asphalt concrete (AC)
- Top-down longitudinal crack in AC
- Plastic deformation in AC
- Transverse cracking in AC
- Plastic deformation in unbound layers
- Crushing/stiffness reduction in stabilized layer

These distresses are compared with the predetermined threshold values set by the design agency. The steps are similar to the AASHTOWare software. However, the models used to determine different distresses are different and locally developed.

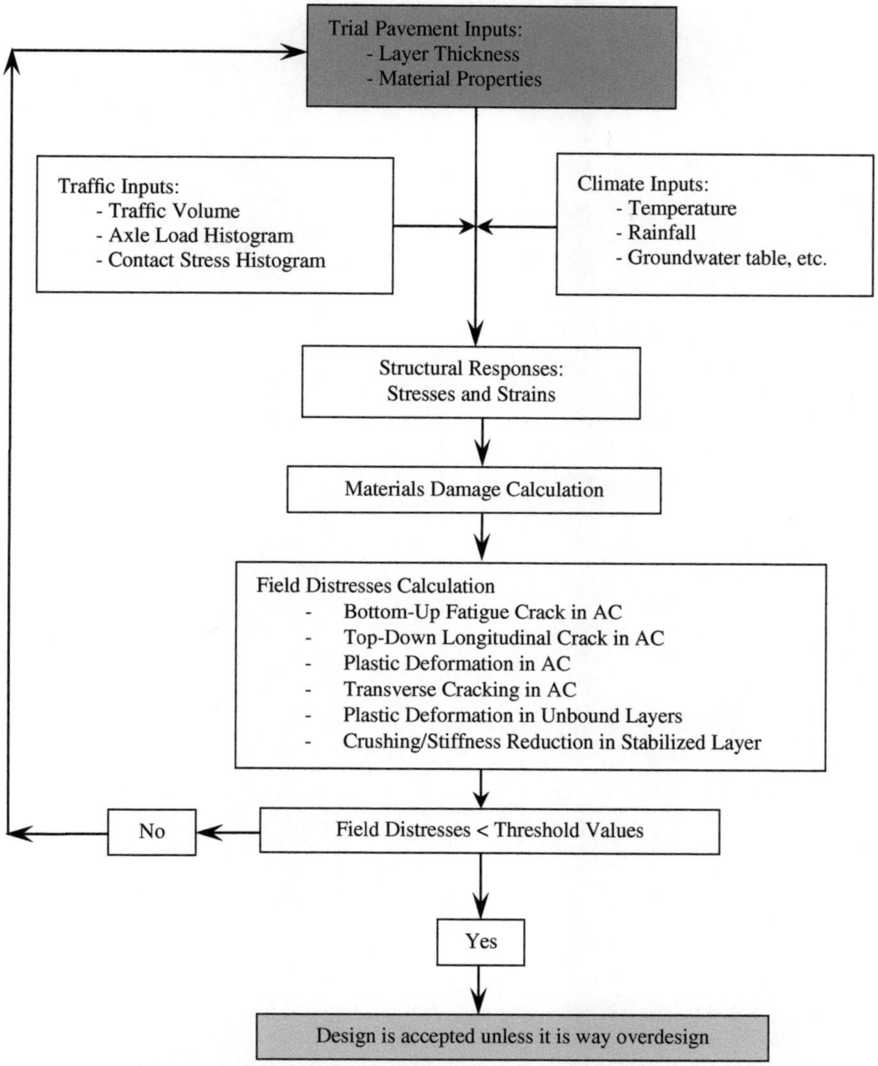

FIGURE A.5 South African pavement design method flowchart.

A.7 Summary

This appendix discusses several international pavement design methodologies. It is seen that pavement design methods vary from region to region. The AASHTOWare method is the most popular design method in the United States and Canada. The African pavement design system is similar to the AASTHOWare. The U.K. pavement design system is quite different from others.

APPENDIX B

Pavement Management System

B.1 General

The pavement management system (PMS) encompasses all activities related to the management of the pavement network for different investment decisions for roads, streets, and highways maintenance. PMS consists of a series of tasks or procedures that help pavement managers prepare over a period of time to maintain pavements in a serviceable condition (Mallick and El-Korchi, 2017). PMS is implemented at two different levels:

1. Network level for an array of pavements
2. Site-specific level for a particular segment

In the network level, PMS is utilized to select the most cost-effective strategies from a number of different alternatives for the maintenance of the pavements within a given analysis period considering the entire network. Only one site is analyzed for its optimal solution at a site-specific level, resulting in the maximum benefit-to-cost ratio over the given analysis period. In the PMS, the pavement maintenance activities can be categorized into three types:

1. Preventive maintenance
2. Corrective maintenance
3. Rehabilitation

Preventive maintenance is performed so that the quality of the pavement does not decrease below a desirable level. For example, application of thin overlay to extend the life of the pavement or keep the skid resistance intact. Corrective maintenance is applied to restore the condition of pavement in response to an existing problem such as potholes. Rehabilitation is conducted when the pavement goes under the quality level service or a threshold condition and a major structural improvement of the pavement is required, such as by recycling, placement of a new overlay, or total reconstruction (Nikolaides, 2015). On the other hand, routine maintenance is carried out to ensure serviceability, such as cleaning drains, checking the road signs, and snow removal, which are parts of the PMS. However, this routine maintenance does not require all the steps of the PMS.

Steps in the PMS can be broadly divided into the sequences listed in Fig. B.1.

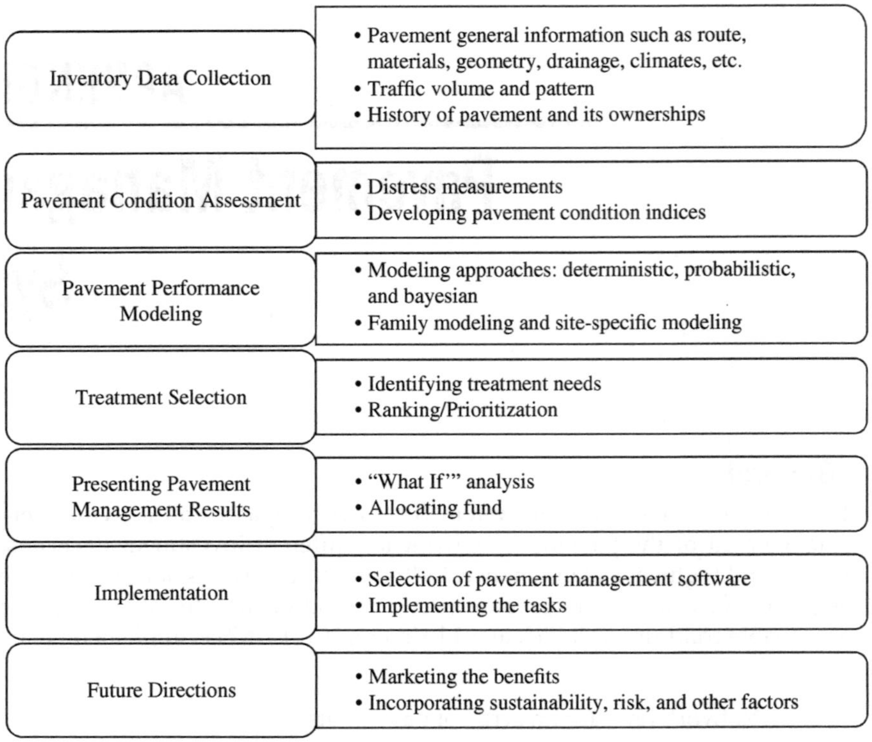

FIGURE B.1 Steps of pavement management systems and its broad tasks.

B.2 Inventory Data Collection

Pavement inventory data collection relevant to the agency's goals is the first and simplest, but very useful, task of the PMS. Some data to be collected are listed below:

- Pavement identity information such as route type, mileposts, functional classification, pavement types, number of lanes, its width, and shoulder information
- Environment and climate data such as freeze-thaw, precipitation, and temperature.
- Traffic volume, load pattern, traffic distribution, etc.
- History of the pavement such as construction dates, maintenance history, and flooding history
- Ownership information such as city or state
- Structural data such as layers, thicknesses, materials, joints characteristics, and subgrade strength

B.3 Pavement Condition Assessment

Pavement condition assessment means the collection of pavement distress and developing pavement condition indices using those distresses.

B.3.1 Distress Measurements

The type of data to be collected is the unique aspects of the agency's network, available resources, demand, etc. Generally, four types of data are collected:

1. Pavement distresses such as cracking and rutting as discussed in Chaps. 8 and 13
2. Structural capacity such as response due to applied load and load transfer quality
3. Surface characteristics such as smoothness, surface texture, and skid resistance
4. Subsurface characteristics such as voids and drainage quality

Different methods are used to collect data, such as manual, automated, and combination of manual and automation (semi-actuated). Some methods are destructive such as coring and boring; some are nondestructive such as the falling-weight deflectometer (FWD) testing. The method to be used is dependent on the agency's budget, needs, manpower, and equipment available to the agency.

Pavement Distresses Measurement

Chapters 8 and 13 discuss different types of major and minor distresses commonly used in pavement.

Structural Capacity Measurement

Pavement structural condition means the structural adequacy of the pavement layers. It is evaluated by conducting the nondestructive testing by measuring the surface deflection/deformation under a static or a dynamic load or both. Few devices are used to measure the surface deflection. Now-a-days, the most common nondestructive testing device is the FWD. The principle of FWD testing has been discussed in Chap. 4.

Pavement structural condition is very often represented by the pavement condition index (PCI). It is a numerical index to indicate the general condition of a pavement between 0 and 100 with 100 being the best possible condition and 0 being the worst possible condition. The PCI is determined by calculating the weighted average of the conditions for different pavement areas in each segment.

Surface Characteristics

Surface characteristics represent pavement's riding quality and the driving safety. Three types of surface characteristics are measured:

1. Longitudinal profile and roughness
2. International roughness index (IRI)
3. Surface texture and friction

Longitudinal Profile and Roughness The unevenness of the pavement's longitudinal profile affecting pavement–vehicle interaction is known as the longitudinal profile and roughness. One indirect way of measuring longitudinal profile and roughness is the present serviceability rating (PSR). PSR is the pavement riding quality based on a panel of observers to ride in a vehicle over the pavement. The scale is 0 to 5, with 5 being excellent and 0 being impassable.

Present serviceability index (PSI) is a substitution of the PSR and is determined based on the measured physical roughness. Again, the scale of 0 to 5 is used with 5 being very good and 0 being very poor. Some of the equipment to measure roughness are listed below:

- Mays Ride Meter
- Bureau of Public Roads Roughometer
- Cox Road Meter

The above-listed equipment does not measure the actual profile of the road, rather it measures the response of the vehicle to surface roughness. The equipment is calibrated to ensure that the response obtained represents true roughness of the pavement. Profilometers can be used to measure the true roughness (profile) and do not need any calibration. Being more sophisticated, profilometers are costly. Examples of profilometers are K. J. Law Profilometers and South Dakota Profilometers.

The PSI can be determined using the following two equations:

$$\text{PSI} = 5.03 - 1.91 \log(1 + \overline{\text{SV}}) - 0.01(C + P)^{0.5} - 1.38\,\overline{\text{RD}}^2 \quad \text{[for flexible pavement]} \tag{B.1}$$

$$\text{PSI} = 5.41 - 1.80 \log(1 + \overline{\text{SV}}) - 0.09(C + P)^{0.5} \quad \text{[for rigid pavement]} \tag{B.2}$$

where PSI = Present serviceability index
$\overline{\text{SV}}$ = Average slope variance on both wheel paths as obtained by a profilometer (this is an expression of the surface irregularities)
C = Major cracking in linear feet per 1,000 ft^2 area of pavement area
P = Asphalt patching in square feet 1,000 ft^2 area of pavement area
$\overline{\text{RD}}$ = Average rut depth of both wheel paths based on a 4-ft straightedge in inches (this is an expression of the pavement deformation)

International Roughness Index (IRI) IRI is a standard method of reporting roughness for both flexible and rigid pavements. IRI is calculated using cracking, rutting, and climate factors at the pavement site. The detailed procedure to calculate IRI is discussed in Chaps. 9 and 14 for flexible and rigid pavements, respectively.

Surface Texture and Friction Surface texture and friction are two sides of a coin. Surface texture is the roughness of the surface and friction is the consequence of the roughness. The term *skid resistance* is often used to describe these two terms of texture and friction on the surface. Skid resistance is the force developed when a tire (that is prevented from rotating) slides along the pavement surface. Skid resistance is an important pavement evaluation parameter because:

- Inadequate skid resistance will lead to higher incidences of skid-related accidents.
- Most agencies have an obligation to provide users with a roadway that is safe.
- Skid resistance measurements can be used to evaluate various types of materials and construction practices.

The basic formula for friction factor (f) is:

$$f = \frac{L}{N} \tag{B.3}$$

where L = Lateral or frictional force required to cause two surfaces to move tangentially to each other

N = Reaction force perpendicular to the surfaces

The skid number SK can be determined as:

$$\text{SK} = 100f \tag{B.4}$$

The SK is usually obtained by measuring the forces obtained with a towed trailer riding on a wet pavement equipped with standardized tires. Skid testing in the United States may occur in a number of ways including:

- The locked wheel tester,
- The spin-up tester, and
- Surface texture measurement

The locked wheel tester speeds up at about 40 mph (64 km/h) and water is sprayed ahead of the test tire to create a wetted pavement surface. Then, brake is applied to lock the test tire, the resulting friction force acting between the test tire and the pavement is measured and reported as the skid number (SN). The standard locked-wheel friction tests are AASHTO T 242 and ASTM E 274.

For a spin-up tester, the vehicle speeds up at about 40 mph (64 km/h) and a locked test wheel is lowered to the pavement surface. Due to its contact with the pavement, the test wheel can spin up to normal travel speed. The force can be computed by knowing the test wheel's moment of inertia and its rotational acceleration.

The simplest surface texture measurement is the sand patch test (ASTM E 965). The test is carried out on a dry pavement surface by pouring a known quantity of sand onto the surface and spreading it in a circular pattern with a straightedge. As the sand is spread, it fills the low spots in the pavement surface. When the sand cannot be spread any further, the diameter of the resulting circle is measured. This diameter can then be correlated to an average texture depth, which can be correlated to skid resistance. Laser or advanced image processing equipment are capable of determining surface macrotexture from a vehicle moving at normal travel speeds. One example is the road surface analyzer (ROSAN) that can be used for measuring texture, aggregate segregation, grooves, joints, and faulting.

Subsurface Characteristics

Subsurface characteristics such as number of layers, thicknesses, and any voids inside can be measured using the ground penetrating radar (GPR). It transmits a pulse of radar energy into the pavement and measures the time required to receive the reflection back by the receiver. Using the reflection time pattern, number of layers and their thicknesses can be interpreted. Pavement cores may be required to calibrate or validate the GPR results. Possible voids inside the pavement can also be determined using the GPR.

B.3.2 Developing Pavement Condition Indices

Different pavement condition indices or condition ratings are frequently used in the PMS to report the pavement conditions and select the appropriate treatment strategies. Two commonly used condition indices and their subdivisions are listed below:

1. Composite indices
   - Subjective composite indices
   - Objective composite indices
2. Individual indices
   - Roughness index
   - Rutting index
   - Structural/fatigue index
   - Nonstructural cracking index
   - Patch index

B.4 Pavement Performance Modeling

Pavement performance modeling is conducted to predict future pavement condition, determine the appropriate timing of action, determine the consequences of different strategies, etc.

B.4.1 Performance Modeling Approaches

Modeling or regression analysis is required to predict the distresses for future years—to evaluate the condition after several years. Four types of modeling approaches are used in the PMS:

1. Deterministic models
2. Probabilistic models
3. Bayesian models
4. Subjective or expert-based models

Deterministic models predict a single parameter (e.g., cracking, rutting, or condition index) by statistical regression analysis using the historical pavement condition information. Simply, it is the statistical way to determine future pavement condition using the current and past survey data. Probabilistic models predict several parameters (e.g., cracking, rutting, or condition index) using the Markov or Semi-Markov probabilistic approach. Bayesian models use both subjective and objective data to develop probabilistic models without using the historical data. In subjective or expert-based models, an individual or an expert group develops equations to determine the pavement deterioration rate throughout pavement life and future pavement condition is predicted accordingly. The agency may set their own predicting model based on their local experience and resources.

B.4.2 Family Modeling

Family-based modeling reduces the number of independent variables and develops pavement performance model for a group of similar pavement sections with similar performance characteristics. The developed single model can be used for any pavement in the family tree.

B.4.3 Site-Specific Modeling

Each pavement is unique and thus two similar pavements are expected to have different performance characteristics. Site-specific model addresses this concept by developing model for a specific pavement site. Sufficient performance data must be collected from that site to develop the site-specific model.

B.5 Treatment Selection

B.5.1 Identifying Treatment Needs

Two components must be defined to determine the current or future treatment needs:

1. Types of treatments to be considered
2. Conditions under which each treatment is considered valuable

For flexible pavements, the following treatment types are commonly considered. Some are shown in Fig. B.2.

Figure B.2 Some primary preventive maintenance activities for HMA pavements. (*From Van Dam, T. J., Harvey, J. T., Muench, S. T., Smith, K. D., Snyder, M. B., Al-Qadi, I. L., Ozer, H., et al. (2015). Towards Sustainable Pavement Systems: A Reference Document. Report FHWA-HIF-15-002. Washington, DC: Federal Highway Administration.*)

- Routine maintenance such as crack sealing
- Surface seal coats
- Milling and inlays
- Thin overlays
- Thick overlays
- Mill and overlay
- Reconstruction

For rigid pavements, the following treatment types are commonly considered. Some are shown in Fig. B.3.

- Slab grinding
- Micro-surfacing
- Full- and partial-depth repairs

FIGURE B.3 Some maintenance works of concrete pavements. (*Van Dam, T. J., Harvey, J. T., Muench, S. T., Smith, K. D., Snyder, M. B., Al-Qadi, I. L., Ozer, H., et al. (2015). Towards Sustainable Pavement Systems: A Reference Document. Report FHWA-HIF-15-002. Washington, DC: Federal Highway Administration.*)

- Crack and seal
- Thin-bonded overlays
- Unbonded overlays
- Slab replacement
- Reconstruction

Sometimes treatment categories are selected instead of selecting the above specific treatment types. Some popular categories are listed below:

- Preventive maintenance
- Surface seal coats
- Minor rehabilitation
- Major rehabilitation
- HMA reconstruction
- PCC reconstruction

B.5.2 Techniques for Treatment Selection

The treatment to be selected must be cost-effective. The following tasks are performed to ensure this objective:

1. Ranking
2. Multi-Year Prioritization
3. Optimization
4. Life-Cycle Cost Analysis

Ranking

Ranking is the method of listing candidate pavements based on their condition, initial cost, life-cycle cost, etc. This method does not consider the cost-effectiveness of alternative strategies. The following steps are commonly followed to rank the candidate pavements:

Step 1. Evaluate the pavements with questionable conditions for the current year.

Step 2. Calculate the probable treatment cost.

Step 3. Prioritize the candidate pavements (the worst condition the first).

Step 4. Select one by one pavement from the prioritized list as many as funding allows.

Step 5. Consider the unfunded pavements in the following year.

Multi-Year Prioritization

A better approach over the ranking method is the multi-year prioritization where the current year or more is considered with multiple alternative treatments, resulting in delay or acceleration of treatment and cost-effectiveness. Two types of cost-effectiveness are considered:

1. Incremental benefit-cost
2. Marginal cost-effectiveness

Incremental benefit-cost approach evaluates if there is any additional benefits such as longer life, riding comfort, and increase in the treatment intervals for incremental increase of investment. Marginal cost-effectiveness considers the cost-effectiveness only, not the benefits.

Optimization

The selection of treatments and their timing of application and the benefits realized from the application of the treatments can be accounted for in the optimization analysis. The optimization can be based on different criteria. For example, either on the concept of minimizing the total cost while keeping all pavements at or above a minimum condition, or on maximizing the total benefit, with the available budget. Every pavement management agency has its own protocol to optimize the benefit and cost.

Life-cycle cost analysis

Life-cycle cost analysis (LCCA) can be used as a decision support tool when selecting treatment selection. LCCA is commonly used while selecting pavement type, determining structure and mix type, construction methods, as well as maintenance and rehabilitation strategy. Typically, LCCA involves the following basic steps:

- Make initial strategy and analysis decisions to establish the parameters under which a LCCA can be carried out.
- Estimate the initial cost and annual costs associated with the owning agency and users for each alternative option.
- Compare alternatives using a common metric such as net present value (NPV) or benefit-cost ratio (B/C).
- Analyze the results and reevaluate alternatives for the most influential costs, factors and assumptions. A sensitivity analysis is often used to do this. Original design strategy alternatives should be reevaluated base on these results analysis in order to improve the cost-effectiveness of each alternative.

The end result of a successful LCCA is not simply the selection of one alternative over the other but the selection of the most cost-effective treatment strategy for a given situation and a greater understanding of the factors that influence cost effectiveness.

B.6 Presenting Pavement Management Results

In the next step of the PMS, "what if" or "trade-off" analysis is conducted to determine the impact of different strategies on the cost-effectiveness, performances, etc. This helps decision-makers to assess the overall condition of the network, such as percentage of the network in poor condition, future costs, and targets. Then funds are allocated to ensure that sufficient funding is available for high-priority roadways and distribute the funding logically among regions, districts, or different jurisdictions.

B.7 Implementation

Once a decision has been made, the following steps are followed to implement the projects:

Step 1. Form a steering committee.

Step 2. Evaluate the goals.

Step 3. Select the software (such as MicroPAVER).

Step 4. Collect the field data.

Step 5. Configure the software accordingly.

Step 6. Test the software.

Step 7. Conduct training to the workforce, if required.

Step 8. Implement the project.

Step 9. Document the implementation and progress.

Step 10. Monitor and control the system on a regular basis.

A steering committee is essentially the first step of implementing the project. The roles of each individual must be clearly declared and known to all. The committee should also be aware of the needs and goals of the agency. A software or computer program is selected to execute the project, record the implementation data, and monitor it. A spreadsheet may be good enough for a small project, whereas MicroPAVERTM may be required for large project. The software may be calibrated or tested based on the field data. Once software is finalized, the workforce will be properly trained if there is a new member. Once the work is started, regular monitoring, documentation, and controlling are required to make it a success. Remember that most projects never run as scheduled or planned. Once finished, a proper documentation must be achieved for future reference.

B.8 Future Directions

Once the project is over, the job is not over. Once the pavement is in operation, the efficacy of the treatments applied must be monitored, and the results disseminated to the stakeholders concerned for learning and future recommendations. How sustainability can be incorporated or improved must be studied from the project. Any catastrophic event that might take place should be judged as well. All these activities will incur better direction for the future.

B.9 Summary

This appendix discusses the basics of the PMS to be followed. However, each agency is unique in terms of needs, available resources, available funding, etc. Thus, most of the agencies practice the PMS as of their own convenience and resources. Details of the PMS can be obtained from the AASHTO (2012) or in the latest edition.

APPENDIX C

Recycling and Rehabilitation of Pavement

C.1 General

With the increase in the population of the world, demand has dramatically increased in the roadway network. Nonetheless, the need to provide a safe, efficient, and economic roadway is challenging due to the reduction in the budgetary fund. The construction cost is considered the most at the time of the extension of the roadway. Nonetheless, less attention is given to future maintenance costs. After a few years of service, two simultaneous problems arise on the roadway:

1. The roadway requires some maintenance due to wear and tear
2. Traffic volume increases

Presence of these two issues in the absence of maintenance will result in rapid damage of the roadway. If the available funds are not sufficient to meet the increased maintenance and demand, there will be a significant reduction in the quality and service level for the roadway network. It eventually results in higher overall preventive maintenance and the higher cost of rehabilitation or reconstruction. It ensures that in earlier stages of pavement damage, it is easier to perform maintenance or rehabilitation. The reactive operation becomes more expensive once the level of damage is severe. Research by the World Bank indicates that each $1.00 spent at the first 40% decrease in roadway quality results in savings of $3.00 to $4.00 relative to the investment expected at the 80% decrease in quality (ARRA, 2015).

Rehabilitation or recycling can be described as measures to improve, reinforce, or rescue existing defective pavements so that only routine maintenance can continue service (ITS, 2000). Recycling of pavements will typically offer the following benefits:

- Conservation of natural resources
- Energy saving
- Conservation of environment
- Money saving

Pavement is made up of aggregates, asphalt binder, and a few additives. Aggregate reuse means saving an aggregate source (say, a mountain). Furthermore, recycled aggregates have inherent asphalt binder. The new aggregates require about 4% to 6% of asphalt binder to satisfy the mix design requirements. However, a recycled aggregate pile requires asphalt binder between 1% and 3% by weight of mix (Nikolaides, 2015). Schwartz (2016) reported that cold-in-place recycling (CIR) technology of asphalt pavement (discussed later in this appendix) saves 60% asphalt binder compared to conventional new asphalt mix.

Energy consumption in pavement construction relates to aggregate and asphalt production, transportation, processing, heating, mixing, placing, and compacting. Less new materials means less energy consumption.

Pollutants such as carbon dioxide (chemically referred to as CO_2) are generated by the use of energy for aggregate and asphalt production, transportation, processing, heating, mixing, placing, and compacting. Recycling of pavement materials saves the emission of this pollutant. For example, the use of asphalt pavement CIR (discussed later in this appendix) is environmentally friendly as it can reduce carbon dioxide emissions by up to 9% compared to traditional HMA mixtures; in the recycling process alone, carbon dioxide emissions are 54% lower (Giani et al., 2015). Schwartz (2016) stated that CIR and cold central plant recycling (CCPR) technologies (discussed later in this appendix) decreased carbon dioxide emissions by 80% and 42%, respectively, compared with conventional HMA applications.

C.2 Asphalt Pavement Recycling

C.2.1 General

Since the late 1970s, recycling of asphalt pavement has grown in popularity, mainly due to higher oil prices. Recycling of old pavements had been more costly than placing new hot-mix asphalt (HMA) (Roberts et al., 1996). The invention of milling machines has improved the cost-effectiveness of asphalt recycling compared to new construction (Cross and Jakatimath, 2007). After that, equipment manufacturers and construction industries were interested in developing the methods and techniques of asphalt recycling and since then it has progressed exponentially (ARRA, 2015).

C.2.2 Asphalt Recycling Types

ARRA (2015) divided various asphalt recycling methods into five broad categories:

1. Hot recycling (HR)
2. Hot in-place recycling (HIR)
3. Cold planing (CP)
4. Full-depth reclamation (FDR)
5. Cold recycling (CR)

Asphalt recycling approaches can be used in some roadway rehabilitation projects in combination with each other. For example, the upper portion of an existing roadway could be removed via CP and the resulting reclaimed asphalt pavement (RAP) could be stored at the asphalt plant. When prepared, the cold planed surface could be overlaid with HMA containing the milled off RAP. Instead, the exposed CP surface could have

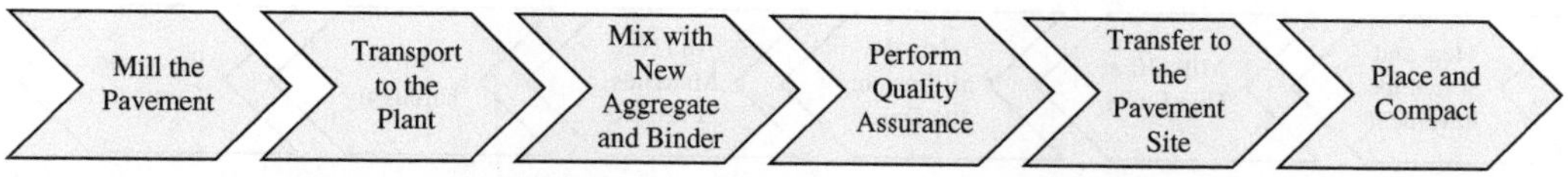

FIGURE C.1 Basic steps of hot recycling.

been HIR, CR, or FDR before the recycled mixture was put to reduce or remove the effects of reflective cracking.

C.2.3 Hot Recycling

Hot recycling (HR) is the world's most widely used method of asphalt recycling. The United States produces more than 100 million tons of RAP. About 15% to 30% of this production is used in hot recycling (ARRA, 2015). Hot recycling is the method of mixing RAP milling with fresh aggregates and a recycled mixture of asphalt binder under heating in a plant. The asphalt binder in the RAP melts first after the RAP has been heated. The new aggregate and binder are then added and thoroughly mixed. It is transported, placed, and compacted once mixed with conventional HMA equipment. Nowadays, this method is the most commonly used practice. Figure C.1 shows the entire HR process.

Some of the major advantages of HR include the following:

- Similar performance as pavements constructed with all new materials, thus can be used in all types of pavements
- Reuse of aggregate and binder, which saves natural aggregates, binder, and energy to produce it
- No need to dispose off the RAP milling

Some of the major disadvantages of HR include the following:

- Problems with existing aggregate gradation and/or asphalt binder may reflect in the new layer.
- Too much RAP produces a stiffer mix and thus leads to thermal cracking.
- Transportation cost is involved while carrying to and from the plant.

C.2.4 Hot In-Place Recycling

Hot in-place recycling (HIR) is conducted for less severely damaged pavement compared to the pavement which requires HR. In HIR, all the recycling of the asphalt pavement is completed on site. The deteriorated top ¾ to 3 in. (19 to 75 mm) is heated and softened, milled off, thoroughly mixed inside the truck, and compacted with conventional HMA paving equipment. If required, it is possible to add virgin aggregates, new asphalt binders, recycling agents, and/or new HMA. Generally, virgin aggregates or additional HMA addition are limited to less than 30% by mass of HIR mix due to equipment restrictions. An analysis of the current asphalt pavement properties and subsequent laboratory mix designs will assess the additional levels of the various additives to ensure compliance with the appropriate mix requirements. Figure C.2 shows the entire HIR process.

Based on the process used, three subcategories of HIR are possible:

1. Surface recycling
2. Remixing
3. Repaving

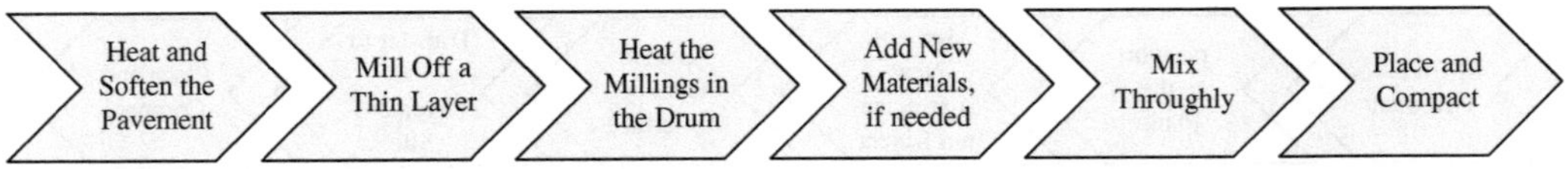

FIGURE C.2 Basic steps of hot in-place recycling.

Surface recycling is the HIR process where the deteriorated surface of the pavement is heated, softened, and scarified, and recycling agent is mixed (if required) and the loose recycled mixture is thoroughly mixed with the standard paver screed. No aggregate or binder except the recycling agent is added. The depth of recycling typically varies from ¾ to 1.5 in. (20 to 40 mm). In a subsequent operation, a surface coat such as a chip seal or HMA overlay is usually placed. Figure C.3 demonstrates the basic HIR surface recycling process.

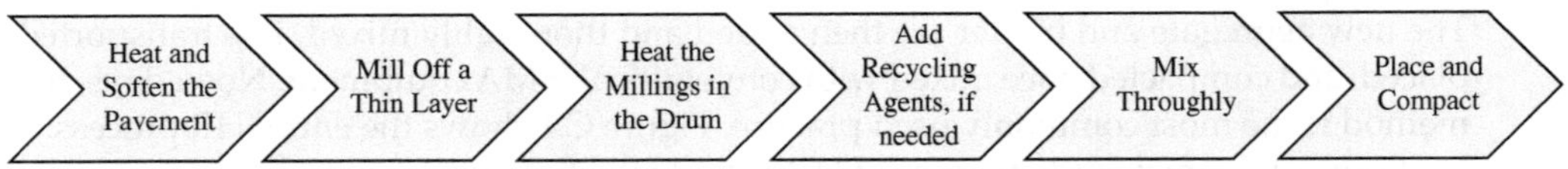

FIGURE C.3 Basic steps of hot in-place surface recycling.

Remixing is similar to the surface recycling but new materials (aggregate and binder) are added in addition to the recycling agent. More specifically, HIR remixing is the process in which the existing deteriorated asphalt pavement is heated, softened, and scarified and virgin aggregate, new asphalt binder, recycling agent, and/or new HMA are added (as required) and the resultant is thoroughly mixed and then paved. Remixing is typically used when existing pavement properties require significant modifications compared to surface recycling. Treatment depths generally range from 1 to 2 in. (25 and 50 mm) for single-stage remixing. A surface coat such as a chip seal or HMA overlay is usually placed in a subsequent operation similar to surface recycling. Figure C.4 shows the basic HIR remixing process.

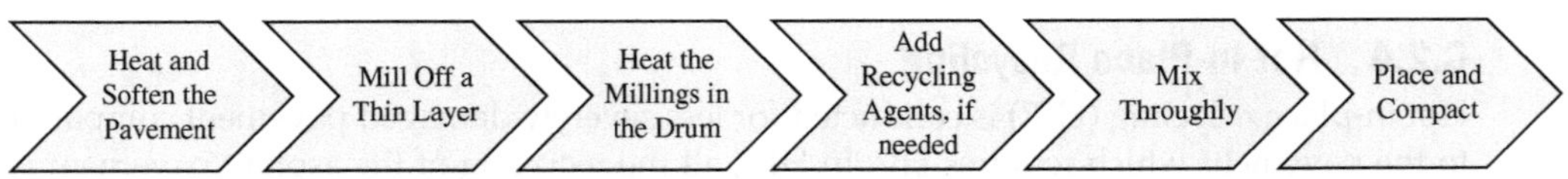

FIGURE C.4 Basic steps of hot in-place remixing.

Remixing is also further classified into two types:

1. Single-stage
2. Multi-stage

In the single-stage method, the above-described process is performed a single time. The process of heating, softening, and scarifying the existing asphalt pavement is repeated several times in the multi-stage remixing method until the full depth of treatment is reached. The scarified material from each stage is placed in a windrow to allow the underlying layer to be heated and scarified. Once the full asphalt layer is scarified, in addition to the recycling agent, new aggregate and binder may be applied. The entire recycled mixture is then thoroughly mixed, placed on site, and compacted.

Repaving is the method of recycling or remixing the surface with the placement of the new HMA overlay. The surface recycled mix functions as a leveling course in the repaving process, while the new HMA acts as the course of surface or wear. The thickness of the pavement can be increased significantly in the course of repaving. Figure C.5 shows the entire hot in-place repaving cycle.

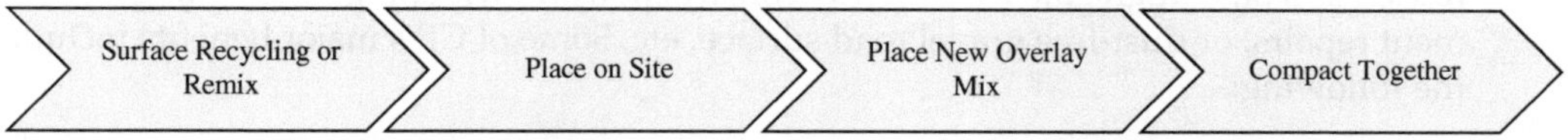

FIGURE C.5 Basic steps of hot in-place repaving.

Repaving is also further classified into two types:

1. Single-stage
2. Multi-stage

One machine fitted with two screeds is used for single-pass repaving. The first screed places the recycled mixture on top of the recycled mix while the second screed places the new HMA layer. Then the two layers are compacted. In the multi-stage method, the surface recycled mix is positioned by its placing and screeding unit to the correct longitudinal profile and cross-slope. The new HMA overlay material is then immediately placed with a traditional asphalt paver on the hot, uncompacted recycled mix. The two layers are then simultaneously compacted.

Some of the major advantages of HIR include the following:

- Reuse of aggregate and binder, which saves natural aggregates, binder, and energy to produce it.
- No need to dispose of the RAP milling.
- No need to transport the milling; thus, no disposal problems.
- Restores the friction and removes minor distresses such as rutting, potholes, raveling, surface irregularities, and oxidized asphalt binder.
- Structural strength and integrity remain intact or increase.
- In-place construction reduces traffic disruptions compared to HR.
- Cost-saving and less hazardous.

Some of the major disadvantages of HIR include the following:

- Large, specially equipped and costly equipment are required.
- There are not as many options for new materials.
- Quality control is not as good as for HR.
- Only suitable for thin layer treatment.

C.2.5 Cold Planing

Cold planing (CP), also called cold millings, is the removal of the required depth of the existing pavement to restore surface friction, correct corrugation, reduce asphalt bleeding, remove shoves, correct slopes (longitudinal profile and cross-slope), etc. This activity is carried out using specially designed equipment called cold planers or pavement profilers. The textured surface can be opened immediately for normal traffic. The pavement can also be treated or overlayed with one of the other forms of asphalt recycling. This method is very quick, and the public is less disturbed.

The RAP created during the CP operation is removed from the site and, like other RAPs, can be used in different ways. The RAP is then further recycled or could be reused as a base aggregate for roadway construction and widening, ditch linings, pavement repairs, or dust-free gravel road surface, etc. Some of CP's major benefits include the following:

- Removal of deteriorated pavement surfaces
- Removal of oxidized asphalt
- Correction of longitudinal profile and cross-slope
- Removal of some distresses such as rutting, shoving, corrugation, and bleeding
- Restore drainage
- Restore friction

Some of the major disadvantages of CP include the following:

- The resulting pavement surface produces noise while using.
- As the thickness of the surface layer decreases, the capacity of pavement structure may decrease.
- The operation may produce dust, which is a nuisance for the surrounding area.
- CP is difficult to operate for stiffer asphalt.

C.2.6 Full-Depth Reclamation

Full-depth reclamation (FDR) means full depth of asphalt layer, and in some cases, some parts of the underlying base/subbase/subgrade layer are pulverized, blended thoroughly, and compacted in place as a new base layer. The blended material is not heated. Rather, it may be stabilized with a wide range of dry or liquid stabilizing agents or may be even untreated base layer. Some of the commonly used additives are portland cement, asphalt emulsion, fly ash, foamed asphalt, lime, calcium chloride, magnesium chloride, etc. or a combination of a few. Treatment depths vary between 4 and 12 in. (100 and 300 mm) in common. FDR is conducted to pavements with severe damage in the underlying base/subbase/subgrade layer and/or satisfies the increased traffic demand.

The combined new layer (treated or untreated) consisting of the bound asphalt layer and the underlying unbound layers can be used as it is if the traffic is low. A new granular layer or wearing HMA layer may be applied as desired. The whole FDR process is shown in Fig. C.6.

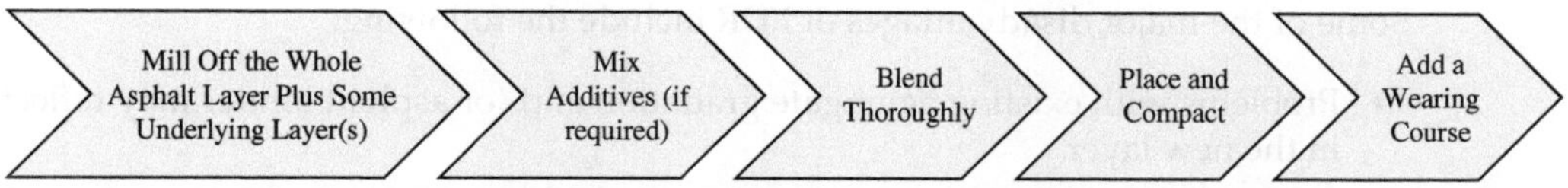

Figure C.6 Basic steps of full-depth reclamation (FDR).

Stabilization of the reclaimed pavement can be done by mechanical, chemical, or bituminous means. Mechanical stabilization methods include the addition of the following:

- Virgin aggregate
- Reclaimed asphalt pavement (RAP)
- Crushed portland cement concrete (PCC)

Chemical stabilization is achieved with the addition of the following:

- Lime
- Portland cement
- Fly ash
- Cement kiln dust
- Calcium/magnesium chloride
- Other proprietary chemical products

Bituminous stabilization can be accomplished with the use of the following:

- Liquid asphalt
- Asphalt emulsion
- Foamed asphalt

For increased stabilization requirements, combinations of all three can also be used. Some of the major advantages of FDR include the following:

- Reuse of aggregate and binder, which saves natural aggregates, binder, and energy to produce it.
- No need to dispose of the RAP milling.
- No need to transport the milling; thus, no disposal problems.
- Elimination of bumps and dips, rutting, potholes, patches, and cracks.
- Base/subbase/subgrade deficiencies can be corrected by stabilization.
- Significant structural improvement with the addition of stabilizing additive(s).
- Produces thick, bound layers that are homogeneous.
- Permits more flexibility in the choice(s) of wearing surface type and thickness.
- Provides significant structural improvement.
- Improves smoothness and riding quality.

Some of the major disadvantages of FDR include the following:

- Problems with existing aggregate gradation and/or asphalt binder may reflect in the new layer.
- Costly as the full depth is reclaimed.
- Not suitable for areas with drainage problems.
- Soil with high plasticity can result in swelling.
- Not recommended for roads with daily traffic of more than 20,000 vehicles.

C.2.7 Cold Recycling

Cold recycling (CR) is the processing and treatment of existing HMA pavements to recover the pavements without heating asphalt materials. A milling machine extracts the degraded top 2 to 4 in. (50 to 100 mm) of HMA, crushes, and screeds the milled material to achieve a required gradation. The milled material is combined with binding additives such as emulsion, cement, lime, or fly ash. Then the mixture is placed back on the roadway, compacted, and graded to the final elevation. If the volume of traffic is relatively high, a fog seal or thin overlay can be added after compaction of the mixture.

Two subcategories within CR are used to further define CR based on the process used:

1. Cold in-place recycling (CIR)
2. Cold central plant recycling (CCPR)

In the CIR, all operations such as milling, screening, and mixing additives are performed on site inside a large truck. After thoroughly mixing, the mixture is placed and compacted. In the CCPR, the milling is transported to a central plant. All the processing of this milling such as screening, gradation, and mixing additives are conducted in the plant. Then the mix is transported to the site, placed, and compacted similar to conventional pavement. Recent studies show that the dynamic modulus of CIR is about half of the conventional mixture at low temperature and is about equal to the conventional mixture at high temperature. Field performance shows that CR and conventional mixtures produce similar performance for low-traffic roadways (Cross and Jakatimath, 2007; Islam et al., 2018; Kim et al., 2010; Schwartz, 2016).

Some of the major advantages of CR include the following:

- Reuse of aggregate and binder, which saves natural aggregates, binder, and energy to produce it.
- No need to dispose of the RAP milling.
- Elimination of bumps and dips, rutting, potholes, patches, and cracks.
- Decrease in carbon dioxide emission.
- Base and subgrade materials are not disturbed.
- Pavement cross-slope and profile can be improved.
- In-place construction reduces traffic disruptions compared to CCPR.

Some of the major disadvantages of CR include the following:

- Problems with existing aggregate gradation and/or asphalt binder may reflect in the new layer.

- Suitable for minor distresses.
- Applicable to low-volume roads.

Compared to CIR, CCPR requires transporting millings to and from the plant but provides better quality control opportunity.

C.2.8 Summary of Rehabilitation Techniques

Various techniques of rehabilitation have been discussed in this appendix. It looks like one technique can be applied to various distresses. In other words, different techniques may be suitable for one form of distress. The decision depends on some factors, including technology availability, skilled labor, traffic volume, etc. Table C.1 lists some general recommendations for the preliminary collection of different rehabilitation techniques.

Pavement distress mode		CP	HIR	CIR	CCPR	FDR
Surface Defects	Raveling	■	■	▒	▒	□
	Potholes	□	▒	▒	▒	▒
	Bleeding	▒	▒	■	■	▒
	Skid Resistance	▒	▒	▒	▒	□
Deformation	Shoulder Drop-Off	▒	□	□	□	□
	Rutting—Wear	▒	■	■	■	□
	Rutting—Mix Instability	▒	▒	▒	▒	■
	Rutting—Deep Structural	□	□	▒	▒	■
	Corrugation	▒	■	■	■	▒
	Shoving	▒	▒	■	■	▒
Load-Related Cracking	Fatigue—Bottom Up	□	□	▒	▒	■
	Fatigue—Top Down	□	▒	■	■	■
	Edge	□	▒	▒	▒	■
	Slippage	□	▒	■	■	□
Non-Load-Related Cracking	Block	□	▒	■	■	▒
	Longitudinal	□	▒	■	■	□
	Transverse	□	▒	■	■	▒
	Reflective	□	▒	■	■	■
Combined	Joint Reflection	□	▒	▒	▒	□
	Discontinuity	□	▒	▒	■	■
Base/Subgrade Deficiencies		□	□	□	▒	■
Rough Ride Quality		■	■	■	■	□

Most Appropriate ■ ▒ □ → *Least Appropriate*

Source: *Adapted from* ARRA. (2015). *Basic Asphalt Recycling Manual.* Glen Ellyn, IL: Asphalt Recycling and Reclaiming Association (ARRA), U.S. Department of Transportation.

Table C.1 Applicability of Recycling Techniques

C.3 Concrete Pavement Recycling

C.3.1 General

Concrete pavement recycling is a simpler process than flexible pavement recycling. Recycling of concrete pavement includes the demolition and crushing of hardened concrete from an existing pavement to produce recycled concrete aggregate (RCA) (ACPA, 2009). RCA can be used in almost any application instead of new granular materials. The aggregate demand for pavement and building construction continues to rise rapidly. In the United States, virgin aggregate production increased from 58 million tons in 1900 (or 0.5 ton/person) to 2.3 billion tons (2.1 billion metric tons) in 1996 (9.6 tons/person) (USGS, 1997). RCA can be a good alternative to meet this high demand. Concrete paving is 100% recyclable (ACPA, 2006). Since the 1940s, concrete recycling has been widely used in Europe and since the 1970s in the United States (NHI, 1998). Concrete recycling is now carried out in at least 41 states for paving applications (FHWA, 2004). Recently, about 140 million tons of annual RCA production in the United States from all sources (both pavements and demolition debris) have been recorded (CDRA, 2018).

Saving natural aggregates, landfill space, energy, and money are some of the advantages of RCA (FHWA, 2002; Hall et al., 2007; Van Dam et al., 2015). Concrete recycling offers 20% to 30% of the cost of pavement with construction materials and supplies and 10% to 15% of total construction costs (Halm, 1980). One case study showed a $5 million savings on a single project, despite cost savings from recycling concrete pavement (CMRA, 2008).

C.3.2 Production of RCA

The basic steps of concrete pavement recycling are listed below and shown in Fig. C.7.

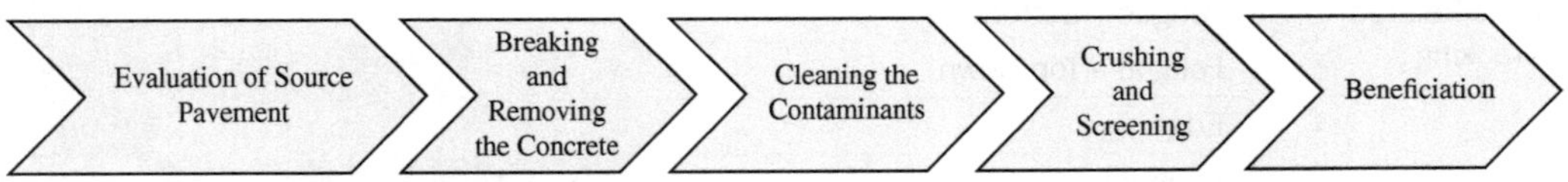

FIGURE C.7 Basic steps of concrete pavement recycling.

- Evaluation of the source concrete
- Breaking and removing the concrete
- Removing any steel mesh, rebar, or dowels
- Crushing and screening the RCA
- Beneficiation or quality control of removing any additional contaminants or improving properties

While producing RCP, effort is made to maximize the production of usable RCA. For example, coarse RCA (material retained on the No. 4 sieve) is typically more valuable and usable than fine RCA (material that passes the No. 4 sieve). Contaminants such as joint sealants, asphalt concrete shoulders, and patching materials, reinforcing steel and dowel bars, and soils and foundation materials should be separated to make the RCA high-quality materials. Any asphalt present in concrete pavement, such as thin layer of asphalt and asphalt repairs, should be separated.

Evaluation of the Source Concrete

In the first step of producing RCA, some basic properties pertaining to the concrete strength and performance such as original aggregate, cement type, and admixture used are evaluated. Based on this information, the proposed use of the RCA is recommended. For example, if the proposed RCA is good-quality material then it could be used for a new structural surface layer of a pavement or other concrete works. On the other hand, inferior RCA may be used for unbound layers of pavement or similar works.

Breaking and Removing the Concrete

The concrete slab is broken into manageable pieces using a heavy scraper or impact breaker and transported to the crushing plant. The first step in the removal process is to loosen the concrete pieces and separate any debonded reinforcing steel. A back hoe or bulldozer with a rhino horn attachment can be used to hook and pull the steel free from the concrete rubble. Some hand work with hydraulic shears may be required to cut the reinforcing steel. Small pieces of embedded steel do not cause problems in the crushing operations and may be removed after crushing.

Removal of Any Steel Mesh, Rebar, or Dowels

The separation of steel materials can be conducted in several phases of the entire process, although the earlier the better. After crushing operations, electromagnets are often used to pick steel from the conveyor belts. Manual labor may be used to expedite the steel removal operations. An operation of steel bar removal is shown in Fig. C.8.

Crushing and Screening the RCA

The transported pieces are crushed and screened to produce desirable aggregate gradation. Care is taken to avoid the production of fine aggregates (passing No. 4 sieve):

1. Primary crushing
2. Secondary crushing

FIGURE C.8 Steel bar removal from an interstate pavement in Pueblo, Colorado.

The primary crusher typically reduces the material size down to about 3 to 4 in. (75–100 mm). The crushed material is then screened and material larger than ⅜ in. (9 mm) is fed into a secondary crusher, which breaks the material to the desired maximum coarse RCA size.

Beneficiation

Beneficiation is the treatment or removal of accidental organic material, excessive dust, soil, etc. to improve its physical or chemical properties prior to further processing or use. The produced RCA may be further graded based on size or density to have some desired properties. This is the last step before the usage of RCA and thus can be treated as a QC/QA check.

C.3.3 Properties of RCA

Properties of RCA largely depend on the proportion of reclaimed aggregate and mortar, properties of original materials, crushed aggregate gradation, etc. Higher amounts of reclaimed mortar result in higher absorption, lower specific gravity, lower particle strength, and lower resistance to abrasion than would be found in the natural aggregate involved. Typical distributions of few natural aggregates and RCA physical properties are listed in Table C.2. It indicates that RCA typically has higher absorption, LA abrasion mass loss, sodium sulfate mass loss, magnesium sulfate soundness mass loss, and chloride content. The actual gravity, however, may be lower than the normal aggregates. In short, RCA's physical properties are different than natural aggregates.

While using RCA, it should be analyzed as an engineered material with appropriate mixture design or construction adjustments. The RCA materials must satisfy the mix design requirements similar to virgin aggregate for the desired application (e.g., pavement surface layer, unbound base layer, etc.). With proper control, RCA can be utilized to satisfy standard quality and gradation.

C.3.4 Properties of Concrete with RCA

The properties of the RCA-prepared concrete mixture depend on RCA's composition, and gradation. The effect of RCA can be minimized with proper mix design and admixture (ACPA, 2009).

Property	Natural Aggregate	RCA
Absorption capacity (%)	0.8–3.7	3.7–8.7
Specific gravity	2.4–2.9	2.1–2.4
LA abrasion test mass loss (%)	15–30	20–45
Sodium sulfate soundness test mass loss (%)	7–21	18–59
Magnesium sulfate soundness test mass loss (%)	4–7	1–9
Chloride content (lb/yd^3)	0–2	1–12

Source: Snyder, M. B., Vandenbossche, J. M., Smith, K. D., and Wade, M. J. (1994). *Synthesis on Recycled Concrete Aggregate*. Interim Report—Task A, DTFH61-93-C00133. Washington, DC: Federal Highway Administration.

Table C.2 Typical Properties of Natural Aggregate and RCA

Fresh RCA Concrete Properties

As RCA particles are angular and rough-textured, RCA produces harsh, fresh concrete. If there is too much fine RCA in the mixture, then workability can be a problem. Therefore, to have adequate workability, fine RCA is usually limited to 30% or less replacement of sand to have sufficient workability. A high RCA absorption rate can also cause problems with workability. It is important to use pozzolanic and chemical admixtures to boost workability.

Hardened RCA Concrete Properties

The ranges of concrete properties resulting from the RCA are listed in Table C.3. In comparison to the mixture with virgin aggregates, the compressive and tensile strength of the RCA mixture can be up to 40% and 20%, respectively. The elasticity modulus may be 40% lower than the virgin aggregate mixture. On the other hand, thermal expansion and contraction, shrinking, and permeability may increase. The degree of difference depends on the amount of fine RCA.

RCA intended for use in concrete paving mixtures must be treated as an engineered material, considering physical and mechanical properties, such as absorption capacity and coefficient of thermal expansion. Consideration of these properties may result in the need to modify the concrete mix design through the use of chemical and/or mineral admixtures, different mix component proportions, and/or aggregate blending. They may also require the consideration of different pavement structural characteristics (i.e., thickness, panel dimensions, reinforcing, etc.).

Since RCA has different physical and mechanical properties than a conventional virgin aggregate, the following considerations should be given while using RCA in pavement design:

- RCA concrete has less strength and elastic modulus. This leads to slightly larger pavement thickness.
- RCA concrete has high shrinkage and thermal expansion/contraction. This leads to more significant joint movements and may require different sealant materials or a small panel span.
- RCA concrete reduces the potential for aggregate interlock. This leads to higher amounts of reinforcing in the mesh-reinforced RCA concrete pavement and continuously reinforced pavement.

Property	Coarse RCA Only	Coarse and Fine RCA
Compressive strength	0–24% lower	15–40% lower
Tensile strength	0–10% lower	10–20% lower
Modulus of elasticity	10–33% lower	25–40% lower
Thermal expansion/contraction	0–30% higher	0–30% higher
Drying shrinkage	20–50% higher	70–100% higher
Permeability	0–500% higher	0–500% higher

TABLE C.3 Typical Properties of RCA Concrete Compared to Similar Mixtures with Natural Aggregate (FHWA, 2007)

C.3.5 Uses of RCA

Since the 1940s in the United States (Epps et al., 1980), RCA has been used widely for roadway concrete layers, shoulders, median barriers, sidewalks, curbs and gutters, building and bridge foundations, and even structural concrete. Since that time, RCA has been used on low-volume roads (e.g., Highway 75 in Iowa) and high-volume roads (e.g., Interstate 10 near Houston, Texas) to create hundreds of concrete pavement construction projects in the United States and around the world (ACPA, 2009). This also included the recycling of pavements that were severely damaged in new concrete pavements due to D-cracking or alkali-silica reactivity (ASR). RCA is common in some European countries (such as Austria) and is increasingly permitted in the United States (such as the Illinois Tollway Reconstruction I-90) (ACPA, 2009).

C.3.6 Considerations for Mix Design Using RCA

RCA materials to be used in new concrete pavements are expected to be free of potentially harmful contaminants. Study shows that in concrete mixtures, the presence of small amounts of joint sealant, traffic engine oil, and other contaminants on the pavement surface cause problems (FHWA, 2007). When designing mixes, RCA materials must satisfy the same quality requirements as virgin aggregate, although there is usually no need for RCA washing prior to batching. Washing, however, helps to reduce the problems of absorption and workability and to strengthen the bond of cement and aggregate. The mix design of concrete containing RCA can be done using the same common procedures for mixing concrete design with virgin aggregate only. The following are some additional activities suggested for mixing with RCA materials:

- A little higher proportion of cement may be necessary to produce the required strength (FHWA, 2007).
- To confirm similar workability to a conventional PCC mixture, 5% to 15% more water and/or a water-reducing admixture may be required (FHWA, 2007).
- FHWA (2007) recommends a water-to-cementitious material ratio of 0.45 or less.
- To prevent the output of a harsh mix, fine RCA should be limited to 30% of the total fine aggregate. The use of coarse RCA in concrete paving mixtures does not have general limitations.
- RCA substitutions for natural aggregate should be conducted volumetrically because of the lower specific gravity of RCA.

C.4 Summary

This appendix addresses asphalt and concrete pavement recycling technology, benefits, procedures, and special consideration requirements. In the *Basic Asphalt Recycling Manual* by Asphalt Recycling and Reclaiming Association (ARRA), U.S. Department of Transportation (referenced as ARRA, 2015), you can find the comprehensive recycling technologies of asphalt pavement. *Recycling Concrete Pavements,* the engineering bulletin EB043P of the American Concrete Pavement Association (referenced as ACPA, 2009), provides the details of concrete pavement recycling technology, additional structural design guidelines, and recommendations.

APPENDIX D

Superpave Asphalt Mix Design

D.1 Background

This textbook explains pavement materials; pavement structures; structural response theories; structural analysis techniques; design of flexible, rigid, and composite pavements; drainage design; and highway geometric design. Although not an integral part of the pavement design, asphalt mix design is commonly performed shortly before the pavement structure and highways are designed. Therefore, asphalt mix design, for the benefit of the readers, is discussed here. This appendix discusses the Superpave asphalt mix design procedure and the associated knowledge required for asphalt mix design. Asphalt mix design determines the following:

- The required quality aggregate and gradation to use
- The required grade (type) of asphalt binder to use
- The required optimum combination of these two ingredients (aggregate and asphalt binder) to have the desired properties and performance

Asphalt concrete (AC) mix consists of asphalt binders and aggregates in general. AC's stiffness and performance depend on types of aggregates, aggregate gradation, binder grade, binder proportion, etc. In the final mix, the following performance is obtained by manipulating the aggregate and asphalt binder variables.

- *Deformation or rutting resistance.* Permanent deformation in AC occurs under wheel loading. The deformation increases as the proportion of round particles, fine particles, binder content, and inappropriate binder grade increase.
- *Fatigue resistance.* Fatigue cracking occurs when the wheel load is applied repeatedly over time. AC fatigue cracking is associated with air void, asphalt binder content, and hard-mixed stiffness. Lower level of asphalt binder leads to less fatigue cracking under repeated loading. However, lower asphalt binder content leads to greater deformation potential (rut) under loading.
- *Low-temperature cracking resistance.* AC cracks when exposed to severe cold temperatures. Low-temperature cracking mainly depends on the grade of the

asphalt binder (low stiffness of temperature). Specifying a suitable grade of asphalt binder with appropriate low-temperature properties will avoid or minimize cracking at low temperatures.

- *Durability.* Chemicals in AC should not age (oxidize) rapidly during production and service life when exposed to air. If the mixture has high air voids (about 8% or more), the permeability of the mixture becomes very high. This high permeability makes it easy for oxygen in the air to access the asphalt binder and accelerates oxidation and volatilization. Also, if the thickness of the asphalt coating surrounding the aggregate is low, the aggregate may become water-accessible through film holes and results in potholes.
- *Skid resistance.* The compacted AC should provide sufficient friction or skid resistance for tires. Smooth, rounded, or polish-susceptible aggregates are less skid resistant.
- *Workability.* During construction, AC must be able to be placed and compacted with reasonable effort. Flat, elongated, or angular particles appear to interlock, making compaction difficult. Excess fines in the mix give rise to tenderness. If the viscosity of the asphalt binder is too high at mixing and lay-down temperatures, the AC becomes difficult to dump, spread, and compact.

The above performance is to be achieved by mix design, choosing the appropriate grade of asphalt binder, optimum binder content, and proposing a simple aggregate ratio and asphalt binder. A variety of methods are available to design asphalt mixtures. The Superpave mix design procedure is the most recent and robust of all methods.

D.2 Superpave Mix Design

Superpave mix design technique has been developed to address the limitations in others' literature. Between 1987 to 1993, Superpave mix design was part of the Strategic Highway Research Program (SHRP) to enhance performance and durability of U.S. roads. The objective of the Superpave program was to develop an asphalt binder specification based on performance, an asphalt mixture specification based on performance, and a mix design system. Over the past few years, Superpave mixes have been commonly used and replacing the Marshall method, which has been used for almost half a century. The following steps are provided by the Superpave mix design method:

Step 1. Selection of aggregates

Step 2. Selection of asphalt binder

Step 3. Selection of design aggregate structure

Step 4. Determination of design binder content

Step 5. Evaluation of moisture susceptibility

Step 1. Selection of Aggregates

In Superpave mix design, aggregate is chosen in a manner that satisfies both the requirements of source and consensus. Source properties are commonly defined by the owner/contractor, which includes the following:

- *Los Angeles Abrasion (AASHTO T 96).* Maximum allowable loss values typically range from approximately 35% to 45%.

- *Soundness (AASHTO T 104).* Maximum allowable loss values typically range from approximately 10% to 20% for five cycles.
- *Deleterious materials (AASHTO T 112).* There is a wide range of maximum allowable percentage of clay lumps and friable particles. Depending on the exact composition of the contaminant, the values range from as little as 0.2% to as high as 10%.

Consensus properties as per the national Superpave specification include:

- Coarse aggregate angularity (AASHTO MP 2 and ASTM D 5821)
- Fine aggregate angularity (AASHTO TP 33)
- Flat and elongated particles (ASTM D 4791)
- Sand equivalency (ASTM D 2419)

The consensus requirements, which depend on traffic level and pavement depth, are listed in Table D.1.

Step 2. Selection of Asphalt Binder

The performance grade (PG) binder required for the project is selected using temperature data, traffic level, and traffic speed. By converting historical air temperatures to pavement temperatures, the temperature data are obtained. Temperatures of the binder laboratory test are defined by the average 7-day maximum pavement temperature (T_{max}) and the minimum one-day pavement temperature (T_{min}). Depending on temperature reliability, a safety factor can be incorporated into the performance grading system. The temperatures of 50% are the average of the weather data.

Design ESAL* (million)	Coarse aggregate angularity (%), minimum		Uncompacted void content of fine aggregate (%), minimum		Sand equivalent (%), minimum	Flat and elongated‡ (%), minimum
	Depth from surface		Depth from surface			
	≤100 mm (4 in.)	>100 mm (4 in.)	≤100 mm (4 in.)	>100 mm (4 in.)		
<0.3	55/–	–/–	–	–	40	–
0.3 to <3	75/–	50/–	40	40	40	10
3 to <10	85/80†	60/–	45	40	45	
10 to <30	95/90	80/75	45	40	45	
≥30	100/100	100/100	45	45	50	

*Design ESALs are the anticipated project traffic level expected on the design lane over 20 years. Regardless of the actual design life of the roadway, determine ESALs for 20 years and choose the appropriate design number of gyration (*Ndes*) level.

†85/80 denotes that 85% of the coarse aggregate has one fractured face, and 80% has two or more fractured faces.

(1)‡A criterion based upon a 5:1 maximum-to-minimum ratio.

Source: *Adapted from FHWA (2017). Superpave Fundamentals Reference Manual. NHI Course #131053. Washington, DC: Federal Highway Administration.*

TABLE D.1 Superpave Aggregate Consensus Property Requirements

Design ESAL (million)	Traffic speed (mile per hour)		
	Less than 12	12 to 42	Above 42
Less than 0.3	–	–	–
0.3 to less than 3	2	1	–
3 to less than 10	2	1	1

Note: 1-grade equivalent is 6°C.
Source: *Adapted from FHWA (2017). Superpave Fundamentals Reference Manual. NHI Course #131053. Washington, DC: Federal Highway Administration.*

TABLE D.2 Increase in High-Temperature Grade for Traffic Load and Speed

The 98% reliability temperatures are determined based on the standard deviations of the low ($\sigma_{\text{Low Temp}}$) and high ($\sigma_{\text{Low High}}$) temperatures data. From statistics, 98% reliability is about two standard deviations from the average value, such that:

$$T_{\text{max at 98\%}} = T_{\text{max at 50\%}} + 2\sigma_{\text{High Temp}} \tag{D.1}$$

$$T_{\text{min at 98\%}} = T_{\text{min at 50\%}} - 2\sigma_{\text{Low Temp}} \tag{D.2}$$

Traffic level and speed are also considered in selecting the PG binder for high temperatures (Table D.2). However, no adjustment is used for low temperatures.

As part of the binder specification, the rotational viscometer is performed at 135°C on the original/unconditioned binder. The specification recommends a viscosity of less than 3 Pascal-seconds (Pa-s) for all binders. At a second temperature, usually 160°C, the rotational viscometer must be run. This is done to determine the appropriate temperatures for mixing and compaction. SHRP adopted the mixing and compaction recommendations of the Asphalt Institute based on the binder's temperature–viscosity relationship, where:

The range for mixing temperature = 150 to 190 centiStokes

The range for compaction temperature = 250 to 310 centiStokes

The rotational viscometer measures viscosity in centipoises (cP), and the values are reported in Pascal-seconds (Pa-s). The conversion from centipoises to Pascal-seconds is as follows:

$$1 \text{ Pa-s} = 1{,}000 \text{ centipoises}$$

The asphalt binder specific gravity (G_b) is determined according to the AASHTO T 228 at 25°C. This specific gravity value is required for the volumetric analysis.

Step 3. Selection of Design Aggregate Structure

To select the design aggregate structure, mathematically combining the gradations of the individual materials into a single blend is used to establish trial blends. The blend is then compared with the design requirements for the appropriate sieves, as stated in Table D.3. Control of gradation is based on four control sieves: maximum sieve, nominal maximum sieve, 2.36-mm sieve, and 0.075-mm sieve.

Trial blends are evaluated by compacting specimens and determining each trial blend's volumetric properties. The compaction is carried out using Superpave gyratory compactor

Sieve size, mm	Nominal maximum aggregate size—control points (% passing)											
	37.5 mm		25.0 mm		19.0 mm		12.5 mm		9.5 mm		4.75 mm	
	Min.	Max.	Min.	Max.	Min.	Max.	Min.	Max.	Min.	Max.	Min.	Max.
50	100	–	–	–	–	–	–	–	–	–	–	–
37.5	90	100	100	–	–	–	–	–	–	–	–	–
25.0	–	90	90	100	100	–	–	–	–	–	–	–
19.0	–	–	–	90	90	100	100	–	–	–	–	–
12.5	–	–	–	–	–	90	90	100	100		100	–
9.5	–	–	–	–	–	–	–	90	90	100	95	100
4.75	–	–	–	–	–	–	–	–	–	90	90	100
2.36	15	41	19	45	23	49	28	58	32	67	–	–
1.18	–	–	–	–	–	–	–	–	–	–	30	55
0.075	0	6	1	7	2	8	2	10	2	10	6	13

Source: *Adapted from FHWA (2017). Superpave Fundamentals Reference Manual. NHI Course #131053. Washington, DC: Federal Highway Administration.*

Table D.3 Aggregate Grading Requirements for Superpave

(Fig. D.1) with a gyration angle of 1.16° with respect to the horizontal at a constant vertical pressure of 600 kPa (87 psi). The number of gyrations used for compaction is determined based on the level of traffic with three critical compaction stages: initial (N_{ini}), design (N_{des}), and maximum (N_{max}) number of gyrations as shown in Table D.4. Initial compaction stage N_{ini} is used to identify the unwanted tenderness of the mix. During construction, a tender mix has less stability and displaces instead of densifying when the roller arrives. Excessive moisture, excessive binder, excessive rounded particles, excessive mixing temperature, insufficient fines, etc. are the causes of the tender mix. N_{des} denote the expected level of compaction that occurs during construction at the design compaction stage. N_{max} corresponds to the maximum compaction stage after years of traffic operation.

Samples are compacted with N_{des} gyrations to determine the design aggregate structure, and the volumetric properties are calculated, as shown in Table D.4, based on the following tasks:

Task 1: Determine the G_{mm} of the loose mix and G_{mb} of the compacted sample using a laboratory test.

Task 2: Calculate VTM: $$\text{VTM} = \left(1 - \frac{G_{mb}}{G_{mm}}\right) \times 100 \qquad \text{(D.3)}$$

VTM should be 4.0%, otherwise adjust other properties.

Task 3: Calculate VMA: $$\text{VMA} = 100 - \left(\frac{G_{mb}P_s}{G_{sb}}\right) \times 100 \qquad \text{(D.4)}$$

Task 4: Calculate VFA: $$\text{VFA} = \left(\frac{\text{VMA} - \text{VTM}}{\text{VMA}}\right) \times 100 \qquad \text{(D.5)}$$

Design ESALs* (million)	Compaction parameters			Typical roadway application†
	N_{ini}	N_{des}	N_{max}	
<0.3	6	50	75	Applications include roadways with very light traffic volumes such as local roads, county roads, and city streets where truck traffic is prohibited or at a very minimal level. Traffic on these roadways would be considered local, not regional, intrastate, or interstate. Special-purpose roadways serving recreational sites or areas may also be applicable to this level.
0.3 to <3	7	75	115	Applications include many collector roads or access streets. Medium-trafficked city streets and the majority of county roadways may be applicable to this level.
3 to <30	8	100	160	Applications include many two-lane, multilane, divided, and partially or completely controlled access roadways. Among these are medium- to highly trafficked city streets, many state routes, U.S. highways, and some rural interstates.
≥30	9	125	205	Applications include the vast majority of the U.S. Interstate system, both rural and urban. Special applications such as truck-weighing stations or truck-climbing lanes on two-lane roadways may also apply to this level.

*Design ESALs are the anticipated project traffic level expected on the design lane over a 20-year period. Regardless of the actual design life of the roadway, determine the design ESALs for 20 years, and choose the appropriate N_{des} level.

†Typical Roadway Applications as defined by *A Policy on Geometric Design of Highway and Streets*, 1994, AASHTO.

Source: *Adapted from FHWA (2017). Superpave Fundamentals Reference Manual. NHI Course #131053. Washington, DC: Federal Highway Administration (FHWA).*

TABLE D.4 Superpave Gyratory Compaction Effort

Task 5: Calculate % G_{mm} at N_{ini}, $$\%G_{mm,N_{ini}} = \%G_{mm,N_{des}}\left(\frac{h_{des}}{h_{ini}}\right) = (100 - VTM)\left(\frac{h_{des}}{h_{ini}}\right) \quad \text{(D.6)}$$

Task 6: Calculate dust to asphalt ratio, $$\frac{F}{A} = \frac{\%\ \text{passing No. 200}}{\%\ P_{be}} \quad \text{(D.7)}$$

where Effective binder, $$P_{be} = (P_b - P_{ba}P_s) \times 100 \quad \text{(D.8)}$$

Absorbed binder, $$P_{ba} = \left(\frac{G_{se} - G_{sb}}{G_{sb}G_{se}}\right)G_b \times 100 \quad \text{(D.9)}$$

Effective specific gravity, $$G_{se} = \frac{P_s}{\dfrac{1}{G_{mm}} - \dfrac{P_b}{G_b}} \quad \text{(D.10)}$$

FIGURE D.1 Superpave gyratory compactor ejecting a compacted specimen.

As mentioned in the Task 2, the volumetric calculations must be made at 4% VTM. If the VTM is not 4%, the results of the volumetric calculations are estimated at 4% as follows:

$$P_{b,\text{est}} = P_b - 0.4(4 - \text{VTM}) \tag{D.11}$$

$$\text{VMA}_{\text{est}} = \text{VMA} + C(4 - \text{VTM}) \tag{D.12}$$

C = 0.1 for VTM < 4% and C = 0.2 for VTM ≥4%.

$$\text{VFA}_{\text{est}} = 100\left(\frac{\text{VMA}_{\text{est}} - 4.0}{\text{VMA}_{\text{est}}}\right) \tag{D.13}$$

$$\%G_{mm,N_{\text{ini}},\text{est}} = \%G_{mm,N_{\text{des}}} - (4 - \text{VTM}) \tag{D.14}$$

$$P_{be,\text{est}} = P_{b,\text{est}} - P_s G_b\left(\frac{G_{se} - G_{sb}}{G_{sb} G_{se}}\right) \tag{D.15}$$

$$\left(\frac{F}{A}\right)_{\text{est}} = \frac{\%\text{ passing No. 200}}{(\%P_{be})_{\text{est}}} \tag{D.16}$$

According to the Superpave mix design criteria listed in Table D.5, the calculated volumetric properties are compared. If more than one aggregate blend meets the criteria, the most economical blend or the choice of the designer could be chosen. If none of the blends fulfill the requirement, you can try another blend of aggregate.

Design ESALs* (million)	Required density (% of theoretical maximum specific gravity)			Voids-in-the mineral aggregate (percent), minimum					Voids filled with asphalt (percent)	Dust-to-binder ratio
				Nominal maximum aggregate size, mm						
	N_{ini}	N_{des}	N_{max}	37.5	25.0	19.0	12.5	9.5		
<0.3	≤91.5	96.0	≤98.0	11.0	12.0	13.0	14.0	15.0	70–80[‡,§]	0.6–1.2
0.3 to <3	≤90.5								65–78[4]	
3 to <10	≤89.0								65–75[†,§]	
10 to <30										
≥30										

*Design ESALs are the anticipated project traffic level expected on the design lane over 20 years. Regardless of the actual design life of the roadway, determine the design ESALs for 20 years, and choose the appropriate N_{des} level.

†For 9.5-mm nominal maximum size mixtures, the specified VFA range shall be 73% to 76% for design traffic levels at $3 million ESALs.

‡For 25.0-mm nominal maximum size mixtures, the specified lower limit of the VFA shall be 67% for design traffic levels at <0.3 million ESALs.

§For 37.5-mm nominal maximum size mixtures, the specified lower limit of the VFA shall be 64% for all design traffic levels.

Source: *Adapted from FHWA (2017). Superpave Fundamentals Reference Manual. NHI Course #131053. Washington, DC: Federal Highway Administration.*

Table D.5 Superpave Volumetric Mixture Design Requirements

Step 4. Determination of Design Binder Content

Design binder content corresponds to the content of the binder, which produces 4% air void in the mix. Once the design aggregate structure is chosen, the specimens are compacted at different asphalt binder contents. The properties of the mixture will then be evaluated to determine a design asphalt binder content. Superpave requires at least two compacted specimens for each of the following asphalt content (FHWA recommends three compacted specimens for each asphalt binder content):

1. Two specimens at the estimated binder content
2. Two specimens 0.5% less than the estimated binder content
3. Two specimens 0.5% more than the estimated binder content
4. Two specimens 1.0% more than the estimated binder content

The volumetric properties are calculated at the design number of gyrations (N_{des}) for each trial asphalt binder content. Optimum asphalt binder content can be arrived at in the following procedures:

1. Plot the following graphs:
   a. *VTM vs. asphalt binder content.* Percent air voids should decrease with increasing asphalt binder content.

b. *VMA vs. asphalt binder content.* Percent VMA should decrease with increasing asphalt binder content, reach a minimum, then increase.
c. *VFA vs. asphalt binder content.* Percent VFA increases with increasing asphalt binder content.
d. G_{mm} *vs. asphalt binder content.* Density generally increases with the asphalt content, reaches a maximum, then decreases. Peak density usually occurs at a high asphalt binder content than peak stability.
e. F/A ratio vs. asphalt binder content.

2. Determine the content of the asphalt binder corresponding to the required amount of air void (typically 4%). This is the optimum content of the asphalt binder.
3. Determine properties at this optimum asphalt binder content by referring to the plots. Compare each of these values to specification values, and the preceding optimum asphalt binder content will be sufficient if all of them are within specification. Otherwise, the mixture should be redesigned if any of these properties are outside the specification range.

Step 5. Evaluation of Moisture Susceptibility

The final step in the volumetric mix design process is to evaluate the moisture sensitivity of the design mixture. This step is accomplished by performing AASHTO T 283 on the design aggregate blend at the design asphalt binder content. Specimens are compacted to approximately 7.0% (±1.0%) air voids. One batch of three specimens is considered the control/unconditioned subset. The other subset of three specimens is the conditioned subset. The conditioned subset is subjected to partial vacuum saturation followed by an optional freeze cycle, followed by a 24-h heating cycle at 60°C. All specimens are tested to determine their indirect tensile strengths, as shown in Fig. D.2.

Deformation rate of 2 in. (50 mm) per minute is applied diametrically until failure, and the peak load is recorded. Then, the indirect tensile strength is calculated as:

$$\text{ITS} = \frac{2P}{\pi D t} \tag{D.17}$$

where P, D, and t are measured for peak load, diameter, and thickness, respectively. The sensitivity to moisture is determined as a ratio of the conditioned subset's tensile strengths divided by the control subset's tensile strengths.

$$\text{Tensile strength ratio} = \frac{\text{Tensile strength after conditioning}}{\text{Tensile strength before conditioning}} \tag{D.18}$$

The minimum criteria for the tensile strength ratio are 0.8 or 80%.

Example D.1: Selection of Asphalt Binder (Step 2 of Superpave Mix Design) The 7-day maximum pavement temperature of a pavement site is 52°C with a standard deviation of 4°C, and the single-day minimum pavement temperature of that pavement site is −12°C with a standard deviation of 5°C. If the design traffic is 5 million and the design speed is 50 mph, then what is the binder grade to be selected at 98% reliability?

Solution

$$T_{\text{max at 98\%}} = T_{\text{max at 50\%}} + 2\sigma_{\text{High Temp}} = 52 + 2(4) = 60^\circ\text{C}$$

$$T_{\text{min at 98\%}} = T_{\text{min at 50\%}} - 2\sigma_{\text{Low Temp}} = -12 - 2(5) = -22^\circ\text{C}$$

The recommended binder at 98% reliability is PG 64–22 without considering the traffic and speed.

For 5 million vehicles at 50 mph, the increase in high-temperature grade, from Table D.5, is 1. Therefore, the final binder is PG 70–22.

Answer *The final binder is PG 70–22.*

(a) Test setup

(b) Failed unconditioned samples

(c) Failed conditioned samples

FIGURE D.2 Indirect tensile strength testing.

Example D.2: Selection of an Aggregate Blend Select the blend using Superpave design aggregate structure for a 5-million ESAL and a 19-mm nominal aggregate size shown in Table D.6.

Data	Blend 1	Blend 2	Blend 3
G_{mb}	2.451	2.341	2.377
G_{mm}	2.528	2.528	2.604
G_b	1.0	1.0	1.0
P_b	5.0	4.7	4.2
P_s	95	95.3	95.8
% Passing No. 200	4.5	4.5	4.5
G_{sb}	2.722	2.625	2.605
H_{ini}	126	132	129
H_{des}	116	117	117

Table D.6 Aggregate Blend Data for Example D.2

Solution

Volumetric analysis to Select an Aggregate Blend (Table D.7)

Computed	Equation	Blend 1	Blend 2	Blend 3
VTM	$\text{VTM} = \left(1 - \frac{G_{mb}}{G_{mm}}\right) \times 100$	3.0	7.4	8.7
VMA	$\text{VMA} = 100 - \left(\frac{G_{mb}P_s}{G_{sb}}\right) \times 100$	14.5	15.0	12.6
VFA	$\text{VFA} = \left(\frac{\text{VMA} - \text{VTM}}{\text{VMA}}\right) \times 100$	78.9	68.1	45.3
% G_{mm} at N_{ini}	$\%G_{mm,N_{ini}} = (100 - \text{VTM})\left(\frac{h_{des}}{h_{ini}}\right)$	89.3	82.1	82.8
G_{se}	$G_{se} = \frac{P_s}{\frac{1}{G_{mm}} - \frac{P_b}{G_b}}$	2.749	2.734	2.801
P_{ba}	$P_{ba} = \left(\frac{G_{se} - G_{sb}}{G_{sb}G_{se}}\right) G_b \times 100$	0.4	1.5	2.7
P_{be}	$P_{be} = (P_b - P_{ba}P_s) \times 100$	4.7	3.3	1.6
F/A	$\frac{F}{A} = \frac{\text{\% passing No. 200}}{\% P_{be}}$	1.0	1.4	2.8

Table D.7 Aggregate Blend Data Analysis for Example D.2

The volumetric calculations must be made at 4% VTM. As the VTM is not 4%, the results of the volumetric calculations are to be estimated at 4%, as shown in Table D.8.

Estimated at 4% air void	Equation	Blend			Criteria	OK
		1	2	3		
$P_{b,est}$	$P_{b,est} = P_b - 0.4(4 - \text{VTM})$	4.6	6.1	6.1		
VMA_{est}	$\text{VMA}_{est} = \text{VMA} + C(4 - \text{VTM})$	14.6	14.3	11.6	Min. 13	1, 2
VFA_{est}	$\text{VFA}_{est} = 100\left(\frac{\text{VMA}_{est} - 4.0}{\text{VMA}_{est}}\right)$	72.7	72.0	67.2	65–75	All
$\% G_{mm}$ at $N_{ini,est}$	$\%G_{mm,N_{ini},est} = \%G_{mm,N_{des}} - (4 - \text{VTM})$	88.3	85.5	87.5	Max. 89	All
$P_{be,est}$	$P_{be,est} = P_{b,est} - P_s G_b\left(\frac{G_{se} - G_{sb}}{G_{sb} G_{se}}\right)$	4.3	4.6	3.5		
$F/A,_{est}$	$\left(\frac{F}{A}\right)_{est} = \frac{\text{\% passing No. 200}}{(\% P_{be})_{est}}$	1.1	1.0	1.3	0.6–1.2	1, 2

Table D.8 Determining the Properties at 4% Air Void for Example D.2

Answer *Both Blend 1 and Blend 2 are satisfactory.*

Example D.3: Selection of Optimum Binder Content Determine the design binder content for a 5-million ESAL and a 19-mm nominal aggregate size, as shown in Table D.9.

Data	Binder content (%)			
	5.0	5.5	6.0	6.5
G_{mb}	2.451	2.491	2.455	2.469
G_{mm}	2.570	2.558	2.530	2.510
G_b	1.023	1.023	1.023	1.023
P_s	95	94.5	94	93.5
% Passing No. 200	4.5	4.5	4.5	4.5
G_{sb}	2.688	2.688	2.688	2.688
H_{ini}	125	131	126	130
H_{des}	115	118	114	112

Table D.9 Mix Properties Data for Example D.3

Solution

Volumetric Analysis to Design Binder Content (Table D.10)

Computed	Equation	Binder content			
		5.0	5.5	6.0	6.5
VTM	$VTM = \left(1 - \frac{G_{mb}}{G_{mm}}\right) \times 100$	7.7	6.3	3.8	6.9
VMA	$VMA = 100 - \left(\frac{G_{mb}P_s}{G_{sb}}\right) \times 100$	15.0	14.9	15.1	15.0
VFA	$VFA = \left(\frac{VMA - VTM}{VMA}\right) \times 100$	48.4	57.7	74.8	54.2
% G_{mm} at N_{ini}	$\%G_{mm,N_{ini}} = (100 - VTM)\left(\frac{h_{des}}{h_{ini}}\right)$	82.4	84.2	86.2	83.9
G_{se}	$G_{se} = \frac{P_s}{\frac{1}{G_{mm}} - \frac{P_b}{G_b}}$	2.857	2.855	2.805	2.966
P_{ba}	$P_{ba} = \left(\frac{G_{se} - G_{sb}}{G_{sb}G_{se}}\right)G_b \times 100$	2.10	2.08	1.46	3.39
P_{be}	$P_{be} = (P_b - P_{ba}P_s) \times 100$	3.01	3.53	4.63	3.33
F/A	$\frac{F}{A} = \frac{\text{\% passing No. 200}}{\%P_{be}}$	1.0	1.1	1.1	1.7

TABLE D.10 Mix Properties Data Analysis for Example D.3

Now the VTM, VMA, VFA, F/A, and G_{mm} ratios are to be plotted to find out the optimum binder content, as shown in Fig. D.3.

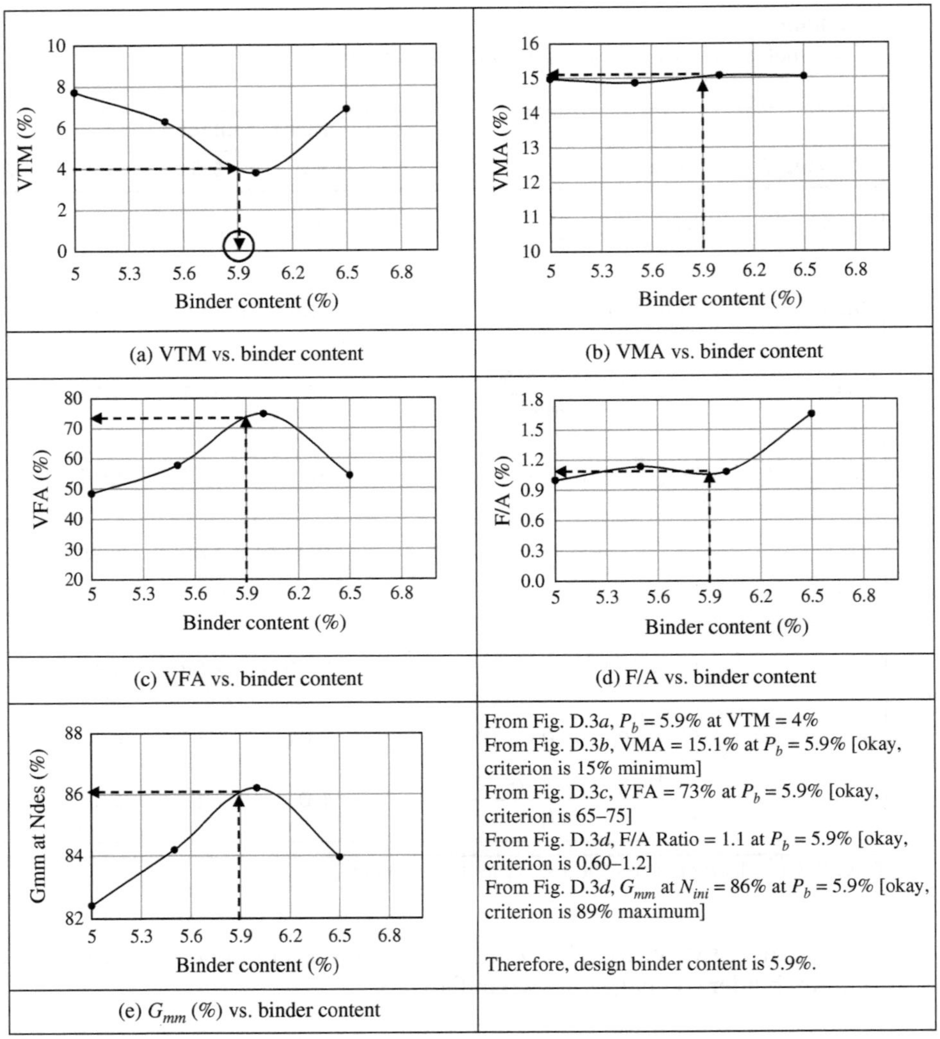

FIGURE D.3 Determining the optimum binder content for Example D.3.

D.3 Summary

The asphalt industry is newer and faster compared to the concrete industry. Federal agencies are largely involved mainly in asphalt, and their uses, characterization, and specifications are rapidly improving daily. This appendix describes the most mixed design procedures currently in use. However, with ongoing research and practice, the design specification may change in the near future. Asphalt mix design procedure is briefly discussed in this appendix, as it is beyond the scope of this textbook. To learn more about the asphalt and its mix design, readers are encouraged to explore another book by the first author, *Civil Engineering Materials: Introduction and Laboratory Testing*, published by CRC Press.

References

AASHTO. (1993). *AASHTO Guide for Design of Pavement Structures*. Washington, DC: American Association of State Highway and Transportation Officials.

AASHTO. (2009). *Transportation and Sustainability Best Practices Background. Sustainability Peer Exchange*. Washington, DC: Center for Environmental Excellence, American Association of State Highway and Transportation Officials.

AASHTO. (2012). *Pavement Management Guide*, 2nd ed. Washington, DC: American Association of State highway and Transportation Officials.

AASHTO. (2015). *Mechanistic-Empirical Pavement Design Guide: A Manual of Practice*. Washington, DC: American Association of State Highway and Transportation Officials.

AASHTO. (2018). *A Policy on Geometric Design of Highways and Streets*, 7th ed. Washington, DC: American Association of State Highway and Transportation Officials.

ACI. (1956). Structural design considerations for pavement joints. Report of the American Concrete Institute (ACI) Committee 325, *ACI Journal*, July.

ACPA. (2006). *Natural Advantage*. QD016P. Skokie, IL: American Concrete Pavement Association.

ACPA. (2009). *Recycling Concrete Pavements*. Engineering Bulletin EB043P. Rosemont, IL: American Concrete Pavement Association.

ARRA. (2015). *Basic Asphalt Recycling Manual*. Glen Ellyn, IL: Asphalt Recycling and Reclaiming Association (ARRA), U.S. Department of Transportation.

Asphalt Institute. (2001). *HMA Construction*. Manual Series No. 22 (MS-22). Lexington, KY: Asphalt Institute.

Austroads. (2012). *Guide to Pavement Technology, Part 2: Pavement Structural Design*. Sydney: Austroads Inc.

Austroads. (2017). *Guide to Pavement Technology, Part 2: Pavement Structural Design*. Sydney: Austroads.

Bonaquist, R. (2011). *Mix Design Practices for Warm Mix Asphalt*. NCHRP Report 691. Washington, DC: Transportation Research Board.

Boussinesq, J. (1985). Application des Potentiels a l'etude de l'equilibre et du Mouvement des Solids Elastiques. Paris: Gauthier-Villars.

Bradbury, R. D. (1938). *Reinforced Concrete Pavements*. Washington, DC: Wire Reinforcement Institute.

Brown E. R., Kandhal, P. S., Roberts, F. L., Kim, Y. R., Lee D-Y., and Kennedy T. W. (2009). *Hot-Mix Asphalt Materials, Mixture Design, and Construction*, 3rd ed. Second Printing 2016. Greenbelt, MD: NAPA Research and Education Foundation.

Burmister, D. M. (1943). *The Theory of Stresses and Displacements in Layered Soil Systems and Applications to the Design of Airport Runways*, Vol. 23: 126–144. Washington, DC: Highway Research Board.

CDOT. (2019). *2019 Pavement Design Manual.*, Denver, CO: Colorado Department of Transportation (CDOT).

CDRA (2018). *Good Economic Sense*. Aurora, IL: Construction and Demolition Recycling Association (CDRA).

Cedegren, H. R. (1974). *Drainage of Highway and Airfield Pavements*. New York: John Wiley & Sons.

Construction and Demolition Recycling Association (CDRA). (2014). *Good Economic Sense*. Aurora, IL: Construction and Demolition Recycling Association.

Cross, S. A., and Jakatimath, Y. (2007). *Evaluation of Cold In-Place Recycling for Rehabilitation of Transverse Cracking on US 412*. Report No. FHWA/OK 07 (04)., Oklahoma City, OK: Oklahoma Department of Transportation, Planning & Research Division.

Darter, M., and Barenberg, E. (1977). *Design of Zero-Maintenance Plain Concrete Pavement*. Report No. FHWA-RD-77-111. Washington, DC: Federal Highway Administration.

Darter, M. I., Titus-Glover, L., Quintus, H., Bhattacharya, B., and Mallela, J. (2014). *Calibration and Implementation of the AASHTO Mechanistic-Empirical Pavement Design Guide in Arizona*. Report No. SPR-606. Phoenix, AZ: Arizona Department of Transportation.

DMRB (2006). *Design Manual for Roads and Bridges (DMRB), Vol. 7, Pavement Design and Maintenance, Section 2, Part 1*. London: Department of Transport, Highway Agency.

DMRB (2009). *Design Manual for Roads and Bridges (DMRB), Vol. 7, Pavement Design and Maintenance, Section 5, Part 2*. London: Department of Transport, Highway Agency.

Dower, R. C., Rand, S. D., and Scodari, P. F. (1985). *The Scrap Tire Problem: A Preliminary Economic Study*. Grant No. CR-811897-01. Washington, DC: U.S. Environmental Protection Agency.

Energy Information Agency (EIA). (2011). *Asphalt and Road Oil Consumption, Price, and Expenditure Estimates*. Washington, DC: Energy Information Agency.

Environmental Protection Agency (EPA). (2009). *Recover Your Resources. Reduce, Reuse, and Recycle Construction and Demolition Materials at Land Revitalization Project*. EPA-560-F09-523. Washington, DC: Environmental Protection Agency.

Epps, J. A., Little, D. N., Holmgren, R. J., and Terrell, R. L. (1980). *Guidelines for Recycling Pavement Materials*. NCHRP Report 224. Washington, DC: Transportation Research Board.

Federal Highway Administration. (FHWA). (2010a). *Highway Statistics, 2010*. Washington, DC: Federal Highway Administration.

Federal Highway Administration. (FHWA). (2010b). *Life-Cycle Cost Analysis: RealCost User Manual*. RealCost Version 2.5. Washington, DC: Federal Highway Administration.

Federal Highway Administration. (FHWA). (2013a). *2013 Pocket Guide to Transportation. Federal Highway Administration*. Washington, DC: Federal Highway Administration.

Federal Highway Administration (FHWA). (2013b). *Improving Transportation Investment Decisions through Life-Cycle Cost Analysis*. Washington, DC: Federal Highway Administration.

FHWA. (1999). *Pavement Subsurface Drainage Design*. NHI Course No. 131026. Washington, DC: Federal Highway Administration.

FHWA. (2002). *Formal Policy on the Use of Recycled Materials*. Washington, DC: Federal Highway Administration.

FHWA. (2004). *Recycled Concrete Aggregate—Federal Highway Administration National Review*. Washington, DC: Federal Highway Administration.

FHWA. (2007). *Use of Recycled Concrete Pavement as Aggregate in Hydraulic-Cement Concrete Pavement*. Technical Advisory T 5040.37. Washington, DC: Federal Highway Administration.

FHWA. (2009). FHWA-HRT-06-106. *Design and Evaluation of Jointed Plain Concrete Pavement with Fiber Reinforced Polymer Dowels*. McLean, VA: Office of Research and Technology Services, Federal Highway Administration.

FHWA. (2010). *Local Calibration of the AASHTOWARE Using Pavement Management Systems*. Report No. HIF-11-026. Washington, DC: Federal Highway Administration Office of Asset Management.

FHWA. (2013). *Traffic Monitoring Guide*. Washington, DC: Federal Highway Administration, U.S. Department of Transportation.

Friberg, B. F. (1940). Design of dowels in transverse joints of concrete pavements. *Transactions of ASCE*, 105, pp. 1076–1095.

Gallivan, F., Ang-Olson, J., and Papson, A. (2010). *Greenhouse Gas Mitigation Measures for Transportation Construction, Maintenance, and Operations Activities*. Final Report. NCHRP Project 25-25(58). Washington, DC: Transportation Research Board.

Garber, N. J., and Hoel, L. A. (2015). *Traffic and Highway Engineering*, 5th ed. Stamford, CT: Cengage Learning.

Ghilani, C. D., and Wolf, P. R. (2014). *Elementary Surveying: An Introduction to Geomatics*, 14th ed. Upper Saddle River, NJ: Pearson.

Giani, M., Dotelli, G., Brandini, N., and Zampori, L. (2015). Comparative life cycle assessment of asphalt pavements using reclaimed asphalt, warm mix technology and cold in-place recycling, *Resources, Conservation and Recycling*, 104, pp. 224–238.

Grass, P. T. (2012). Asphalt consumption trends. *Technical Presentation, South East Asphalt User/Producer Group Annual Meeting*. South East Asphalt User/Producer Group, Ridgeland, MS.

Guo, X. (2013). *Local Calibration of the AASHTOWare Using Test Track Data*. Graduate dissertation. Auburn, AL: Auburn University.

Haider, S. W., Harichandran, R. S., and Dwaikat, M. B. (2010). Effect of axle load measurement errors on pavement performance and design reliability. *Transportation Research Record*, no. 2160, Data Systems and Travel Survey Methods 2010, pp. 107–117.

Hall, K. T., Dawood, D., Vanikar, S., Tally, Jr., R., Cackler, T., Correa, A., Deem, P., et al. (2007). *Long-Life Concrete Pavements in Europe and Canada*. FHWA-PL-07-027., Washington, DC: Federal Highway Administration.

Hall, K., Xiao, D., and Wang, K. (2011). Calibration of the AASHTOWARE for Flexible Pavement Design in Arkansas. 2011 Annual Meeting of the Transportation Research Board, Washington, DC.

Halm, H. J. (1980). Concrete recycling. *Transportation Research News*. 89, p. 6–10.

Halstead, W. (1985). Relation of asphalt chemistry to physical properties and specifications. *Proceedings of AAPT*, No. 54.

Hansen, K., and Copeland, A. (2013). *Second Annual Asphalt Pavement Industry Survey on Reclaimed Asphalt Pavement, Reclaimed Asphalt Shingles, and Warm-Mix Asphalt Usage: 2009–2011*. Information Series 138. Lanham, MD: National Asphalt Pavement Association.

Hansen, K., and Newcomb, D. (2011). *Asphalt Pavement Mix Production Survey on Reclaimed Asphalt Pavement, Reclaimed Asphalt Shingles, and Warm-Mix Asphalt Usage. 2009-2010.* Information Series 138. Lanham, MD: National Asphalt Pavement Association.

Hasan, M., Islam, M., and Tarefder, R. (2016b). Clustering vehicle class distribution and axle load spectra for mechanistic-empirical predicting pavement performance. *Journal of Transportation Engineering*, https://doi.org/10.1061/(ASCE)TE.1943-5436.0000876, 05016006.

Hasan, M. A., Islam, M. R., and Tarefder, R. A. (2016a). Site-specific versus pavement mechanistic-empirical (ME) default traffic data on interstate pavement performances. *Transportation Research Board (TRB) 95th Annual Meeting*, January 10–14, 2016, Washington, DC. Paper ID. 16-5619.

Hasan, M. M., Islam, M. R., and Tarefder, R. A. (2018). Characterization of subgrade soil mixed with recycled asphalt pavement (RAP). *Journal of Traffic and Transportation Engineering*, 5(3), pp. 207–214.

Holtz, R. D., Christopher, B. R., and Berg, R. R. (1998). *Geosynthetic Design and Construction Guidelines*. FHWA Technical Report No. FHWA-HI-95-038. Washington, DC: Federal Highway Administration, pp. 460.

International Organization for Standardization (ISO). (2006). *ISO 14040: 2006 Environmental Management—Life Cycle Assessment—Principles and Framework.* Geneva, Switzerland: International Organization for Standardization.

Ioannides, A. M., Thompson, M. R., and Barenberg, E. J. (1985). Westergaard solutions reconsidered. *Journal of Transportation Research Record*, 1043, pp. 13–23.

Ishak, S., Shin, H. C., Sridhar, B. K., and Zhang, Z. (2010). Characterization and Development of Truck Axle Load Spectra for Future Implementation of Pavement Design Practices in Louisiana. *Journal of Transportation Research Record*, no. 2153, 0361-1981, pp. 121–129.

Islam, M. R. (2015). *Thermal Fatigue Damage of Asphalt Pavements*. Doctoral dissertation. Department of Civil Engineering, University of New Mexico, Albuquerque, NM.

Islam, M. R., Ahmed, M., and Tarefder, R. A. (2016). Measuring the cross-anisotropy of hot-mix asphalt. *International Journal of Pavement Engineering*, 17(3), pp. 265–274.

Islam, M. R., Faisal, H., and Tarefder, R. A. (2015a). Determining temperature and time dependent Poisson's ratio of asphalt concrete using indirect tension test. *Journal of Fuel*, 146(2015), pp. 119–124.

Islam, M. R., Hossain, M. I., and Tarefder, R. A. (2015b). A study of asphalt aging using Indirect Tensile Strength test. *Construction and Building Materials*, 95(2015), pp. 218–223.

Islam, M. R., Kalevela, S. A., and Rivera, J. (2018a). *Dynamic Modulus of Cold-in-Place Recycling Asphalt Materials.* Final Report, 2018-10. Denver, CO: Colorado Department of Transportation (CDOT): pp. 1–158.

Islam, M. R., Mannan, U. A., Rahman, A., and Tarefder, R. A. (2014). Effects of reclaimed asphalt pavement on hot-mix asphalt. *ASTM Journal of Advances in Civil Engineering Materials*, 3(1), pp. 291–307.

Islam, M. R., Rahman, A., and Tarefder, R. A. (2018b). Open graded friction course in resisting transverse cracking in asphalt pavement. *ASCE Journal of Cold Regions Engineering*, 32(2), pp. 1–7.

Islam, M. R., and Tarefder, R. (2015). Moisture variations in unbound layers in an instrumented pavement section. *International Journal of Environmental, Ecological, Geological and Geophysical Engineering*, 9(4), pp. 167–171.

Islam, M. R., Tarefder, R. A., and Syed, I. (2014). Measurements of lateral distribution of vehicle wheels and its effect on fatigue life of asphalt concrete. In: *Sustainability, Eco-efficiency, and Conservation in Transportation Infrastructure Asset Management, 3rd International Conference on Transportation Infrastructure (ICTI)*, April 22–25, 2014, Pisa, Italy, pp. 379–383.

ITS. (2000). *University of California Institute of Transportation Studies, South Africa/United States Pavement Technology Workshop, Module 3: Pavement Structural Design*. Videotape series of seminar, March 20–23, 2000. Richmond, CA: University of California Field Station.

Kasperick, T., and Ksaibati, K. (2015). *Calibration of the Mechanistic-Empirical Pavement Design Guide for Local Paved Roads in Wyoming*. Report No. MPC 15-294. Laramie, WY: Mountain-Plains Consortium, Department of Civil and Architectural Engineering, University of Wyoming.

Kim, J., Lee, H., Jahren, C. T., Heitzman, M., and Chen, D. (2010). Long-term field performance of cold in-place recycled roads in Iowa. *Journal of Performance of Constructed Facilities*, 24(3), pp. 265–274.

Mallick, R. B., and El-Korchi, T. (2017). *Pavement Engineering*, 2nd ed. Boca Raton, FL: CRC Press.

McGennis, R., Shuler, S., and Bahia, H. (1994). *Background of Superpave Asphalt Binder Test Methods*. FHWA Report No. FHWA-SA-94-069. Washington, DC: FHWA.

Moulton, L. K. (1980). *Highway Subsurface Design*. Report No. FHWA-TS-80-224. Washington, DC: Federal Highway Administration.

Mukherjee, A., Stawowy, B., and Cass, D. (2013). Project emission estimator: tool for contractors and agencies for assessing greenhouse gas emissions of highway construction projects. *Transportation Research Record*, 2366, pp. 3–12.

National Asphalt Pavement Association (NAPA), European Asphalt Pavement Association (EAPA). (2012). *The Asphalt Paving Industry, a Global Perspective, Production, Use, Properties, and Occupational Exposure Reduction Technologies and Trends*. Global Series 101, 3rd ed. Lanham, MD: National Asphalt Pavement Association.

Natural Resources Conservation Service (NRCS). (2010). Time of concentration. *National Engineering Handbook. Part 630: Hydrology*. U.S. Department of Agriculture, Chap. 15. 210-VI-NEH.

NCHRP. (2004). *Guide for Mechanistic-Empirical Design of New and Rehabilitated Pavement Structures*. Final report, NCHRP Project 1-37A. Washington, DC: Transportation Research Board of the National Academies.

NHI. (1998). *Techniques for Pavement Rehabilitation: A Training Course—Participant's Manual*. FHWA-HI-99-006. Washington, DC: National Highway Institute, Federal Highway Administration.

Nikolaides, A. (2015). *Highway Engineering—Pavements, Materials and Control of Quality*. Boca Raton, FL: CRC Press.

NRCS. (1986). *Urban Hydrology for Small Watersheds*. Technical Release (TR) 55. Conservation Engineering Division, Natural Resources Conservation Service (NRCS), U.S. Department of Agriculture.

Nunn, M. (2004). *Development of a Mode Versatile Approach to Flexible and Flexible Composite Pavement Design*. Report TRL 615, Crowthorne, UK.

Odemark, N. (1949). *Undersokning av Elasticitetegenskaperna hos Olika Jordarter samt Teori for Berakning av Bel¨agningar Enligt Elasicitetsteorin*. Meddelande 77: Statens Vaginstitut [in Swedish].

ODOT. (2014). *Hydraulics Design Manual*. Salem, OR: Highway Division, Oregon Department of Transportation.

Papagiannakis, A. T., and Masad, E. A. (2008). *Pavement Design and Materials*. Hoboken, NJ: John Wiley & Sons.

Papagiannakis, A. T., Bracher, M., and Jackson, N.C. (2006). Utilizing clustering techniques in estimating traffic data input for pavement design. *Journal of Transportation Engineering*, 132, pp. 872–879.

Peterson, J. (1984). Chemical composition of asphalt as related to asphalt durability: State of the art. *Transportation Research Record*, no. 999. Retrieved from http://onlinepubs.trb.org/Onlinepubs/trr/1984/999/999-004.pdf.

Rahman, M. (2014). *Local Calibration of the AASHTOWARE Prediction Models for Pavement Rehabilitation and Evaluation of Top-Down Cracking for Oregon Roadways*. Doctoral dissertation, Iowa State University, Iowa.

Ray, E., Kandhal, P. S., Roberts, F. L., Kim, Y. R., Lee, D.-Y., et al. (2009). *Hot-Mix Asphalt Materials, Mixture Design, and Construction*, 3rd ed. Second Printing 2016. Greenbelt, MD: NAPA Research and Education Foundation.

Ridgeway, H. H. (1976). Infiltration of water through the pavement surface. *Transportation Research Record*, 616, pp. 98–100.

Rivera, J., Islam, M. R., and Mincic, M. (2017). Statistical analysis of pavement system in the United States. CSU-Pueblo Student Symposium, April 7, 2017.

Roberts, F. L., Kandhal, P. S., Brown, E. R., Lee, D-Y., and Kennedy, T. W. (1996). *Hot Mix Asphalt Materials, Mixture Design, and Construction*. Greenbelt, MD: NAPA Research and Education Foundation.

Romanoschi, S., A., Momin, S., Bethu, S., and Bendana, L. (2011). Development of traffic inputs for the new ME pavement design guide: a case study. *Journal of Transportation Engineering*, 2256, pp. 142–150.

Roque, R., Yan, Y., and Lopp, G. (2018). *Impact of Recycled Asphalt Shingles (RAS) on Asphalt Binder Performance*. Final Report BDV31-977-36. Tallahassee, FL: Florida Department of Transportation.

SAPDM. (2013). *South African Pavement Design Manual (SAPDM)*. Pretoria, South Africa.

Sayers, M. W., Gillespie, T. D., and Queiroz, C. A. V. (1986). *The International Road Roughness Experiment: Establishing Correlation and a Calibration Standard for Measurements*. World Bank Technical Paper No. 45. Washington, DC: The World Bank.

Schwartz, C. W. (2016). *Structural Characteristics and Environmental Benefits of Cold-Recycled Asphalt Paving Materials*. National Pavement Preservation Conference, October 12–14, 2016, Nashville, TN.

Smith, B., and Nair, H. (20015). *Calibration Factors and Design Criteria Values for Mechanistic-Empirical Pavement Design*. Report No. VCTIR 16-R1., Charlottesville, VA: Virginia Center for Transportation Innovation and Research.

Tarefder, R., and Islam, M. (2015). *Study and Evaluation of Materials Response in Hot Mix Asphalt Based on Field Instrumentation*. Final Report, Project No. NM11MSC-03., Albuquerque, NM: New Mexico Department of Transportation (NMDOT).

Tran, N. H., and Hall, K. D. (2007a). Development and influence of statewide axle load spectra on flexible pavement performance. *Journal of Transportation Research Record*, no. 2037, pp. 106–114.

Tran, N. H., and Hall, K. D. (2007b). Development and significance of statewide volume adjustment factors in mechanistic-empirical pavement design guide. *Journal of Transportation Research Record*, no. 2037, pp. 97–106.

United States Census Bureau. (2007a). *EC0731SX1: Manufacturing: Subject Series: Industry-Product Analysis: Industry Shipments by Products: 2007 Economic Census*. Washington, DC: United States Census Bureau.

United States Census Bureau. (2007b). *Manufacturing: Industry Series: Detailed Statistics by Industry for the United States: 2007 Economic Census. Ready Mixed Concrete Manufacturing*. Washington, DC: United States Census Bureau.

United States Census Bureau. (2013). *Number of Firms, Number of Establishments, Employment, and Annual Payroll by Enterprise Employment Size for the United States, All Industries: 2011*. Washington, DC: United States Census Bureau.

United States Geological Survey. (USGS). (2013a). *Mineral Commodity Summaries 2013.*, Reston, VA: United States Department of the Interior, United States Geological Survey.

United States Geological Survey. (USGS). (2013b). *2011 Minerals Yearbook, Cement*. Advance Release. Reston, VA: United States Department of the Interior, United States Geological Survey.

USGS. (1997). *Natural Aggregates—Foundation of America's Future*. USGS Fact Sheet FS 144-97. Washington, DC: U.S. Geological Survey, U.S. Department of the Interior.

Van Dam, T. J., Harvey, J. T., Muench, S. T., Smith, K. D., Snyder, M. B., Al-Qadi, I. L., Ozer, H., et al. (2015). *Towards Sustainable Pavement Systems: A Reference Document*. FHWA-HIF-15-002. Washington, DC: Federal Highway Administration.

Wallace, H., and Martin, J. (1967). *Asphalt Pavement Engineering*. New York: McGraw Hill.

Westergaard, H. M. (1926). Stresses in concrete pavements computed by theoretical analysis. *Public Roads*, 7, pp. 25–35.

World Highways. (2001). Foamed asphalt takes hold. *World Highways*, April 2001.

Index

Note: Page numbers followed by *f* refer to figures; page numbers followed by *t* refer to tables.

A

D

E

I

J

K

L

M

N

O

P

S

T

U

V

W

Y

Z

Printed by Libri Plureos GmbH in Hamburg,
Germany